EXTREME ENVIRONMENTS
Unique Ecosystems – Amazing Microbes

T0273450

Editors

Anita Pandey

Department of Biotechnology
Graphic Era (Deemed to be University)
Dehradun 248002, Uttarakhand, India

Avinash Sharma

National Centre for Cell Science
Savitribai Phule Pune University Campus, Ganeshkhind,
Pune 411007, India

CRC Press
Taylor & Francis Group
Boca Raton   London   New York

CRC Press is an imprint of the
Taylor & Francis Group, an **informa** business

A SCIENCE PUBLISHERS BOOK

Cover credit
Photo caption: View of Himalayan Mountain range from Spiti Valley, Himachal Pradesh, India
Photo credit: Dr Sanjeeva Pandey

First edition published 2021
by CRC Press
6000 Broken Sound Parkway NW, Suite 300, Boca Raton, FL 33487-2742

and by CRC Press
2 Park Square, Milton Park, Abingdon, Oxon, OX14 4RN

© 2021 Taylor & Francis Group, LLC

CRC Press is an imprint of Taylor & Francis Group, LLC

Library of Congress Cataloging-in-Publication Data

Names: Pandey, Anita, 1959- editor. | Sharma, Avinash, 1984- editor.
Title: Extreme environments : unique ecosystems – amazing microbes /
editors, Anita Pandey, Centre for Environmental Assessment & Climate
Change, G.B. Pant National Institute of Himalayan Environment and
Sustainable Development (An Autonomous Institute of Environment, Forest
& Climate Change), Almora, Uttarakhand, India, Avinash Sharma, National
Centre for Cell Science, Pune, India and SBP Pune University Campus,
Pune, India.
Description: First edition. | Boca Raton : CRC Press, 2021. | Includes
bibliographical references and index.
Identifiers: LCCN 2020046701 | ISBN 9780367350161 (hardcover)
Subjects: LCSH: Extreme environments–Microbiology.
Classification: LCC QR100.9.E94 2021 | DDC 577.5/8–dc23
LC record available at https://lccn.loc.gov/2020046701

ISBN: 978-0-367-35016-1 (hbk)
ISBN: 978-0-367-64904-3 (pbk)
ISBN: 978-0-429-34345-2 (ebk)

Typeset in Times
by TVH Scan

I dedicate this work to my father, Late Dr. Brahma Prakash Pandey, for his inspiration to learn the nuances of science and research.

Anita Pandey

Preface

Extremophiles are the unique organisms that survive in extreme ecosystems, like hot springs, volcanos, glaciers, alkaline, and acidic environments, based on which these are classified into different classes. Over the last few decades, extremophiles are getting attention because of their unique mechanisms to adhere to extreme environments; the other important aspect of studying these organisms is the linkage of these organisms to the evolution of life. These organisms not only have the unique mechanisms to survive in harsh conditions, but they also produce unique enzymes called 'extremozymes' which are highly stable in harsh conditions and are very useful in harsh industrial production procedures.

There are many success stories about the usage of the enzymes isolated from extremophiles in industries; one of those success stories is the discovery of *Taq* DNA *Polymerase* without which the automated version of the PCR would not have been possible. Although extremophiles have been recognized over the last two decades, they are not yet explored much for their functionality and biotechnological applications. Comparing extremophiles with the other groups of microorganisms, it is clear that not many species from extreme environments are isolated. Based on the high throughput sequencing data of the microbial communities from the extreme environment, it is believed that these organisms perform complex processes in the ecosystem and have huge potential for the production of secondary metabolites. It has been estimated that the majority of the microorganisms associated with extreme environments are not yet cultivated because of the difficulties in offering complex environmental conditions to the microbial cells in the laboratory.

In general, the phylogenetic diversity of extremophiles is very high and complex to study. Some orders or genera contain only extremophiles, whereas other orders or genera contain both extremophiles and non-extremophiles. Interestingly, extremophiles that adapted to the same extreme condition may be broadly dispersed in the phylogenetic tree of life. This is the case for different psychrophiles or barophiles for which members may be found dispersed in three domains of life. There are also groups of organisms belonging to the same phylogenetic family that has adapted to very diverse extreme or moderately extreme conditions.

This issue *Unique Ecosystems: Amazing Microbes*, under the series *Extreme Environments*, aims to consolidate the research highlights and identify the gaps within the diverse aspects of extremophiles. A total of twenty-one chapters discuss the biotechnological and industrial applications, adaptation strategies, and exploration of microbial diversity using classical and next-generation sequence approaches. All the chapters are contributed by well-known researchers performing quality research in the field of extremophiles.

Citing the example of the Venezuelan Andes, Chapter 1 discusses the importance of the vanishing extreme ecosystems, in particular the tropical glaciers. The chapter recognizes the potential of psychrophilic microorganisms with their value as biotechnological tools and is fundamental to understand the origin of life on Earth and the possible existence of extraterrestrial life. The concern shown over the melting glaciers in recent decades, through the knowledge of the microbial composition of tropical glaciers along with the associated biotechnological applications, is a relevant topic to understand the recent advances in climate change research. Another extreme temperature environment, discussed in Chapter 2, is addressed by the remarkable example of the thermophilic microorganisms that colonize the Himalayan Geothermal Belt. While these hot ecosystem thermophiles have already made their way to biotechnological, industrial, agricultural, and pharmaceutical applications, these are also contributing to the studies of phylogenetic associations, among various prokaryotic domains, and the origin of life. The chapter is concluded with a concept to hypothesize the Himalayan Geothermal Belt as "Himalayan *Geobacillus* Corridor".

Chapter 3 narrates the story of The Dead Sea, one of the most saline water ecosystems on Earth. It deals with many types of halophilic and halotolerant microorganisms, including Archaea, Bacteria, and different types of eukaryotes, highlighting the ones that served as model organisms for the study of life at high salt concentrations. It further highlights the importance of metagenomics that helped in a better understanding of the microbial diversity of the lake. Complementing to the subject line, Chapter 4 describes the distribution of the halophilic microorganisms in peculiar environments like hypersaline lakes, solar salterns, saline soil, evaporation ponds, and marine environments along with the unique property of thriving in an environment characterized by high salt concentration. The chapter emphasizes on the biotechnological applications of the most useful halophilic products produced by bacterial, archaeal, and eukaryal halophiles.

Chapter 5 is focused on another extreme environment, the Copahue geothermal system, with unique characteristics such as the presence of the still active Copahue volcano and the geochemical composition of the area. It discusses the rich microbial diversity of the extreme environment, Rio Agrio, with outstanding high tolerance to heavy metals and metalloids along with their role in bioleaching and bioremediation strategies. Chapter 6 further discusses the mechanisms acquired by metallophilic microorganisms that inhabit metal-rich or metal-contaminated sites and are likely to have applications in the bioremediation of heavy metals. It highlights the potential and advantage of native metal tolerant and genetically modified microorganisms in bioremediation approaches.

Chapters 7-10 address the bioprospection of microbial diversity of Antarctica, a plethora of harsh habitats spanning from marine to terrestrial and lacustrine ones. It provides an environment that stimulates the development of communities of organisms which, individually or in symbiotic relationships, possesses very peculiar adaptation strategies and produces specified biomolecules that may not be detected in other environments. The focus of Chapter 7 is on the bioprospection of cold-adapted bacteria associated with biotic and abiotic Antarctic matrices as a potential resource of novel bioactive molecules, such as antimicrobials and extracellular polymeric substances. It also gives a glimpse of the ethical issues of Antarctic bioprospecting. Chapter 8 focuses on the diversity of cold-adapted yeasts of King George Island (Maritime Antarctica) given their relevance as a source of cold-active exoenzymes and their biotechnological applications in food, biofuel, and detergent industries. Chapter 9 discusses the bacterial microbiota associated with Antarctic macroalgae which are otherwise a rarely studied research topic. It is important in view of the high marine richness and the high number of endemic macroalgal species that are colonized by a variety of bacterial species with particular reference to their taxonomy, ecology, and biotechnological potential. Chapter 10 refers to a case study on the bacterial diversity in microbial mats from Fildes Peninsula in Maritime Antarctica and their genetic potential for nitrogen (N) acquisition, an essential element that often limits microbial growth.

Microorganisms, which thrive under extreme environments, possess unique adaptation strategies to cope with the prevalent adverse climatic conditions. While many chapters include this concern partly, Chapter 11 presents a comprehensive view of the microbial adaptation strategies that are functional under extreme limits of various environmental factors. It discusses the specialized physiological adaptations and the unique metabolic machinery with specifically modified biomolecules that confer the extremophilic microorganisms with the ability to thrive under stressful environments. Chapter 12 gives an example of a microbial adaptation process elaborating the man-made hypersaline environments, referred to as solar salterns. It highlights the adaptation strategy of microbes to salinity gradients along with the associated specific applications. Chapter 13 discusses the mechanisms involved in metabolic pathways in biodegrading psychrotrophic bacteria under low-temperature environments with particular reference to the cold-active extracellular enzymes.

Chapter 14 presents an overview of the microbial diversity that harbors various extreme environments of Armenia, namely geothermal springs, saline-alkaline soils, subterranean salt deposits, polymetallic mines, and karst caves. It gives a diversified account of extremophilic microbes, isolated from these environments belonging to different phylogenetic groups of Bacteria and Archaea with respect to their distribution, ecological significance, and biotechnological potential. Similarly, the following chapter (Chapter 15) gives an overview of the microbial diversity associated with the extreme environments of Egypt, namely soda lakes, solar salterns, hot springs, deep sea, and desert.

Chapter 16 elaborates the microbial life in deep terrestrial continental crust, describing the subsurface of the planet Earth which is hot, aphotic, and an extreme habitat for life. It provides in-depth knowledge of the deep terrestrial biosphere, highlighting its extremities and nutrient sources for microbial life, microbial diversity, and function along with the implications of deep life in astrobiological research. Chapter 17 recognizes the deep subsurface aquifers as a unique source of biodiversity and presents an insight into deep subsurface microbial life in the therapeutic waters of Poland. It highlights the practical acquisition of microorganisms, their genes, and the products relevant to the use of therapeutic waters.

Bioprospection of extremophiles has taken space in frontline research along with the challenging difficulties in culturing and analyzing these organisms under laboratory conditions. With the background that the use of classic microbiological methods does not allow culturing of as many as 99% of the microorganisms from environmental samples, metagenomics emerged as a powerful tool to explore unculturable microbes through the sequencing and analysis of DNA extracted from the environmental samples. Chapter 18 addresses the basis of advanced methods including metagenomics and other omics approaches applicable in the exploration of the extremophilic communities that link phylogenetic and functional information about the environmental microbial diversity including unculturable microbiota. The metagenomic approach has been discussed with various groups of extremophiles including thermophiles, psychrophiles, halophiles, and acidophiles. Chapter 19 briefly discusses the impact of microbial genome sequencing in understanding extremophiles. It highlights the advancement in sequencing technologies that provide high throughput genomic data with its uses in genomic analysis of extremophiles. Then, Chapter 20 gives an overview of the application of integrated multiomics approaches in microbial ecotoxicology, giving the example of the use of extremophilic isolates of *Pseudomonas aeruginosa* as relevant biomarkers.

Chapter 21, giving a general account of actinobacteria, emphasizes the scope of bioprospection of new species isolated from extreme environments of Morocco. Studies on Actinobacteria from extreme environments, a rich microbial resource of diverse biologically active compounds, remain scanty and need the attention of researchers.

We are grateful to the distinguished authors-Luis Andrés Yarzábal, Nagendra Thakur, Aharon Oren, Alessandra Morana, Edgardo Donati, SR Joshi, Lo Giudice Angelina, Silvana Vero, Sergio Leiva, Silvia Batista, Praveen Kumar, Alyssa Carré-Mlouka, Rakshak Kumar, Panosyan Hovik, Samy Selim, Pinaki Sar, Agnieszka Kalwasińska, Rajesh K. Sani, Rajpal Srivastav, Ivanka M. Karadžić, Ahmed Nafis and their co-authors for their contribution.

We acknowledge the GB Pant National Institute of Himalayan Environment, Almora, India; National Centre for Cell Science, Pune, India, and Graphic Era (Deemed to be University), Dehradun, India, for supporting this endeavor.

Grateful thanks to Dr Sanjeeva Pandey for providing the View of Himalayan Mountain range from Spiti Valley, Himachal Pradesh, India for the Cover page.

Anita Pandey, Avinash Sharma

Contents

1

Bioprospecting Extreme Ecosystems Before They Vanish: The (Poorly Studied) Microbiology of Tropical Glaciers

Luis Andrés Yarzábal[1,2]

Introduction

Earth is a cold place to live in. Glaciers, ice caps, ice sheets, and more than 90% of the oceans' waters, whose temperature remains below 5°C, make the Earth's biosphere a place dominated by cold environments.

According to the National Snow and Ice Data Center, glaciers cover approximately 10 percent of the Earth's land area. If glaciers and ice sheets from Greenland and Antarctica are subtracted from the figure, still around 706,000 km^2 of land remains covered by glaciers (RGI Technical Report 2017). The total volume of this glacial ice has been estimated to reach 170,000 km^3 which, if completely melted, would increase the sea level by 0.4 meters worldwide (Huss and Farinotti 2012).

The glacierized component of the cryosphere is by far the most extreme in terms of constraining the development of life, as we know it. And this is not only because cold temperatures impose serious threats to life, limiting vital processes like solute diffusion rates, and macromolecular interactions, or else modifying membrane fluidity and enzyme kinetics (D'Amico et al. 2006, Rodrígues and Tiedje 2008). In glacier environments, liquid water is scarce which makes them cold deserts. However, strange as it may seem, many microorganisms thrive under these conditions; these are collectively known as psychrophiles (i.e. cold-loving organisms) and psychrotolerants (i.e. cold-tolerant organisms), the most abundant and diverse among all living beings as some authors may argue (Margesin and Feller 2010).

In fact, it has been estimated that as many as 10^{29} microbial cells are immured in glaciers and ice sheets around the globe (Anesio et al. 2017). However, their abundance varies in the range of several orders of magnitude, if we consider ice associated with impurities in the surface of glaciers (10^{11} cells per L) (Lutz et al. 2014, Stibal et al. 2015) or subglacial sediments and runoff waters (0.87 to 7.9×10^6 cells per g) (Stibal et al. 2012). Many of these psychrophiles (and their derived metabolites) have caught attention as they could be useful to develop biotechnological processes and/or products, some of which have already been protected by patents or trade secrecy (Margesin and Feller 2010, Feller 2013, Collins and Margesin 2019).

[1] Unidad de Salud y Bienestar, Universidad Católica de Cuenca, Av. Las Américas y Calle Humboldt, Cuenca, Ecuador.
[2] Laboratorio de Microbiología Molecular y Biotecnología, Facultad de Ciencias, Universidad de Los Andes, Av. Alberto Carnevalli, Mérida, Estado Mérida, Venezuela.
E-mail: yarzabalandres@gmail.com

With a few exceptions, many of the glaciers that contain this rich diversity of psychrophiles have been retreating rapidly since the beginning of the 20th century. In fact, some glaciers have completely disappeared and many others are projected to disappear in the near future (Zemp et al. 2019). As a direct consequence of glacier melting, microbes immured in ice may enter new environments and be disseminated globally, bringing unknown consequences (Rogers et al. 2004), or worst, go extinct and be lost forever.

Here, I review current knowledge on the microbial composition of tropical glaciers, including the potential use of this biological resource to develop environmentally-friendly biotechnological tools to support a variety of applications in different fields.

Tropical Glaciers

According to the U.S. Geological Survey, a glacier can be defined as a large, perennial accumulation of crystalline ice, snow, rock, sediment and often liquid water that originates on land and moves—under the influence of gravity and its weight—downslope. Glaciers naturally form in areas where the accumulation of snow exceeds its ablation (melting and sublimation) over many years, often centuries.

In 1999, Georg Kaser defined tropical glaciers, from a glaciological point of view as those that occur in regions where three geographical zones coincide: (1) the astronomical tropics (radiative delimitation), (2) any area in which the daily air temperature amplitude exceeds the annual temperature amplitude (thermal delimitation), and (3) the area of oscillation of the Intertropical Convergence Zone (ITCZ) (hygric delimitation). Glaciers from New Guinea (Irian Jaya), East Africa (Mt. Kenya, Mt. Kilimanjaro, and Rwenzori Mountains), and South America (Andes Mountains between Venezuela and Bolivia) fall within this category (Figure 1). However, approximately 99% of them are found in the Tropical Andes region of South America.

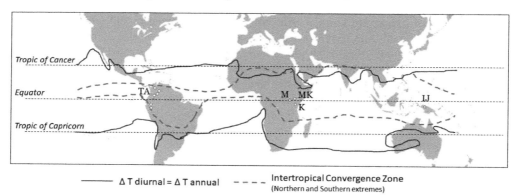

Figure 1. **Tropical Glaciers:** The figure shows the tropics and their limits from a glaciological perspective. Some of the tropical glaciers discussed in this review are shown (dots). K: Kilimajaro; MK: Mount Kenya; M: Margherita Glacier; IJ: Irian Jaya; TA: Tropical Andes glaciers (Adapted from Kaser 1999 and Veettil et al. 2017).

In the last decades, glaciological studies have shown that tropical glaciers can be particularly useful to understand current and past fluctuations of climate since they exhibit dramatic modifications owing to an increased sensitivity to these fluctuations as compared to mid- and high-latitude glaciers (Albritton et al. 2001, Kaser 2001, Kaser et al. 2004a). While air temperature is the most important parameter affecting the mass balance of the latter (Ohmura 2001), atmospheric moisture content has been proposed to strongly modulate the fluctuations of tropical glaciers (Kaser et al. 2004a). This implies that parameters like precipitation, cloudiness, and incoming solar radiation—strongly impacting the glacier surface—are of paramount importance in terms of tropical glacier retreat, as proposed several decades ago (Hastenrath 1984, Kaser and Noggler 1991).

Besides their importance as proxies for climate change studies, tropical glaciers are also natural reservoirs of freshwater and tourist attractions. Therefore, their rapid melting can lead to more problems on several fronts than previously anticipated.

Glacier Retreat in the Tropics

Glaciers around the world are rapidly retreating as the Earth's climate warms. According to the most recent information provided by the American Meteorological Society, glaciers across the world lost mass for the 38th consecutive year since 1980 (Pelto et al. 2018). This mass balance loss is equivalent to slicing off 22 meters from the top of an average glacier. In a recent report, Bamber et al. (2018) estimated a global mass loss from glaciers and ice caps of -227 ± 31 Gt/yr solely from 2012 to 2016. This estimation did not include losses from other peripheral glaciers, i.e. glaciers from Greenland and Antarctica.

Tropical glaciers are no exception to this global trend, mostly because they are particularly sensitive to climatological variations as mentioned above (Thompson et al. 2011). In fact, it has been clearly shown that increases in both global temperatures and the height of the 0°C isotherm in the tropical atmosphere (i.e. the freezing level) over the past 60 years coincided with the retreat of the tropical ice caps (Thompson et al. 2006; Bradley et al. 2009). This retreat is more pronounced at higher elevations since warming in the mid- to upper troposphere is twice the one recorded at the Earth's surface (Bradley et al. 2006). On the other hand, small glaciers shrink faster than bigger ones owing to the fact that their rate of melting is inversely proportional to their size. Altogether, these are the main reasons explaining why low-latitude glaciers—like those that still exist in the Tropical Andes, Eastern Africa, or Indonesia—have lost a significant amount of their mass in the last decades.

Currently available data confirm that glacier retreat in the tropics is an undeniable fact that unfortunately will end in their disappearance in decades to come (Figure 2). For instance, small glaciers on top of Kilimanjaro lost approximately 85% of their ice cover from 1912 to 2011 (Cullen et al. 2013). A similar trend has been recorded in the case of Indonesian glaciers. According to the NASA Earth Observatory, by 1989, five glaciers were present on top of Mount Punkaj Jaya (New Guinea) at 4,884 m.a.s.l.; 20 years later, in 2009, two of these glaciers were gone (Kincaid 2007). Most recent reports confirm this trend: in 27 years, from 1988 to 2015, the glacierized area in these mountains was reduced by 84.9% (Veettil and Wang 2018).

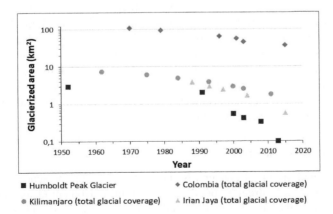

■ Humboldt Peak Glacier   ◆ Colombia (total glacial coverage)

● Kilimanjaro (total glacial coverage)   ▲ Irian Jaya (total glacial coverage)

Figure 2. Glacier Retreat in the Past 50 years: (plotted with data from Veettil et al. 2017, Cullen et al. 2013, Veettil and Wang 2018).

In the Tropical Andes—a vast area in South America that runs through seven countries following the path of the Andes Mountains from Southwestern Venezuela to Northern Chile and

Argentina—glaciers also retreated rapidly during the last decades. This is particularly significant if one considers that almost 99% of the remaining tropical glaciers are located in this region of South America (Kaser 1999). As highlighted above, small glaciers located below 5,400 m.a.s.l. are prone to experience more negative mass balances (i.e. they shrink more rapidly) than those located at higher altitudes. In 2013, Rabatel et al. calculated that in approximately 50 years, Bolivian small glaciers had lost between 50% and 90% of their surface area.

By far, the most dramatic example of glacier disappearance is the one recorded in the Venezuelan Andes. As documented by Braun and Bezada (2013), the area covered by glacial ice in these mountains decreased from approximately 10 km^2 in 1919 to less than 0.1 km^2 by 2011. Today, only one ice patch remains on top of Humboldt Peak i.e. at 4,940 m.a.s.l. This ice patch was projected to disappear by 2015; however, due to a combination of several favorable topographic factors, a few remnants still exist but are considered a glaciological anomaly.

Venezuela is likely to be the first country to lose all of its glaciers (Braun and Bezada 2013, NASA Global Climate Change 2018). Unfortunately, it will not be the last since Indonesia will also likely follow suit by the end of this decade (NASA Earth Observatory). The consequences of glacier loss are already being felt at different levels and encompass a variety of impacts that ranges from socio-economic to ecologic (Vuille et al. 2018). For instance, the tree line in the Andes is moving upwards (Zimmer et al. 2018) and with it, invasive species, pests, and disease vectors gain access to new niches (Siraj et al. 2014, Wiens 2016, Freeman et al. 2018). Unfortunately, the consequences of elevational range shifts are even worst, with studies confirming the widespread extinction of hundreds of species at a local range (Wiens 2016). Even though these pieces of evidence are alarming, they are strictly related to plants and animals. In the case of glacier-microbes, the melting of their habitat will likely have an even deeper impact: total extinction.

Glacier Ecology

Once considered to be devoid of life, too extreme and inhospitable, glaciers are nowadays acknowledged not only as repositories of life but as true ecosystems harboring remarkable biodiversity (Miteva 2008, Boetius et al. 2017). In 2004, Priscu and Christner calculated that Antarctica and Greenland contained approximately 9.61×10^{25} bacteria immured in their ice sheets. That explains why glacial ice environments are currently considered major habitable ecosystems of Earth's biosphere.

Following the discovery of viable extremophilic microorganisms in almost all the ice phases known to occur in glacial environments (Margesin and Collins 2019) and having in mind their horizontal stratification, Hodson et al. (2008) proposed the existence of at least three ecosystems in a glacier: a supraglacial ecosystem, a subglacial ecosystem and an englacial ecosystem (Figure 3). A fourth ecosystem has been also proposed to be as prominent as the formers: proglacial streams.

Since the four ecosystems differ in parameters, like solar radiation, water content, nutrient abundance, ionic strength, rock-water contact, pressure, pH conditions, and redox potential, they sustain significantly different microbial communities.

The Supraglacial Ecosystem

This ecosystem, located on top of glaciers, is characterized by the presence of five main habitats: snowpack, cryoconite holes, streams, ponds, and moraines (Hodson et al. 2008). In the glacier surface, incident solar radiation is barely reflected or absorbed by clouds. This radiation is absorbed by dark organic- or inorganic matter present, causing the rise in the local temperature and the melting of ice. Meltwater dissolves mineral- and organic nutrients—transported to- and deposited in the surface by winds and precipitation—which concentrate in sun cups upon snow and in cryoconite holes, flow in streams and/or accumulate in moraines (Tranter et al. 2004). Under these conditions, the growth of photosynthetic (photoautotrophic) microorganisms like cyanobacteria,

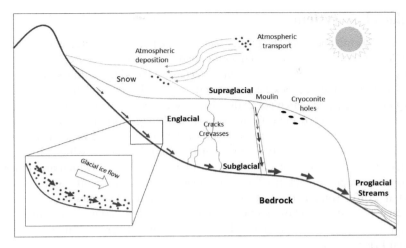

Figure 3. Glacial Ecosystems: (arrows indicate water, meltwater, and nutrient flow under the glacier). Inset: rock comminution due to the downhill flow of the glacier releases small particles and debris and liberates energy sources that sustain chemolithoautotroph growth. Nutrients and water are also delivered into englacial and subglacial ecosystems through moulins and cracks/crevasses (Adapted from Boetius et al. 2015 and Hotaling et al. 2017b).

microalgae, and diatoms is favored. However, other microorganisms that obtain energy through the oxidation of inorganic substrates (= *chemolithotrophic*) also thrive in this environment. Finally, many heterotrophic bacteria, fungi, and microeukaryotes rely on nutrients provided by primary producers and thus are also found in this ecosystem (Boetius et al. 2015, Mieczen et al. 2013).

Many of the microorganisms colonizing supraglacial environments typically synthesize pigments, so to not only capture the light energy necessary for photosynthesis but also to resist UV radiation and frequent freeze-thaw cycles (Dieser et al. 2010, Marizcurrena et al. 2019). Under some circumstances these microorganisms bloom, giving rise to phenomena known as green-snow or red-snow (Lutz et al. 2014, 2016, Anesio et al. 2017). These pigments can also contribute to accelerate ice melting since they are also able to absorb solar heat and reduce snow albedo.

The Englacial Ecosystem

From a general point of view, englacial ecosystems are defined as deep, immured environments, where conditions reach extreme values of temperature, pressure, acidity, and darkness (Boetius et al. 2015, Hotaling et al. 2017b). This is where the bulk of the glacial ice lies, a vast environment running from a few meters down the surface to hundreds (or even thousand) meters deep. Englacial ecosystems lack significant volumes of liquid water but also energy sources (Christner et al. 2006); nevertheless, water flows periodically from the surface through crevasses and channels, delivering nutrients, gases, and viable cells to the ice bed (Hotaling et al. 2017b). That is why the redox potential in this particular ecosystem can vary greatly from well-oxygenated to completely anoxic environments (Anesio et al. 2017).

In these ecosystems, living microorganisms are mostly confined to small veins and interstices located between ice crystals, where scarce nutrients and dissolved ions concentrate (Price 2000). Owing to these conditions, microorganisms immured in englacial ecosystems are scarce (typically present at a density of 10^1 to 10^3 cells per ml) and mostly chemolitoautotrophs (Boyd et al. 2014); nevertheless, a few chemoheterotrophs are also found, which rely for their nutrition on organic nutrients released by primary producers or provided by water streams from the surface. Some anaerobes and methanogens can be found at great depths (Tung et al. 2005, Price 2007, Hodson et al. 2008).

The Subglacial Ecosystem

At the interphase formed between the glacier base and the bedrock, located from dozens to hundreds of meters below the surface, the environmental conditions change drastically. The combination of glacier movement and the enormous pressure exerted by the gigantic mass of the ice atop results in the fine grinding of the bedrock and other sediments. This basal debris—rich in minerals and sedimentary organic carbon which act as the energy sources for many microbial species—is mixed with liquid water, transported there from the surface through moulins, fractures, and cavities, or originating from the basal ice melted as a result of the high pressure of the overlying ice plus the heat produced by geothermal activity (Weertman and Birchfield 1983). Although light is completely absent, under these conditions some hotspots of microbial life arise which are dominated mainly by chemolithoautotrophs or autotrophic methanogens (Skidmore et al. 2005, Boyd et al. 2010, Hamilton et al. 2013). Some heterotrophic bacteria and microeukaryotes have also been recovered from subglacial ecosystems, whose metabolism is supported by organic molecules, either released by primary producers, deposited in preglacial times, or delivered from the supraglacial zone (Mikucki and Priscu 2007).

Proglacial Streams

As mentioned before, streams of actively running water are important components of glacierized ecosystems. In high-elevation glaciers, these streams link glacial ecosystems and processes (Battin 2003 2004), playing a paramount role as biogeochemical conduits whose importance has not been properly addressed to date (Hotaling et al. 2017a).

Studies conducted in European and North-American glaciers show that Proteobacteria, Bacteroidetes, Actinobacteria, Cyanobacteria, and algae are the dominant taxa in proglacial stream water and biofilms (Wilhelm et al. 2013, Fegel et al. 2016). As expected, microbial communities in proglacial stream water are similar to those found in subglacial ecosystems, showing a direct link between both habitats (Mitchell et al. 2013).

Glacial-Ice Microbes

Cold environments are harsh places to live for any microorganism. This is because reductions in ambient temperature are deleterious to cell structures and functions. For instance, the ability of membranes to take nutrients or to release metabolites is negatively affected by the rigidity imposed on the phospholipid bilayer by low temperatures. Similarly, the diffusion of solutes is also limited at low temperatures, not only because membranes are less fluid but also because water increases its viscosity (D'Amico et al. 2006, Rodríguez and Tiedje 2008). In the particular case of frozen environments, the formation of ice crystals is a dangerous threat to cell integrity since they can damage the fragile cytoplasmic membranes.

On the other hand, from a kinetic point of view, the functioning of enzymes is also dramatically impaired by low temperatures. As Arrhenius law dictates, the catalytic activity of an enzyme is exponentially dependent on temperature; that is why, for every 10°C decrease, the catalytic rate of an enzymatic reaction drops approximately two-fold (Feller 2013).

Therefore, in order to resist the environmental challenges imposed by frozen environments, many microbial species evolved different strategies. This ability to grow at low temperatures, named *psychrophily*, was first noticed by Forster (1887) more than a century ago. Psychrophiles evolved a panoply of mechanisms to cope with coldtemperatures among which stand out are regulation of membrane fluidity, production of cryoprotectors and antifreeze proteins, regulation of ion channel permeability, synthesis of cold-shock chaperones, seasonal dormancy and others (Georlette et al. 2004, D'Amico et al. 2006).

On the other hand, many bacterial species found in frozen environments belong to bacterial taxa that form spores (e.g., *Bacillus* and *Actinomyces*) while others have thick cell walls and polysaccharide capsules. These structures are supposed to allow bacterial cells to endure the challenges imposed by extreme conditions (Priscu et al. 2007).

However, contrary to what expected, microbes found in glacial ice are far from mere survivors, trying to resist the prevailing stressful conditions and expecting environmental changes to multiply. Indeed, many microbial species are actively metabolizing and growing, although at low rates. Generally, these microbes are found in association with liquid water, commonly found inside liquid veins, at sub-freezing temperatures (Price 2007). Others colonize water either at the bedrock or atop the ice (i.e., inside cryoconite holes).

By far the most important adaptation evolved by microorganisms to multiply at low temperatures is the modulation of enzyme structure (Feller 2013). This strategy permitted to optimize the enzymatic activity at low temperatures by destabilizing their structure throughout the entire molecule and/or at specific domains, including the catalytic one. In other words, psychrophilic enzymes are much more flexible than their mesophilic counterparts because their structure is close to the lowest possible stability (Feller 2013), and this flexibility compensates for the reduction of molecular movements at low temperatures (D'Amico et al. 2006).

Destabilization of psychrophilic enzymes can be reached by a combination of genetic modifications that result in a reduction of the intramolecular interactions, mainly of the weak type, and subsequent improvement of the dynamics of the active site residues in the cold. Incidentally, these features are at the basis of the thermolability of psychrophilic enzymes and explain why they are so easily denatured when the temperature increases (Feller 2013).

Colonization of Mountain Glaciers

How did microbes reach the top of high mountains like the Andes or the Himalayas? Did they evolve *in situ* or were they transported and deposited there from other distant ecosystems? In fact, both hypotheses are plausible. Glaciers are considered repositories of microbial forms of life that were transported by winds from distant areas and further immured under multiple layers of ice. In some cases, there is a proximity between the mountain glaciers and other ecosystems; for instance, Andean glaciers are very close (i.e., a few hundred kilometers) to heavily colonized environments (like the Amazon forest or the exposed soils from the Pacific coast) from which atmospheric particles arise (Priscu et al. 2007). According to some authors, this explains why glaciers from low-latitude regions contain more recoverable bacteria than those located in the Earth poles (Priscu et al. 2007).

Microbes can also travel long distances, sometimes several thousand kilometers, transported on top of small particles that can originate, for instance, in the Sahara desert and be deposited on top of European mountains (Barberán et al. 2014, Weil et al. 2017); they can also cross the Atlantic ocean and reach the Amazon basin (Griffin 2007) as well as the top of Andes mountains (Boy and Wilcke 2008, González-Toril et al. 2009).

It has been recently shown that the rising bubbles from the ocean can also be a way to scavenge bacteria and diatoms from saline water, which are ejected inside microscopic droplets into the air (Marks et al. 2018). The survival of these microbes under the extreme conditions prevailing in the troposphere (cold temperatures, high doses of UV radiation, and absence of water) has been related to the production of pigments (Meola et al. 2008) and the ability to sporulate (Nicholson et al. 2000). In addition, Liu et al. (2019) proposed that the ability to grow over a wide range of low temperatures (a trait known as *eurypsychrophily*) is a paramount strategy evolved by microbes to survive the long journey from source habitat to glacier surface. This also explains why many of the microorganisms found in glacial ice samples display such an ability.

As mentioned above, once deposited on the surface of glaciers, these microbes are covered by successive layers of snow and ice year after year. Consequently, glaciers become historical archives of microbial communities that can help us understand both climatic and environmental changes from a chronological perspective (Priscu and Christner 2004). This is because these microbes are not only able to survive but are also able to adapt, proliferate, and evolve in the ice core while remaining viable for thousand and even million years (Bidle et al. 2007, D'Elia et al. 2008).

Microbiology of Tropical Glaciers

In spite of their importance and biotechnological value—as well as their value as analogs of potential extraterrestrial life forms (Ponce et al. 2011)—microbes immured in tropical glaciers remain to be studied in depth. In fact, only a few works have been published to date concerning this topic (Table 1). As can be seen, almost no studies have depicted the details of the microbial communities colonizing such types of environments contrary to what happens in the case of glaciers located at higher latitudes. Only a few have been published, concerning the culturable fraction of bacteria colonizing Venezuelan glaciers and the characteristics of some particular strains (Table 1).

TABLE 1: Summary of Studies Mentioned in This Chapter and the Main Features/Results Reported

Glacier	Features	Reference
Humboldt Peak Glacier (Venezuelan Andes)	Isolation and characterization of glacial-ice bacteria	Ball et al. (2014)
Humboldt Peak Glacier (Venezuelan Andes)	Bioprospection of glacial ice for PGP-bacteria	Balcázar et al. (2015)
Bolívar Peak Glacier (Venezuelan Andes)	Diversity of culturable bacteria	Rondón et al. (2016)
Humboldt Peak and Bolívar Peak Gaciers (Venezuelan Andes)	Plant-growth promoting ability of glacial-ice microbes	Rondón et al. (2019)
Nevado Illimani (Bolivian Andes)	Surface snow blocks; cultivation in liquid and solid media	Elster et al. (2007)
Quelccaya Glacier (Peruvian Andes)	Diatoms entrapped in glacial ice (visualized through Scanning Electron Microscopy)	Fritz et al. (2015)
Quelccaya, Sajama and Coropuna Glaciers (Peru and Bolivia Tropical Andes)	Diatoms entrapped in glacial ice (visualized through Scanning Electron Microscopy)	Weide at al. (2017)
Mount Kilimanjaro Glaciers (East Africa)	Bacteria immured in glacial ice (detected by 16S rDNA sequencing)	Ponce et al. (2011a)
Mount Kilimanjaro Periglacial Soils (Tanzania)	No relevant information published	Ponce et al. (2011b)
Mount Kilimanjaro Glacier (Tanzania)	Detection of Actinobacteria, Gammaproteobacteria, Betaproteobacteria, Alphaproteobacteria, and Bacteroidetes (by 16S rDNA sequencing)	NASA 2012 Annual Report
Mount Kilimanjaro Periglacial Soils (Tanzania)	*In situ* real-time quantification of microbial communities (viable cells, non-viable cells, and spores)	Powers et al. (2018)
Lewis Glacier (Mount Kenya)	Isolation and identification of culturable bacteria from glacial ice samples	Kuja et al. (2018b)
Lewis Glacier (Mt. Kenya, Kenya)	Preliminary studies on Bacteria and Archaea from ice samples	Kuja et al. (2018a)
Margherita Glacier (Rwenzori Mountains, Uganda and D.R. of Congo)	*Ceratodon purpureus* (Moss Gemmae) aggregation	Uetake et al. (2014)

This is disturbing since, as we have highlighted above, there is not much time for scientists to gain access to glacial ice samples. This dramatic reality is perfectly exemplified by the disappearance, in less than four years, of one of the last two Venezuelan glaciers, just after some samples were collected by this author and his team (Rondón et al. 2017) (Figure 4). Therefore, we ignore most details concerning the peculiarities of the microbial communities immured in these glaciers.

Figure 4. Glacier Retreat in Humboldt Peak, Venezuelan Andes: (a: ca. 1950; b: 2007; c: 2015) (Reprinted with permission from Archivo Diario Ultimas Noticias, Marco Cayuso and Jota García respectively).

Kilimanjaro Glaciers (East Africa)

Initial studies, conducted mainly by NASA teams, reported on the microbiology of Kilimanjaro glaciers. In one of the published works, two dust-rich layers embedded within the ice, located at approximately 1.5 m above the base of a 40 m-deep glacier and dating from around 1,200 AD, were shown to harbor a higher proportion of bacteria than any other microorganisms, as revealed by gene cloning and sequencing procedures (Ponce et al. 2011a). Actinobacteria-related 16S rDNA sequences were the most frequently detected in both samples followed by Gamma-proteobacteria (in the upper layer) or β- and α-proteobacteria and Bacteroidetes (in the lower layer) (NASA Report 2012). The majority of OTUs identified in these samples (12 out of 14) were closely related to species previously isolated from cold-water environments. In another brief report, periglacial soils of Mount Kilimanjaro, considered among the most extreme soils on Earth and analogs to Martian soils, were also investigated by NASA teams (Ponce et al. 2011b). Unfortunately, no data were included in this report.

Margherita Glacier (Rwenzori Mountains, Eastern Africa)

An unprecedented form of a biogenic aggregation was found by Uetake et al. (2014) at the forefront of Margherita Glacier, located on top of the Rwenzori Mountains at 5,109 m (Uganda and Democratic Republic of Congo border). This aggregation contained long filaments of the rhizoidal moss *Ceratodon purpureus* (Hedw.), a cosmopolitan species found worldwide from tropical to polar regions.

Lewis Glacier (Mt. Kenya, Kenya)

Only two reports have been published to date concerning the microbiology of this rapidly melting tropical glacier. In one of them, the authors claimed that Lewis glacier is colonized by a rich and diverse microbial community (Kuja et al. 2018a). Even though no data was provided concerning the details of such prospection, the following bacterial taxons were (allegedly) identified by both culture-dependent and culture-independent methods: *Cyanobacterium* sp., *Cryobacterium* sp., *Agreria* (sic), Bacteroidetes, Actinobacteria, Acidobacteria, Chlamydiae, Chloroflexi, Firmicutes, Nitrospirae, Fibrobacteres, Planctomycetes, Proteobacteria, and *Spirochaeta* sp. Some fungi were also claimed to be present in these communities, including members of *Penicillium* and *Fusarium* genera.

The second report depicted the bacterial diversity present on the snow cover of Lewis Glacier (Kuja et al. 2018b). Bacteria were isolated from snow samples and identified by nucleotidic sequencing of their 16S rDNA genes. The isolates belonged to three bacterial phyla, namely Firmicutes, Proteobacteria, and Actinobacteria. *Bacillus* (53%) and *Stenotrophomonas* (23.4%) were the most frequent genera followed by *Cryobacterium* (5.9%), *Paenibacillus* (5.9%), *Subtercola* (5.9%), and *Arthrobacter* (5.9%). Most of these isolates are frequent soil inhabitants and were probably transported on top of the glacier by atmospheric winds; however, a few isolates were closely related to psychrophilic or psychrotolerant species of *Cryobacterium*, *Subtercola/Agreia*, and *Arthrobacter* genera, previously isolated from permanently cold environments.

Quelccaya Glacier (Peru)

In 2015, Fritz and co-workers reported on the serendipitous discovery of freshwater diatoms in ice-core samples dating from the late Holocene, collected at the Quelccaya Summit Dome in the tropical Andes of Peru. By means of scanning electron microscopy, the most frequently observed diatoms were identified as members of the *Hantzschia*, *Pinnularia*, and *Aulacoseira* genera. According to the authors, these diatoms were probably transported to the glacier by winds from nearby high elevation lakes and/or wetlands. Two years later, the same group extended their findings to another three glaciers—namely Sajama, Quelccaya, and Coropuna glaciers, located in Bolivian and Peruvian tropical Andes—showing that the diatom assemblages included cosmopolitan and aerophilic species (Weide et al. 2017). From 44 taxa identified, only 11 were common to all three glaciers with *Pinnularia* cf. *borealis* Ehrenberg as the dominant species.

Illimani Nevado (Bolivia)

In 2007, Elster and co-workers reported on the isolation of bacteria from surface snow blocks—collected seven years before—at the summit of Illimani Nevado, Bolivia (Bolivian Cordillera Real, 6,350 m). The samples were inoculated in mineral media and further used to inoculate agarized media. Although the microorganisms were only identified from a morphological point of view (as coccoids, rods, or red clusters of bacteria; unusual prokaryotes; hyphae and spores of fungi; yeasts and diatoms), culturable heterotrophs were highly abundant in snow samples. Their abundance was related to dust transportation by winds, especially from nearby ecosystems. Crustal aerosols (i.e., microorganisms and biotic remnants found in aerosol and snow samples) were also present in the samples.

La Corona Glacier (Humboldt's Peak, Venezuelan Andes) and Bolivar Peak's Glacier (Venezuelan Andes)

Since 2012, the research conducted by my scientific team, at the University of Los Andes (Mérida, Venezuela), focused on the bacteria immured in the last two remnants of Venezuelan glaciers. In the first report published in 2014, Ball and co-workers confirmed the presence of an abundant community of bacteria (>10^6 UFC/ml) in ice samples collected at La Corona glacier (Humboldt's Peak) at approximately 5,000 m.a.s.l. (Figure 5a). Many of these isolates were able to grow in agarized media *in vitro* (Figure 5b) and were identified as members of the following phyla/classes: Proteobacteria (α, β, and γ), Actinobacteria, and Flavobacteria. *Pseudomonas* was the most frequent genus detected and many isolates related to previously identified psychrophilic or psychrotolerant strains. Most isolates were psychrophilic and synthesized cold-active proteases and amylases. Strikingly, a high percentage of them were resistant to several antibiotics and heavy-metals.

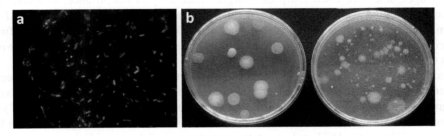

Figure 5. Glacial Ice Microbes: (a) Epifluorescence microscopy at × 1,000 magnification of living cells stained with propidium iodide and SYTO9 using the LIVE/DEAD BacLight bacteria viability kit (Molecular Probes, Eugene, Oregon); (b) Colonies of heterotrophic bacteria grown in Nutrient Agar (left) and R2A medium (right). (Reprinted with permission from Luis Andrés Yarzábal R.).

Rondón and co-workers (2016) obtained similar results when they prospected glacial ice from another Venezuelan glacier, located on top of Bolivar Peak at approximately 4,850 m.a.s.l. Again, many isolates belonged to the *Pseudomonas* genus, and many displayed resistance to several antibiotics and heavy metals. Also, as previously described, many isolates were related to well-known psychrophilic strains that were previously isolated from cold environments.

In 2015, Balcázar and co-workers reported on the ability of some of the *Pseudomonas* spp. isolates to display, *in vitro* and low temperatures, several traits related to plant-growth promotion (PGP). Among these, the ability to dissolve mineral phosphates and to produce secondary metabolites, such as indol-acetic acid, siderophores, and hydrogen cyanide. These last two volatiles are highly toxic to other microbes. In line with these findings, the *Pseudomonas* spp. isolates inhibited the growth of three phytopathogens (*Pythium ultimum*, *Fusarium oxysporum*, and *Phytophthora infestans*). Consequently, we claimed that some of these strains would be useful to develop cold-active biofertilizers.

The confirmation of this hypothesis arrived in 2019 when Rondón et al. reported that wheat (*T. aestivum*) seedlings developed longer roots and shoots when inoculated with some of the abovementioned *Pseudomonas* spp. isolates. Similarly, wheat plantlets grown in sterile sand or soil, at 15°C, were protected by the bacterization of their seeds. Noticeably, when the plantlets were challenged with a well-known pathogenic oomycete (*P. ultimum*), they were less damaged when protected by *Pseudomonas* isolates. This report was the first one showing, not only the PGP-abilities of ice-immured bacteria but the possibility of using this microbiological resource to develop cold-active biofertilizers and/or biocontrol agents.

Glacial Ice as Reservoirs of Microbes and Their Genes: The 'Genome Recycling' Paradigm

Glacial ice is considered a huge microbial reservoir. As said before, microorganisms can remain viable for millennia when entrapped inside ice crystals or concentrated in microscopic liquid veins. When their habitat melts down, glacier-immured microbes are released and spread everywhere. Therefore, the transportation of microbes can happen both ways: from the Amazon forest to the glaciers (transported by winds) and from the Andean glaciers back to the Amazon basin (transported by glacier water fed streams and rivers) (da Rocha Riveiro et al. 2017).

According to some authors, as much as 4×10^{21} cells are released per year into the environment when non-Antarctic glacial systems meltdown (Irvine-Fynn and Edwards 2013). Since many of these microbes are ancient pathogens, it has been proposed that their spreading (and of their genes) can be considered a threat to other ecosystems and even to human and animal life (Rogers et al. 2004, Smith et al. 2004, Edwards 2015).

Even though the information concerning the diversity of viruses in glacial systems is scarce, a few studies are beginning to shed light on this subject (Rassner 2017). In addition to several bacteriophages, other viruses are also present in glacial ice samples. For instance, some circumstantial evidence supports the presence of caliciviruses, influenza viruses, and enteroviruses—all of which are human pathogens—in glacial ice (Smith et al. 2004). However, other microbes–fungi, like *Cryptococcus* sp. and bacteria, like many coliforms, were also detected in both glacial ice or glacial streams (Goodwin et al. 2012, Turchetti et al. 2015).

As said before, in addition to the release of these microbes, there is also a serious concern about the potential spreading of their genes and genomes. Some authors call this the 'genome recycling' paradigm. Indeed, it is well known that in nature, the horizontal transfer of genes (HGT) between microorganisms belonging either to the same or to different domains of life, occurs at a high rate (Soucy et al. 2015). This HGT influences the evolution of all kinds of living beings, eukaryotes included.

When addressing HGT, one aspect is of particular relevance: spreading of antibiotic resistance determinants. These genes are frequently located in mobile genetic modules, prone to be transferred quickly and efficiently from bacteria to other microbes. It has been shown that glacial ecosystems, many of which are considered 'pristine', contain an abundant repertoire of antibiotic resistance determinants. This includes genes conferring resistance to antibiotics used to fight human, animal, and agriculturally important diseases, such as tetracycline, chloramphenicol, streptomycin, and beta-lactams (Ushida et al. 2010, Segawa et al. 2013).

In the case of tropical glaciers, this hot topic has been barely addressed. Antibiotic-resistant bacteria were shown to be abundant in glacial ice from two Venezuelan glaciers (Ball et al. 2014, Rondón et al. 2016). In fact, more than 65% of the strains isolated in La Corona glacier (Humboldt's Peak) were resistant to high doses of ampicillin (>100 µg/ml) whereas more than half of the strains were resistant to nalidixic acid, penicillin, or chloramphenicol (Ball et al. 2014). Resistance to kanamycin and streptomycin was less frequent (approximately 25% of the strains). Multi-resistance (i.e., the ability to grow in the presence of three or more antibiotic classes) was also frequent among those strains with almost 60% able to resist at least three antibiotics and more than 20% able to grow in the presence of five different antibiotics. Similar results were obtained when characterizing bacterial isolates from glacial ice samples collected at Pico Bolivar Glacier (Rondón et al. 2016). We also showed that many isolates grew in the presence of high doses (i.e. 100 ppm) of several toxic metals, such as Ni^{++}, Zn^{++}, and Cu^{++}.

In accordance with the previous results, my colleagues and I detected the presence of low- and high-molecular-weight plasmids in approximately 47% of the strains tested. This was not completely unexpected since it is well documented that antibiotic-resistance determinants are frequently located inside mobile genetic elements, like plasmids.

Given the absence of information concerning this topic and the role of glacial ecosystems as reservoirs of antibiotic resistance and other potentially harmful genes (such as those coding for virulence traits in pathogenic bacteria), it is evident that more efforts should be made to understand the downstream effects of microbial- and genetic release on natural ecosystems due to tropical glaciers' runoff.

Conclusions

As we have seen, topical glaciers are certainly 'endangered species' on the road to extinction, a concept already emphasized—although criticized—by Carey in 2007. There is seldom doubt that we, scientists, have a very limited timeframe to learn whatever we can from these masses of ice before they disappear forever (or at least, for a long period until the next glacial age arrive). As Kennedy and Hanson stated in 2006, studying glacier ice is important "not only because we are losing it but also because it is an archive that has told us much about past climates". It is tempting to add that glaciers are also important because they represent a huge reservoir of a still-unknown microbial diversity.

In this race against time, many aspects deserve to be considered. Perhaps the most important is to depict the microbial diversity that remains entrapped inside tropical glaciers. Indeed, to date, there is not one single work published on this matter, which is rather strange. With the use of current methods for characterizing the microbial diversity of natural ecosystems (e.g., NGS) it would be feasible at a low cost to describe this diversity. Among the many questions that remain to be answered, it is important to assess not only how many microbial species lie there, but how diverse these communities are from a functional point of view. This can help us understand how different (or similar) ancient microorganisms were to extant species; this can also shed light on the strategies evolved by these microbes to deal with the challenges imposed by such an extreme environment.

Studying tropical glaciers can also help to get a glimpse into the future by allowing us to understand which threats we can face when all these microbes and their genes will be finally released into the natural environment, or how they could modify ecosystem functioning. In this sense, it is important to focus on the glacial virome, i.e., the collection of all viruses contained inside glacial ice. To date, no one study has addressed this aspect in the case of tropical glaciers. This is not surprising since very little is known about the genetic diversity of viruses in glacial habitats. As recently stated by Rassner (2017), "The number of studies of the viral component of (glacial) communities is small and typically limited to measuring specific parameters in relation to the prokaryotic community". Among the reasons explaining this lag, the absence of 'universal' genetic markers allowing surveying the abundance and diversity of viruses in natural microbial communities. Quite recently, a promising software tool was developed to identify viral genomes from raw reads obtained through massive parallel sequencing of metagenomes, representative of viral or mixed (viral and bacterial) communities (Garreto et al. 2019).

But, when observing from a different perspective, there are also more positive outcomes that can result from this quest. For instance, the possibility of finding useful microbes (or genes) that are relevant in the development of processes or products with biotechnological value. This was recently shown to be a reality since a few of these glacier microorganisms promoted plant-growth and development at low temperatures (Rondón et al. 2019). Even though the results are preliminary, they provide a glimpse of what could be a vast and diverse toolbox for future biotechnological applications.

Acknowledgments

Dr. Eduardo Chica (Facultad de CienciasAgropecuarias, Universidad de Cuenca, Ecuador) is gratefully acknowledged for helpful comments and suggestions.

References

Albritton, D.L., L.G. Meira Filho, U. Cubasch, X. Dai, Y. Ding, D.J. Griggs, et al. 2001. Technical summary. *In*: Houghton, J.T., Y. Ding, D.J. Griggs, M. Noguer, P.J. van der Linden, X. Dai, K. Maskell and C.A. Johnson [eds.]. Climate Change 2001: The Scientific Basis. Contribution of Working Group I to the Third Assessment Report of the Intergovernmental Panel on Climate Change. Cambridge University Press, Cambridge, United Kingdom and New York, NY, USA, 881 pp.

Anesio, A.M., S. Lutz, N.A.M. Chrismas and L.G. Benning. 2017. The microbiome of glaciers and ice sheets. npj Biofilms and Microbiomes, 3, Article No. 10. doi: http://doi.org/10.1038/s41522-017-0019-0.

Balcázar, W., J. Rondón, M. Rengifo, M. Ball, A. Melfo, W. Gómez, et al. 2015. Bioprospecting glacial ice for plant growth promoting bacteria. Microbiol. Res. 177: 1–7.

Ball, M.M., W. Gómez, X. Magallanes, R. Moreno, A. Melfo and L.A. Yarzábal. 2014. Bacteria recovered from a high-altitude, tropical glacier in Venezuelan Andes. World J. Microbiol. Biotechnol. 30: 931–941.

Bamber, J.L., R.M. Westaway, B. Marzeion and B. Wouters. 2018. The land ice contribution to sea level during the satellite era. Environ. Res. Lett. 13: 063008.

Barberán, A., J. Henley, N. Fierer and E.O. Casamayor. 2014. Structure, inter-annual recurrence, and global-scale connectivity of airborne microbial communities. Sci. Total Environ. 487: 187–195. Available from: http://dx.doi.org/10.1016/j.scitotenv.2014.04.030.

Battin, T.J., L.A. Kaplan, J.D. Newbold and C.M. Hansen. 2003. Contributions of microbial biofilms to ecosystem processes in stream mesocosms. Nature 426: 439–442.

Battin, T.J., A. Wille, R. Psenner and A. Richter. 2004. Large-scale environmental controls on microbial biofilms in high-alpine streams. Biogeosciences 1: 159–171.

Bidle, K.D., S.H. Lee, D.R. Marchant and P.G. Falkowski. 2007. Fossil genes and microbes in the oldest ice on Earth. PNAS 104: 13455–13460. doi: 10.1073/pnas.0702196104.

Boetius, A., A.M. Anesio, J.W. Deming, J.A. Mikucki and J.Z. Rapp. 2015. Microbial ecology of the cryosphere: sea ice and glacial habitats. Nature Rev. Microbiol. 13: 677–690.

Boyd, E.S., M. Skidmore, A.C. Mitchell, C. Bakermans and J.W. Peters. 2010. Methanogenesis in subglacial sediments. Environ. Microbiol. Rep. 2: 685–692.

Boyd, E.S., T.L. Hamilton, J.R. Havig, M.L. Skidmore and E.L. Shock. 2014. Chemolithotrophic primary production in a subglacial ecosystem. Appl. Environ. Microbiol. 80: 6146–6153.

Boy J. and W. Wilcke. 2008. Tropical Andean forest derives calcium and magnesium from Saharan dust. Global Biogeochem. Cycles 22: GB1027. doi:10.1029/2007GB002960.

Bradley, R.S., M. Vuille, H.F. Diaz and W. Vergara. 2006. Threats to water supply in the tropical Andes. Science 312: 1755–1756.

Bradley, R.S., F.T. Keimig, H.F. Diaz and D.R. Hardy. 2009. Recent changes in freezing level heights in the tropics with implications for the deglacierization of high mountain regions. Geophys. Res. Lett. 36: L17701. (10.1029/2000GL037712).

Braun, C. and M. Bezada. 2013. The history and disappearance of glaciers in Venezuela. J. Latin Am. Geogr. 12: 85–124.

Carey, M. 2007. The history of ice: how glaciers became an endangered species. Environ. History 12: 497–527. Retrieved from http://www.jstor.org/stable/25473130.

Christner, B.C., G. Royston-Bishop, C.M. Foreman, B.R. Arnold, M. Tranter, K.A. Welch, et al. 2006. Limnological conditions in subglacial Lake Vostok, Antarctica. Limnol. Oceanogr. 51: 2485–2501.

Collins, T. and R. Margesin. 2019. Psychrophilic lifestyles: mechanisms of adaptation and biotechnological tools. Appl. Microbiol. Biotechnol. 103: 2857–2871.

Cullen, N.J., P. Sirguey, T. Mölg, G. Kaser, M. Winkler and S.J. Fitzsimons. 2013. A century of ice retreat on Kilimanjaro: the mapping reloaded. The Cryosphere 7: 419–431.

da Rocha Ribeiro, R., J. Cardia Simoes and E. Ramirez. 2017. The Amazon glaciers. In: Glacier Evolution in a Changing World (DaniloGodoneed). IntechOpen. doi: 10.5772/intechopen.70490.

Dieser, M., M. Greenwood and C.M. Foreman. 2010. Carotenoid pigmentation in Antarctic heterotrophic bacteria as a strategy to withstand environmental stresses. Arctic Antarct. Alp. Res. 42: 396–405.

Edwards, A. 2015. Coming in from the cold: potential microbial threats from the terrestrial cryosphere. Front. Earth Sci. 3: 12. doi: 10.3389/feart.2015.00012.

Elster, J., R.J. Delmas, J.-R. Petit and K. Reháková. 2007. Composition of microbial communities in aerosol, snow and ice samples from remote glaciated areas (Antarctica, Alps, Andes). Biogeosciences Discuss. 4: 1779–1813.

Fegel, T.S., J.S. Baron, A.G. Fountain, G.F. Johnson and E.K. Hall. 2016. The differing biogeochemical and microbial signatures of glaciers and rock glaciers. J. Geophys. Res. Biogeosci. 121: 919–932.

Feller, G. 2013. Psychrophilic enzymes: from folding to function and biotechnology. Scientifica, Article ID 512840, 28 pages. http://dx.doi.org/10.1155/2013/512840.

Freeman, B.G., J.A. Lee-Yaw, J.M. Sunday and A.L. Hargreaves. 2018. Expanding, shifting and shrinking: the impact of global warming on species' elevational distributions. Global Ecol. Biogeogr. 27: 1268–1276.

Fritz, S.C., B.E. Brinson, W.E. Billups and L.G. Thompson. 2015. Diatoms at >5,000 meters in the Quelccaya Summit Dome Glacier, Peru. Arctic Antarct. Alp. Res. 47: 369–374.

Garretto, A., T. Hatzopoulos and C. Putonti. 2019. Vir mine: automated detection of viral sequences from complex metagenomic samples. Peer J. 7: e6695. http://doi.org/10.7717/peerj.6695.

González-Toril, E., R. Amils, R.J. Delmas, J.-R. Petit, J. Komarek and J. Elster. 2009. Bacterial diversity of autotrophic enriched cultures from remote, glacial Antarctic, Alpine and Andean aerosol, snow and soil samples. Biogeosciences 6: 33–44.

Goodwin, K., M.G. Loso and M. Braun. 2012. Glacial transport of human waste and survival of fecal bacteria on Mt. McKinley's Kahiltna Glacier, Denali National Park, Alaska. Arctic Antarct. Alp. Res. 44: 432–445. doi: 10.1657/1938-4246-44.4.432.

Hamilton, T.L., J.W. Peters, M.L. Skidmore and E.S. Boyd. 2013. Molecular evidence for an active endogenous microbiome beneath glacial ice. ISME J. 7: 1402–1412.

Hastenrath, S. 1984. The Glaciers of Equatorial East Africa. Dordrecht, Boston, and Lancaster: Reidel.

Hodson, A., A.M. Anesio, M. Tranter, A. Fountain, M. Osborn, J. Priscu, et al. 2008. Glacial ecosystems. Ecol. Monogr. 78: 41–67.

Hotaling, S., D.S. Finn, J.J. Giersch, D.W. Weisrock and D. Jacobsen. 2017a. Climate change and alpine stream biology: progress, challenges, and opportunities for the future. Biol. Rev. CambPhilos. Soc. 92: 2024–2045. doi: 10.1111/brv.12319.

Hotaling, S., E. Hood and T.L. Hamilton. 2017b. Microbial ecology of mountain glacier ecosystems: biodiversity, ecological connections and implications of a warming climate. Environ. Microbiol. 19: 2935–2948. doi: 10.1111/1462-2920.13766.

Huss, M. and D. Farinotti. 2012. Distributed ice thickness and volume of all glaciers around the globe. J. Geophys. Res. 117: F04010.

Irvine-Fynn, T.D.L. and A. Edwards. 2013. A frozen asset: the potential of flow cytometry in constraining the glacial biome. Cytometry A. 85: 3–7. doi: 10.1002/cyto.a.22411.

Kaser, G. and B. Noggler. 1991. Observations on Speke Glacier, Rwenzori Range, Uganda. J. Glaciology 37: 313–318.

Kaser, G. 1999. A review of the modern fluctuations of tropical glaciers. Glob. Planet. Change 22: 93–103.

Kaser, G. 2001. Glacier-climate interaction at low latitudes. J. Glaciology 47: 195–204.

Kaser, G., C. Georges, I. Juen, T. Mölg, P. Wagnon and B. Francou. 2004a. The behavior of modern low-latitude glaciers. PAGES News 12: 15–17.

Kennedy, D. and B. Hanson. 2006. Ice and History. Science 311: 1673.

Kincaid, J.L. 2007. An assessment of regional climate trends and changes to the Mt. Jaya glaciers of Irian Jaya. (MSc thesis, Texas A&M University).

Kuja, J.O., H.M. Makonde, A.T. Muigai, A. Omire, H.I. Boga and J. Uetake. 2018a. The status of Lewis Glacier of Mount Kenya and the threat to Novel microbial communities. Int. J. Microbiol. Mycol. 7: 6–13.

Kuja, J.O., H.M. Makonde, H.I. Boga, A.T.W. Muigai and J. Uetake. 2018b. Phylogenetic diversity of prokaryotes on the snow-cover of Lewis Glacier in Mount Kenya. Afr. J. Microbiol. Res. 12: 574–579.

Liu, Y., J. Priscu, T. Yao, T. Vick-Majors, A. Michaud and L. Sheng. 2019. Culturable bacteria isolated from seven high-altitude ice cores on the Tibetan Plateau. J. Glaciol. 65: 29–38.

Lutz, S., A.M. Anesio, S.E.J. Villar and L.G. Benning. 2014. Variations of algal communities cause darkening of a Greenland glacier. FEMS Microbiol. Ecol. 89: 402–414.

Lutz, S., A.M. Anesio, R. Raiswell, A. Edwards, R.J. Newton, F. Gill and L.G. Benning. 2016. The biogeography of red snow microbiomes and their role in melting arctic glaciers. Nature Communications volume 7, Article number: 11968.

Margesin, R. and G. Feller. 2010. Biotechnological applications of psychrophiles. Environ. Technol. 31: 835–844. doi: 10.1080/09593331003663328.

Margesin, R. and T. Collins. 2019. Microbial ecology of the cryosphere glacial and permafrost habitats: current knowledge. Appl. Microbiol. Biotechnol. 103: 2537. https://doi.org/10.1007/s00253-019-09631-3.

Marizcurrena, J.J., M.F. Cerdá, D. Alem and S. Castro-Sowinski. 2019. Living with pigments: the colour palette of antarctic life. pp. 65–82. *In*: S. Castro-Sowinski [ed.]. The Ecological Role of Microorganisms in the Antarctic Environment, Springer Polar Sciences. Springer Nature Switzerland AG 2019.

Marks, R., E. Górecka, K. McCartney and W. Borkowski. 2017. Rising bubbles as mechanism for scavenging and aerosolization of diatoms. Ocean Sci. Discuss. https://doi.org/10.5194/os-2017-82.

Meola, M., A. Lazzaro and J. Zeyer. 2015. Bacterial composition and survival on Sahara dust particles transported to the European Alps. Front. Microbiol. 6: 1–17.

Mieczen, T., D. Gorniak, A. Swiatecki, M. Zdanowski, M. Tarkowska-Kukuryk and M. Adamczuk. 2013. Vertical microzonation of ciliates in cryoconite holes in Ecology Glacier, King George Island. Polish Polar Res. 34: 201–212.

Mikucki, J.A. and J.C. Priscu. 2007. Bacterial diversity associated with Blood Falls, a subglacial outflow from the Taylor Glacier, Antarctica. Appl. Environ. Microbiol. 73: 4029–4039.

Mitchell, A.C., M.J. Lafreniere, M.L. Skidmore and E.S. Boyd. 2013. Influence of bedrock mineral composition on microbial diversity in a subglacial environment. Geology 41: 855–858.

Miteva, V. 2008. Bacteria in snow and glacier ice. pp. 31–50. *In*: R. Margesin, F. Schinner, J.C. Marx and C. Gerday [eds.] Psychrophiles: from Biodiversity to Biotechnology. Springer, Berlin, Heidelberg.

NASA Annual Report. 2012. Survability of icy worlds. Available at: https://nai.nasa.gov/annual-reports/2012/jpl-icy-worlds/survivability-of-icy-worlds/.

NASA Earth Observatory. Ice loss on Puncak Jaya. Available at: https://earthobservatory.nasa.gov/images/79084/ice-loss-on-puncak-jaya.

NASA Global Climate Change. 2018. Last glacier standing in Venezuela. Available at: https://climate.nasa.gov/news/2792/last-glacier-standing-in-venezuela/.

Nicholson, W.L., N. Munakata, G. Horneck, H.J. Melosh and P. Setlow. 2000. Resistance of *Bacillus* endospores to extreme terrestrial and extraterrestrial environments. Microbiol. Mol. Biol. Rev. 64: 548–572. Available from: http://dx.doi.org/10.1128/MMBR.64.3.548-572.2000.

Pelto, M. and W.G.M.S. Network. 2018. Alpine glaciers in state of the climate in 2017. Bull. Am. Meteorol. Soc. 99: S23–S25.

Ponce, A., S.M. Beaty, C. Lee, A.C. Noell, C.N. Stam and S.A. Connon. 2011a. Microbial Habitat on Kilimanjaro's Glaciers. 42nd Lunar and Planetary Science Conference, held March 7–11, 2011 at The Woodlands, Texas. LPI Contribution No. 1608, p. 2645. Available at: https://www.researchgate.net/publication/252351729_Microbial_Habitat_on_Kilimanjaro's_Glaciers.

Ponce, A., R.C. Anderson and C.P. McKay. 2011b. Microbial Habitability in Periglacial Soils of Kilimanjaro. Analogue Sites for Mars Missions (6018.pdf). Available at: https://www.lpi.usra.edu/meetings/analogues2011/pdf/6018.pdf.

Price, P.B. 2000. A habitat for psychrophiles in deep Antarctic ice. Proc. Natl. Acad. Sci. USA. 97: 1247–1251.

Price, P.B. 2007. Microbial life in glacial ice and implications for a cold origin of life. FEMS Microbiol. Ecol. 59: 217–231.

Priscu, J. and B. Christner. 2004. Earth's icy biosphere, pp. 130–145. *In*: A. Bull [ed.]. Microbial Diversity and Bioprospecting. ASM Press, Washington, DC. doi: 10.1128/9781555817770.ch13.

Priscu, J.C., B.C. Christner, C.M. Foreman and G. Royston-Bishop. 2007. Ice core methods. Biological Material. pp. 1156–1167. *In*: S.A. Elias [ed.]. Encyclopedia of Quaternary Science, Elsevier, Amsterdam. https://doi.org/10.1016/B0-44-452747-8/00335-5.

Rassner, S.M.E. 2017. Viruses in glacial environments. pp. 111–131. *In*: R. Margesin [ed.]. Psychrophiles: From Biodiversity to Biotechnology. doi: 10.1007/978-3-319-57057-0_6.

RGI Consortium Randolph Glacier Inventory (v.6.0): A Dataset of Global Glacier Outlines. Global Land Ice Measurements from Space, Boulder, Colorado USA (RGI Technical Report, 2017). https://doi.org/10.7265/N5-RGI-60.

Rogers, S.O., W.T. Starmer and J.D. Castello. 2004. Recycling of pathogenic microbes through survival in ice. Med. Hypotheses. 63: 773–777.

Rondón, J., W. Gómez, M.M. Ball, A. Melfo, M. Rengifo, W. Balcázar, et al. 2016. Diversity of culturable bacteria recovered from Pico Bolívar's glacial and subglacial environments, at 4950 m, in Venezuelan tropical Andes. Can. J. Microbiol. 62: 1–14. doi: 10.1139/cjm-2016-0172.

Rondón, J., M.M. Ball, L.T. Castro and L.A. Yarzábal. 2019. Eurypsychrophilic *Pseudomonas* spp. isolated from Venezuelan tropical glaciers as promoters of wheat growth and biocontrol agents of plant pathogens at low temperatures. Environ. Sust. https://doi.org/10.1007/s42398-019-00072-2.

Segawa, T., N. Takeuchi, A. Rivera, A. Yamada, Y. Yoshimura, G. Barcaza, et al. 2013. Distribution of antibiotic resistance genes in glacier environments. Environ. Microbiol. Reports 5: 127–134. doi: 10.1111/1758-2229.12011.

Siraj, A.S., M. Santos-Vega, M.J. Bouma, D. Yadeta, D. Ruiz Carrascal and M. Pascual. 2014. Altitudinal changes in malaria incidence in highlands of Ethiopia and Colombia. Science 343: 1154–1158.

Skidmore, M., S.P. Anderson, M. Sharp, J. Foght and B.D. Lanoil. 2005. Comparison of microbial community compositions of two subglacial environments reveals a possible role for microbes in chemical weathering processes. Appl. Environ. Microbiol. 71: 6986–6997.

Smith, A.W., D.E. Skilling, J.D. Castello and S.O. Rogers. 2004. Ice as a reservoir for pathogenic human viruses: specifically, caliciviruses, influenza viruses, and enteroviruses. Med. Hypotheses. 63: 560–566.

Soucy, S.M., J. Huang and J.P. Gogarten. 2015. Horizontal gene transfer: building the web of life. Nat. Rev. Genet. 16: 472–482. doi: 10.1038/nrg3962.

Stibal, M., J. Telling, J. Cook, K.M. Mak, A. Hodson and A.M. Anesio. 2012. Environmental controls on microbial abundance and activity on the Greenland ice sheet: a multivariate analysis approach. Microb. Ecol. 63: 74–84.

Stibal, M., E. Gözdereliler, K.A. Cameron, J.E. Box, I.T. Stevens, J.K. Gokul, et al. 2015. Microbial abundance in surface ice on the Greenland Ice Sheet. Front. Microbiol. 6: 225. doi: 10.3389/fmicb.2015.00225.

Thompson, L.G., E. Mosley-Thompson, H. Brecher, M. Davis, B. León, D. Les, et al. 2006. Abrupt tropical climate change: past and present. Proc. Natl. Acad. Sci. USA (PNAS) 103: 10,536–10,543.

Thompson, L.G., E. Mosley-Thompson, M.E. Davis and H.H. Brecher. 2011. Tropical glaciers, recorders and indicators of climate change, are disappearing globally. Ann. Glaciol. 52: 23–34.

Tranter, M., A. Fountain, C. Fritsen, B. Lyons, J. Priscu, P. Statham, et al. 2004. Extreme hydrochemical conditions in natural microcosms entombed within Antarctic ice. Hydrol. Process. 18: 379–387.

Tung, H.C., N.E. Bramall and B.P. Price. 2005. Microbial origin of excess methane in glacial ice and implications for life on Mars. Proc. Natl. Acad. Sci. USA (PNAS) 102: 18292–18296.

Turchetti, B., L. Selbmann, R.A. Blanchette, S. DiMauro, E. Marchegiani, L. Zucconi L., et al. 2015. *Cryptococcus vaughanmartiniae* sp. nov and *Cryptococcus onoforii* sp. nov.: two new species isolated from worldwide cold environments. Extremophiles 19: 149. doi: 10.1007/s00792-014-0692-3.

U.S. Geological Survey. What is a glacier? Available at: https://www.usgs.gov/faqs/what-a-glacier?qt-news_science_products=0#qt-news_science_products.

Uetake, J., S. Tanaka, K. Hara, Y. Tanabe, D. Samyn, H. Motoyama, et al. 2014. Novel biogenic aggregation of moss gemmae on a disappearing African Glacier. PLoS ONE 9(11): e112510. pmid:25401789.

Ushida, K., T. Segawa, S. Kohshima, N. Takeuchi, K. Fukui, Z. Li, et al. 2010. Application of real-time PCR array to the multiple detection of antibiotic resistant genes in glacier ice samples. J. Gen. Appl. Microbiol. 56: 43–52. doi: 10.2323/jgam.56.43.

Veettil, B.K., S. Wang, S. Florêncio de Souza, U.F. Bremer and J.C. Simões. 2017. Glacier monitoring and glacier-climate interactions in the tropical Andes: a review. J. South Am. Earth Sci. doi: 10.1016/j.jsames.2017.04.009.

Veettil, B.K. and S. Wang. 2018. State and fate of the remaining tropical mountain glaciers in australasia using satellite imagery. S. J. Mt. Sci. 15: 495. https://doi.org/10.1007/s11629-017-4539-0.

Weertman, J. and G.E. Birchfield. 1983. Basal water film, basal water pressure, and velocity of traveling waves on glaciers. J. Glaciol. 29: 20–27.

Weide, D.M., S.C. Fritz, B.E. Brinson, L.G. Thompson and W.E. Billups. 2017. Freshwater diatoms in the Sajama, Quelccaya, and Coropuna glaciers of the South American Andes, Diatom Research, doi: 10.1080/0269249X.2017.1335240.

Weil, T., C. De Filippo, D. Albanese, C. Donati, M. Pindo, L. Pavarini, et al. 2017. Legal immigrants: invasion of alien microbial communities during winter occurring desert dust storms. Microbiome 5: 32. doi: 10.1186/s40168-017-0249-7.

Wiens, J.J. 2016. Climate-related local extinctions are already widespread among plant and animal species. PLoS Biol. 14: e2001104.

Wilhelm, L., G.A. Singer, C. Fasching, T.J. Battin and K. Besemer. 2013. Microbial biodiversity in glacier-fed streams. ISME J. 7: 1651–1660.

Zemp, M., M. Huss, E. Thibert, N. Eckert, R. McNabb, J. Huber, et al. 2019. Global glacier mass changes and their contributions to sea-level rise from 1961 to 2016. Nature 568: 382–386.

Zimmer, A., R.I. Meneses, A. Rabatel, A. Soruco, O. Dangles and F. Anthelme. 2018. Time lag between glacial retreat and upward migration alters tropical alpine communities. Perspect. Plant Ecol. Evol. Syst. 30: 89–102. http://dx.doi.org/10.1016/j.ppees.2017.05.003.

2

The Microbial Diversity of Hot Springs Located in Himalayan Geothermal Belts (HGB)

Sayak Das and Nagendra Thakur[*]

Introduction

The Himalayan Geothermal Belt (HGB) have canaliculated topography, diversified contours, psychedelic valleys, pulchritudinous landscapes and is regarded as among the largest geothermal areas in our Earth's domain. The HGB banquets the entire Himalayas covering more than 3,500 km spanning from Afghanistan, Tajikistan, Pakistan, North West Indian Himalayas (regions comprising of Jammu-Kashmir, Ladakh, Himachal Pradesh, and Uttarakhand), Nepal, Bhutan, North East Indian Himalayas (regions comprising of Sikkim, Assam, Meghalaya, and Arunachal Pradesh), Tibet, and Yunnan Provinces of China, Myanmar, and Thailand. This belt is reminiscent of the great geotectonic collision of the Indian Plate while merging to the Eurasian Plate. The term "Himalayan Geothermal Belt" was first defined by Tong and Zhang in 1981 (Hochstein and Regenauer-Lieb 1998).

Geological Setting

Afghanistan lounges in the collision contact zone of the Indo-Pakistan-Afghan block (Treolar and Izatt 1993). This province is known as the Hindu Kush Geothermal Province. The merging of Gondwana Plate into the Eurasian Plate elated the Hindu Kush region. This neotectonic shift created fracture all along the Herat-Panjshir fault allowing in water seepage into the Earth's crust, superheated zones, creating geothermal fluids that are spread around the geothermal active areas in Afghanistan (Saba et al. 2004). This geothermal province is spread across Herat, Panjao, Ghorband, Panjshir, Badakhstan to Pamirs and even extending till Tajikstan (SAARC Energy Centre 2011). Many thermal and hot springs have been found in this region. For example, Herat province hosts a N_2-SiO_3 hot spring at Obe district (45°C-55°C); the Panjao-Bande Amir region contains many bicarbonate hot springs (24°C-35°C); Kalu Valley hot springs (35°C-65°C); Ghorband Valley hot springs (25°C-40°C); Khwaja Qeech hot springs (35°C-47°C); Ghorghauri hot springs (32°C-43°C); Qala-e-Saraab nitrogen hot springs (20°C-42°C) of Andarab region; sulfur-rich Bobe-Tangi hot springs (25°C-41°C) of Wakhan region and Sarghaliyan hot springs (34°C-40°C) of Badakhshan region (Kurenoe and Belianin 1969).

Department of Microbiology, School of Life Sciences, Sikkim University, 6th Mile, Samdur, Gangtok, 737102-Sikkim.
[*]Corresponding author: nthakur@cus.ac.in

The Pakistan Himalayan Geothermal provinces host many hot springs in the Himalaya-Karakoram-Hindu Kush region. More than 20 hot springs lie in this fault line. The hot springs of Murtazabad are rich in solfataric mud and emit a sulfur smell. Here, seven hot springs are reported to exist along the fault line of Main Karakoram Thrust, and their temperature ranges from 45°C to ~100°C. There are hot springs near the confluence of river Bundelas (34°C to 46°C), Gilgit (27°C to 56°C), Rakhiot Valley (34°C to 60°C), Moshkin Valley (45°C to 52°C), Darkot hot spring (21°C to 43°C) of Yasen Valley, Chitral region (41°C to 60°C) and Choutron hot spring (40°C to 50°C) of Basho Valley (SAARC Energy Centre 2011).

The Indian Himalayan Geothermal Belt can be classified into two groups: North-western India Himalayan Geothermal Belt (nwIHGB) and North-eastern India Himalayan Geothermal Belt (neIHGB).

North West Indian Geothermal Province covers the hot springs of Jammu, Kashmir, Ladakh, Himachal Pradesh, and Uttarakhand. Puga and Chumathang are the most famous geothermal provinces of Ladakh, which are being planned to harness the geothermal energy and utilize it to produce electricity. The hot springs of Himachal Pradesh lie mostly in the Beas Valley (Bashist, Kalath, Rampur, and Kulu) and Parbati Valley (Jan, Kasol, Manikaran, Pulga, and Khirganga) of Kulu district. The temperature of Beas Valley hot springs ranges from 20°C to 67°C, and in Parbati Valley it ranges from 20°C to ~100°C. The Uttarakhand Geothermal Provinces lie along the confluence of the river Alaknanda in the Dhauliganga Valley of Chamoli district. Many hot springs are located at the Alaknanda valley and Dhauliganga Valley, where their temperature varies from 23°C to 70°C.

The geothermal provinces of Nepal constitute the most diverse fault lines and varied topographical clusters. Majorly it contains Main Frontal Thrust, Main Boundary Thrust, Main Central Thrust, the South Tibetan Detachment System, and the Indo-Tsangpo Suture Zone tectonic plates (SAARC Energy Centre 2011). Most of the hot springs lie along the Main Central Thrust and few are scattered at Main Boundary Fault. The famous hot springs of Nepal are located at various districts of Dharchula, Bajhang, Jumla, Dhanchauri area, Riar, Mayangdi, Surai Khola, Mustang, and western regions (Bhurung Tatopani, Dokhola, Singha Tatopani) respectively.

The Geothermal Provinces of Bhutan falls under Main Boundary Thrust (MBT) and has several areas characterized by geothermally superheated hot springs. The thermal springs can be found at Punakha, Gasa, Zhembang, Bumthang, and Gelpehu areas (SAARC Energy Centre 2011).

The North East Indian Himalayan Geothermal provinces lie in the states of Sikkim, Assam, Meghalaya, and Arunachal Pradesh. Toward further east, Tibetan and Yunnan Geothermal Provinces host more than 1,000 hot springs and are a very significant geothermal area of the Himalayan Geothermal Belt. The ending of this belt touches northern Myanmar and Thailand, where the hottest geothermal zones can be found.

Hot Spring Microbiology and Their Significance

The most prodigious gift of nature if anyone has to consider, then undoubtedly, it has to be the extreme environmental conditions. The word "extreme" has been concocted by mankind as the abiotic parameters governing these niches or ecosystem is beyond human adaptive physiological capabilities. Our cognitive functioning and metabolomics cannot explain or survive the ruthlessness of nature's extreme ecosystems, be it on the basis of temperature, pH, or salinity or atmospheric pressure. The unexpected serendipity of *Thermus aquaticus* fueled a new dimension to biology and gave birth to numerous academic fields and researches. The countenance of hot springs is their invaluable microflora and fauna, which has gained impetus in recent decades. Now, the hot spring microbiology is regarded as the hotspot of research in the arena of microbial ecology (López-López et al. 2013).

The microorganisms present in such hot springs can thrive under extreme temperatures. A major term conferring microorganisms surviving in harsh or extreme ecosystems are known as extremophiles. These environments hold various combinations of extreme conditions, such as enormously high or low pH or enormously high or low temperature, high pressure, high salinity, or other mishmashes thereof (Aanniz et al. 2015, Elleuche et al. 2015). These extreme or exhilarating environments and the microorganisms living in them have opened the doors for the search of life exterior to Earth and also given leeway to transfer life from one planet to another (Rothschild and Mancinelli 2001). The discovery of extremophiles has also poured vivacity into the biotech and other industries (Madigan and Marrs 1997). Mac Elroy coined the term "extremophiles" (Greek "philos" means lovers) i.e., lovers of extreme environments (Mac Elroy 1974). The resulting environments, based on the elevated or low conditions, such as high-temperature, low temperature, low pH, high pH, and salinity, are then known as thermophilic, psychrophilic, acidophilic, alkaliphiles, and halophilic respectively (Rothschild and Mancinelli 2001). As the hot springs are high-temperature ecosystems, thus our focus will be mainly on thermophiles. Thermophile also carried a diction from Greek ("thermos" means heat and "philos" means lovers) and thus are the lovers of heat or high-temperature. Thermophiles are generally designated as the organisms, which can grow above 45°C (Madigan and Oren 1999). These possess an optimum range of temperature from 55-80°C; however, the microorganisms growing above 80°C are referred to as hyperthermophiles (Bertoldo and Antranikian 2002).

However, currently, thermophiles are classified into moderate thermophiles (organisms possessing optimal growth temperatures between 50-64°C with maximum growth at a temperature below 70°C), extreme thermophiles (organisms possessing optimal growth temperatures between 65-85°C with maximum growth at a temperature above 70°C), and hyperthermophiles (organisms possessing optimal growth temperatures above 85°C with maximum growth at a temperature above 90°C) (Wiegel 2001).

Thermophiles have a place within two phylogenetically, altogether different spaces of life, bacteria, and archaea (Stetter 1999). It is generally confined that thermophilic bacteria is dominating the community at temperatures between 50°C and 90°C in most hydrothermal environments. Among bacteria, there are just a couple of species that can be called hyperthermophiles, for example, *Thermotoga* and *Aquifex* that have an ideal temperature in the scope of 90 to 95°C (Huber and Stetter 2000). In environments with a temperature above 90°C Archaea are dominating (Reysenbach and Shock 2002), such as *Pyrolobus fumarii* that has a temperature optimum of 106°C and can live at temperature up to 113°C. However, the first hyperthermophilic archaea found in enormously hot and acidic hot spring was *Sulfolobus acidocaldarius* (Brock et al. 1972).

In view of their ideal temperature, some thermophilic microorganisms having a place with direct thermophiles include *Bacillus caldolyticus*, *Geobacillus stearothermophilus*, *Thermoactinomyces vulgaris*, *Clostridium thermohydrosulfuricum*, *Thermoanaerobacter ethanolicus*, *Thermoplasma acidophilum*, etc. The group termed extreme thermophiles includes both bacterial and archaeal thermophiles. Among aerobic bacteria, several species include *Bacillus caldolyticus*, *B. caldotenax*, *Bacillus Schlegelii*, *Hydrogenobacter thermophilus*, *Thermothrix thiopara*, *Thermus thermophilus*, *Thermus filiformis*, *Thermomicrobium roseum*, and *Calderobacterium hydrogenophilum*; some anaerobic thermophilic bacteria include *Dyctioglomus thermophilum*, *Thermosipho africanus*, *Thermotoga maritima* and *Thermotoga neapolitana*, *Fervidobacterium pennavorans*, *Acetomicrobium faecalis* and some archaea include *Thermodesulfobacterium commune*, *Sulfolobus acidocaldarius*, *Methanococcus vulcanicus*, and *Sulfurococcus mirabilis*. The hyperthermophiles include *Methanoccus jannaschii*, *Acidianus infernos*, *Archaeoglobus profundus*, *Methanopyrus kandleri*, *Pyrobaculum islandicum*, *Pyrococcus furiosus*, *Pyrodictiumoccultum*, *Pyrolobus fumarii*, *Thermococcus littoralis*, *Ignicoccus islandicum*, etc. (Ghosh et al. 2003, Kristjansson et al. 1986, Reysenbach and Shock 2002, Stetter 1999, Wiegel 2001).

The thermophilic microorganisms are perceived by their metabolic thermo-security, which are floated by their thermophilic proteins. The thermostability of the thermophilic catalysts has been built up as important biocatalysts for different biotechnological and modern purposes (Niehaus et al. 1999). Taq-polymerase from *Thermus aquaticus* laid the cornerstone for polymerase chain reaction (PCR) technique (Brock and Freeze 1969, Brock 1997). From the biotechnological standpoint, the thermophilic microorganisms are the most attractive organisms due to their ability to produce enzymes proficient to catalyze industrial relevant process at a higher temperature than corresponding enzymes from mesophiles. Extracellular-polymer-degrading enzymes and DNA-modifying enzymes have several applications in food, chemical and pharmaceutical industries, and environmental biotechnology (Ladenstein and Antranikian 1998, Synowiecki 2010).

Here, in this chapter, a panned view of the Himalayan Geothermal Belt and their microbial diversity has been presented. These areas, due to their rugged topography and unavailability to access the hot springs, make it a major challenge for the researchers to decipher the microbial diversity of these hot springs. Thus, the future provides a unique opportunity to understand these hot springs, and their study will help the scientific community to hunt biotechnologically and industrially important microbes.

A. Pamir Geothermal Province Hot Springs (Pam-GPHS) and Its Microbial Diversity

This geothermal province is located in Tajikistan. They are extremely mountainous terrains and geographically isolated. It is sandwiched in between China at the east front, Pakistan at the southern front, Afghanistan at the western front, and Kyrgyzstan at the northern front (Breu and Hurni 2003). The geological features of Pamir Mountains begin at the Himalaya-Hindukush massif of HGB. Due to the tectonic collision, orogeny and denudation are common characteristic phenomena.

There are supposedly many unaccounted hot springs of both cold as well as thermal in nature. There are approximately 72 thermal discharging spots of various physicochemical characteristics, and some also emit sulfur and nitrogen gases. On the basis of carbonate content of the thermal springs are classified into three types: (a) Essentuki type (comprising of hot springs located at Vyazdara, Iniv, Churzh, Darshan, Garm Chaashma, Udit, Anderob, Avci, Daraistazh, and Barshor), (b) Narzanov type (comprising of hot springs located at Shirgin, Dustiroz, Vrang, Rivak, Saryshitharv, Nemats, Nerhun, Hozhuni, and Zhunt), and (c) Borjomi type (comprising of hot springs located at Bahmyr, Achiktash, Mihmandzhuly, Bakhtiar, and Dzhartygumbez). Among them, temperature ranging from 20°C to 62°C have been observed at the hot springs of Bahmyr, Garm Chaasma, Darshan, Dzhartygumbez, Avci, and Shirgin.

Microbiology: There have been very early studies on the bacteriology of these hot springs that suggested that they might harbor *Thermus flavus* and *Thermus ruber* and its closely associated-like species (Egorova and Loginova 1975, Loginova and Bogdanova 1977). However, most of the studies have been on the algal communities so far (Barinova and Niyatbekov 2017, Barinova and Niyatbekov 2018, Jumaeva 2008, Niyatbekov and Barinova 2018). A huge diversity of species (243 taxa) belonging to the algal genus has been reported, such as *Achnanthes* sp., *Achnanthidium* sp., *Actinella* sp., *Bertalot* sp., *Amphora* sp., *Aneumastus* sp., *Anomoeoneis* sp., *Brachysira* sp., *Caloneis* sp., *Cavinula* sp., which were some of the major findings (Niyatbekov and Barinova 2018).

B. North Pakistan Geothermal Province Hot Springs (NorPak-GPHS) and Its Microbial Diversity

This geothermal province contains many hot springs. The microbial world of these hot springs is concealed and most of them are unexplored. Very few recent studies are going on the various hot springs of HGB, which are located in Chu Teran, Kotli, Mashkin, Gilgit, Murtazabad, Karachi, and Budelas areas (Zaigham et al. 2009). Most of the culture-dependent studies have reported of diversified bacteria, like *Pseudomonas* sp., *Bacillus* sp., *Thermus* sp., *Geobacillus* sp., which sowed industrially important biotechnological properties (Asad et al. 2011, Ghumro et al. 2011, Javed et al. 2012, Khan et al. 2011, Khan et al. 2018, Niaz et al. 2010, Qadar et al. 2009, Rafique et al. 2010, Saleem et al. 2012, Siddiqui et al. 2014, Zahoor et al. 2012, Zaidi 2007).

The hot springs from Kotli district (Azad Kashmir, Pakistan), Gilgit, and Karachi have been discussed here.

i. Kotli Geothermal Hot Springs

The hot springs in these regions are locally called as "Tatta Pani", meaning "hot water". The Kotli district of Azad Kashmir (Pakistan) hosts sulfuric hot spring – Tatta Pani. There are approximately eight discharging areas located around 33°36'43.57"N-73°56'49.35"E in the vicinity of the river Poonch (six thermal springs) and river Kotli (two thermal springs) confluence (Anees et al. 2017). These geothermal zones are situated 26 km north from Kotli's main city. The Tatta Pani area lies on the southeastern slope of the Hazara-Kashmir Syntaxis (HKS), which is part of HGB (Anees et al. 2017). It is located on the Poonch river confluence at its right bank side. The geothermal heat produced might be due to the MBT friction of the Himalayan Geothermal Belt (HGB). The *in situ* temperature of this hot spring ranges from 79°C-86°C and is neutral in pH (6.93). The temperature and pH of the hot spring has a seasonal variation (Khan et al. 1999) and was reported to have a lower temperature earlier. The hot spring is of bicarbonate type (Anees et al. 2015) and had high amounts of Na^+ cations (208 mgL^{-1}), Cl$^-$ ions (58.3 mgL^{-1}), and NO_3^{2-} (26.23 mgL^{-1}) (Zahoor et al. 2016). It had a high electrical conductance of 462.6 μScm^{-1} as it showed a considerable amount of total dissolved solids (TDS) value of 231.3 mgL^{-1} (Zahoor et al. 2016). Thus, it can be understood that this hot spring is rich in minerals, and radon and tritium studies were also conducted that showed the presence of naturally occurring radioactivity among the samples (Anees et al. 2015).

Microbiology: From this hot spring, a novel Gram-positive thermophilic bacterium was reported. The 16S rRNA gene (GeneBank Accession No.: JQ284017) of this rod-shaped bacterium TP-2 was claimed to have only 89% similarity with *Geobacillus debilis* (Zahoor et al. 2016). It was able to hydrolyze gelatin and metabolize ortho nitrophenyl-β-D-galactopyranosidase (ONPG) at 65°C. Zahoor and his co-workers demonstrated that this bacterium could also produce many significant and industrially important enzymes, like xylanase, lipase, FPase, protease, and CMCase.

A thermophilic amylase producing actinomycetes *Thermoactinomyces sacchari* and few unidentified strains of thermophilic amylase producing *Bacillus* sp. were also reported from Tatta Pani (Jadoon et al. 2014). Similarly, Erum and his co-workers reported a thermoalkalophilic bacterium strain KP10, showing 88% similarity to *Bacillus clausii* (GenBank Accession No.: KX013388) from this hot spring (Erum et al. 2017).

ii. Gilgit Geothermal Hot Springs

This geothermal province of Northern Pakistan hosts many hot springs along the valleys of Gilgit, Yasin, and the Hunza region. The Gilgit Tatta Pani is situated at the Indus river confluence at its right bank side on the Karakoram Highway. They are located 1,200 m above the mean sea level. Murtazabad has two alkaline hot springs featuring 80°C temperature near the Khunjerab river confluence in the Hunza Valley; one at Murtazabad Zairen and another at Murtazabad Balai, where the latter is located above the former geologically (Javed et al. 2012). Another alkaline hot spring can be found at Darkut Pass having 62°C temperature in the Yasin valley above the Rawat basecamp at an altitude of 4,650 above mean sea level.

Microbiology: From these hot springs, three small Gram-negative facultative thermophilic bacterial rod strains GCTP-1 (Gilgit Tatta Pani), GCMB-1 (Murtazabad), and GCDP-1 (Darkut Pass) were reported from the geothermal province of Gilgit region (Javed et al. 2012). A lipase producing bacterium *Geobacillus* sp. SB-4S was found from Gilgit Hot Spring province (Tayyab et al. 2011), which was metabolically active at 75°C. Another novel bacterium *Geobacillus thermopakistaniensis* strain MAS1 was discovered from this hot spring (Siddiqui et al. 2014). These hot springs favored more *Geobacillus* sp. in its niche.

iii. Karachi Geothermal Hot Springs

This geothermal province marks the ending of HGB at the western front. The most famous hot spring found in this zone is located at the foothills of Hallar mountains, which is approximately 13 km north to Karachi. This hot spring is known as Mangho Pir Hot Spring and is renowned for its balneotherapeutic properties. The mud bath is a very common practice at this hot spring. The temperature of this hot spring ranges from 40°C to 49°C and is alkaline. It is rich in sulfur content ranging from 210-250 mgL^{-1}. This hot spring is a thermally mineral-rich and has the highest Na$^+$ content (350 mgL^{-1}-600 mgL^{-1}) among all other hot springs (Jahangir et al. 2001). The conductivity ranges from 2,700-3,500 mgL^{-1} on an average (Jahangir et al. 2001). This hot spring is a bicarbonate type of thermal spring. Due to its rich mineral content, the balneotherapy is very prevalent in this spring and possibly has cured many ortho-related diseases (Javed et al. 2009).

Microbiology: Earlier reports of Gram-positive thermophilic rods, mostly belonging to *Bacillus* genus comprising of *B. badius, B. pumilus, B. circulans, B. pulvifacience*, were most prevalent (Chagtai and Siddiqui 1980). Other early microbiological reports suggest it host various microorganisms, like *Beggiato alba, B. minima, Thioploca* sp., Cynaophyta (comprising of *Aphanocapsa* sp., *Chroococcus* sp., *Cylindrospermum stagnale, Merismopedia thermalis, M. glauca, Anabaena* sp., *Nostoc carneum, Oscillatoria amphigranulata, O. subtilissima, O. acuminata, O. tenius, O. limosa, O. angusta, O. okeni, O. amoena, O. chalybea, Phormidium molle, P. laminosum, P. tenue, P. subcapitatum, P. frigidum, Pleurocapsa caldaria, Synechocystis salina, Synechococcus cedyorum*), Cholorophyta (comprising of *Chlorella vulgaris, Chlorococcum humicola, Kirchneriella contorata, Scenedesmus* sp.), and Bacillariophyta (comprising of *Cymbella turgiduia, C. ventricosa, Navicula viridula*) among other algal communities. Most recently, a thermoalkalophilic xylanase producing bacterium *Bacillus pumilus* K22 was reported (Ullah et al. 2019) from this hot spring.

North West Indian Himalayan Geothermal Belt (nwIHGB)

i. Jammu-Kashmir and Ladakh Geothermal Province Hot Springs (JK-Lad-GPHS) and Its Microbial Diversity

This geothermal province has many unexplored hot springs that are recently being studied. On the basis of their geographical distribution, they are divided into five zones: (a) Nubra Valley hot springs of Ladakh (Craig et al. 2013) that are located at Changlung (temperature ranging from 55°C-61°C), Pulthang (temperature ranging from 25°C-30°C), and Panamik (temperature ranging from 75°C-78°C); (b) Chenab Valley hot springs (Craig et al. 2013) are located at Atholi (temperature ranging from 53°C-58°C), Chinkah (50°C-56°C), Galhan (temperature ranging from 55°C-63°C), Gul (38°C-45°C), Kiar (55°C-58°C), Kurah (50°C-55°C), Mahogala (45°C-52°C), Sidhu (58°C-65°C), Sweed (25°C-30°C), Loharna (37°C-42°C), Honzur (35°C-45°C), Tattapani (60°C-75°C), Tatwain (50°C-55°C), and Yurdu (40°C-50°C); (c) Jammu Valley hot spring is located at Rajouri Tatta Pani (50°C-100°C); (d) Kashmir valley hot spring (Craig et al. 2013) is located at Gandakh Nag (20°C-28°C), and (e) Puga-Chumathang Valley hot springs (Craig et al. 2013) are located at Puga (76°C-88°C) and Chumathang (80°C-90°C). Usually, the water of northwest Himalayan geothermal areas is Na-Ca-HCO$_3$-SO$_4$-Cl type.

Microbiology: From the Tatta Pani (33°25′N-74°25′E) of Rajouri district, Jammu and Kashmir, India eight different Gram-positive cellulose-degrading thermophilic bacterial rod strains were reported (Priya et al. 2016). The surface temperature of the hot spring ranges from 50°C-100°C depending on the seasonal variation. The isolates were isolated from the sediment *Bacillus licheniformis* IP_WH4 (GenBank Accession No.: KP842612), *B. aerius* IP_WH3 (GenBank Accession No.: KP842611), *B. licheniformis* IP_WH2 (GenBank Accession No.: KP842610), *B. licheniformis* IP_60Y (GenBank Accession No.: KP842613), *Geobacillus thermodenitrificans* IP_WH1 (GenBank Accession No.:

KP842609), *G. thermodenitrificans* IP_60A1 (GenBank Accession No.: KP842614), *Geobacillus* sp. IP_60A2 (GenBank Accession No.: KP842615), and *Geobacillus* sp. IP_80TP (GenBank Accession No.: KP842616) (Priya et al. 2016).

Similarly, three plant growth-promoting thermophilic Gram-positive rods, (*Bacillus subtilis* BHUJP-H1 (GenBank Accession No.: KU312403); *Bacillus* sp. BHUJP-H2 (GenBank Accession No.: KU312404); *B. licheniformis* BHU-H3 (GenBank Accession No.: KU312405), were described from Chumathang hot spring (Verma et al. 2018). From the Kishtwar district Tatta Pani, a Gram-negative large thermophilic rod, *Flavobacterium thermophilum* (GenBank Accession No.: NR104891.1), and a Gram-positive medium thermophilic rod, *Anoxybacillus* sp. (GenBank Accession No.: EU621359.1), were reported and they could produce acetoin (Sharma et al. 2018a).

ii. Himachal Pradesh Geothermal Province Hot Springs (HP-GPHS) and Its Microbial Diversity

This geothermal province contains many hot springs located at the river confluence of Beas and Sutlej. The major hot springs present in Beas and Parbati valley include Manikaran, Khirganga, Kasol, and Awas. Whereas Satluj and Spiti Valley include Tapri, Chuza-Sumdo, Tattapani, Garam Kund, and Vasisht. The temperature of these hot springs also varies and ranges from 30-100°C. Mannikaran hot spring has the temperature of 94°C, Tattapani has 65°C, Kasol's ranges from 69-89°C, and Vasisht has a temperature of 47°C (Cinti et al. 2009). Most of these hot springs emit gas, like CO_2, CH_4, H_2, and N_2 (Chandrasekharam et al. 2005). The water type of thermal springs of Himachal is mainly Na-Ca-Cl-HCO_3.

Microbiology: Among all the hot springs, the majority of the microbial studies have been on Manikaran hot spring as it has the highest temperature, and also the balneotherapy is very much prevalent in these hot springs. The Manikaran hot springs are located at 1,760 m above the mean sea level between 32°01′N-77°20′48″E., has an average temperature ranging from 70°C to 110°C depending on the seasonal variation with slightly acidic pH as the water is of Na-Ca-Cl-HCO_3. Thermotolerant fungal isolates—*Penicillium citrinum, Paecilomyces variotii, Pichia guilliermondii, Paecilomyces* sp., *Aspergillus sydowii, Myceliophthora thermophila*—from Manikaran hot spring water and soil samples were reported (Verma et al. 2012, Sharma et al. 2013, Suman et al. 2015). Industrially important thermophilic enzymes producing bacteria were found—*Klebsiella* sp., *Lysinibacillus* sp., *Enterobacter cloacae, Exiguobacterium indicum, Stenotrophomonas maltophilia, Acinetobacter baumannii, Rhodococcus qingshengii, Paenibacillus pabuli, P. tylopili, Bacillus* sp., *B. licheniformis, B. subtilis, Microbacterium oxydans, Micrococcus indicus, Pseudomonas fragi, P. reactans, Brevibacillus* sp., *Thermusparvatiensis* (Dwivedi et al. 2012, Dwivedi et al. 2015, Kumar et al. 2014, Sharma et al. 2012, Sharma et al. 2013, Suman et al. 2015, Verma et al. 2014a,b)—from the Manikaran hot springs. A novel species was also discovered—*Lampropedia cohaerens* (Tripathi et al. 2016)—from here. Metagenomic studies showed that Firmicutes were the most predominant type followed by Aquificae and Deinococcos-Thermus group of bacteria, and Crenarchaeota was the most dominant archaeal phylum in the hot spring (Bhatia et al. 2015).

iii. Uttarakhand Geothermal Province Hot Springs (Uttar-GPHS) and Its Microbial Diversity

This geothermal province is the residence of many famous hot springs, like Yamunotri hot spring, Soldhar, Ringigad, Suryakund, Tapt Kund, Gangnani, Bhukki, etc. The temperature of these hot springs ranges from 27°C-100°C. Soldhar (Figure 1a) has the highest recorded temperature of 110°C, Suryakund (57°C-92°C), Ringigad (75°C-95°C), Gangani (45°C-65°C), and Bhukki (35°C-50°C). The average temperature of the northwest hot springs in India is highest among all the other Himalayan geothermal hot spring provinces with slightly alkaline pH and Na-Ca-HCO_3-Cl type of water hydrochemistry. There are also reports of sulfur cold spring Sahastradhara (20°C to

23°C) near river Baldi, Dehradun, which has numerous balneotherapeutic properties (Bhatt and Mir 2015).

District Chamoli of Uttarakhand hosts three important and famous hot springs, Soldhar, Tapovan Kund, and Ringigad, which are situated at Garhwal Himalaya. Soldhar is located near Tapovan between 39°29′25″N and 79°39′29″E at an altitude of 1,900 m above mean sea level at a roadside distance from Joshimath-Malari road. The maximum temperature of Soldhar (Figure 1b) can reach above 95°C on seasons (Kumar and Sharma 2019). Tapovan Kund is near Joshimath, Tapovan (Ranawat and Rawat 2017). Ringigad lies 30°33′14″N and 79°40′0.06″E at an altitude of 1,850 m above mean sea level. The maximum temperature of Ringigad (Figure 1c) is around 90°C (Kumar and Sharma 2019).

Figure 1. Hot Springs of Uttarakhand – (a) Soldhar hot spring, showing algal maps on top (b) Soldhar hot spring with hot water reservoir (c) Ringigad hot spring. Hot Springs of Sikkim – (d) Polok hot spring, West Sikkim (e) Borong hot spring, West Sikkim (f) Reshi hot spring, West Sikkim (g) Yumthang hot spring, North Sikkim (h) Dzongu hot spring, North Sikkim (i) Takrum hot spring, North Sikkim (j) New Yume Samdung hot spring, North Sikkim (k) Old Yume Samdung hot spring, North Sikkim.

Microbiology: From the hot springs of Garhwal, mostly the culture-dependent bacteria belonged to genus *Bacillus*, like *B. sonorensis*, *B. tequilensis*, *B. licheniformis*, and *Paenibacillus ehimensis*, and its closely related branched family genus, like *Paenibacillus* sp., *Geobacillus* sp., *Brevibacillus* sp., *Lysinibacillus* sp., *Aneurinibacillus* sp., and few *Deinococcus* sp., *Staphylococcus epidermidis*, and *Pseudomonas* sp. were reported (Arya et al. 2015, Kumar et al. 2004, Pandey et al. 2014a,b, Ranawat and Rawat 2017, Sharma et al. 2014a, 2009, Trivedi et al. 2006). Many industrially important thermozyme producing bacteria were reported (Dhyani et al. 2017) from the hot springs. The Sahastradhara cold sulfur spring also had many industrially potential bacteria (Rawat et al. 2018).

The algal diversity predominantly belonged to Cyanophyta from the hot springs of Soldhar and Ringigad belonging to *Myxosarcina* sp., *Phormidium bohaneri*, *Synechocystis sallensis*, *Spirulina subsalsa*, *Chlorogloeopsis* sp., *Chroococus turgidus*, *Hydrococcus rivularis*, *Lyngbya hieronyamus*, *Pseudanabaena galeata*, *Spirulina meghiniana*, *Gloeocapsalivida* sp., *Lyngbya hieronyamusii*, and many various *Oscillatoria* species, like *Oscillatoriapseudogeminata*, *O. animalis*, *O. cruenta*, *O. princeps*, among other species (Bhardwaj et al. 2010, 2011, Ranawat and Rawat 2017). A mycelial yeast *Saccharomycopsis fibuligera* was also reported from these hot springs (Kumar et al. 2005).

Culture-independent studies on the Soldhar hot spring through metagenomics revealed the predominance of many bacterial taxonomical families, such as Actinobacteria, Bacteroidetes, Deinococcus, Thermus, Firmicutes, Gammaproteobacteria, Betaproteobacteria, Xanthomonadaceae, Moraxellaceae, Enterobacteriaceae, Chromatiaceae, Alteromonadaceae, Comamonadaceae, Erythrobacteraceae, Rhodobacteraceae, Thermaceae, Cyclobacteriaceae, Microbacteriaceae, Acidobacteria, Aquificae, Verrucomicrobia, Deltaproteobacteria, Methylophilaceae, Sphingomonadaceae, Caulobacteraceae, Bacillaceae, Chitinophagaceae, Propionibacteriaceae, and Micrococcaceae (Sharma et al. 2017). The metagenomics of the soil samples of Soldhar hot spring had common hot spring microflora (like *Anoxybacillus, Bacillus, Clostridium, Flavobacterium, Geobacillus, Meiothermus, Pseudomonas, Symbiobacterium, Thermus*, and *Ureibacillus* as the predominant genus) found in the geothermal areas of the Himalayas (Rawat and Joshi 2018).

During the recent studies by Kumar and his co-workers, it was reported that the Ringigad hot spring was the ecological niche to few cultivable bacterial strains of *Bacillus cerus, B. cibi, Aeromonas veronii, Strenotrophomonas maltophila, Paenibacillus dendritiformis, Brevibacillus borstelensis*, and *Streptococcus pyogenes*; actinomycetes strains of *Streptomyces albus, S. canescens; Thermoactinomyces candidus* and *T. thalopophilum*; and fungal strains of *Fusarium oxysporum, Sclerotium rolfsii*, and *S. sclerotiorum*) (Kumar and Sharma 2019).

The genomic DNA of some significant thermophilic bacteria have been sequenced and reported by some researchers, such as *Thermus* sp. strain RL (Dwivedi et al. 2012), *Deinococcus* sp. strain RL (Mahato et al. 2014), *Cellulosimicrobium* sp. strain MM (Sharma et al. 2014b), and *Fictibacillus halophilus* (Sharma et al. 2016). Few bacterial and archeal viral genomes from this hot spring were also revealed (Sharma et al. 2018b). From Khirganga hot spring, 45 unidentified bacterial isolates, capable of producing industrially relevant thermozymes, were reported (Shirkot and Verma 2015).

C. Nepal Himalayan Geothermal Belt (NHGB)

This geothermal province consists of many hot springs throughout its region. There are reports of more than 20 well documented hot springs spread among ten geothermal provinces. The hot spring water is usually Na-K-HCO$_3$ type. The major geothermal provinces of Nepal are:

i. Dharchula Geothermal Province Hot Springs (Dhar-GPHS)

This geothermal province of the Dharchula district contains three major hot springs—Sribagar, Sina Tatopani, and Chamaliya. The Sribagar hot springs are located at 80.6°E-29.9°N coordinates, and the average temperature ranges from 55°C-75°C. The hot springs are located near the river confluence where there is tectonic contact. The Sina Tatopani hot springs are located at 80.7°E-29.9′N coordinates, and the average temperature ranges from 30°C-37°C. These hot springs are located at the thrust contact region between augen gneiss and sericitic schist and quartzite. The Chamaliya hot springs are located at 80.6°E-29.7°N coordinates, and the average temperature ranges from 35°C-40°C.

ii. Bajhang Tapovan Geothermal Province Hot Springs (BajhTap-GPHS)

This geothermal province hosts only one thermal spring at 81.2°E-29.6°N near the Major Thrust. The average temperature of this hot spring is 37°C-40°C.

iii. Jumla Geothermal Province Hot Springs (Jumla-GPHS)

Jumla district hosts two hot springs at its geothermal province—Dhanchauri and Tila Nadi. Dhanchauri hot spring is located between 82.3°E-29.6°N, and the average temperature of this hot spring is 20°C-25°C. Characteristic dolomites and silica deposits are found at Dhanchauri hot spring. Here, three more hot springs can be found in this vicinity. The Tila Nadi hot spring is located between 82.1°E-29.2°N, and the average temperature of this hot spring is 34°C-45°C. There are seven different hot springs are located along the right side of the river bank confluence of Tila Nadi, which is located just below the Tatopani village.

iv. Jomson Geothermal Province Hot Springs (Jom-GPHS)

This geothermal province hosts only one thermal spring at 83.7°E-29.8°N. The average temperature of this hot spring is 20°C-25°C. The springs can be found at the confluence of river Kaki Gandaki banks.

v. Mustang Tatopani Geothermal Province Hot Springs (Must-GPHS)

This geothermal province is located between 83.7°E-28.5°N. The average temperature of this hot spring is 66°C-72°C. There are more five hot springs in this vicinity.

vi. Sadhu Khola Geothermal Province Hot Springs (Sadhu-GPHS)

This geothermal province hosts only one thermal spring at 84.2°E-28.4°N. The average temperature of this hot spring is 63°C-70°C.

vii. Mayangdi Geothermal Province Hot Springs (Maya-GPHS)

This geothermal province is situated between 83.5°E-28.4°N. The average temperature of this hot spring is 33°C-40°C. Here, four more hot springs can be found.

viii. Rior Geothermal Province Hot Springs (Rior-GPHS)

This geothermal province hosts only one thermal spring at 82.7°E-27.9°N. The average temperature of this hot spring is 30°C-35°C.

ix. Surai Khola Geothermal Province Hot Springs (Surai-GPHS)

This geothermal province hosts only one thermal spring at 83.3°E-27.8°N. The average temperature of this hot spring is 33°C-38°C.

x. Chilime Geothermal Province Hot Springs (Chili-GPHS)

This geothermal province hosts only one thermal spring at 85.3°E-28.3°N. The average temperature of this hot spring is 50°C-58°C.

xi. Kodari Geothermal Province Hot Springs (Koda-GPHS)

This geothermal province hosts only one thermal spring at 83.9°E-27.9°N. The average temperature of this hot spring is 38°C-45°C.

Microbiology: Very few studies have been reported with respect to microbial diversity from the hot springs of Nepal. From Bhurung Tatopani located at Myagdi district, few Gram-positive thermophilic rod-shaped bacteria showing >99% similarity to *Bacillus licheniformis*, *B. subtilis*, and *B. pumilus* were reported (Adhikari et al. 2015). Also, different Gram-positive thermophilic bacterial rod strains belonging to various genus, like *Geobacillus* sp., *Bacillus* sp., *Aeribacillus* sp., *Anoxybacillus* sp., *Brevibacillus* sp., etc., were found from various hot springs of Ratopani, Bhurung, Paudwar, Sinkosh, and Singha, geographically located at Myagdi district, Nepal (Yadav et al. 2017). This hot springs is one of the hottest among all other hot springs of Nepal and emits a heavy sulfur smell. It is situated at 2,262 m above the mean sea level. Similarly, from Paudwar hot spring, a thermoalkalophilic xylanase producing bacterium *Anoxybacillus kamchatkensis* NASTPD13 was isolated (Yadav et al. 2017, Yadav et al. 2018).

D. Bhutan Himalayan Geothermal Belt (BHGB)

This geothermal belt hosts numerous hot springs but their microbiology has not been explored or studied. All these hot springs are frequently visited by the patients and healthy individuals owing to their immense religious beliefs and also to practice balneotherapy (Wangchuk and Dorji 2007, Wangdi et al. 2014).

i. Punakha Geothermal Province Hot Springs (Puna-GPHS)

This geothermal province consists of two main hot springs—Chuboog and Koma Tsha Chu.

1. Chuboog Tsha Chu: The word "Chuboog" is a Bhutanese terminology, where "Chu" means water and "boog" means center. The name of this hot spring has been derived from its origin point, i.e., it is geographically located in between the confluence of river Pho Chu and Mo Chu. It is situated 1,737 m above the mean sea level. People bathe in this hot spring and also worship the local deity Khachaep Dralay Gyelpo. This hot spring has two separate bathing ponds—Gongma (which means upper pond) and Wogma (which means lower pond). The average temperature of Gongma pond ranges from 37°C to 45°C depending on the seasons. It is believed to cure ailments like tuberculosis, gastric, and skin-related diseases. Similarly, the average temperature of Wogma pond ranges from 40°C to 48°C depending on the seasons. It is believed to cure ailments, like tuberculosis, gastric problems, muscular sprain, and ortho-related diseases (Wangdi et al. 2014).

2. Koma Tsha Chu: It is located 1,839 m above the mean sea level. This hot spring has three bathing ponds and is believed to cure different ailments, like anal fistula, arthritis, accident wounds, dermal diseases, osteo-related pains, and tuberculosis (Wangdi et al. 2014). The average temperature of this spring is 40°C-45°C.

Microbiology: None reported to date.

ii. Gasa Province Hot Springs (Gasa-GPHS)

This geothermal province consists of three main hot springs—Gasaand Koma Tsha Chu.

1. Gasa Tsha Chu: It is located at the banks on the river Mo Chu confluence of Gasa Dzongkhang at 2,100 m elevation above the mean sea level. There are four different bathing ponds at this hot spring. This hot spring is believed to cure ailments, like chronic arthritis, paralysis, syphilis, and headaches (Wangdi et al. 2014). The temperature of the ponds on an average varies from 50°C-60°C.

Microbiology: None reported to date.

2. Gayza Tsha Chu: It is located in the northern areas of Laya Geog of Gasa town at 3,826 m above the mean sea level. This hot spring is believed to cure conjunctivitis, arthritis, and inflammations (Wangdi et al. 2014). The average temperature of the hot springs ranges from 35°C-42°C.

Microbiology: None reported to date.

iii. Bumthang Province Hot Springs (Bum-GPHS)

This geothermal province consists of only one main hot spring—Dhur Tsha Chu.

Dhur Tsha Chu: It is located at Bumthang under the Wangchuk Centennial Park at 3,522 m above the mean sea level. This geothermal province has seven different hot spring bathing ponds. They are all famous for their respective balneotherapeutic properties (Wangdi et al. 2014). The various hot springs under this province are Beken Menchu Tsha Chu (35°C-42°C; believes to cure tuberculosis), Chenrezig Tsha Chu (25°C-30°C; believes to cure vision-related ailments), Dangwa Phochu Tsha Chu (45°C-50°C; believes to cure sexually transmitted diseases, like gonorrhea and syphilis for males only), Dangwa Mochu Tsha Chu (47°C-52°C; believes to cure sexually transmitted diseases, like gonorrhea and syphilis for females only), Gunay Tsha Chu (40°C-45°C; chronic headaches), Guru Tsha Chu (45°C-50°C; believes to cure dermal and osteo-related ailments), and ZekhamTsha Chu (35°C-40°C; believes to cure tetanus patients).

Microbiology: None reported to date.

iv. Zhemgang Province Hot Springs (Bum-GPHS)

This geothermal province consists of only one main hot spring—Duenmang Tsha Chu.

Duenmang Tsha Chu: It is located at the foothills of Kamjong hill of the Khyeng region and on the river banks confluence of the river Mangde Chu. It is also popularly known as Khyeng Tsha Chu. It is situated at an elevation of 385 m above the mean sea level. There are six different bathing ponds at this geothermal province, whose average temperature ranges from 35°C to 55°C. These hot springs are believed to cure chronic joint pains, dermal diseases, sinusitis, headaches, and tuberculosis (Wangdi et al. 2014).

Microbiology: None reported to date.

v. Gelephu Province Hot Springs (Gele-GPHS)

This geothermal province consists of only one main hot spring—Gelephu Tsha Chu.

Gelephu Tsha Chu: It is located at 15 km uphill from Gelephu town. It is situated at an elevation of 332 m above the mean sea level. There are five different bathing ponds at this geothermal province whose average temperature ranges from 35°C to 40°C. These hot springs are believed to cure chronic skin diseases, gastric problems, anal fistula, chronic arthritis, wounds, osteo-related diseases, dermatitis, hemorrhoids, conjunctivitis, fever, hypertension, ulcers (Wangdi et al. 2014).

Microbiology: None reported to date.

vi. Lhuentse Gelephu Province Hot Springs (Lhu-GPHS)

This geothermal province consists of three main sulfur hot springs—Khambalung gNey Tsha Chu, Yoenten Kuenjung Tsha Chu, and Pasalum Tsha Chu.

1. Khambalung gNey Tsha Chu: It is located at 15 km uphill from Gelephu town. It is situated at an elevation of 2,472 m above the mean sea level. There are three different bathing ponds—Guru Tsha Chu (50°C to 60°C), Tshepameg Tsha Chu (35°C to 40°C), and Khandro Yeshey Thsogyal Tsha Chu (45°C to 55°C). Guru Tsha Chu is the largest among the others located in a nearby waterfall. It is a bicarbonate type of spring rich in sulfur (Wangchuk and Dorji 2007). Tshepameg Tsha Chu and Khandro Yeshey Thsogyal Tsha Chu contain limestone (Wangchuk and Dorji 2007). The latter hot spring is situated in a stone bowl-shaped zone just below the Guru Tsha Chu. Guru Tshachu is believed to cure various gastric problems, dermal diseases, sexually transmitted diseases, rheumatoid diseases, osteo-related problems, muscular pains, gout, paralysis, and urinary tract infections (Wangchuk and Dorji 2007, Wangdi et al. 2014). Tshepameg Tsha Chu helps in the ailment of dermal diseases. Khandro Yeshey Thsogyal Tsha Chu is believed to cure gout, arthritis, paralysis, and pains (Wangchuk and Dorji 2007, Wangdi et al. 2014).

Microbiology: None reported to date.

2. Yoenten Kuenjung Tsha Chu: It is situated at an elevation of 2,761 m above the mean sea level. It is believed to cure memory disorders. There are two bathing ponds; one under a sacred cave which is considered to be the main source pond, and the other a bathing pond (Wangdi et al. 2014). At this geothermal province, the average temperature ranges from 35°C to 40°C.

Microbiology: None reported to date.

3. Pasalum Tsha Chu: It is situated at an elevation of 4,795 m under Lhuentse Dzongkhag in Gangzur Gewogabove the mean sea level. It is believed to cure many diseases, like backaches, headaches, and gastric problems (Wangdi et al. 2014). At this geothermal province, the average temperature ranges from 37°C to 45°C.

Microbiology: None reported to date.

North East Indian Himalayan Geothermal Belt (neIHGB)

1) Sikkim Geothermal Province Hot Springs (Sikkim GP-HS) and Its Microbial Diversity

Sikkim naturally hosts many hot springs. It is a major tourist attractive state of India where nature is in its juvenile form and a refreshing season greets its visitors. In local languages, these hot springs are called as Tatopani or Tsha chu. Tatopani is a Nepali word where "Tato" means Hot and "pani" means water, whereas "Tsha chu" is a Tibetan word where "Tsha" means hot and "chu" means water (Das et al. 2012a). Here, at Sikkim, hot springs are sociologically very relevant and hold a prime importance (Das et al. 2012a). It is regarded as an elixir and it is believed that bathing in it can cure many bone-related diseases and drinking it can also cure gastric problems (Das et al. 2012b). Located at various places, Yumthang, Yume Samdung, Tarum, Polok, Borong, Reshi, etc., they are major tourist attractions.

Polok Tatopani (Ralang Tchu/Ralang Tsha chuu/Rabong Tatopani) (Figure 1d) is located at the base of Gangyab, West Sikkim, by the banks of the river Rangeet (Thakur et al. 2013, Das et al. 2016). Borong Tatopani (Figure 1e) is located at lower Borong and the ponds are situated at the banks of river Rangit in West Sikkim. There are three ponds for bathing, but it depends on seasons (Das et al. 2016). Reshi Tatopani/Phur Tsha Chu ("Phur" means bubble in the Tibetan language) (Figure 1f) and is located approximately 25 km from Jorethang to the east of Reshi (Tinkitam) (Sherpa et al. 2013). Hot spring source is located near the bank of river Testa. One can feel a strong sulfurous smell from going closer to the hot spring vicinity (Das et al. 2016). Yumthang Tatopani (Figure 1g) is located on the base of the mountain across the river Lachung Chu in the town of Lachung (Das et al. 2016). Dzongu Tatopani (Figure 1h) is located in the valley of Upper Dzongu, Sikkim. Dzongu is closely associated with three terms—Land of Lepcha, Natural Hub of medicinal plants, and interaction of nature and culture (Das et al. 2016). Ponds are like modern pools for bathing purposes. Two separate bathing ponds are present; one for males and the other for females in separate two rooms. The water is used only for the bathing purpose (Das et al. 2016). Takrum Tatopani (Figure 1i) is located at Lachen valley in the North Sikkim district (Das et al. 2016). Yume Samdong Tatopani (Figure 1j; Figure 1k) is located in the North Sikkim district at Yume Samdong valley. It is above zero-point and is located at the highest altitude (Das et al. 2016).

Polok Tatopani and Borong Tatopani lie at the crucial dividing junction or the border zone of West Sikkim and South Sikkim. Rocks observed at the Tatopani study area of West Sikkim showed the presence of Sillimanite Granite Gneiss, Migmatite, Biotite Gneiss, Mica Schist, and Kyanite geological features. They belong to major rock types of Darjeeling Gneiss or Kanchenjunga Gneiss. And rocks observed at the Tatopani study area of South Sikkim showed the presence of Interbedded Schist, Phyllites, and Mica Schist geological features. They belong to major rock types of Gneiss of Gorubathan Formation under Daling Group. All the rocks most probably were evolved during the Proterozoic era. As the rocks are usually carried down by the river action, hence its origin cannot be deciphered accurately and further rock biochemistry needs to be carried out.

Microbiology: The culture-dependent studies showed the complete dominance of phylum *Firmicutes* in the hot springs of Sikkim. *Geobacillus* was the predominant genus along with few representatives of *Anoxybacillus* and *Bacillus*. The culture-dependent study showed that *G. stearothermophilus* XTR25, *G. kaustophilus* YTPR1, *G. subterraneus* 17R4, *G. lituanicus* TP11, *G. kaustophillus* YTPB1, *Parageobacillus toebii* 10PHP2, *G. toebii* strains, *Anoxybacillus caldiproteolyticus* TRB1, *Anoxybacillus gonensis* TP9, *Bacillus smithii* 17R6, *Bacillus* sp. 17R5, were the bacterial flora present, respectively (Najar et al. 2018a). A novel culturable bacterium was also reported for the first time, which was isolated from Yumthang hot spring—*G. yumthangensis* (Najar et al. 2018b, c).

Culture-independent analysis through metagenomics of the hot springs of Sikkim showed various phylum diversity, like *Proteobacteria* (~63%), *Bacteroidetes* (~15%), *Acidobacteria* (~4%), *Nitrospirae* (~4%), and *Firmicutes* (~3%) in Borong Tatopani; Polok Tatopani had *Proteobacteria* (~47%), *Bacteroidetes* (~4%), *Firmicutes* (~3%), *Parcubacteria* (~3%), and *Spirochaetes* (~3%); Yumthang Tatopani had *Actinobacteria* (~98%) and *Proteobacteria* (~2%) in the majority; Reshi Tatopani had *Proteobacteria* (~76%), *Actinobacteria* (~23%), *Firmicutes* (~1%), and *Cyanobacteria* (0.03%) (Najar 2018).

At genus level, there was a distinct variation in hot springs. The genuses present in Borong Tatopani had *Acinetobacter* (~8%), *Flavobacterium* (~4%), *Vogesella* (~4%), *Ignavibacterium* (~3%), *Sediminibacterium* (~3%), *Thermodesulfovibrio* (~3%), and *Acidovorax* (~2%); Polok Tatopani had *Flavobacterium* (~3%), *Sediminibacterium* (~3%), *Pseudomonas* (~2%), *Treponema* (~2%), and *Opitutus* (~1%); Yumthang Tatopani had *Rhodococcus* (~98%), *E. coli* (~0.7%), *Serratia* (~0.5%), *Nocardiopsis* (~0.5%), *Brevundimons* (~0.2%), and *Acinetobacter* (~0.2%); Reshi Tatopani had *Pseudomonas* (~85%), *Rhodococcus* (~4%), *Dietzia* (~4%), *Arthobacter* (~4%), *Staphylococcus* (~1%), and *Paracoccus* (~0.3%), respectively (Najar 2018).

The diversity at species level varied significantly in all the four hot springs. Polok Tatopani had *Sediminibacterium goheungense*, *Opitutus terrae*, *Treponema caldarium*, *Ignavibacterium album*, *Desulfobulbus mediterraneus* and *Thermodesulfovibrio hydrogeniphilus*, *T. yellowstoni*, *Hydrogenobacter thermophiles*, *Thermoanaerobacter uzonensis*, *Thermoanaerobaculum aquaticum*, *Thermolithobacter ferrireducens*, *Thermus arciformis*, *T. caliditerrae*, etc. Borong Tatopani had *Ignavibacterium album*, *Rheinheimera aquatic*, *Flavobacterium cheonhonense*, *Thermodesulfovibrio yellowstonii*, *Thiovirga sulfuroxydans*, *Meiothermus hypogaeus*, etc. ReshiTatopani had *Microbacterium species* (~67%), *Arthrobacter phenanthrenivorans* (~3%), and *Rhodococcus erythropolis* (~2%) and Yumthang Tatopani had *Rhodococcus ruber* (~98%) and *Escherichia coli* (~1%), respectively. Polok and Borong Tatopani had a lesser amount of archaeal communities. Borong Tatopani had *Crenarchaeota* (~1%), whereas Polok Tatopani had *Euryarchaeota* (~0.6%). *Desulfurococcales* and *Desulfurococcus* were the major order and genus under *Crenarchaeota*, respectively, whereas *Methanomicrobiales* and *Methanospirillum* were the major order and genus under *Euryarchaeota*, respectively (Najar 2018).

2) Assam Geothermal Province Hot Springs (AhomGP-HS) and Its Microbial Diversity

The Ahom-GP-HS consists of two major alkaline thermal springs: (a) Garampani area in Garampani Wildlife Sanctuary, Karbi-Anglong district (25°30′N and 92°37′E) of the Narmada-Sone Dauki geothermal province lineament and (b) Nambor Garampani area in Nambor Wildlife Sanctuary, Golaghat district, Assam. In the local dialect, "Garampani" refers to "hot spring".

Karbi-Anglong Garampani consists of three clusters of hot springs along with the main source, Ansuya Kund. The hot spring temperature ranges between ~49°C to 58°C, which lies at the right bank of the Kopili River confluence. Due to the presence of hot springs, the village is known as "Garampani village". The three hot spring clusters within the approximate range of 100 m have numerous Eocene sandstone and nummulitic limestones stretched throughout the river confluence. Here, bubbles can be seen emerging beneath the ponds releasing H_2S gas odor. These zones have pyrites and chalcopyrites mineralogy. People have been bathing at these waters for ages for their balneotherapeutic properties. They are rich in bicarbonates, sulfides, and chlorine. These thermal springs are reported to be of $Na-HCO_3-SO_4$ type (Sharma et al. 1982).

The Nambor Garampani area lies in the Nambor wildlife sanctuary, which is located at 26°40'-26°45'N latitude and 94°20'-94°25'E longitude (Hazarika et al. 2014). Here, also some three to four clusters of hot springs can be found within this 100 m area around the sanctuary confluence of rivulet Nambor. The temperature of these slightly alkaline thermal springs ranges from 40°C to 46°C depending on the season (Hazarika et al. 2014, Hazarika and Gogoi 1985).

Microbiology: There has been a very limited study of the microbial population of these hot springs. From the Karbi-Anglong Garampani, many *Pseudomonas* sp. were reported which can be used for bioremediation of lead from environments due to their biosorption activity (Kalita and Joshi 2017). Predominantly, industrially important bacterial isolates, like *Pseudomonas aeroginosa* 2474, *P. ficuserectae* PKRS11, *P. alcaligenes* MJ7, and *Pseudomonas* sp. W6, were found from this hot spring area (Kalita and Joshi 2017).

The predominant bacteria found in the Nambor Garampani area are *Staphylococcus aureus*, *S. epidermis*, *Citrobacter diversus*, *Micrococcus luteus*, and few unidentified species of *Bacillus* (Hazarika et al. 2014). Until recently, many cellulolytic bacterial strains were also reported—*Stenotrophomonas maltophilia*, *Bacillus cereus*, *B. pumilus*, and *B. thuringiensis*. The crude enzymes extracted from these bacteria were found to be enzymatically thermostable at 100°C, whereas commercially available enzymes are functionally active till 60°C (Parveen et al. 2016).

There has been also reports of numerous cyanobacterial species from Nambor hot springs—*Aphanothece saxicola* Näg, *Chroococcus minor* (Kütz.) Näg, *Chlorogloca microcystoides* Geitler, *Lyngbya martensiana* Menegh. ex. Gomont, *Lyngbya putealis* Mont. ex. Gomont, *Oscillatoria acuminata* f. *tenius* Parukutty, *O. chlorine* Kütz. ex. Gomont, *O. fremyii* De Toni, J., *O. proboscidea* Gomont, *O. subtilissima* Kütz, *Phormidium subincrustatum* Fritch et Rich., *Schizothrix lacustrics* A. Br. ex. Gomont., *Spirulina subsalsa* Öerst. ex. Gomont, and *Anabaena vaginicola* f. *fertilissima* Prasad (Hazarika et al. 2014).

To date, no culture-independent strategies have been used to explore the microbial diversity of these hot springs of Assam.

3) Meghalaya Geothermal Province Hot Springs (MeghaGP-HS) and Its Microbial Diversity

This part of the northeastern region hosts two hot spring areas, one at Jakrem and another at Mendipathar. Jakrem is located about 64 km west of Shillong, in the East Khasi Hills district of Meghalaya. This geothermal province lies in the central Shillong plateau composting mostly of granite mineralogy. The average temperature of the hot spring water ranges from 45°C – 49°C depending on the seasonal variations and is slightly neutral in pH. There is no such foul smell of sulfur from this water but is very well known for its balneotherapeutic properties among the locals.

The Mendipathar hot spring is situated north of Bakrapara village of Resubelpara, North Garo Hills district. There are two to three discharging points for the thermal springs and the temperature of them ranges from 30°C to 32°C on an average.

Microbiology: Among Jakrem and Mendipathar hot springs, the later has not been explored in terms of microbiological activity or microbial diversity. Jakrem hot spring has been thoroughly studied by many researchers (Kumar et al. 2013, Panda et al. 2016). A novel thermophilic bacteria was discovered by Kumar and his co-workers from this geothermal province, which was nomenclatured as *Caldimonas meghalayensis* AK31[T] (Kumar et al. 2013). There has been extensive study on the archeal and bacterial metagenomics by Panda and his co-workers, where it was found that among the major bacterial phyla, Firmicutes, Chloroflexi, and Thermi were predominant (Panda et al. 2016), whereas in the case of archaea, phyla *Euryarchaeota* and *Crenarchaeota* were predominant. *Clostridium, Chloroflexus,* and *Meiothermus* were the majority of the bacterial genus, and among archaeal genus, *Methanoculleus*, and *Methanosaeta* were reported (Panda et al. 2016).

4) Arunachal Pradesh Geothermal Province Hot Springs (ArunGP-HS) and Its Microbial Diversity

This geothermal province lies within the Bomdila Gneiss and Darjeeling Gneiss (Sinha 1980). The initial report of hot springs of this province was done by Bakliwal et al. (1971) and they are situated at the west of Dirang near the river banks of Diggin, whereas another was reported from Bishum village of Sangti Valley (Sinha 1980). The geological set up of Dirang is due to shattered; jointed and phyllitic quartzite rocks that emitted high sulfuric odor and gave the characteristic smell to the hot springs, and for Bishum it is fractured and shattered chlorite mica schist with pyrite specks (Sinha 1980). A cold spring was also reported from the same strata during the geological survey. Researches on the geological front are being carried out presently by a few researchers (Taye and Chutia 2016). Overall, the hot springs can be divided into three zones: a) Dirang Hot Springs, b) Bishum Hot Springs, and c) Tawang hot springs. There are reports of four various geothermal ponds at Dirang, two at Bishum and four at Tawang. The problem of the research areas is that these are very closely situated at the river banks and hence owing to natural calamities like flash floods, earthquakes, or floods in river destroys these ponds, and hence it is very difficult to study them. Only proper bathing ponds are the best sampling sites for seasonal study. The hot springs of Tawang and Dirang are frequently visited by the people for balneotherapeutic purposes as they believe it to cure various skin diseases.

The four hot springs of Tawang are located at Braksar, Thimbu, Tsachu, and Kitpi. Braksar hot spring is located at the right-hand side of the Mago Chu river bank vicinity, New Melling village. The Thimbu hot spring is located at the right-hand side of the Luguthang river bank of Thimbu village near the confluence of Mago-Luguthang Rivers. The Tsachu hot spring is situated at the left-hand side of the near the confluence of Tsachu-Nyukcharong Chu river of Tsachu river. Kitpi hot spring is located at the river banks of Jong of Greng Khar village of Kitpi (Taye and Chutia 2016).

The Tawang hot springs have varied surface temperatures from 33°C to as high as 70°C, viz., Braksar (33°C), Kitpi (43°C), Tsachu (65°C), Thimbu (70°C), whereas in case of Dirang there are four different spots with varied surface temperature ranging from 21°C to 40°C. In Bishum, there is a thermal pond with a 40°C surface temperature and a cold spring of 2°C temperature. All the physicochemical characterization reported so far has been tabulated in Table 1.

Microbiology: An extracellular alkalophilic lipase producing Gram-positive bacteria *Bacillus* sp. LBN2 was reported by Bora and Bora (2012), which was isolated from Dirang hot spring. It showed lipolytic activity at pH ranging from 8 to pH11 and is stable at 70°C (Bora and Bora 2012). The environmental conditions, i.e., the temperature, Ph, and other geological parameters enable the dominance of Firmicutes in the hot spring. The hot spring was found to be predominant with various isolates of *Bacillus* sp. (unidentified) having amylolytic, cellulolytic, and proteolytic activities respectively (Bora and Kalita 2006).

To date, no culture-independent strategies have been used to explore the microbial diversity of these hot springs of Arunachal Pradesh.

E. Tibet Himalayan Geothermal Belt (TibHGB) and Their Microbial Diversity

Tibetan plateau is the ecological cluster that houses numerous hydrothermal geysers and hot springs and other characteristic geothermal fields. The hydrothermal areas include Dangxiong, Dongweng, Gariqiao, Gulu, Laduogang, Luoma, Qucai, Tuoma, Xumai, Yangbajing, and Yangyi. These geothermal fields are mainly hosted on seven prefectures of Tibet, such as Ngari, Xigaze, Lhasa city, Nagqu, Shannan, Nylngchi, and Qamdo prefectures. It is assumed that in Tibet there are more than 300 hot springs present. The average temperature of the Tibet geothermal hot springs ranges from 59°C-64°C with slightly neutral to alkaline in nature. Most of them are rich in Na and K ions and are of generally HCO_3 type. Acidobacteria, Bacteroidetes, Proteobacteria, Nitrospirae, Firmicutes, Aquificae, Cyanobacteria, Chloroflexi, Planctomycetes, and Thermodesulfobacteria were the most

TABLE 1: Physicochemical Properties of Arunachal Pradesh Geothermal Province Hot Springs (ArunGP-HS)

Parameters	Dirang Hot Springs				§Bishum Hot Springs			Tawang Hot Springs		
	DR H-5§	DR H-6§	DR H-7§	Hot spring 1¥,€	SB-116A§	SB116C (cold)§	Kitpi Hot spring¥	Braksar hot spring€	Thimbu hot spring€	Tsachu hot spring€
Surface Temperature (in °C)	n.r.	n.r.	n.r.	35-40	n.r.	2	43	33	70	65
Ph	7.3	7.9	7.1	8.1-8.25	6.0	6.0	8.16	7.85	6.87	7.25
Conductivity (in Mhos cm^{-1})	425	1,700	1,150	1,853.7	n.r.	n.r.	979.1	n.r.	n.r.	n.r.
Ca	27	81	58	97.9	524	8	122.8	n.r.	n.r.	n.r.
Mg	8	12	9	n.r.	5	n.r.	n.r.	n.r.	n.r.	n.r.
Na	50	250	150	269.5	n.r.	n.r.	88.2	n.r.	n.r.	n.r.
K	9	45	35	4.6	n.r.	n.r.	4.4	n.r.	n.r.	n.r.
Cl	20	123	85	40.4-76	6	5	44.6	70	52	61
HCO$_3$	77	316	240	n.r.	75	31	n.r.	n.r.	n.r.	n.r.
CO$_3$	n.r.	n.r.	n.r.	n.r.	n.r.	n.r.	n.r.	n.r.	n.r.	n.r.
TDS	–	n.r.	n.r.	766	1,992	50	n.r.	833	703	901
Total Hardness (as CaCO$_3$)	100	253	182	162.51	1,340	20	204.1	n.r.	n.r.	n.r.
Total Alkalinity	n.r.	n.r.	n.r.	56	n.r.	n.r.	216	n.r.	n.r.	n.r.
SO$_4$	n.r.	n.r.	n.r.	28-56.4	1,240	n.r.	60.6	23	76	60
B	n.r.	n.r.	n.r.	n.r.	<0.5	<0.5	n.r.	n.r.	n.r.	n.r.
Na/K	9.35	9.35	7.31	n.r.	n.r.	n.r.	n.r.	n.r.	n.r.	n.r.
Suspended matters (in mgL^{-1})	n.r.	n.r.	n.r.	24	n.r.	n.r.	n.r.	26	28	21
Organic matters (in mgL^{-1})	n.r.	n.r.	n.r.	220	n.r.	n.r.	n.r.	247	27	249
Inorganic matters (in mgL^{-1})	n.r.	n.r.	n.r.	546	n.r.	n.r.	n.r.	586	676	652

Symbols used: § = as per Sinha, 1980; ¥ = as per Bora et al. 2006; € = as per Taye and Chutia, 2016; Abbreviations used: n.r. = not reported dominant group of bacteria reported in these hot springs, and among the archaea, Thaumarchaeota, Euryarchaeota, Halobacteriales, Methanomicrobiales, Crenarchaeota, and Desulfurococcales were the most predominant groups (Huang et al. 2011) in Tibetan geothermal provinces hot springs.

Sichuan is located in the eastern Mediterranean-Himalayan geothermal activity zone near Tibet with intermediate to high-temperature geothermal fields spread across it. Many hot springs are found in Batang, Litang, Ganzi, Daofi, and Kangding region, where more than 250 hot springs exist having an average temperature ranges from 55°C-80°C. The most famous boiling hot springs (Batang) can be found in this region.

Microbiology: The most predominant genus reported from Sichuan geothermal hot springs are *Acinetobacter, Aquaspirillum* sp., *Azospira* sp., *Chloroflexus* sp., *Chryseobacterium* sp., *Clostridium* sp., *Dechloromonas* sp., *Duganella* sp., *Flavobacterium* sp., *Hydrogenobacter* sp., *Massilia* sp., *Novosphingobium* sp., *Paludibacter* sp., *Pseudomonas* sp., *Rhodobacter* sp., *Sulfurihydrogenibium* sp., *Sulfurovum* sp., *Tepidimonas* sp., *Thermodesulovibrio* sp., *Thermus* sp. (Tang et al. 2018).

F. Yunnan Himalayan Geothermal Belt (YHGB)

Yunnan Province hosts high-temperature hydrothermal features where some boiling hot springs and hot springs of low to moderate temperatures occur. It is situated in the Indian Ocean Plate and Eurasian Plate collision zone. In this geothermal province, there are more than 850 hot springs alone. The thermal springs in Yunnan Province are dispersed around Lincang City, Baoshan City, Dehong Dai, Xishuangbanna Dai Autonomous Prefecture, the Puer City, the Dali Bai Autonomous Prefecture, Honghe Hani, etc. The average temperature of the hot springs residing in these prefectures of Yunnan ranges from 52°C-64°C.

Microbiology: At Hehuawenquan hot springs—*Micromonospora* sp., *Nocardiospis* sp., *Nonomurea* sp., *Streptomyces* sp.; at Dagunguo hot springs—*Streptomyces* sp.; at Diretiyanqu hot springs—*Micrococcus* sp., *Streptomyces* sp.; at Gumingquan hot springs—*Micromonospora* sp., *Streptomyces* sp., *Verrucosispora* sp.; at Hamazui hot springs—*Actinomadura* sp.; at ShuiReBaoZhaqu hot springs—*Actinomadura* sp., *Microbispora* sp., *Micromonospora* sp., *Nonomurea* sp., *Pseudonocardia* sp., *Streptomyces* sp.; at Zhenzhuquan hot springs—*Verrucosispora* sp.; at Zimeiquan hot springs—*Micromonospora* sp., *Promicromonospora* sp.; at Gongxiaoshe hot springs—*Micrococcus* sp.; at Jinze hot springs—*Verrucosispora* sp.—were the most predominant genus reported from the Yunnan Geothermal Province (Liu et al. 2016).

Myanmar Himalayan Geothermal Belt (MyanHGB)

Myanmar is situated at the conjunction of the Alpine–Himalayan Orogenic Belt and the Indonesian Island Arc System. It is also regarded as the end-point of the Himalayan Geothermal Belt on the eastern side. In the northern areas of Myanmar, the geological setting of the geothermal zone is arched around the Eastern Himalayan Syntaxis and passes through the Indo-Myanmar Ranges (Barber et al. 2017, Bertrand et al. 2001, Khin et al. 2017). Northern Myanmar is reminiscent of the tectonic collision between India–Eurasian plates. Thus, due to the seismic movements, this region has some of the hottest geothermal fields and is being harnessed to produce clean and green energy for domestic as well as industrial uses.

Myanmar hosts many hot springs throughout its length and breadth of the country due to the rigorous tectonic movements and hot geothermal fields. Kachin State (2 hot springs; 35°C-50°C); Kayah State (5 hot springs; 30°C-40°C); Kayin State (15 hot springs; 34°C-63°C); Sagaing region (10 hot springs; 28°C-50°C); Taninthayi region (19 hot springs; 36°C-53°C), Magway region (5 hot springs; 31°C-49°C); Mandalay region (3 hot springs; 30°C-42°C); Mon State (19 hot springs; 36°C-67°C); Rakhine State (1 hot spring; 30°C-35°C), and Shan State (17 hot springs; 23°C-64°C) (Tun et al. 2019) are the few well documented geothermal hot springs in terms of their energy assets.

Microbiology: Here, in these hot springs of Myanmar very limited studies on microbial diversity has been done and need more exploration. Only a few studies have been on the thermophilic Cyanophyta *Mastigocladus laminosus* (Soe et al. 2019) and other algal communities comprising of *Ulothrix cylindricum, Ulothrix variabilis, Geminella crenulatocollis, Binuclearia tatrana, Microspora crassior, M. floccose, M. stagnorum, Stigeoclonium longipilum, Protoderma* sp., *Cladophora crispate, C. fracta, C. glomerata, Pithophora oedogonia, Hormidium reticulatum, Spirogyra rhizobrachiats, S. jugalis, Closterium baillyanum, C. pachyderma, Cosmarium rectosporum,* and *Hormidium subtile* were some of the reported Cyanophyta members found in the hot springs of Yenwe village located in Taunggyi, southern Shan state (Win and New 2019).

Thailand Himalayan Geothermal Belt (ThaiHGB)

The collision of Indo-Eurasian tectonic plates has resulted in the formation of hot springs in Northern Thailand. There are 114 hot springs spread across Thailand whose surface temperature ranges from 30°C-99°C (Wood and Singharajwarapan 2014). The hot springs here are classified on the basis of temperature (65% are hyperthermal springs ranging temperature between 45°C to 100°C, and 35% are thermal springs whose surface temperature varies from 30°C to 45°C) and pH [65% are weak alkaline (pH 7.7-8.7); 20% are slightly neutral (pH 6.8-7.2), and 15% are moderately alkaline (pH 8.9-10.2)] springs (Subtavewung et al. 2005).

In northern Thailand, the HGB ends at the eastern side of the Indian peninsular region. There are 16 hot springs in this geothermal province, and they are located at Kok River, Yang Pa Kael, Ban Pa Suert, Mae Chan, Pong Phu Fuang, Sop Pong, North of Mae Chan Hot Springs, Na Pong, Huai Sai Khao, San Kamphaeng, Ban Pong Kum, Fang, Along the Mae Chan Fault, Ban Pong Nam Ron, Wiang Nong Lom Swamp, and Mae Jok (Wood and Singharajwarapan 2014). This geothermal province's hot springs have an average temperature of more than 80°C with slightly neutral to alkaline pH. This region contains granitic rocks. The Fang Hot Springs and San Kamphaeng Hot Springs are the famous hot spring of Myanmar, whose temperatures are in the range of 110°C-130°C. The hot springs at Fang evolved from the granite foliated fractures of the active Mae Chan fault (Wood and Singharajwarapan 2014).

Microbiology: Many Cynaobacteria have been reported from the various hot springs of North Thailand, and the most predominant genus belongs to *Synechococcus* sp., *Cyanobacterium* sp., *Phormidium* sp., *Lyngbya* sp. *Chroococcus* sp., *Oscillatoria* sp., *Mastigocladus* sp., *Symploca* sp., *Aphanothece* sp., *Cyanosarcina* sp., *Calothrix* sp., *Scytonema* sp., *Gloeocapsa* sp., *Bacularia* sp., *Onkonema* sp., *Calothrix* sp., and *Merismopedia* sp. among other species (Sompong et al. 2005). The Bor Khleung hot spring in Ratchaburi province, Thailand, bears around 50°C-60°C temperature with mild acidic pH. It has been extensively studied and the major predominant groups of this hot spring are Actinobacteria, Chloroflexi, Cytophagales, Thermaceae, Nostocales, Chryseobacterium, Acidobacteria, Achromobacter, Bacteroidetes, Planctomycetales, Betaproteobacteria, Verrucomicrobiales, Chlorobi, and Enterobacteriaceae (Kanokratana et al. 2004).

Conclusion

Here in this chapter, we have tried to give a sneak peek into the Himalayan Geothermal Belt, and the various microbial diversity of their hot springs. A representation of Pamirs, North Pakistan, Indian Himalayan (North-west and North-east), Bhutan, Nepal, China (Tibet, Sichuan, and Yunnan), Myanmar, and Thailand Geothermal Provinces have been presented. With respect to microbial studies, it was found that only a small percentage of hot springs have been studied throughout the above mentioned geothermal areas. The data is still very scarce, and thus there is a great scope of research in these extreme environments.

In the case of Indian Himalayan Geothermal areas, including Ladakh, Himachal, Uttrakhand, and Northeast Himalayas, the prominent microflora of these hot springs was found to be Proteobacteria,

Firmicutes, and Actinobacteria. However, there is a distinction in the presence of various other phyla, such as Chloroflexi, which was only found in hot springs of North-east geothermal areas of India. Similarly, *Deinococcus thermi* and Bacteriodetes were only found to be present in Ladakh and Uttrakhand-Himachal Pradesh hot springs respectively. However, Actinobacteria was only found in Ladakh and North-east hot springs.

In the case of North Pakistan, the major phyla were found to be Proteobacteria, Chloroflexi, and Thrermotogae. These hot springs also possess the lower abundances of phyla Firmicutes, Bacteriodetes, and Actinobacteria. However, Spirochetes, Deinococcus Thermi, Acidobacteria, and Chlorobi were also found to be present but were least abundant. Proteobacteria was found to be abundant in all three major geothermal areas, such as Tibet and Sichuan except Yunnan. However, in the case of Tibet, Firmicutes, and Chloroflexi were abundant following Proteobacteria. In the case of Sichuan Cyanobacteria, Aquificae and Bacteroidetes were abundant following Proteobacteria. In contrast to Tibet and Sichuan, the most abundant phylum in Yunnan was Aquificae followed by Proteobacteria, Deinococcus Thermi, and Thermotogae.

In the case of all the hot springs, one aspect of microflora was common and it was Cyanophyta. All the hot springs had various species of algal communities either in microbial mats or in benthic sediments of thermal pools. Another interesting aspect was the isolation of *Geobacillus* sp. from most of the hot springs of Pakistan, Jammu-Kashmir, Uttarakhand, Nepal, Sikkim, Tibetan, and Yunan provinces. However, the microbiology of most of the hot springs of the Himalayan Geothermal Belt has not been studied, and it is not known that these hot springs also harbors *Geobacillus* sp. Based on the available data from some of the hot springs of the Himalayan Geothermal Belt, we would like to hypothesize that the Himalayan Geothermal Belt can be called as "Himalayan *Geobacilli* Corridor".

Acknowledgements

The author SD would like to thank the Department of Science and Technology, DST INSPIRE for providing the INSPIRE FELLOWSHIP (IF130091) for the research work. We heartily thank and express our most sincere gratitude to the Department of Forest, Govt. of Sikkim for giving us permission and kind cooperation during field research work and support. The authors are thankful to the Department of Biotechnology, Government of India (DBT-NER/Health/45/2015 & BT/PR25092/NER/95/1009/2017) for providing funds for research work.

References

Aanniz T., M. Ouadghiri, M. Melloul, J. Swings, E. Elfahime, J. Ibijbijen, et al. 2015. Thermophilic bacteria in Moroccan hot springs, salt marshes and desert soils. Braz. J. Microbiol. 46(2): 443–453. doi:10.1590/S1517-838246220140219.

Adhikari H., S. Ghimire, B. Khatri and K.C. Yuvraj. 2015. Enzymatic screening and molecular characterization of thermophilic bacterial strains isolated from hot spring of Tatopani, Bhurung, Nepal. Int. J. App. Sci. Biotech. 3(3): 392–397. doi: 10.3126/ijasbt.v3i3.12724.

Anees M., M.M. Shah and A.A. Qureshi. 2015. Isotope studies and chemical investigations of Tattapani hot springs in Kotli (Kashmir, NE Pakistan): implications on reservoir origin and temperature. Procedia Environ. Sci. 13: 291–295.

Anees M., M.M. Shah, A.A. Qureshi and S. Manzoor. 2017. Multi proxy approach to evaluate and delineate the potential of hot springs in the Kotli District (Kashmir, Pakistan). Geologica. Acta. 15(3): 217–230, I–III. doi:10.1344/GeologicaActa2017.15.3.5.

Arya M., G.K. Joshi, A.K. Gupta, A. Kumar and A. Raturi. 2015. Isolation and characterization of thermophilic bacterial strains from Soldhar (Tapovan) hot spring in central Himalayan region, India. Ann. Microbiol. 65: 1457–1464.

Asad W., M. Asif and S.A. Rasool. 2011. Extracellular enzyme production by indigenous thermophilic bacteria: partial purification and characterization of α–amylase by *Bacillus* sp. WA21. Pak. J. Bot. 43: 1045–1052.

Barber A.J., K. Zaw and M.J. Crow. 2017. The Pre-Cenozoic tectonic evolution of Myanmar. pp. 687–712. *In*: A.J. Barber, K. Zaw and M.J. Crow [eds.]. Myanmar: Geology, Resources and Tectonics. Geological Society, London, Memoirs. doi: 10.1144/M48.31.

Barinova S. and T.P. Niyatbekov. 2017. Algal diversity of the Pamir high mountain mineral springs in environmental variables gradient. Int. J. Environ. Sci. Nat. Resour. 14(3): 555706.

Barinova S. and T.P. Niyatbekov. 2018. Diatom species richness in algal flora of Pamir, Tajikistan. Eur. Sci. J. 14(3): 301–323.

Bertoldo C. and G. Antranikian 2002. Starch-hydrolyzing enzymes from thermophilic archaea and bacteria. Curr. Opin. Chem. Biol. 6: 151–160.

Bertrand G., C. Rangin, H. Maluski, H. Bellon and The GIAC Scientific Party. 2001. Diachronous cooling along the Mogok metamorphic belt (Shan Scarp, Myanmar): trace of the northward migration of the Indian Syntaxis. J. Asian Earth Sci. 19: 649–659.

Bhardwaj K.N. and S.C. Tiwari. 2010. Cyanobacterial diversity of two hyperthermal springs, Ringigad and Soldhar in Tapovan geothermal field, Uttarakhand Himalaya. Curr. Sci. 99(11): 1513–1515.

Bhardwaj K.N., A. Kainthola and S.C. Tiwari. 2011. Reassessment of taxonomic diversity of cyanobacteria in microbial mats and physicochemical characteristics of Badrinath thermal spring, Garhwal Himalaya. India. Proc. Natl. Acad. Sci. India 81: 235–241.

Bhatia S., N. Batra, A. Pathak, S.J. Green, A. Joshi and A. Chauhan. 2015. Metagenomic evaluation of bacterial and archaeal diversity in the geothermal hot springs of Manikaran, India. Genome Announc. 3(1): e01544–14. doi:10.1128/genomeA.01544-14.

Bhatt B.B. and R.A. Mir. 2015. Impact of tourism on water qualities of Sahastradhara. J. Agri. Forest. Environ. Sci. 1: 60–66.

Bora L. and M.C. Kalita. 2006. Occurrence and Extracellular enzymatic activity profiles of bacterial strains isolated from hot springs of West Kameng district of Arunachal Pradesh, India. Internet J. Microbiol. 4: 1–4.

Bora L., A. Kar, I. Baruah and M.C. Kalita. 2006. Hot springs of Tawang and West Kameng districts of Arunachal Pradesh. Curr. Sci. 91(8): 1011–1013.

Bora L. and M. Bora. 2012. Optimization of extracellular thermophilic highly alkaline lipase from thermophilic *Bacillus* sp. isolated from hot spring of Arunachal Pradesh, India. Braz. J. Microbiol. 43(1): 30–42.

Breu T. and H. Hurni 2003. The Tajik Pamirs: Challenges of sustainable development in an isolated mountain region; Centre for Development and Environment (CDE), University of Berne: Berne, Germany, 2003; ISBN 3-906151-74-3.

Brock T.D. and H. Freeze. 1969. *Thermus aquaticus* gen. n. and sp. n., a nonsporulating extreme thermophile. J. Bacteriol. 98: 289–297.

Brock T.D., K.M. Brock, R.T. Belly and R.L. Weiss. 1972. *Sulfolobus*: a new genus of sulfur-oxidizing bacteria living at low pH and high temperature. Arch. Microbiol. 84: 54–68.

Brock T.D. 1997. The value of basic research: Discovery of *Thermus aquaticus* and other extreme thermophiles. Genetics 146: 1207–1210.

Chagtai H.U. and P.M.A. Siddiqui. 1980. Systematic bacteriological study of Manghopir hot spring water. J. Pak. Med. Assoc. 30(2): 34–40.

Chandrasekharam D., M.A. Alam and A. Minissale. 2005. Thermal discharges at Manikaran, Himachal Pradesh, India. Proceed. World Geotherm. Congress Antalya, Turkey: pp. 1-4.

Cinti D., L. Pizzino, N. Voltattorni, F. Quattrocchi and V. Walia. 2009. Geochemistry of thermal waters along fault segments in the Beas and Parvati valleys (north-west Himalaya, Himachal Pradesh) and in the Sohna town (Haryana), India. Geochem. J. 43: 65–76.

Craig J., A. Absar, G. Bhat, G. Cadel, M. Hafiz, N. Hakhoo, et al. 2013. Hot springs and the geothermal energy potential of Jammu and Kashmir State, N.W. Himalaya, India. Earth Sci. Rev. 126: 156–177. doi:10.1016/j.earscirev.2013.05.004.

Das S., M.T. Sherpa, S. Sachdeva and N. Thakur. 2012a. Hot springs of Sikkim (Tatopani): A Socio medical conjuncture which amalgamates religion, faith, traditional belief and tourism. Asian Acad. Res. J. Social Sci. Human. 1(4): 80–93.

Das S., M.T. Sherpa and N. Thakur. 2012b. Sikkim's Tatopani - A balneotherapeutic prospect for community health in North East India. Int. J. Agri Food Sci. Tech. 3(2): 149–152.

Das S., I.N. Najar, M.T. Sherpa and N. Thakur. 2016. Biotechnological and sociological importance of hot springs of Sikkim. pp. 149–181. *In*: N. Bag, R. Murugan and A. Bag [eds.]. Biotechnology in India: Initiatives and Accomplishments. New India Publishing Agency, New Delhi, India.

Dhyani A., R. Gururani, S.A. Selim, P. Adhikari, A. Sharma and V. Pande. 2017. Production of industrially important enzymes by thermobacilli isolated from hot springs of India. Res. Biotechnol. 8: 19–28. doi: 10.25081/Rib.2017.V8.3594.

Dwivedi V., N. Sangwan, A. Nigam, N. Garg, N. Niharika, P. Khurana, et al. 2012. Draft genome sequence of *Thermus* sp. Strain RL, isolated from a hot water spring located atop the Himalayan ranges at Manikaran, India. J. Bacteriol. 194(13): 3534.

Dwivedi V., K. Kumari, S.K. Gupta, R. Kumari, C. Tripathi, P. Lata, et al. 2015. *Thermus parvatiensis* RL(T) sp. nov., isolated from a hot water spring, located atop the Himalayan ranges at Manikaran, India. Indian J. Microbiol. 55(4): 357–365. doi:10.1007/s12088-015-0538-4.

Egorova L.A. and I.G. Loginova. 1975. Distribution of highly thermophilic, nonsporulating bacteria in the hot springs of Tadzhikistan. Mikrobiologia 44(5): 938–942.

Egorova L.A. and I.G. Loginova. 1977. Distribution of highly thermophilic, nonsporulating bacteria in the hot springs of Tadzhikistan. Mikrobiologia 46(2): 342–345.

Elleuche S., C. Schäfers, S. Blank, C. Schröder and G. Antranikian. 2015. Exploration of extremophiles for high temperature biotechnological processes. Curr. Opin. Microbiol. 25: 113–119. doi:10.1016/j.mib.2015.05.011.

Erum N., Z. Mushtaq and A. Jamil. 2017. Isolation and identification of a catalase producing thermoduric alkalotolerant *Bacillus* sp. strain KP10 from hot springs of Tatta Pani, Azad Kashmir. J. Anim. Plant Sci. 27(6): 2056–2062.

Ghosh D., B. Bal, V.K. Kashyap and S. Pal. 2003. Molecular phylogenetic exploration of bacterial diversity in a Bakreshwar (India) Hot Spring and culture of *Shewanella* - related thermophiles. Appl. Environ. Microbiol. 69: 4332–4336.

Ghumro P.B., M. Shafique, M.I. Ali, I. Javed, B. Ahmad, A. Jamal, et al. 2011. Isolation and screening of protease producing thermophilic *Bacillus* strains from different soil types of Pakistan. Afr. J. Microbiol. Res. 5: 5534–5539.

Hazarika A., T. Sarmah and A. Hazarika. 2014. Study of the prokaryotes from hot springs and its neighbour area from Nambor wildlife sanctuary, Assam. Indian J. Fundamental Appl. Life Sci. 4(4): 62–67.

Hazarika D. and P. Gogoi. 1985. Thermal algae from Hot Springs of Nambor Forest, Assam. Geobios News Report 4: 187-190.

Hochstein M.P. and K.R. Lieb. 1998. Heat generation associated with collision of two plates: the Himalayan geothermal belt. J. Volcanol. Geotherm. Res. 83: 75–92.

Huang Q., C.Z. Dong, R.M. Dong, H. Jiang, S. Wang, G. Wang, et al. 2011. Archaeal and bacterial diversity in hot springs on the Tibetan Plateau, China. Extremophiles 15(5): 549–563. doi:10.1007/s00792-011-0386-z.

Huber R., H. Huber and K.O. Stetter. 2000. Towards ecology of hyperthermophiles biotopes new isolation strategies and novel metabolic properties. FEMS Microbiol. Rev. 24: 615–623.

Jadoon M.A., T. Ahmad, M.M.U. Rehman, A. Khan and A. Majid. 2014. Isolation and identification of thermophillic actinomycetes from hot water springs from Azad Jammu and Kashmir Pakistan for the production of thermophillic amylase. World Appl. Sci. J. 30(3): 350–354. doi: 0.5829/idosi.wasj.2014.30.03.82338.

Jahangir T.M., M.Y. Khuhawar, S.M. Laghari and A. Laghari. 2001. Physico-chemical and Biological study of Mangho Pir euthermal springs, Karachi, Sindh Pakistan. Online J. Biol. Sci. 1(7): 636–639.

Javed A., J. Iqbal, U. Asghar, F.A. Khan, A.B. Munshi and I. Siddiqui. 2009. A study to evaluate therapeutic properties of minerals of Mangho Pir hot spring, Karachi. J. Chem. Soc. Pak. 31: 396–401.

Javed M.M., S. Zahoor, H. Sabar, I.U. Haq and M.E. Babar. 2012. Thermophilic bacteria from the hot springs of Gilgit (Pakistan). J. Anim. Plant Sci. 22: 83–87.

Jumaeva G.R. 2008. Algal Flora of Major Thermal and Mineral Springs of Pamir. PhD Thesis, Tajikistan, Dushanbe.

Kalita D. and S.R. Joshi. 2017. Study on bioremediation of Lead by exopolysaccharide producing metallophilic bacterium isolated from extreme habitat. Biotechnol. Rep. (Amst) 16: 48–57.

Kanokratana P., S. Chanapan, K. Pootanakit and L. Eurwilaichitr. 2004. Diversity and abundance of *bacteria* and *archaea* in the Bor Khlueng Hot Spring in Thailand. J. Basic Microbiol. 44(6): 430–444. doi: 10.1002/Jobm.200410388.

Khan M.A., N. Ahmad, A.U. Zafar, I.A. Nasir and M.A. Qadir. 2011. Isolation and screening of alkaline protease producing bacteria and physio–chemical characterization of the enzyme. Afr. J. Biotech. 10: 6203–6212.

Khan R., S.H. Shah and N.A. Khan. 1999. Investigation of the geothermal springs of the Tatta Pani area, district Kotli, Azad Jammu and Kashmir. Geol. Surv. Pak. 701: 1–11.

Khan R., M.I. Khan, A. Zeb, N. Roy, M. Yasir, I. Khan. et al. 2018. Prokaryotic diversity from extreme environments of Pakistan and its potential applications at regional levels. bioRxiv preprint dx. doi. org/10.1101/342949.

Khin K., Z. Khin and T.A. Lin. 2017. Geological and tectonic evolution of the Indo-Myanmar Ranges (IMR) in the Myanmar region. pp. 65–79. *In*: A.J. Barber, Z. Khin and M.J. Crow [eds.]. Myanmar: Geology, Resources and Tectonics. Geological Society, London, Memoirs. doi:10.1144/M48.4.

Kristjansson J.K., G.O. Hreggvidsson and G.A. Alfredsson. 1986. Isolation of halotolerant *Thermus* spp. from submarine hot springs in Iceland. Appl. Environ. Microbiol. 52: 1313–1316.

Kumar B., P. Trivedi, A.K. Mishra, A. Pandey and L.M.S. Palni. 2004. Microbial diversity of soil from two hot springs in Uttaranchal Himalaya. Microbiol. Res. 159: 141–146.

Kumar B., A. Pandey and L.M.S. Palni. 2005. Extracellular amylase activity of *Saccharomycopsis fibuligera*, a mycelial yeast isolated from a hot spring site in Garhwal Himalayas. Indian J. Microbiol. 45: 211–215.

Kumar M., A.N. Yadav, R. Tiwari, R. Prasanna and A.K. Saxena. 2014. Deciphering the diversity of culturable thermotolerant bacteria from Manikaran hot springs. Ann. Microbiol. 64: 741–751.

Kumar R., K. Ravinder, Nupur, T.N. Srinivas and P.V. Kumar. 2013. *Caldimonas meghalayensis* sp. nov., a novel thermophilic betaproteobacterium isolated from a hot spring of Meghalaya in northeast India. Antonie van Leeuwenhoek 104(6): 1217–1225.

Kumar R. and R.C. Sharma. 2019. Microbial diversity and physico-chemical attributes of two hot water springs in the Garhwal Himalaya, India. J. Microbiol. Biotechnol. Food Sci. 8(6): 1249–1253. doi: 10.15414/ Jmbfs.2019.8.6.1249-1253.

Kurenoe V.V. and V.I. Belianin. 1969. Mineral Waters of Afghanistan, Dept. Geological Survey, Kabul, Afghanistan.

Ladenstein R. and G. Antranikian. 1998. Proteins from hyperthermophiles: stability and enzymatic catalysis close to the boiling point of water. Adv. Biochem. Eng. Biotechnol. 61: 37–85.

Liu L., N. Salam, J.Y. Jiao, H.C. Jiang, E.M. Zhou, Y.R. Yin, et al. 2016. Diversity of culturable thermophilic actinobacteria in hot springs in Tengchong, China and studies of their biosynthetic gene profiles. Microb. Ecol. 72(1): 150–162. doi:10.1007/s00248-016-0756-2.

López-López O., M.E. Cerdán and M.I. González-Siso. 2013. Hot spring metagenomics. Life 3(2): 308–320. doi:10.3390/life3020308.

Mac Elroy R.D. 1974. Some comments on the evolution of extremophiles. BioSystems 6: 74–55.

Madigan M.T. and B.L. Marrs. 1997. Extremophiles. Sci. Am. 276: 82–87.

Madigan M.T. and A. Oren. 1999. Thermophilic and halophilic extremophiles. Curr. Opin. Microbiol. 2: 265–269.

Mahato N.K., C. Tripathi, H. Verma, N. Singh and R. Lal. 2014. Draft genome sequence of *Deinococcus* sp. strain RL isolated from sediments of a hot water spring. Genome Announc. 2: e703–e714. doi: 10.1128/ genomeA.00703-14.

Najar I.N. 2018. Bacterial Diversity and Antibiotic Resistance Profile of Four Hot Springs of Sikkim. PhD Thesis, Sikkim University.

Najar I.N., M.T. Sherpa, S. Das and N. Thakur. 2018a. Microbial ecology of two hot springs of Sikkim: predominate population and geochemistry. Sci. Total Environ. 637(1): 730–745.

Najar I.N., M.T. Sherpa. S. Das, K. Verma, V.K. Dubey and N. Thakur. 2018b. *Geobacillus yumthangensis* sp. nov., a thermophilic bacterium isolated from a north-east Indian hot spring. Int. J. Syst. Evol. Microbiol. 68: 3430–3434.

Najar I.N., M.T. Sherpa, S. Das and N. Thakur. 2018c. Draft genome sequence of *Geobacillus yumthangensis* AYN2 sp. nov., a denitrifying and sulfur reducing thermophilic bacterium isolated from the hot springs of Sikkim. Gene Reports 10: 162–166.

Niaz M., T. Iftikhar, R. Tabassum, M.A. Zia, H. Saleem, S.Q. Abbas, et al. 2010. α-amylase production by *Bacillus licheniformis* under solid state fermentation conditions and its cross linking with metalo-salts to confer thermostability. Int. J. Agric. Biol. 12: 793–795.

Niehaus F., F. Niehaus, C. Bertoldo and G. Antranikian. 1999. Extremophiles as a source of novel enzymes for industrial application. Appl. Microbiol. Biotechnol. 51: 711–729.

Niyatbekov T. and S.S. Barinova. 2018. Bioindication of aquatic habitats with diatom algae in the Pamir Mountains, Tajikistan. MOJ Ecol. Environ. 3(3): 117–120.

Panda A.K., S.S. Bisht, S.D. Mandal and N.S. Kumar. 2016. Bacterial and archeal community composition in hot springs from Indo-Burma region, North-east India. AMB Exp. 6(111): 1–12.

Pandey A., K. Dhakar, A. Sharma, P. Priti, P. Sati and B. Kumar. 2014a. Thermophilic bacteria, that tolerate wide temperature and pH range, colonize the Soldhar (95°C) and Ringigad (80°C) hot springs of Uttarakhand, India. Ann. Microbiol. 65(2): 809–816.

Pandey A., K. Dhakar, P. Sati, A. Sharma, B. Kumar and L.M.S. Palni. 2014b. *Geobacillus stearothermophilus* a resilient hyperthermophile isolated from an autoclaved sediment sample. Proc. Natl. Acad. Sci. 84(2): 349–356.

Parveen A., A. Deka, G. Goswami, M. Barooah and R.C. Boro. 2016. Cellulase producing thermophilic bacteria from hot spring of Assam. Int. J. Microbiol. Res. 8(8): 776–780.

Priya I., M.K. Dhar, B.K. Bajaj, S. Koul and J. Vakhlu. 2016. Cellulolytic activity of thermophilic bacilli isolated from Tattapani hot spring sediment in North West Himalayas. Indian J. Microbiol. 56(2): 228–231. doi:10.1007/s12088-016-0578-4.

Qadar S.A.U., E. Shereen, S. Iqbal and A. Anwar. 2009. Optimization of protease production from newly isolated strains of *Bacillus* sp. PCSIR EA-3. Indian J. Biotechnol. 8: 286–290.

Rafique N., M.S. Awan, H.A. Rathore, D.A. Gardezi, M. Ahmad, M. Tariq, et al. 2010. Kinetics of endoglucanase and cellobiohydrolase production by parent and mutant derivative of moderately thermotolerant *Bacillus subtilis* GQ 301542 on optimized medium. Afr. J. Biotechnol. 9: 7531–7538.

Ranawat P. and S. Rawat. 2017. Characterization of bacterial diversity of hot springs of Chamoli region of Garhwal Himalaya. ENVIS Bull. Himalayan Ecol. 25: 120–128.

Rawat N. and G.K. Joshi. 2018. Bacterial community structure analysis of a hot spring soil by next generation sequencing of ribosomal RNA. Genomics 111(5): 1053–1058. doi:10.1016/j.ygeno.2018.06.008.

Rawat R., R. Rautela, S. Rawat and A.B. Bhatt. 2018. Hydrolytic enzyme production potential of bacterial population from Sahastradhara cold sulfur spring, Uttarakhand. Plant Arch. 18(2): 1313–1316.

Reysenbach A. and E. Shock. 2002. Merging genomes with geochemistry in hydrothermal ecosystems. Science 296: 1077–1083.

Rothschild L.J. and R.L. Mancinelli. 2001. Life in extreme environments (nature). Nature 409: 1092–1101.

SAARC Energy Centre. 2011. Study on Geothermal resources of South Asia. www.saarcenergy.org.

Saba D.S., M.E. Najaf, A.M. Musazai and S.A. Taraki. 2004. Geothermal energy in Afghanistan: prospects and potential centre on internal cooperation, New York University, New York, USA, and Afghanistan Centre for Policy and Development Studies, Kabul, Afghanistan.

Saleem F., M. Asif, M. Ajaz and S.A. Rasool. 2012. Studies on extracellular protein metabolites of thermophilic bacterial strains isolated from local hot spring. J. Chem. Soc. Pak. 34: 1–5.

Sharma A., A. Pandey, Y.S. Shouche, B. Kumar and G. Kulkarni. 2009. Characterization and identification of *Geobacillus* spp. isolated from Soldhar hot spring site of Garhwal Himalaya, India. J. Basic. Microbiol. 49: 187–194.

Sharma A., K. Jani, Y.S. Souche and A. Pandey. 2014a. Microbial diversity of Soldhar, hot spring, India, assessed by analyzing 16S rRNA and protein coding genes. Ann. Microbiol. 65(3): 1323–1332.

Sharma A., P. Hira, M. Shakarad and R. Lal. 2014b. Draft genome sequence of *Cellulosimicrobium* sp. strain MM, isolated from arsenic-rich microbial mats of a himalayan hot spring. Genome Announc. 2: e01020–14. doi: 10.1128/genomeA.01020-14.

Sharma A., P. Kohli, Y. Singh, P. Schumann and R. Lal. 2016. *Fictibacillus halophilus* sp. nov., from a microbial mat of a hot spring atop the himalayan range. Int. J. Syst. Evol. Microbiol. 66: 2409–2416. doi: 10.1099/ijsem.0.001051.

Sharma A., D. Paul, D. Dhotre, K. Jani, A. Pandey and Y.S. Shouche. 2017. Deep sequencing analysis of bacterial community structure of Soldhar hot spring, India. Microbiology 86(1): 136–142.

Sharma A., M. Schmidt, B. Kiesel, N.K. Mahato, L. Cralle, Y. Singh, et al. 2018b. Bacterial and archaeal viruses of Himalayan hot springs at Manikaran modulate host genomes. Front. Microbiol. 9: 3095. doi: 10.3389/fmicb.2018.03095.

Sharma B., R. Verma, K. Dev and R. Thakur. 2012. Molecular characterization of Manikaran hot spring microbial community by 16S rRNA and RAPD analysis. BioTechnology: An Indian J. 6: 254–266.

Sharma N., G. Vyas and S. Pathania. 2013. Culturable diversity of thermophilic microorganisms found in hot springs of Northern Himalayas and to explore their potential for production of industrially important enzymes. Scholars Acad. J. Biosci. 1(5): 165–178.

Sharma S., P. Sharma, A. Sourirajan, D.J. Baumler and K. Dev. 2018a. Identification of thermophilic *Flavobacterium* and *Anoxybacillus* in unexplored Tatapani hot spring of Kishtwar district of Jammu and Kashmir: A North Western Himalayan state. Curr. Trends Biotechnol. Pharm. 12(3): 245–256.

Sherpa M.T., S. Das and N. Thakur. 2013. Physicochemical analysis of hot water springs of Sikkim – Polok Tatopani, Borong Tatopani and Reshi Tatopani. Recent Res. Sci. Tech. 5(1): 63–67.

Shirkot P. and A. Verma. 2015. Assessment of thermophilic bacterial diversity of thermal springs of Himachal Pradesh. ENVIS Bull. Himalayan Ecol. 23: 27–34.

Siddiqui M.A., N. Rashid, S. Ayyampalayam and W.B. Whitman. 2014. Draft genome sequence of *Geobacillus thermopakistaniensis* strain MAS1. Genome Announc. 2: e00559–14. doi: 10.1128/genomeA.00559-14.

Sinha R.K. 1980. Some thermal springs in Kameng District, Arunachal Pradesh. J. Geol. Soc. India 21: 464–467.

Soe K.M., A. Yokoyama, J. Yokoyama and Y. Hara. 2011. Morphological and genetic diversity of the thermophilic cyanobacterium, *Mastigocladus laminosus* (Stigonematales, Cyanobacteria) from Japan and Myanmar. Phycol. Res. 59: 135–142.

Sompong U., P.R. Hawkins, C. Besley and Y. Peerapornpisal. 2005. The distribution of cyanobacteria across physical and chemical gradients in hot springs in northern Thailand. FEMS Microbiol. Ecol. 52: 365–376.

Stetter KO. 1999. Extremophiles and their adaptation to hot environments. FEBS Lett. 452: 22–25.

Subtavewung P., M. Raksaskulwong and J. Tulyatid. 2005. The characteristic and classification of hot springs in Thailand. Proceed. World Geotherm. Cong. 2: pp. 1–7.

Suman A., P. Verma, A.N. Yadav and A.K. Saxena. 2015. Bioprospecting for extracellular hydrolytic enzymes from culturable thermotolerant bacteria isolated from Manikaran thermal springs. Res. J. Biotechnol. 10(4): 33–42.

Synowiecki J. 2010. Some applications of thermophiles and their enzymes for protein processing. J. Biotechnol. 9: 7020–7025.

Tang J., Y. Liang, D. Jiang, L. Li, Y. Luo, M.M.R. Shah, et al. 2018. Temperature-controlled thermophilic bacterial communities in hot springs of western Sichuan, China. BMC Microbiol. 18: 134. https://doi.org/10.1186/s12866-018-1271-z.

Taye C.D. and A. Chutia. 2016. Physical and chemical characteristics of a few hot springs of Tawang and West Kameng district, Arunachal Pradesh, Northeast India. J. Assam Sci. Soc. 57(1&2): 47–55.

Tayyab M., Rashid N. and M. Akhtar. 2011. Isolation and identification of lipase producing thermophilic *Geobacillus* sp. SBS-4S: cloning and characterization of the lipase. J. Biosci. Bioeng. 111(3): 272–278. doi:10.1016/j.jbiosc.2010.11.015.

Thakur N., S. Das, M.T. Sherpa and R. Ranjan 2013. GPS mapping and physical description of Polok, Borong and Reshi Tatopani – Hot Springs of Sikkim. J. Int. Acad. Res. Multidiscip. 1(10): 637–648.

Treloar P.J. and C.N. Izatt. 1993. Tectonics of the Himalayan collision between the Indian Plate and the Afghan Block: a synthesis, Geological Society, London, Special Publications 74: 69–87.

Tripathi C., N.K. Mahato, A.K. Singh, K. Kamra, S. Korpole and R. Lal. 2016. *Lampropedia cohaerens* sp. nov., a biofilm-forming bacterium isolated from microbial mats of a hot water spring, and emended description of the genus *Lampropedia*. Int. J. Syst. Evol. Microbiol. 66: 1156–1162. doi:10.1099/ijsem.0.000853.

Trivedi P., B. Kumar and A. Pandey. 2006. Conservation of soil microbial diversity associated with two hot springs in Uttaranchal Himalaya. Natl. Acad. Sci. Lett. 29: 185–188.

Tun M.M. 2019. An overview of renewable energy sources and their energy potential for sustainable development in Myanmar. Eur. J. Sustain. Dev. 3(1): em0071. https://doi.org/10.20897/ejosdr/3951.

Ullah S., M. Irfan, W. Sajjad, Q.U.A. Rana, F. Hasan, S. Khan, et al. 2019. Production of an alkali-stable xylanase from *Bacillus pumilus* K22 and its application in tomato juice clarification. Food Biotechnol. 33(4): 353–372. doi: 10.1080/08905436.2019.167 4157.

Verma A., M. Gupta and P. Shirkot. 2014a. Isolation and characterization of thermophilic bacteria in natural hot water springs of Himachal Pradesh (India). The Bioscan 9(3): 947–952.

Verma A., K. Dhiman, M. Gupta and P. Shirkot. 2014b. Bioprospecting of thermotolerant bacteria from hot water springs of Himachal Pradesh for the production of Taq DNA polymerase. Proc. Natl. Acad. Sci. India Sect. B. Biol. Sci. doi: 10.1007/s40011-014-0412-x.

Verma J.P., D.K. Jaiswal, R. Krishna, S. Prakash, J. Yadav and V. Singh. 2018. Characterization and screening of thermophilic *Bacillus* strains for developing plant growth promoting consortium from hot spring of Leh and Ladakh region of India. Front. Microbiol. 9: 1293. doi: 10.3389/fmicb.2018.01293.

Verma P., A.N. Yadav, A. Suman and A.K. Saxena. 2012. Isolation and molecular characterization of thermotolerant lignocellulose producing fungi from Manikaran thermal springs. *In*: Proceeding of National Symposium on Microbes in Health and Agriculture, at School of Life Sciences, Jawaharlal Nehru University New Delhi, 82.

Wangchuk P. and Y. Dorji. 2007. Historical roots, spiritual significance and the health benefits of mKhempa-l Jong gNyes Tshachu (hot spring) in Lhuntshe. J. Bhutan Studies 112–128.

Wangdi N., T. Dorji and K. Wangdi. 2014. Hot springs and mineral springs of Bhutan. Ugyen Wangchuck Institute for Conservation and Environment. Lamai Gompa: Bhumtang. UWICE Press.

Wiegel J. 2001. Extreme Thermophiles. eLS. 2: 1–12.

Win O.Y. and L.T. Nwe. 2019. Taxonomic study on some members of Chlorophyta in Yenwe village, Taunggyi Township. J. Myanmar Acad. Arts Sci. XVII(4): 433–450.

Wood S.H. and F.S. Singharajwarapan. 2014. Geothermal systems of Northern Thailand and their association with faults active during the Quaternary. GRC Trans. 38: 607–615.

Yadav P., S. Korpole, G.S. Prasad, G. Sahni, J. Maharjan, L. Sreerama, et al. 2017. Morphological, enzymatic screening and phylogenetic analysis of thermophilic bacilli isolated from five hot springs of Myagdi. Nepal. J. Appl. Biol. Biotechnol. 6: 1–8. doi: 10.7324/JABB.2018.60301.

Yadav P., J. Maharjan, S. Korpole, G.S. Prasad, G. Sahni, T. Bhattarai, et al. 2018. Production, purification, and characterization of thermostable alkaline xylanase from *Anoxybacillus kamchatkensis* NASTPD13. Front. Bioeng. Biotechnol. 6: 65. doi: 10.3389/fbioe.2018.00065.

Zahoor S., M.M. Javed, M.N. Aftab and U.H. Ikram. 2012. Isolation and molecular identification of a facultatively anaerobic bacterium from the hot spring of Azad Kashmir. Pak. J. Bot. 44: 329–333.

Zahoor S., M.M. Javed and E. Babarh. 2016. Characterization of a novel hydrolytic enzyme producing thermophilic bacterium isolated from the hot spring of Azad Kashmir-Pakistan. Brazilian Braz. Arch. Biol. Technol. 59: e16150662. dx.doi.org/10.1590/1678-4324-2016150662.

Zaidi NS. 2007. Cloning and overexpression of cellulase genes of thermophillic bacterial species. Dissertation, University of the Punjab.

Zaigham N.A., Z.A. Nayyar and N. Hisamuddin. 2009. Review of geothermal energy resources in Pakistan. Renew. Sust. Energ. Rev. 13: 223–232.

3

The Dead Sea: Lessons Learned
From the Study of Microorganisms
Isolated From a Dying Lake

Aharon Oren

Introduction

The Dead Sea is one of the most saline bodies of water on Earth. Today, the total dissolved salts concentration is around 350 g/l. It is an 'athalassohaline' lake, meaning that the ionic composition of the brines differs greatly from that of seawater. Divalent cations (~2.2 M Mg^{2+}, ~0.5 M Ca^{2+}) dominate over monovalent cations (~1.2 M Na^+, ~0.2 M K^+). Cl^- is the main anion (~99% of the anion sum) followed by Br^-. The pH of the brine is ~6.

Because of its high salinity and the abundance of toxic divalent cations, the Dead Sea was for long considered sterile. The first evidence of life in the lake was obtained during the studies of Benjamin Elazari-Volcani (Wilkansky) in the 1930s (Wilkansky 1936, Elazari-Volcani 1940, Volcani 1944). In the past decades, the Dead Sea has been drying out rapidly; the lake level is decreasing by more than a meter annually. Due to the resulting increase in salinity, the conditions in the water column have now become so extreme that even the best salt- and divalent cation-adapted microorganisms can no longer grow there. But from the times of Volcani's studies until the late 1990s, conditions for life were more favorable in the less saline upper layers of the water column. Before 1979, the lake was 'permanently' stratified (meromictic). As a result of the negative water balance, the complete mixing of the water column occurred in February 1979. Since that time the lake has been generally holomictic, but short meromictic episodes occurred (1980-1982 and 1992-1996) following massive inflow of freshwater from the Jordan River and rain floods after exceptionally wet winters.

Volcani's studies in the late 1930s and early 1940s were qualitative only and were based on enrichment cultures and characterization of the organisms that had developed. Some of his isolates were named as new species (further information is provided in the sections below). The first quantitative information about microbial community densities was collected only in 1963-1964 when up to 8.9×10^6 prokaryotes (most probably being red halophilic Archaea) and up to 4×10^4 cells per ml of the unicellular green alga *Dunaliella* were reported. The highest densities of organisms were found in the surface layers (Kaplan and Friedmann 1970). Following the rainy winter of 1979-1980, a diluted surface layer of about 5 m depth was formed, initiating a short meromictic episode that lasted until the end of 1982. *Dunaliella parva* reached population densities of up to 8,800 cells per ml. Both dilution of the brine and the availability of phosphate are required for algal development in the Dead Sea (Oren and Shilo 1982). Following the algal bloom, a bloom of red halophilic Archaea developed with population densities of up to 1.9×10^7 cells per ml, which imparted a red color to the water (Oren 1983a). In addition to bacterioruberin carotenoids, this

The Institute of Life Sciences, The Hebrew University of Jerusalem, The Edmond J. Safra Campus, 9190401 Jerusalem, Israel.

E-mail: aharon.oren@mail.huji.ac.il

bloom contained the purple light-driven proton pump bacteriorhodopsin (Oren and Shilo 1981). A renewed microbial bloom developed in the lake in the spring of 1992 in the upper 5-10 m of the water column with up to 1.5×10^4 Dunaliella cells per ml (Oren et al. 1995a) and up to 3.5×10^7 prokaryotic cells per ml that once more imparted a red coloration to the Dead Sea due to their high content of bacterioruberin. Bacteriorhodopsin was not detected in this bloom (Oren and Gurevich 1993).

Currently, the lake is saturated with NaCl. Massive amounts of halite are therefore precipitated and sink to the bottom as a result of the drop in the water level. The ratio of toxic divalent cations to the better tolerated Na^+ ions is therefore increasing, making the lake an even more extreme environment for life than it had been in the past (Oren 2010).

Microorganisms Isolated From the Dead Sea

Over the years, many cultivation-dependent studies have been performed in the Dead Sea and surrounding hypersaline environments, and these have yielded a large number of novel species of Archaea, Bacteria as well Eukaryotes that were described in the literature.

Archaea

- *Haloarcula marismortui* (ex Volcani 1940) Oren et al. 1990) was first used in a physiological study by Ginzburg et al. (1970). The type strain (ATCC 43049, DSM 6131, JCM 7785, NBRC 101032, VKM B-2009) (Oren et al. 1990) closely resembles the lost culture of '*Halobacterium marismortui*', first described by Elazari-Volcani (1940).
- *Haloferax volcanii* Torreblanca et al. 1986 (basonym: *Halobacterium volcanii* Mullakhanbhai and Larsen 1975; Approved Lists 1980) was isolated from surface sediment. Type strain is DS2 (ATCC 29605, DSM 3757, NBRC 14742, JCM 8879, NBBC 85050, NCIMB 2012, VKM B-1768). It is a moderate halophile with a high magnesium tolerance.
- *Halorubrum sodomense* (Oren 1983) McGenity and Grant 1996 (basonym *Halobacterium sodomense* Oren 1983) with type strain RD-26 (ATCC 33755, CIP 105330, DSM 3755, NBRC 14740, JCM 8880, NCIMB 2197, VKM B-1771). It has an extremely high magnesium requirement and tolerance and was isolated from the archaeal bloom that developed in the Dead Sea in 1980.
- *Halobaculum gomorrense* Oren et al. 1995, the first and thus far the only member of the genus *Halobaculum* with type strain DS2807 (ATCC 700876, DSM 9297, JCM 9908). It was isolated from the 1992 archaeal bloom in the lake. Like *Hrr. sodomense*, it requires relatively low sodium for growth but needs a very high level of magnesium; optimal growth occurs in the presence of 0.6-1.0 M Mg^{2+} (Oren et al. 1995b).
- *Haloplanus natans* Elevi Bardavid et al. 2007, the first described species of the genus *Haloplanus* with type strain RE-101 (DSM 17983, JCM 14081). This little-pigmented species of pleomorphic flat cells, which contain gas vesicles, was isolated from Dead Sea-Red Sea water mixtures in the experimental outdoor ponds (Elevi Bardavid et al. 2007).

In addition to these taxa described as new species, many isolates of the class *Halobacteria* recovered from the Dead Sea were characterized in-part. These included strains affiliated with the genera *Halobacterium*, *Haloferax*, and *Haloarcula* that were isolated in the 1990s from enrichments prepared by Volcani from Dead Sea water samples collected in the 1930s (Arahal et al. 1996, 2000a).

Aerobic Bacteria

- *Halomonas halmophila* (Elazari-Volcani 1940) Franzmann et al. 1989 (basonym: *Flavobacterium halmophilum* corrig. Elazari-Volcani 1940) with type strain ACAM 71

(ATCC 19717, CIP 105455, DSM 5349, NBRC 15537, JCM 21222, LMG 4023, NCIMB 1971). This species was isolated by Volcani in the late 1930s. Dobson et al. (1990) provided an emended description of the taxon.

- *Chromohalobacter israelensis* (Huval et al. 1996) Arahal et al. 2001 (basonym: *Halomonas israelensis* Huval et al. 1996) with type strain Ba1 (ATCC 43985, CECT 5287, CCM 4920, CIP 106853, DSM 6768, LMG 19546, NCIMB 13766). It was isolated from a crude salt sample collected at the Dead Sea evaporation ponds (Rafaeli-Eshkol 1968). Stain Ba1 was later named *Halomonas israelensis* (Huval et al. 1995) and subsequently assigned to the genus *Chromohalobacter* as *Chromohalobacter israelensis* (Arahal et al. 2001).

- *Chromohalobacter marismortui* (ex Elazari-Volcani 1940) Ventosa et al. 1989. This species, originally described as 'Chromobacterium marismortui' by Volcani, was described as a member of the genus *Chromohalobacter* as it differs from the genus *Chromobacterium* in many aspects, including the nature of its pigment and the arrangement of the flagella (Ventosa et al. 1989).

- *Virgibacillus marismortui* (Arahal et al. 1999) Heyrman et al. 2003 (basonym: *Bacillus marismortui* Arahal et al. 1999; other synonyms: *Salibacillus marismortui* (Arahal et al. 1999) Arahal et al. 2000) with type strain 123 (ATCC 700626, CECT 5066, CIP 105609, DSM 12325, LMG 18992). This Gram-positive, rod-shaped organism was isolated in the late 1990s from enrichments prepared from Dead Sea water samples in the late 1930s. It was first described as *Bacillus marismortui*, then renamed *Salibacillus marismortui*, and finally *Virgibacillus marismortui* (Arahal et al. 1999, 2000b; Heyrman et al. 2003).

Five more *Chromohalobacter* strains were retrieved from a water sample collected in November 2008 from a depth of 75 m (Atanasova et al. 2012). A novel halophilic actinobacterium described as *Amycolatopsis flava*, a name not yet validly published, with proposed type strain AFM 10111 (DSM 46658, CGMCC 4.7123) was isolated from Dead Sea sediment (Wei et al. 2015).

Anaerobic Chemotrophic Bacteria

- *Halobacteroides halobius* Oren et al. 1984, with type strain MD-1 (ATCC 35273, DSM 5150) is a moderately halophilic anaerobic, long rod-shaped moderately halophilic bacterium from the bottom sediments of the Dead Sea. It ferments simple sugars to ethanol, acetate, hydrogen, and carbon dioxide. It is the type species of the type genus of the family *Halobacteroidaceae* (order *Halanaerobiales*) (Oren et al. 1984).

- *Sporohalobacter lortetii* (Oren 1984) Oren et al. 1988 (basonym: *Clostridium lortetii* Oren 1984) with type strain DH-1 (ATCC 35059, DSM 3070), also isolated from Dead Sea sediments, is a halophilic anaerobic bacterium that produces endospores with attached gas vesicles. Products of amino acid fermentation include acetate, butyrate, and hydrogen. Like *Halobacteroides halobius*, it belongs to the family *Halobacteroidaceae* (Oren 1983c, Oren et al. 1987).

- *Orenia marismortui* (Oren et al. 1988) Rainey et al. 1995 (basonym: *Sporohalobacter marismortui* Oren et al. 1988) with type strain DY-1 (ATCC 35420, DSM 5156) is an endospore-forming halophile that ferments sugars to ethanol, acetate, butyrate, formate, hydrogen, and carbon dioxide (Oren et al. 1987, Rainey et al. 1995).

- *Selenihalanaerobacter shriftii* Switzer Blum et al. 2001 with type strain DSSe-1 (ATCC BAA-73), also a member of the *Halobacteroidaceae*, is a halophilic isolate from Dead Sea sediment that respires selenate. It oxidizes glucose to acetate + CO_2 with concomitant reduction of selenate to selenite plus elemental selenium. Other electron acceptors that supported anaerobic growth on glycerol are nitrate and trimethylamine-N-oxide (Switzer Blum et al. 2001).

Halophilic Anoxygenic Phototrophs

- *Ectothiorhodospira marismortui* Oren et al. 1990 with type strain EG-1 (DSM 4180) (proposed to be a later heterotypic synonym of *Ectothiorhodospira mobilis* Pelsh 1936; Approved Lists 1980), was isolated from Hamei Mazor, a warm (39°C) hypersaline sulfur spring on the shore of the Dead Sea. It is a moderate halophile that can grow photoautotrophically with sulfide as an electron donor or photoheterotrophically using acetate, succinate, fumarate, malate, or pyruvate as carbon sources (Oren 1989, Oren et al. 1989).

- *Rhodovibrio sodomensis* (Mack et al. 1996) Imhoff et al. 1998 (basonym: *Rhodospirillum sodomense* Mack et al. 1996), of which the type strain DSI (ATCC 51195, DSM 9895) is a pink-colored halophilic anoxygenic purple bacterium isolated from water and sediment of the Dead Sea. It grows between 6% and 21% NaCl with an optimum at 12% NaCl (Mack et al. 1993, Imhoff et al. 1998).

Eukaryotic Microorganisms

- *Gymnascella marismortui* strain UHA 1632, KW 2005, described as a new species of the ascomycete genus *Gymnascella* (Buchalo et al. 1998).

The best-known eukaryote in the biota of the Dead Sea is the unicellular flagellated green alga *Dunaliella parva*, documented from the algal blooms in the lake in 1980 and 1992 (Oren and Shilo 1992, Oren et al. 1995a). A Dead Sea strain of *D. parva* was isolated in pure culture in the mid-1960s and served in a number of physiological studies (e.g., Ginzburg 1969). The *Dunaliella* cells observed in the lake in 1964 were identified as *D. viridis* (Kaplan and Friedmann 1970).

There is no evidence that fungi are quantitatively important part of the Dead Sea biota. Still, a great diversity of fungi of different groups (*Oomycota, Zygomycota, Ascomycota,* mitosporic fungi) were cultivated from water samples collected from different depths and in different seasons. Species that were found included *Aspergillus terreus, A. sydowii, A. versicolor, Eurotium herbariorum, Penicillium westlingii, Cladosporium cladosporoides,* and *C. sphaerospermum.* To what extent the fungal diversity recovered was present as dormant spores or as vegetative mycelia is unknown (Buchalo et al. 1998, 2000, Kis-Papo et al. 2001). Using a medium, without salt that contained 18% glycerol, different species of yeasts were cultivated from Dead Sea water, including *Trichosporon mucoides, Rhodotorula laryngis,* and a *Candida* sp. resembling *Candida glabrata* (Butinar et al. 2005).

Dead Sea Microorganisms as Models for the Study of Life at High Salt Concentrations

Some Dead Sea isolates have been extensively studied and have become popular model organisms for study. As the Archaea of the class *Halobacteria* are much easier to grow and handle than methanogens or hyperthermophiles, they also have become models for the study of the properties in which Archaea differ from Bacteria. The development of methods for the genetic manipulation of *Haloferax volcanii* made that organism especially popular for study. A search in the ISI Web of Knowledge (www.webofknowledge.com) (June 6, 2019) yielded 402 articles in which '*volcanii*' features in the title and 1,125 articles in which the word is found in the topic. All these deal with *Haloferax volcanii* or its basonym *Halobacterium volcanii.* For '*marismortui*' the numbers are 166 and 983, respectively (including genera *Haloarcula/Halobacterium, Bacillus, Chromohalobacter, Ectothiorhodospira, Orenia, Salibacillus, Sporohalobacter,* and *Virgibacillus*).

Haloferax volcanii

Haloferax volcanii (basonym: *Halobacterium volcanii*) was isolated in the early 1970s from shallow sediment at the Jordanian side of the Dead Sea (Mullakhanbhai and Larsen 1975). *Hfx. volcanii* has several advantages as a model organism for studies in many fields of biology such as it grows on simple defined media, it has a relatively stable genome, and it can be genetically manipulated relatively easily.

Its polar lipids differ significantly from those of the genus *Halobacterium* (Mullakhanbhai and Francis 1972), and this property was one of the main reasons for the reclassification in the newly established genus *Haloferax* (Torreblanca et al. 1986). Like in the other members of the class *Halobacteria*, osmotic adaptation is mainly based on the accumulation of KCl inside the cells. Intracellular K^+ concentrations as high as 3.6 M were measured, creating a 500– to 1,000-fold potassium gradient across the cell membrane. The cytoplasmatic K^+ concentration is not in equilibrium with the membrane potential so that K^+ transport cannot be accounted for solely by a passive uniport process. An ATP-dependent K^+ transport system was characterized in *Hfx. volcanii* that had properties similar to the Trk potassium transport system in *Escherichia coli* (Meury and Kohiyama 1989). *Hfx. volcanii* has great potential for survival under unfavorable conditions. This is in part due to the formation of dormant persister cells as a stress response. The addition of spent culture media to planktonic cells showed that *Hfx. volcanii*-conditioned media stimulated persistence, whereas conditioned media in which other haloarchaea or halophilic bacteria had been grown did not; thus, a species-specific signal may be involved. Homoserine lactone-like signal molecules may be involved in the process (Megaw and Gilmore 2017).

The genome of the type strain of *Hfx. volcanii* consists of a main 2.848 Mb chromosome, three smaller chromosomes of 636, 438, and 85 kb, and a 6.4 kb plasmid (Hartman et al. 2010). The genome encodes a type I-B CRISPR/Cas system, and it carries three CRISPR loci and eight Cas proteins. All three CRISPR loci are transcribed and processed into mature crRNAs (Maier et al. 2013). The system relies on more than 50 different crRNAs, whose stability and maintenance depend on the proteins Cas5 and Cas7, which bind the crRNA. *Hfx. volcanii* is the first known example of an organism that can tolerate autoimmunity via the CRISPR-Cas system while maintaining a constitutively active system (Maier et al. 2019). An experimental system was established in which a non-essential gene, required for pigment production responsible for the reddish colony color, is targeted by the CRISPR-Cas I-B system. When both the self-targeting and native crRNAs are expressed, self-targeting by CRISPR-Cas causes no reduction in transformation efficiency of the plasmid encoding the self-targeting crRNA so that the reddish colony phenotype due to mutations in the targeted region could be observed. In cells lacking the pre-crRNA processing gene *cas6*, and only the self-targeting crRNA exists as mature crRNA, self-targeting leads to moderate toxicity and the emergence of deletion mutants (Stachler et al. 2017).

The structure and biosynthesis of the glycoprotein S-layer cell wall have been investigated in-depth for over thirty years. High-resolution electron microscopy complemented with image analysis and processing revealed the presence of a near-hexagonal lattice of unit cells with a center-to-center spacing of 15.5 nm. Negatively stained samples showed the unit cell to be composed of six protomers (Kessel et al. 1988a). The three-dimensional structure could be elucidated from the electron micrographs to a resolution of 2 nm. The glycoprotein forms 4.5 nm high, dome-shaped complexes with a narrow pore at the apex, widening to a funnel-like opening toward the cell membrane. Six radial protrusions emanate from each complex and provide lateral connectivity (Kessel et al. 1988b). Based on investigations of the secondary and tertiary organization of the S-layer protein using circular dichroism, fluorescence spectroscopy, dynamic light scattering, and transmission electron microscopy, it was shown that the detailed protein structure depended on environmental factors, especially on pH. The β-sheet structure is affected by environmental pH with higher disorder under more alkaline conditions. The concentrations of Na^+, Mg^{2+}, and Ca^{2+}

also affect the protein structures, causing small changes in α-helix and β-sheet content (Rodriguez-Oliveira et al. 2019).

The gene for the S-layer protein encodes a mature polypeptide of 794 amino acids, preceded by a signal sequence of 34 amino acid residues. The S-layer protein is a glycoprotein containing both N- and O-glycosidic bonds (Sumper et al. 1990). The *Hfx. volcanii* S-layer glycoprotein and the mode of its biosynthesis and transport through the cell membrane were investigated in-depth (Kandiba and Eichler 2014). It has seven putative N-glycosylation sites, decorated with a trisaccharide (glucuronic acid (GlcA)-β1,4-GlcA-β1,4-glucose-β1-Asn), a tetrasaccharide (methyl-O-4-GlcA-β-1,4-galacturonic acid-α1,4-GlcA-β1,4-glucose-β1-Asn), and a pentasaccharide (hexose-1,2-[methyl-O-4-]GlcA-β-1,4-galacturonic acid-α1,4-GlcA-β1,4-glucose-β1-Asn) (Kandiba et al. 2016). The newly synthesized S-layer glycoprotein undergoes a maturation step following translocation across the plasma membrane (Kandiba et al. 2013). The processing step, detected as an increase in the apparent molecular mass, is apparently unrelated to glycosylation of the protein. Maturation on the external face of the plasma membrane requires the presence of magnesium ions that are involved in membrane association of the S-layer glycoprotein (Eichler 2001).

The genes involved in the N-glycosylation process are today well known. The gene *alg11* is essential, while the two versions of *alg5* are not. However, the deletion of *alg5* led to slower growth and interfered with the glycosylation process. Deletion of *stt3*, the only component of the oligosaccharide transferase complex, did not affect cell viability. N-glycosylation is thus not essential in *Hfx. volcanii*, but in its absence cells grow more slowly (Abu-Qarn and Eichler 2006). AglE is an integral membrane protein that acts as a glycotransferase and plays a role in the addition of a sugar subunit of the pentasaccharide (Abu-Qarn et al. 2008). AglF and AglI are involved in the addition of the hexuronic acid found at position three of the pentasaccharide, while AglG contributes to the addition of the hexuronic acid found at position two (Yurist-Doutsch et al. 2008). AglP is an S-adenosyl-L-methionine-dependent methyltransferase that adds a 14 Da moiety to a hexuronic acid found at position four of the pentasaccharide (Magidovich et al. 2010). The genes encoding all but one of the Agl proteins are sequestered into a single gene island. AglS is necessary for the addition of the final mannose subunit of the pentasaccharide. It acts as a dolichol phosphate-mannose mannosyltransferase (Cohen-Rosenzweig et al. 2012). Distinct glycan-charged C_{55} and C_{60} phosphodolichol carriers are required for the assembly of the pentasaccharide. The oligosaccharides are sequentially assembled from glycans originating from distinct phosphodolichol carriers (Guan et al. 2010). In *Hfx. volcanii*, the first four pentasaccharide subunits are initially transferred from a common dolichol phosphate carrier to the target protein, and only then the final pentasaccharide subunit is delivered from a distinct dolichol phosphate to the N-linked tetrasaccharide. *Haloarcula marismortui* decorates the S-layer glycoprotein with the same N-linked pentasaccharide and employs dolichol phosphate as lipid glycan carrier, but in that species, the complete pentasaccharide is first assembled on dolichol phosphate and only then is the glycan transferred to the target protein (Calo et al. 2011).

Replacement of *Haloferax volcanii* AglD, the glycosyltransferase involved in adding the final hexose of the pentasaccharide, with homologous glycosyltransferases from other halophilic Archaea, led to the appearance of an S-layer glycoprotein similar to the protein from the native strain. Using this approach, it will be possible to generate glycoproteins with more diverse N-linked glycans (Calo et al. 2010).

The tetrasaccharide attached to Asn-498 of the S-layer glycoprotein comprises a sulfated hexose, two hexoses, and a rhamnose. The Agl11-Agl14 proteins catalyze the stepwise conversion of glucose-1-phosphate to dTDP-rhamnose, the final sugar of the tetrasaccharide glycan (Kaminski and Eichler 2014).

N-glycosylation is important for the stability and function of the *Hfx. volcanii* S-layer. Cryo-electron microscopy examination of membrane vesicles, prepared from cells of a parent strain and from strains lacking genes encoding glycosyltransferases involved in assembling the N-linked pentasaccharide, showed that vesicles from mutant cells were only partially covered. Moreover, cells lacking AglD involved in adding the final pentasaccharide sugar were easily degraded by proteases (Tamir and Eichler 2017). Cell surface glycosylation is also required for efficient mating and lateral gene transfer using the fusion-based mating system. Differences in glycosylation of the S-layer glycoprotein affected mating success (Shalev et al. 2017). S-layer glycoprotein glycosylation in *Hfx. volcanii* depends on medium salt concentration. When grown in 3.4 M NaCl, S-layer glycoprotein Asn-13 and Asn-83 are modified by a pentasaccharide, while dolichol phosphate is modified by a tetrasaccharide comprising the first four pentasaccharide residues. In cells grown in 1.75 M NaCl, substantially less pentasaccharide was detected, and a dolichol phosphate modified by a distinct tetrasaccharide was found that was absent in cells grown at high salinity. The same tetrasaccharide modified S-layer glycoprotein Asn-498 in cells are grown in low salt, a residue that is not decorated in cells grown at high salt (Guan et al. 2012, Kaminski et al. 2013a). Manipulation of the Agl pathway, together with the capability of *Hfx. volcanii* to N-glycosylate non-native proteins forms the basis for establishing this species as a glyco-engineering platform (Kaminski et al. 2013b).

Besides the S-layer glycoprotein, *Hfx. volcanii* contains several other glycosylated membrane proteins (Eichler 2000). A 98-kDa glycoprotein was partially characterized. This glycoprotein is not associated with the membrane in a Mg^{2+}-dependent manner, but its mode of glycosylation appears to similar to that of the S-layer protein (Konrad and Eichler 2002).

Already, more than three decades ago, it was discovered that *Hfx. volcanii* can be transformed with foreign DNA. High-molecular-weight DNA can be taken up by spheroplasts in the presence of polyethylene glycol. Auxotrophic mutants could be transformed into prototrophs with genomic DNA from wild-type cells (Cline et al. 1989). Using auxotrophic mutants generated by chemical mutagenesis, the presence of a native genetic transfer system was demonstrated (Mevarech and Werczberger 1985). Selectable markers based on the *leuB* gene encoding 3-isopropylmalate dehydrogenase and the *trpA* genes encoding tryptophan synthase—lack of which make the cells auxotrophic for leucine and tryptophan, respectively—are available as a positive selection system for genetic manipulation (Allers et al. 2004). Using the immobile plasmids pHV2 and pHV11 as cytoplasmic markers, it was shown that the cytoplasms of the parental types do not mix during the mating process. Each parental type can serve both as a donor and a recipient. Cytoplasmic bridges are formed between the parental types, and these bridges are apparently used for DNA transfer between cells (Rosenshine et al. 1989). *Hfx. volcanii* thus became a model organism for the study of recombination in Archaea (Allers and Ngo 2003). An efficient gene knockout system was developed, which used the *pyrE2* gene that coded for orotate phosphoribosyl transferase as a selectable marker (Bitan-Banin et al. 2003).

Interspecies genetic transfer between cells *Hfx. volcanii* and *Hfx. mediterranei* is also possible, as shown by the transfer of selectable shuttle vectors (Tchelet and Mevarech 1994). This enabled the study of the effect of mutations in the major gas vesicle protein GvpA of *Hfx. mediterranei* expressed in *Hfx. volcanii*, a species that does not form gas vesicles (Knitsch et al. 2017). *Hfx. volcanii* was even used as a host system for expressing mammalian genes in a transgenic strain that produces mammalian olfactory receptors embedded in the archaeal lipid bilayer (Lobasso et al. 2015).

The original species description of *Hfx. volcanii* stated the organism to be non-motile (Mullakhanbhai and Larsen 1975). Still, the genome contains the *flgA1-flgA2* genes that encode flagellins and genes that encode proteins involved in flagellar assembly. Indeed, under special conditions, motility can be observed. Biosynthesis of the archaeal flagella (*'archaella'*) resembles the mode of production of bacterial type IV pili rather than the assembly of bacterial flagella (Tripepi et al. 2010). The *Hfx. volcanii* flagellins are N-glycosylated by the same glycosylation (Agl) components responsible for glycosylation of the S-layer protein described above. The deletion of any of the *agl* genes impairs swimming motility to various extents. The archaella consist of a major flagellin, FlgA1 that has three predicted N-glycosylation sites, and a minor flagellin, FlgA2 (Tripepi et al. 2012). Cells that lack FlgA2 but express FlgA1 are hypermotile, have an increased number of flagella per cell, and increased flagellum length. FlgA2 may play both structural and regulatory roles in *Hfx. volcanii* motility (Tripepi et al. 2013).

Hfx. volcanii can adhere to surfaces and form biofilms. A nonflagellar type IV pilus-like structure plays a critical role in the process (Tripepi et al. 2010) and at least 20 genes, including potential novel regulatory genes, affect the adhesion process (Legerme et al. 2016). Microcolonies formed on surfaces can form flake-like towers exceeding 100 µm in height after seven days. The extracellular matrix may contain polysaccharides, extracellular DNA, and amyloid protein. When biofilm development is induced, a subpopulation of cells may differentiate into chains of long rods sometimes exceeding 25 µm in length. Frequent gene exchange occurs between cells within the biofilm (Chimileski et al. 2014). N-glycosylation of pilins plays an important role in regulating the transition between planktonic to sessile cell states. Cells that lack the *aglB* gene have thick pili bundles and form microcolonies, whereas the surface of wild-type cells is decorated with discrete pili; such cells form a dispersed layer of cells on a plastic surface (Esquivel et al. 2016).

Polyploidy is common among the halophilic Archaea, and the phenomenon was best studied in *Hfx. volcanii*. The chromosome copy number ranges from two following phosphate starvation to more than 40 when phosphate is abundant and is highly regulated (Maurer et al. 2017). A heterozygous strain was constructed that contained two different types of genomes: one that contained the wild-type *leuB* gene and one that had the *leuB:trpA* gene introduced by gene replacement. Growth in the absence of both leucine and tryptophan is only possible when both types of genomes are simultaneously present. The heterozygous strain was grown in the presence of tryptophan, selecting for the presence of *leuB*, or in the presence of leucine selecting for *leuB:trpA*. The first condition led to a complete loss of *leuB:trpA*-containing genomes, while under the second condition *leuB*-containing genomes were lost. In the absence of selection, gene conversion led to a fast equalization of genomes and resulted in homozygous *leuB*-containing cells (Lange et al. 2011). The *Hfx. volcanii* chromosome has multiple origins of replication (Maurer et al. 2017). Homologous recombination can easily generate two chromosomes, each possessing replication origins, rRNA loci, and essential genes (Ausiannikava et al. 2018).

A few enzymes of *Hfx. volcanii* were studied in-depth. The organism has two different dihydrofolate reductases encoded by the genes *hdrA* and *hdrB*. The deletion of *hdrA* did not affect growth in minimal medium. The *hdrB* gene is located immediately downstream of the thymidylate synthase gene. The *hdrB* alone can support growth in minimal medium, whereas *hdrA* alone can support growth only when the medium is supplemented with thymidine (Ortenberg et al. 2000). *Hfx. volcanii* dihydrofolate reductase is a moderately halophilic enzyme; it retains its activity and secondary structure at salt concentrations as low as 0.5 M, and it lacks the typical structural features of truly halophilic proteins. The enzyme was crystallized and its structure was resolved at 2.6 Å resolution (Pieper et al. 1998). *Hfx. volcanii* dihydrolipoamide dehydrogenase and citrate synthase were expressed in *E. coli*, and protocols for refolding and reactivation were established (Connaris et al. 1999). *Hfx. volcanii* contains genes for three group II chaperonins that encode complexes made up of hetero-oligomers with eightfold symmetry. At least one of these must be present for growth (Kapatai et al. 2006).

Hfx. volcanii synthesizes three proteins (α1, α2, and β) that are classified in the proteasome superfamily. Two types of proteasomes can be found: α1-β and α1-α2-β (Kaczowka and Maupin-Furlow 2003). Proteasomes degrade proteins in an ATP-dependent manner following post-translational modification of the proteins by sampylation, a process that resembles ubiquitinylation in eukaryotes (Prunetti et al. 2014). Eichler and Maupin-Furlow (2013) presented a review of post-translational modifications and protein degradation, mainly based on studies with *Hfx. volcanii*. A caspase-like protease, closely resembling caspase-4, which may initiate programmed cell death was identified in *Hfx. volcanii* (Seth-Pasricha et al. 2013).

Hfx. volcanii possesses the twin-arginine translocation (Tat) pathway to export partially or fully folded proteins from the cytoplasm (Dilks et al. 2005) as well as the Sec-mediated cotranslational protein translocation and membrane protein insertion. The signal recognition particle, a ribonucleoprotein complex involved in the recognition and targeting of nascent extracytoplasmic proteins, is required for viability (Rose and Pohlschröder 2002). Assembly of the *Hfx. volcanii* signal recognition particle does not require high salt concentrations (Tozik et al. 2002). Like in Bacteria and Eukaryotes, binding of ribosomes to the membrane occurs at SecYE-based sites (Ring and Eichler 2004).

Thousands of small non-coding RNAs (sRNAs), both intergenic and antisense, were discovered using strand-specific sRNA sequencing of *Hfx. volcanii*, comprising 25% to 30% of the total transcriptome under no-challenge and oxidative stress conditions, respectively. Hundreds of differentially expressed sRNAs were found in response to hydrogen peroxide-induced oxidative stress (Gelsinger and DiRuggiero 2018). The *Hfx. volcanii* genome contains 14 annotated genes that encode initiation factors or their subunits. Nine of these are non-essential as single deletion mutants could be constructed, but the genes encoding initiation factors aIF1, aIF2γ, aIF5A, aIF5B, and aIF6 were found to be essential (Gäbel et al. 2013). The majority of transcripts in *Hfx. volcanii* is leaderless; most transcripts with a leader sequence do not make use of the Shine-Dalgarno mechanism for translation initiation in untranslated regions (Kramer et al. 2014).

To further understand its metabolic versatility, many of the metabolic pathways encoded by the *Hfx. volcanii* genome has been investigated in detail. Glucose is degraded via the semiphosphorylative Entner-Doudoroff pathway is catalyzed by glucose dehydrogenase, gluconate dehydratase, and 2-keto-3-deoxy-6-phosphogluconate aldolase. Detailed studies were reported of the glucose dehydrogenase and the 2-keto-3-deoxy-6-phosphogluconate aldolase (Sutter et al. 2016). The genome also encodes an inducible 2-keto-3-deoxygalactonate kinase, which is likely involved in galactose catabolism (Pickl et al. 2014). Fructose can be taken up by a fructose-specific phosphotransferase system producing fructose-1-phosphate, which is converted via fructose-1,6-bisphosphate to triose phosphates by a class II fructose-1,6-bisphosphate aldolase (Pickl et al. 2012). D-xylose and L-arabinose are converted to α-ketoglutarate, the initial step being the formation of the corresponding lactones, D-xylono-γ-lactone, and L-arabino-γ-lactone, rather than the respective sugar acids (Sutter et al. 2017). D-Xylose is degraded to α-ketoglutarate via D-xylose dehydrogenase and a novel 2-keto-3-deoxyxylonate dehydratase in a pathway that differs from the classical pathway used by most bacteria with xylulose 5-phosphate as an intermediate (Johnsen et al. 2009). The inducible L-arabinose dehydrogenase is a homotetrameric protein (Johnsen et al. 2013). The oxidative pentose phosphate pathway involves a novel type of glucose-6-phosphate dehydrogenase catalyzing the oxidation of glucose-6-phosphate to ribulose-5-phosphate (Pickl and Schönheit 2015). *Hfx. volcanii* possesses two distinct glyceraldehyde-3-phosphate dehydrogenases, one being involved in sugar catabolism in glycolysis and one in gluconeogenesis (Tästensen and Schönheit 2018). When glycerol is available as a nutrient, glucose catabolism is repressed and glycerol is activated by a glycerol kinase (Sherwood et al. 2009). Dihydroxyacetone, an overflow product from glycerol utilization by organisms such as *Salinibacter ruber*, can also be used as a sole carbon source for growth; both dihydroxyacetone kinase and glycerol kinase play a role in this metabolism (Ouellette et al. 2013). *Hfx. volcanii* further contains two alcohol dehydrogenases with different substrate range. Both are haloalkaliphilic and thermoactive for the oxidative reaction and catalyze the reductive reaction at a slightly acidic pH (Timpson et al. 2013). The *Hfx. volcanii* NADP-dependent isocitrate dehydrogenase is a dimeric enzyme that is highly resistant to chemical denaturation (Madern et al. 2004). When grown on acetate, high activities of malate synthase, one of the two enzymes unique to the glyoxylate cycle were detected (Serrano et al. 1998). Malate synthase is found in three isoforms that differ in size and sequence conservation (Bracken et al. 2011).

According to the original species description (Mullakhanbhai and Larsen 1975), *Hfx. volcanii* does not grow anaerobically by denitrification. However, later studies showed that anaerobic growth in the presence of nitrate is possible. Transcription of the genes encoding the denitrifying enzymes nitrate reductase and nitrite reductase is activated under anaerobic conditions. A putative DNA-binding protein, NarO, is encoded upstream of the respiratory nitrate reductase gene. It may have an oxygen-sensing function and also work as a transcriptional activator of the denitrifying genes (Hattori et al. 2016). *Hfx. volcanii* can also grow anaerobically by respiration of dimethylsulfoxide. An inducible DMSO reductase encoded by the *dmsR* gene is a key enzyme in the process (Qi et al. 2016). Hydroxymethylglutaryl-CoA reductase, a 46 kDa protein that is a key enzyme in the mevalonate pathway of isoprenoid biosynthesis for lipid and carotenoid production, is more abundant during growth at high salinity. Cultures grown at low salt contained lower amounts of *hmgR* transcript compared to high salt-grown cells, suggesting that the regulation occurs at the level of transcription (Bidle et al. 2007). The membrane protease LonB was recently identified as a regulator of carotenogenesis controlling degradation of phytoene synthase in *Hfx. volcanii* (Cerletti et al. 2018). A genetically modified hyperpigmented stain was proposed as a producer strain for the carotenoid bacterioruberin that has strong antioxidant activity (Zalazar et al. 2018).

Haloarcula marismortui

The red extremely halophilic archaeon from the Dead Sea isolated by Ginzburg et al. (1970) and later formally described as a new species *Haloarcula marismortui* (Oren et al. 1990) closely resembles the organism described by Volcani during his early studies of the Dead Sea as *Halobacterium marismortui* and was later lost. *Har. marismortui* has a relatively large genome (4.27 Mb), organized into nine circular replicons with G+C ratios ranging from 54% to 62%. Putative functions could be assigned to at least 58% of the 4,242 predicted proteins, including an unusually large number of environmental response regulators, which suggests that *Har. marismortui* may have the potential of exploiting diverse environments. Analysis of the genome shows a high potential for the biosynthesis of cellular components, explaining its low nutritional requirements (Baliga et al. 2004). The genome encodes six retinal proteins. These include a dual-bacteriorhodopsin system composed of HmbR1 and HmbRII, enabling light-driven proton pumping over a wide range of pH (Fu et al. 2013a). The photocycle of HmbRI mainly follows the conventional pathway, including intermediates M, N, and O; however, the photocycle of HmbRII does not include the O intermediate (Tsai et al. 2014).

Several aspects of the metabolic pathways of *Har. marismortui* was investigated in-depth. Acetate is activated via an inducible AMP-forming acetyl-CoA synthetase that catalyzes the reaction Acetate+ATP+ CoA $\rightarrow$ Acetyl-CoA+AMP+ Phosphate. The 72-kDa monomeric enzyme requires between 1 M and 1.5 M KCl for optimal activity. It was functionally overexpressed in *E. coli* and could be reactivated from inclusion bodies following solubilization in urea and refolding in the presence of salts (Bräsen and Schönheit 2005). *Har. marismortui* also contains an unusual ADP-forming acetyl-coenzyme A synthetase. The homodimeric 166 kDa enzyme, catalyzing the reversible ADP- and Pi-dependent conversion of acetyl-CoA to acetate, is salt-dependent. Also, this enzyme was functionally expressed in *E. coli* (Bräsen and Schönheit 2004).

Accumulation of poly-β-hydroxyalkanoate (PHA) by *Har. marismortui* was first reported by Kirk and Ginzburg (1972), and it was the first report of the occurrence of this storage polymer in a member of the Archaea. When cultured in a minimal medium with excess glucose, PHA can be accumulated to up to 21% of the cellular dry weight. The *phaE* and *phaC* genes, probably encoding two subunits of a class III PHA synthase, were identified. The introduction of these genes into *Haloarcula hispanica*, an organism that harbors highly homologous *phaEC* genes, could enhance PHA synthesis in the recombinant strains (Han et al. 2007).

Har. marismortui can be grown anaerobically using nitrate as the electron acceptor. Induction of both dissimilatory nitrate and nitrite reductases depends on the presence of nitrate and only occurs under anaerobic conditions. The final products of denitrification are nitrogen gas and nitrous oxide. A 2Fe-2Sferredoxin, present in large quantities in the cells, can serve as an electron donor for the nitrite reduction by nitrite reductase (Werber and Mevarech 1978a). This ferredoxin has been characterized in-depth (see below). Genes encoding the NarG and NarH subunits of the dissimilatory nitrate reductase, a molybdo-iron-sulfur enzyme, were identified and the properties of the NarGH nitrate reductase were described (Yoshimatsu et al. 2002). The *nirK* gene encoding a Cu-containing dissimilatory nitrite reductase was cloned and sequenced (Ichiki et al. 2001).

The first physiological studies with *Har. marismortui* dealt with its ion metabolism and mode of osmotic adaptation. Potassium was identified as the main intracellular cation. In cells grown in medium with 3.9 M NaCl and 7 mM K^+, the intracellular K^+ concentration decreased from 5.5 to 3.8 moles per kg cell water during the exponential growth phase and remained at the latter value during the stationary phase. Most of the potassium was estimated to be bound within the cytoplasm (Ginzburg et al. 1970). Neutron scattering experiments suggested that about three-quarters of the intracellular water has extremely slow mobility. It was hypothesized that the slow mobility of a large part of *Har. marismortui* cell water indicates a specific water structure responsible for the large amounts of K^+ bound within the cells (Tehei et al. 2007).

Proteins of halophilic Archaea of the class *Halobacteria* that function in the cytoplasm containing molar concentrations of KCl generally require molar concentrations of salts for structural stability and activity. Halophilic proteins are typically characterized by a large excess of acidic over basic amino acid residues. The high negative surface charge of such proteins makes them more soluble and renders them more flexible at high salt concentrations, while non-halophilic proteins tend to aggregate and become rigid under these conditions. The high surface charge is neutralized mainly by tightly bound water dipoles. High salt concentrations are required for stabilization because of the low-affinity binding of the salt to specific sites on the surface of the folded polypeptide, thus stabilizing the active conformation of the protein (Eisenberg and Wachtel 1987, Mevarech et al. 2000). Much of our in-depth understanding of the structure and functioning of halophilic proteins were derived from proteins of *Har. marismortui*, notably malate dehydrogenase and 2Fe-2S ferredoxin.

The *Har. marismortui* malate dehydrogenase was purified and characterized using many different methods. The enzyme is a dimer that requires salt for stability and activity. At NaCl concentration below 2 M, the enzyme is inactivated in a first-order reaction probably due to the dissociation of the dimeric enzyme. When the salt concentration is increased the inactivated enzyme is reactivated in a second-order reaction, determined by the reassociation rate of the subunits to form dimers. At even lower salt concentrations, an irreversible (first-order) denaturation process occurs (Mevarech and Neumann 1977). Minor conformational changes were observed over the whole range between 1 M and 5 M NaCl. The special properties of the halophilic enzyme seem to be related to its capacity of associating with unusually large amounts of water and salts (Pundak and Eisenberg 1981). The three-dimensional structure of the enzyme was determined at high resolution by X-ray crystallography. In addition to the large excess of acidic residues at the protein surface, *Har. marismortui* malate dehydrogenase showed a much larger number of salt bridges compared with similar enzymes from non-halophilic organisms (Dym et al. 1995).

At least one of the physiological functions of the *Har. marismortui* 2Fe-2S ferredoxin may be to serve as an electron donor for nitrite reduction. Also, this protein has an extremely high excess of acidic amino acid residues in comparison with ferredoxins of non-halophiles (Werber and Mevarech 1978b). The protein has been crystallized and its three-dimensional structure was solved to high resolution by X-ray diffraction (Sussman et al. 1979). The protein differs from plant-type 2Fe-2S ferredoxins in two important aspects: its surface is coated with acidic residues except for the vicinity of the iron-sulfur cluster, and there is an insertion of two amphipathic helices near the N-terminus, and these form a separate hyperacidic domain that may provide extra surface carboxylates for solvation. The bound surface water molecules have on the average 40% more hydrogen bonds than in a typical non-halophilic protein crystal structure (Frolow et al. 1996).

Other *Har. marismortui* enzymes that were characterized further are an esterase and a lipase (Camacho et al. 2009, Müller-Santos et al. 2009) and a catalase-peroxidase (Yamada et al. 2002). The gene encoding the esterase was cloned and expressed in *E. coli*. The enzyme is salt-dependent and folds optimally in the presence of 3 M KCl (Müller-Santos et al. 2009). The structure of the catalase-peroxidase was resolved to a resolution of 2.0 Å. The enzyme is a dimer of two identical subunits and has heme b as a cofactor (Yamada et al. 2002).

Much of the detailed information on the three-dimensional structure of the ribosome and its function was obtained from studies of crystals of ribosomal particles of *Har. marismortui*. For this work, Thomas Steitz and Ada Yonath shared the 2009 Nobel Prize in Chemistry with Venkataraman Ramakrishnan. Good-quality crystals of the large ribosomal subunit of *Har. marismortui* diffracting to 2.9 Å resolution were reported already in 1994, and chimeric complexes could be reconstituted with *E. coli* ribosomal components, indicating rather a high homology despite the evolution distance (Franceschi et al. 1994). A few years later, X-ray crystallography allowed resolution of the structure to 2.4 Å, while high-resolution cryo-electron microscopy and single-particle reconstruction techniques enabled solving the structure to 19 Å. A comparison of the two methods enabled the detection of small differences in the mass distribution in the flexible regions of the 50S subunit. Most of the proteins stabilize the structure by interacting with RNA domains (Penczek et al. 1999, Ban et al. 2000, Gabdulkhakov et al. 2013). These studies also led to the recognition of the binding sites of different antibiotics that interact with the ribosome and inhibit protein synthesis (Schroeder et al. 2007).

The genome of *Har. marismortui* encodes three rRNA operons with considerable sequence heterogeneity: *rrnA*, *rrnB*, and *rrnC*. Operons A and C are nearly identical, whereas operon B shows a high divergence in nucleotide sequence, having up to 135 nucleotide polymorphisms among the 16S, 23S, and 5S ribosomal RNA genes (López-López et al. 2007). The 16S rRNA genes *rrnA* and *rrnB* differ by nucleotide substitutions at 74 positions. The 16S rRNAs transcribed from both operons are equally represented in the functional 70S ribosome population (Mylvaganam and Dennis 1992). Fluorescence *in situ* hybridization experiments with operon-specific probes showed that both operons were expressed in each cell (Amann et al. 2000). The intragenomic 16S rRNA gene divergence may be an adaptation to different temperatures. The expression level of operon B was about four times higher than the other two together in cells incubated at 50°C and was three times lower at 15°C. The predicted secondary structure of these genes indicated that they have distinct stabilities at different temperatures. A mutant strain lacking the *rrnB* operon grew slower at high temperatures. Variation in salt concentration did not affect the expression of the different operons (López-López et al. 2007).

According to the original description of *Haloarcula marismortui*, the species is non-motile. However, the genome encodes archaeal flagella (*'archaella'*). Archaella are assembled similarly to type IV pili. Flagellar filaments can be observed in the electron microscope. The genome encodes two archaellin genes—*flaA2* and *flaB*. *Har. marismortui* strains have been obtained with archaella consisting of FlaA2 archaellin with a minor FlaB fraction or FlaB only. Both strains were motile and produced functional helical archaella. FlaA2 archaellin functions better at more severe conditions of high temperature and/or low salinity, while FlaB has an advantage at increasing salinity (Syutkin et al. 2014). Thus, functional differences exist that enable the cells to cope with varying environmental conditions (Fu et al. 2013b). Under suitable experimental conditions, the organism displays phototactic behavior (Lin et al. 2010). Six genes for retinal proteins were annotated, three being ion transporters, and three probably are sensory rhodopsins. The sensory rhodopsins, HmSRI and HmSRII mediate photoattractant and photorepellent responses, respectively (Lin et al. 2010). A transducer for microbial sensory rhodopsin was also identified (Syutkin et al. 2018).

Four major polar lipids were identified in *Har. marismortui*, all being derivatives of 2,3-di-*O*-phytanyl-sn-glycerol: (1) TGD (2,3-di-*O*-phytanyl-1-i-[β-D-glucopyranosyl-(1'-6')-*O*-α-D-mannopyranosyl-(1'-2')-*O*-α-D-glucopyranosyl]-sn-glycerol (11 mol%); (2) phosphatidylglycerol (11 mol%); (3) phosphatidylglycerophosphate (62 mol%); (4) phosphatidylglycerosulfate (17 mol%). Squalenes, vitamin MK-8, and bacterioruberins were the dominant non-polar lipids with traces of β-carotene, lycopene, and retinal (Evans et al. 1980).

Chromohalobacter israelensis

Chromohalobacter israelensis (in early studies designated strain Ba1) was isolated from a crude salt sample collected at the Dead Sea evaporation ponds (Rafaeli-Eshkol 1968) and was later described as *Halomonas israelensis* (Huval et al. 1995) until it was finally renamed *Chromohalobacter israelensis* (Arahal et al. 2001).

The importance of glycine betaine as an osmotic solute in heterotrophic halophilic or halotolerant bacteria, including many that cannot produce the compound themselves, was described in the mid-1980s (Galinski and Trüper 1982, Imhoff and Rodriguez-Valera 1984). However, it is seldom realized that the first evidence for the function of betaine in osmotic adaptation in bacteria was obtained already in the late 1960s from studies with *C. israelensis*. Energy-dependent accumulation of labeled choline was documented followed by its conversion to a derivative, probably betaine, that increased the salt resistance of the respiratory system (Rafaeli-Eshkol 1968). Accumulation of labeled betaine in a salt-dependent manner was also demonstrated. Respiratory activity of cells that had lost their salt resistance as a result of washing in the absence of sodium chloride could be restored by the addition of glycine betaine (Rafaeli-Eshkol and Avi-Dor 1968). Betaine was accumulated by respiring cells, and the maximum amount taken up was correlated with the osmolarity of the medium. The claim that betaine may exert its stimulatory action on respiration at the outside of the cellular membrane (Shkedy-Vinkler and Avi-Dor 1975) deserves to be further investigated.

In the 1980s, *C. israelensis* was also used as a model organism to investigate ion metabolism in moderately halophilic bacteria. The cell membrane was found permeable to K^+ and Rb^+ ions but not to Na^+ and Li^+ and contains ion pumps powered by the electrochemical proton gradient without the mediation of ATP (Shnaiderman and Avi-Dor 1982). Respiratory activity is stimulated by Na^+, especially in the alkaline pH range where it induced acidification of the intracellular milieu, but respiratory stimulation induced by K^+ was most prominent in the acidic pH range and led to alkalinization of the internal pH (Ken-Dror and Avi-Dor 1985). The presence of a Na^+-dependent NADH:quinone oxidoreductase was documented (Ken-Dror et al. 1986).

Salt-dependent regulation of the lipid composition of cell membranes of moderately halophilic bacteria was extensively investigated in *Halomonas elongata*, *Salinivibrio costicola*, and other model organisms since the 1980s (Ventosa et al. 1998). Those studies were preceded by studies of glycolipid and phospholipid metabolism in *C. israelensis*. When grown in a liquid medium that contained 2 M NaCl, the following lipids (in reducing order) were found: phosphatidylethanolamine, phosphatidyl glycerol, two glycolipids tentatively identified as glucuronosyl-diglyceride and glucosyl phosphatidylglycerol, and cardiolipin. The lipids had an unusually high content of 11,12-methylene octadecanoic acid, both in cells grown in 2 M and 4 M NaCl (Peleg and Tietz 1971). Cells grown in high salt-nutrient broth contained approximately equal amounts of glucosylphosphatidylglycerol and a glucuronic acid-containing glycolipid. In the stationary growth phase, the phosphatidylglycerol content was lower and the content of cardiolipin increased. Phosphatidylethanolamine isolated from 18-h cultures contained 12 and 25% of hexadecenoic and octadecenoic acids, respectively, while in 48-h cultures only palmitic and cyclopropane acids were found. All other polar lipids contained palmitic and cyclopropane acids, while diglycerides contained palmitic and monoenoic acids (Stern and Tietz 1973a). The glucosyl phosphatidylglycerol was isolated and subjected to a detailed structure elucidation (Peleg and Tietz 1973). It could be synthesized *in vitro* by cell-free particles from UDP-[^{14}C]glucose and ^{32}P-labeled phosphatidylglycerol. The system required Mg^{2+} or Ca^{2+} for activity, while KC1 and NaCl were inhibitory (Stern and Tietz 1978). Biosynthesis of the glucuronosyldiglyceride from UDP-[^{14}C]glucuronic acid and externally added diglycerides was also observed in cell-free particles. Only bacterial 1,2-diglycerides could serve as a substrate for the enzyme. Also, here Mg^{2+} was required for activity, while KC1 and NaCl were inhibitory (Stern and Tietz 1973b). The major fatty acids are 16:0, 18:0, 16:1cisΔ9, and 18:1Δ11. At sub-optimal salt concentrations, the degree of saturation decreased (Mutnuri et al. 2005).

   C. israelensis can utilize nitrate or ammonium as a nitrogen source. Nitrate is taken up only when no ammonium is present in the medium. The nitrate reductase is associated with the respiratory electron transport chain on the cell membrane. It serves for assimilatory purposes only, and nitrate is not used as an electron acceptor for anaerobic respiration (Hochman et al. 1988).

Halorubrum sodomense

After it was discovered that *Halorubrum sodomense* required starch for growth (which could be replaced by the addition of clay minerals such as bentonite in the medium) and that starch was degraded (Oren 1983b), the nature of the starch-utilizing enzymatic activity was investigated in greater depth. Starch is degraded to glucose by an extracellular amyloglucosidase. The enzyme requires at least 1 M salt for optimal activity and is thermophilic, having its temperature optimum at 65°C and 75°C in the presence of 1.4 M and 3.9 M NaCl, respectively (Oren 1983d). The enzyme has an apparent molecular mass of 175 kDa and is a dimer of two different subunits with apparent molecular masses of 82 and 72 kDa (Chaga et al. 1993).

   The glycolipid of *Hrr. sodomense* is a sulfated C$_{20}$, C$_{20}$ mannosyl glucosyl glycerol diether. Its structure differs from the glycolipids of most other halophilic Archaea in having a sulfate substitution at the C-2 of the mannose residue and in being an α-D-(1→4)-linked disaccharide (Trincone et al. 1990).

Halobacteroides halobius and Other Members of the *Halanaerobiales* From Dead Sea Sediments

Halobacteroides halobius (Oren et al. 1984), the second characterized member of the *Halanaerobiales*, an order of halophilic, mostly fermentative anaerobes (class *Clostridia*), was tested for its mode of osmotic adaptation. Cells grown in medium containing 1.56 M NaCl and 0.013 M KCl contained ~0.53 M Na$^+$ and ~0.92 M K$^+$ in the cytoplasm. This, together with the finding of only very low concentrations (< 0.01 M) of free amino acids and glycine betaine, suggested that the organism uses inorganic salts rather than organic solutes for osmotic adaptation (Oren 1986). This was later confirmed for other members of the *Halanaerobiales*.

   The related *Orenia marismortui* can reduce different nitro-substituted aromatic compounds (nitrobenzene, *o*-nitrophenol, *m*-nitrophenol, *p*-nitrophenol, nitroanilines, 2,4-dinitrophenol, 2,4-dinitroaniline) to the corresponding amines (Oren et al. 1991a).

   When grown with selenium oxyanions as the electron acceptor, *Selenihalanaerobacter shriftii* (Switzer Blum et al. 2001) produces nanospheres of ~300 nm diameter of elemental selenium, both inside and outside the cells (Oremland et al. 2004).

   No further attempts toward physiological and biochemical characterization were reported for the phylogenetically related Dead Sea isolates *Sporohalobacter lortetii* and *Sporohalobacter marismortui* beyond the properties reported in the species descriptions (Oren 1983c, Oren et al. 1987).

Ectothiorhodospira marismortui

Ectothiorhodospira marismortui, a halophilic anaerobic purple photosynthetic sulfur bacterium, was isolated from a warm hypersaline sulfur spring on the western shore of the Dead Sea (Oren 1989, Oren et al. 1989). Studies of its mode of osmotic adaptation resulted in the purification and structure elucidation of a novel organic osmotic solute that thus far has not been documented from any other microorganism; N-α-carbamoyl-L-glutamine 1-amide (CGA), an unusual amino acid derivative with no previous reference in the chemical literature. It is accumulated as a compatible solute in the cytoplasm together with two other osmotic solutes known also from other photosynthetic prokaryotes: glycine betaine and sucrose. The intracellular concentration of CGA was calculated to be up to 0.5 M, suggesting that the compound may serve an important function as an osmoprotectant (Galinski and Oren 1991). The estimated CGA concentration in the cells rose from about 0.2 M to 0.5 M in cells grown from 0.85 to 2.56 M NaCl, showing that its intracellular concentration is salt-dependent. Under the same conditions, intracellular glycine betaine increased from 0.47 M to 1.29 M, while the sucrose concentrations remained constant at about 0.05 M (Oren et al. 1991b). When cells grown at high salinity were subjected to dilution stress, part of the glycine betaine was rapidly excreted into the medium and was later taken up again by the cells; the intracellular concentrations of CGA and sucrose decreased within 1-2 h to new levels corresponding with the lowered salinity. As no CGA appeared in the medium, it was concluded that the compound is degraded intracellularly (Fischel and Oren 1993).

Physiological and Genomic Studies of Fungi Isolated From the Dead Sea

After a large number of fungal strains were obtained from the Dead Sea water column (Buchalo et al. 2000, Kis-Papo et al. 2001), in-depth studies were performed with selected isolates. Since fungi can be consistently recovered from the Dead Sea brines, the questions were asked whether growth as vegetative mycelia in the lake is possible and would low long fungal spores and mycelia may retain their viability in the hostile Dead Sea environment. Mycelia of Dead Sea isolates of *Aspergillus versicolor* and *Chaetomium globosum* could survive up to eight weeks in the undiluted Dead Sea water. Other strains (*Aspergillus versicolor, Eurotium herbariorum, Gymnascella marismortui*) retained their viability after suspension for three months in 80% Dead Sea water–20% distilled water mixtures. Strains obtained from the Dead Sea generally showed better survival than closely related isolates from other locations. Except for *Emericella* spores, spores of Dead Sea fungi survived a three-months exposure to the Dead Sea brines (Kis-Papo et al. 2003a).

A comparative genomic study was made of strains of the ascomycete *Aspergillus versicolor* obtained from the Dead Sea and the other saline as well as non-saline environments. The screening was based on more than 600 amplified fragment length polymorphism markers (equal to loci). Genomic diversity was positively correlated with stress and was highest in isolates from the Dead Sea surface but dropped drastically in Dead Sea samples from depths of 50 to 280 m (Kis-Papo et al. 2003b).

Putative presence of the high-osmolarity glycerol response (HOG) pathway, important in osmotic regulation, heat stress, freeze stress, and oxidative stress, was demonstrated in the Dead Sea strain of *Eurotium herbariorum*. The deduced amino acid sequence of a mitogen-activated protein kinase (MAPK) encoded by the *EhHOG* gene indicated high similarity with homologous genes from other fungi and yeasts, including *Aspergillus nidulans, Saccharomyces cerevisiae*, and *Schizosaccharomyces pombe* (Jin et al. 2005).

Cultivation-Independent Studies – Environmental Genomics and Metagenomics

Cultivation-independent studies of the nature of the microbial communities in the Dead Sea, based on 16S rRNA gene libraries and metagenomic analyses (fosmid clones) prepared from DNA extracted from the communities, were performed on biomass samples collected during the 1992 microbial bloom in the lake and preserved as frozen cell pellets for more than 15 years prior to analysis and were compared with the sparse biomass found in water samples collected from the lake in 2007, 2010, and 2015 (Bodaker et al. 2010, 2012, Rhodes et al. 2012, Jacob et al. 2017). The diversity of 16S rRNA genes retrieved from the 1992 bloom sample was low: a single sequence distantly related to the genus *Halobacterium* was recovered with a high frequency (Bodaker et al. 2010); analysis of ~400 base-pair fragments suggested the presence of organisms related to *Halosarcina* and the alkaliphilic genus *Natronococcus* (Rhodes et al. 2012). No genes encoding retinal proteins, such as bacteriorhodopsin and halorhodopsin, were recovered from the 1992 bloom DNA (Bodaker et al. 2012).

The metagenomes of the communities collected in 2007 and 2010 were much more diverse in 16S rRNA sequences, which lineages related to *Halorhabdus*, *Haloplanus*, the alkaliphilic genus *Natronomonas*, and other groups of the class *Halobacteria*, some of which still lacked cultivated representatives (Bodaker et al. 2010, Rhodes et al. 2012). Analysis of 16S rRNA amplicons from a surface water sample collected in 2015 (<300 bases, covering a different part of the molecule than that analyzed by Rhodes et al., 2012) showed 52% archaeal sequences (mainly related to *Halorhabdus* and *Natronomonas*) and 45% Bacteria (mainly related to *Acinetobacter* (*Gammaproteobacteria*) and *Bacillus* [*Firmicutes*]) (Jacob et al. 2017). Novel bacteriorhodopsin and sensory rhodopsin genes were found in samples collected in 2007 and 2010 (Bodaker et al. 2012).

Metagenomic analysis showed enrichment in clusters of orthologous genes related to transposable elements, intein, and homing endonucleases, Na^+/H^+ and K^+/H^+ antiporters as well as channels for Mg^{2+} and Co^{2+} (Bodaker et al. 2010).

Conclusions

The Dead Sea has been a source of isolation for diverse halophilic microorganisms: Archaea, Bacteria as well as different types of eukaryotes (fungi, *Dunaliella*). As documented above, some of these have become important model organisms for the study of life at high salt concentrations. A prominent case is *Haloferax volcanii*, fittingly named in honor of Benjamin Elazari-Volcani, the pioneer of the Dead Sea microbiology studies. It is one of the few members of the Archaea for which convenient methods for genetic manipulation have been developed, and as a result, many basic features of the life of Archaea and haloadaptation are studied in this organism. *Haloarcula marismortui*, earlier known as *Halobacterium marismortui* and first isolated by Volcani (Oren et al. 1990), became a model organism for the study of the structure of halophilic proteins. Studies on this organism led to the elucidation of the three-dimensional structure of the ribosome, awarded with the 2009 Nobel Prize in Chemistry shared by Venkataraman Ramakrishnan, Thomas Steitz, and Ada Yonath. A few other Dead Sea isolates have also contributed important insights into the nature of life at high salt concentrations.

Due to the rapid decrease in water level, resulting in an overall increase in salinity and an increasing dominance of toxic divalent cations magnesium and calcium, the lake's brines are nowadays too extreme to support large communities of microorganisms as they are too extreme even for the best known salt-adapted organisms. Still, use of cultivation-independent environmental genomic and metagenomic approaches in the past decade showed that a small but diverse microbial community still inhabits the lake and that most of the diversity is due to organisms that are still awaiting cultivation (Bodaker et al. 2010, Rhodes et al. 2012, Jacob et al. 2017). The finding of dense communities of microorganisms in underwater fresh to brackish water springs in the lake shows that the microbiota in the Dead Sea sediments may be much more diverse than previously recognized (Ionescu et al. 2012). Thus, a wealth of yet unknown microorganisms may be waiting to be characterized and studied.

In a recent essay, Paul and Mormile (2017) published a call for the protection of saline and hypersaline environments from a microbiological perspective. They rightfully argued that the organisms isolated from hypersaline lakes provide valuable information about the adaptation, evolutionary history, and potential environmental and biotechnological applications of salt-adapted organisms and that it is critical to conserve these unique environments and limit the damage inflicted by anthropogenic influences. In the past decades, the Dead Sea as an ecosystem has deteriorated as a result of increased salinization due to water diversions, mineral extraction, and climatic change. It may be expected that in the near future the current trend of decreasing water levels and increasing salinity will continue, making the Dead Sea an even more hostile environment for life than it is today.

Plans are currently considered to dramatically interfere with the Dead Sea ecosystem by introducing massive amounts of water from the Red Sea (the Red Sea-Dead Sea water conveyance program) (Gavrieli et al. 2005, Glausiusz 2010). If this project will indeed be implemented in the future, it may be predicted that the Dead Sea will once more become populated by dense microbial communities, and these may well differ from those that developed in the lake in the past.

References

Abu-Qarn, M. and J. Eichler. 2006. Protein N-glycosylation in Archaea: defining a *Haloferax volcanii* genes involved in S-layer glycoprotein glycosylation. Mol. Microbiol. 61: 511–525.

Abu-Qarn, M., A. Giordano, F. Battaglia, A. Trauner, P.G. Hitchen, H.R. Morris, et al. 2008. Identification of AglE, a second glycosyltransferase involved in N glycosylation of the *Haloferax volcanii* S-layer glycoprotein. J. Bacteriol. 190: 3140–3146.

Allers, T. and H.-P. Ngo. 2003. Genetic analysis of homologous recombination in Archaea: *Haloferax volcanii* as a model organism. Biochem. Soc. Trans. 31: 706–709.

Allers, T., H.-P. Ngo, M. Mevarech and R.G. Lloyd. 2004. Development of additional selectable markers for the halophilic archaeon *Haloferax volcanii* based on the *leuB* and *trpA* genes. Appl. Environ. Microbiol. 70: 943–953.

Amann, G., K.O. Stetter, E. Llobet-Brossa, R. Amann and J. Antón. 2000. Direct proof for the presence and expression of two 5% different 16S rRNA genes in individual cells of *Haloarcula marismortui*. Extremophiles 4: 373–376.

Arahal, D.R., F.E. Dewhirst, B.J. Paster, B.E. Volcani and A. Ventosa. 1996. Phylogenetic analyses of some extremely halophilic archaea isolated from Dead Sea water, determined on the basis of their 16S rRNA sequences. Appl. Environ. Microbiol. 62: 3779–3786.

Arahal, D.R., M.C. Márquez, B.E. Volcani, K.H. Schleifer and A. Ventosa. 1999. *Bacillus marismortui* sp. nov., a new moderately halophilic species from the Dead Sea. Int. J. Syst. Bacteriol. 49: 521–530.

Arahal, D.R., M.C. Gutiérrez, B.E. Volcani and A. Ventosa. 2000a. Taxonomic analysis of extremely halophilic archaea isolated from 56-years-old Dead Sea brine samples. Syst. Appl. Microbiol. 23: 376–385.

Arahal, D.R., M.C. Márquez, B.E. Volcani, K.H. Schleifer and A. Ventosa. 2000b. Reclassification of *Bacillus marismortui* as *Salibacillus marismortui* comb. nov. Int. J. Syst. Evol. Microbiol. 50: 1501–1503.

Arahal, D.R., M.T. García, W. Ludwig, K.H. Schleifer and A. Ventosa. 2001. Transfer of *Halomonas canadensis* and *Halomonas israelensis* to the genus *Chromohalobacter* as *Chromohalobacter canadensis* comb. nov. and *Chromohalobacter israelensis* comb. nov. Int. J. Syst. Evol. Microbiol. 51: 1443–1448.

Atanasova, N.S., E. Roine, A. Oren, D.H. Bamford and H. Oksanen. 2012. Global network of specific virus-host interactions in hypersaline environments. Environ. Microbiol. 14: 426–440.

Ausiannikava, D., L. Mitchell, H. Marriott, V. Smith, M. Hawkins, K.S. Makarova, et al. 2018. Evolution of genome architecture in Archaea: spontaneous generation of a new chromosome in *Haloferax volcanii*. Mol. Biol. Evol. 35: 1855–1868.

Baliga, N.S., R. Bonneau, M.T. Facciotti, M. Pan, G. Glusman, E.W. Deutsch, et al. 2004. Genome sequence of *Haloarcula marismortui*: a halophilic archaeon from the Dead Sea. Genome Res. 14: 2221–2234.

Ban, N., P. Nissen, J. Hansen, P.B. Moore and T.A. Steitz. 2000. The complete atomic structure of the large ribosomal subunit at 2.4 Å resolution. Science 289: 905–934.

Bidle, K.A., T.E. Hanson, K. Howell and J. Nannen. 2007. HMG-CoA reductase is regulated by salinity at the level of transcription in *Haloferax volcanii*. Extremophiles 11: 49–55.

Bitan-Banin, G., R. Ortenberg and M. Mevarech. 2003. Development of a gene knockout system for the halophilic archaeon *Haloferax volcanii* by use of the *pyrE* gene. J. Bacteriol. 185: 772–778.

Bodaker, I., I. Sharon, M.T. Suzuki, R. Feingersch, M. Shmoish, E. Andreishcheva, et al. 2010. Comparative community genomics in the Dead Sea: an increasingly extreme environment. ISME J. 4: 399–407.

Bodaker, I., M.T. Suzuki, A. Oren and O. Béjà. 2012. Dead Sea rhodopsins revisited. Environ. Microbiol. Rep. 4: 617–621.

Bracken, C.D., A.M. Neighbor, K.K. Lamlenn, G.C. Thomas, H.L. Schubert, F.G. Whitby, et al. 2011. Crystal structures of a halophilic archaeal malate synthase from *Haloferax volcanii* and comparisons with isoforms A and G. BMC Struct. Biol. 11: 23.

Bräsen, C. and P. Schönheit. 2004. Unusual ADP-forming acetyl-coenzyme a synthetases from the mesophilic halophilic euryarchaeon *Haloarcula marismortui* and from the hyperthermophilic crenarchaeon *Pyrobaculum aerophilum*. Arch. Microbiol. 182: 277–287.

Bräsen, C. and P. Schönheit. 2005. ATP-forming acetyl-CoA synthetase from the extremely halophilic archaeon *Haloarcula marismortui*: purification, identification, and expression of the encoding gene, and phylogenetic affiliation. Extremophiles 9: 355–365.

Buchalo, A., E. Nevo, S.P. Wasser, H.P. Molitoris, A. Oren and P.A. Volz. 2000. Fungi discovered in the Dead Sea. Mycol. Res. 104: 129–133.

Buchalo, A.S., E. Nevo, S.P. Wasser, A. Oren and H.P. Molitoris. 1998. Fungal life in the extremely hypersaline water of the Dead Sea: first records. Proc. R. Soc. B. 265: 1461–1465.

Butinar, L., S. Santos, I. Spencer-Martins, A. Oren and N. Gunde-Cimerman. 2005. Yeast diversity in hypersaline habitats. FEMS Microbiol. Lett. 244: 229–234.

Calo, D., Y. Eilam, R.G. Lichtenstein and J. Eichler. 2010. Towards glycoengineering in Archaea: replacement of *Haloferax volcanii* AglD with homologous glycosyltransferases from other halophilic Archaea. Appl. Environ. Microbiol. 76: 5684–5692.

Calo, D., Z. Guan, S. Naparstek and J. Eichler. 2011. Different routes to the same ending: comparing the N-glycosylation process of *Haloferax volcanii* and *Haloarcula marismortui*, two halophilic archaea from the Dead Sea. Mol. Microbiol. 81: 1166–1177.

Camacho, R.M., J.C. Mateos, O. González-Reynoso, L.A. Prado and J. Córdova. 2009. Production and characterization of esterase and lipase from *Haloarcula marismortui*. J. Ind. Microbiol. Biotechnol. 36: 901–909.

Cerletti, M., R. Paggi, C. Troetschel, M.C. Ferrari, C.R. Guevara, S. Albaum, et al. 2018. LonB protease is a novel regulator of carotenogenesis controlling degradation of phytoene synthase in *Haloferax volcanii*. J. Proteome Res. 17: 1158–1171.

Chaga, G., J. Porath and T. Illéni. 1993. Isolation and purification of amyloglucosidase from *Halobacterium sodomense*. Biomed. Chromatogr. 7: 256–261.

Chimileski, S., M.J. Franklin and R.T. Papke. 2014. Biofilms formed by the archaeon *Haloferax volcanii* exhibit cellular differentiation and social motility, and facilitate horizontal gene transfer. BMC Biology 12: 65.

Cline, S.W., L.C. Schalkwyk and W.F. Doolittle. 1989. Transformation of the archaebacterium *Halobacterium volcanii* with genomic DNA. J. Bacteriol. 171: 4987–4991.

Cohen-Rosenzweig, C., S. Yurist-Doutsch and J. Eichler. 2012. AglS, a novel component of the *Haloferax volcanii* N-glycosylation pathway, is a dolichol phosphate-mannose mannosyltransferase. J. Bacteriol. 194: 6909–6916.

Connaris, H., J.B. Chaudhuri, M.J. Danson and D.W. Hough. 1999. Expression, reactivation, and purification of enzymes from *Haloferax volcanii* in *Escherichia coli*. Biotechnol. Bioengin. 64: 38–45.

Dilks, K., M.I. Giménez and M. Pohlschröder. 2005. Genetic and biochemical analysis of the twin-arginine translocation pathway in halophilic archaea. J. Bacteriol. 187: 8104–8113.

Dobson, S.J., S.R. James, P.D. Franzmann and T.A. McMeekin. 1990. Emended description of *Halomonas halmophila* (NCMB 1971[T]). Int. J. Syst. Bacteriol. 40: 462–463.

Dym, O., M. Mevarech and J.L. Sussman. 1995. Structural features that stabilize halophilic malate dehydrogenase from an archaebacterium. Science 267: 1344–1346.

Eichler, J. 2000. Novel glycoproteins of the halophilic archaeon *Haloferax volcanii*. Arch. Microbiol. 173: 445–448.

Eichler. J. 2001. Post-translational modification of the S-layer glycoprotein occurs following translocation across the plasma membrane of the haloarchaeon *Haloferax volcanii*. Eur. J. Biochem. 268: 4366–4373.

Eichler, J. and J. Maupin-Furlow. 2013. Post-translation modification in Archaea: lessons from *Haloferax volcanii* and other haloarchaea. FEMS Microbiol. Rev. 37: 583–606.

Eisenberg, H. and E.J. Wachtel. 1987. Structural studies of halophilic proteins, ribosomes, and organelles of bacteria adapted to extreme salt concentrations. Ann. Rev. Biophys. Biophys. Chem. 16: 69–92.

Elazari-Volcani B. 1940. Studies on the Microflora of the Dead Sea. Ph.D. Thesis, the Hebrew University of Jerusalem.

Elevi Bardavid R.E., L. Mana and A. Oren. 2007. *Haloplanus natans* gen. nov., sp. nov., an extremely halophilic, gas-vacuolate archaeon isolated from Dead Sea-Red Sea water mixtures in experimental outdoor ponds. Int. J. Syst. Evol. Microbiol. 57: 780–783.

Esquivel, R.N., S. Schulze, R. Xu, M. Hipper and M. Pohlschroder. 2016. Identification of *Haloferax volcanii* pilin *N*-glycans with diverse roles in pilus biosynthesis, adhesion, and microcolony formation. J. Biol. Chem. 291: 10602–10614.

Evans, R.C., S.C. Kushwaha and M. Kates. 1980. The lipids of *Halobacterium marismortui*, an extremely halophilic bacterium in the Dead Sea. Biochim. Biophys. Acta 619: 533–544.

Fischel, U. and A. Oren. 1993. Fate of compatible solutes during dilution stress in *Ectothiorhodospira marismortui*. FEMS Microbiol. Lett. 113: 113–118.

Franceschi, F., I. Sagi, N. Böddeker, U. Evers, E. Arndt, C. Paulke, et al. 1994. Crystallographic, biochemical and genetic studies on halophilic ribosomes. Syst. Appl. Microbiol. 16: 697–705.

Franzmann, P.D., U. Wehmeyer and E. Stackebrandt. 1988. *Halomonadaceae* fam. nov., a new family of the class *Proteobacteria* to accommodate the genera *Halomonas* and *Deleya*. Syst. Appl. Microbiol. 11: 16–19.

Frolow, F., M. Harel, J.L. Sussman, M. Mevarech and M. Shoham. 1996. Insights into protein adaptation to a saturated salt environment from the crystal structure of a halophilic 2Fe-2S ferredoxin. Nat. Struct. Biol. 3: 452–457.

Fu, H.-Y., H.-P. Yi, Y.-H. Lu and C.-S. Yang. 2013a. Insight into a single halobacterium using a dual-bacteriorhodopsin system with different functionally optimized pH ranges to cope with periplasmic pH ranges associated with continuous light illumination. Mol. Microbiol. 88: 556–561.

Fu, H.-Y., Y.-H. Lu, H.-P. Yi and C.-S. Yang. 2013b. A transducer for microbial sensory rhodopsin that adopts GTG as a start codon is identified in *Haloarcula marismortui*. J. Photochem. Photobiol. B. Biol. 121: 15–22.

Gabdulkhakov, A., S. Nikonov and M. Garber. 2013. Revisiting the *Haloarcula marismortui* 50S ribosomal subunit model. Acta Crystall. D69: 997–1004.

Gäbel, K., J. Schmitt, S. Schulz, D.J. Nather and J. Soppa. 2013. A comprehensive analysis of the importance of translation initiation factors for *Haloferax volcanii* applying deletion and conditional depletion mutants. PLoS One 8: e77188.

Galinski, E.A. and H.G. Trüper. 1982. Betaine, a compatible solute in the extremely halophilic phototrophic bacterium *Ectothiorhodospira halochloris*. FEMS Microbiol. Lett. 13: 357–360.

Galinski, E.A. and A. Oren. 1991. Isolation and structure determination of a novel compatible solute from the moderately halophilic purple sulfur bacterium *Ectothiorhodospira marismortui*. Eur. J. Biochem. 198: 593–598.

Gavrieli, I., A. Bein and A. Oren. 2005. The expected impact of the "Peace Conduit" project (the Red Sea – Dead Sea pipeline) on the Dead Sea. Mitig. Adapt. Strateg. Glob. Change 10: 759–777.

Gelsinger, D.R. and J. DiRuggiero. 2018. Transcriptional landscape and regulatory roles of small noncoding RNAs in the oxidative stress response of the haloarchaeon *Haloferax volcanii*. J. Bacteriol. 200: e00779-17.

Ginzburg, M. 1969. The unusual membrane permeability of two halophilic unicellular organisms. Biochim. Biophys. Acta 173: 370–376.

Ginzburg, M., L. Sachs and B.Z. Ginzburg. 1970. Ion metabolism in a *Halobacterium*. I. Influence of age of culture on intracellular concentrations. J. Gen. Physiol. 55: 187–207.

Glausiusz, J. 2010. New life for the Dead Sea. Nature 464: 1118–1120.

Guan, Z., S. Naparstek, L. Kaminski, Z. Konrad and J. Eichler. 2010. Distinct glycan-charged phosphodolichol carriers are required for the assembly of the pentasaccharide N-linked to the *Haloferax volcanii* S-layer glycoprotein. Mol. Microbiol. 78: 1294–1303.

Guan, Z., S. Naparstek, D. Calo and J. Eichler. 2012. Protein glycosylation as an adaptive response in Archaea: growth at different salt concentrations leads to alterations in *Haloferax volcanii* S-layer glycoprotein N-glycosylation. Environ. Microbiol. 14: 743–753.

Han, J., Q. Lu, L. Zhou, J. Zhou and H. Xiang. 2007. Molecular characterization of the $phaEC_{Hm}$ genes, required for biosynthesis of poly(3-hydroxybutyrate) in the extremely halophilic archaeon *Haloarcula marismortui*. Appl. Environ. Microbiol. 73: 6058–6065.

Hartman, A.L., C. Norais, J.H. Badger, S. Delmas, S. Haldenby, R. Madupu, et al. 2010. The complete genome sequence of *Haloferax volcanii* DS2, a model archaeon. PLoS One 5: e9605.

Hattori, T., H. Shiba, K.-i. Ashiki, T. Araki, Y.-k. Nagashima, K. Yoshimatsu, et al. 2016. Anaerobic growth of haloarchaeon *Haloferax volcanii* by denitrification is controlled by the transcription regulator NarO. J. Bacteriol. 198: 1077–1086.

Heyrman, J., N.A. Logan, H.-J. Busse, A. Balcaen, L. Lebbe, M. Rodriguez-Diaz, et al. 2003. *Virgibacillus carmonensis* sp. nov., *Virgibacillus necropolis* sp. nov. and *Virgibacillus picturae* sp. nov., three novel species isolated from deteriorated mural paintings, transfer of the species of the genus *Salibacillus* to *Virgibacillus*, as *Virgibacillus marismortui* comb. nov. and *Virgibacillus salexigens* comb. nov., and emended description of the genus *Virgibacillus*. Int. J. Syst. Evol. Microbiol. 53: 501–511.

Hochman, A., A. Nissany and M. Amizur. 1988. Nitrate reduction and assimilation by a moderately halophilic, halotolerant bacterium Ba_1. Biochim. Biophys. Acta 965: 82–89.

Huval, J.H., R. Latta, R. Wallace, D.J. Kushner and R.H. Vreeland. 1995. Description of two new species of *Halomonas*: *Halomonas israelensis* sp. nov. and *Halomonas canadensis* sp. nov. Can. J. Microbiol. 41: 1124–1131.

Ichiki, H., Y. Tanaka, K. Mochizuki, K. Yoshimatsu, T. Sakurai and T. Fujiwara. 2001. Purification, characterization, and genetic analysis of Cu-containing dissimilatory nitrite reductase from a denitrifying halophilic archaeon, *Haloarcula marismortui*. J. Bacteriol. 183: 4149–4156.

Imhoff, J.F. and F. Rodriguez-Valera. 1984. Betaine is the main compatible solute of halophilic eubacteria. J. Bacteriol. 184: 478–479.

Imhoff, J.F., R. Petri and J. Süling. 1998. Reclassification of species of the spiral-shaped phototrophic purple non-sulfur bacteria of the α-*Proteobacteria*: description of the new genera *Phaeospirillum* gen. nov., *Rhodovibrio* gen. nov., *Rhodothalassium* gen. nov. and *Roseospira* gen. nov. as well as transfer of *Rhodospirillum fulvum* to *Phaeospirillum fulvum* comb. nov., of *Rhodospirillum molischianum* to *Phaeospirillum molischianum* comb. nov., of *Rhodospirillum salinarum* to *Rhodovibrio salinarum* comb. nov., of *Rhodospirillum sodomense* to *Rhodovibrio sodomensis* comb. nov., of *Rhodospirillum salexigens* to *Rhodothalassium salexigens* comb. nov. and of *Rhodospirillum mediosalinum* to *Roseospira mediosalina* comb. nov. Int. J. Syst. Bacteriol. 48: 793–798.

Ionescu, D., C. Siebert, L. Polerecky, Y.Y. Munwes, C. Lott, S. Häusler, et al. 2012. Microbial and chemical characterization of submarine freshwater springs in the Dead Sea, harboring rich microbial communities. PLoS One 7: e38319.

Jacob, J.H., E.I. Hussein, M.A.K. Shakhatreh and C.T. Cornelison. 2017. Microbial community analysis of the hypersaline water of the Dead Sea using high-throughput amplicon sequencing. MicrobiologyOpen 6: e500.

Jin, Y., S. Weining and E. Nevo. 2005. A MAPK gene from Dead Sea fungus confers stress tolerance to lithium salt and freezing-thawing: prospects for saline agriculture. Proc. Natl. Acad. Sci. USA 52: 18992–18997.

Johnsen, U., M. Dambeck, H. Zaiss, T. Fuhrer, J. Soppa, U. Sauer, et al. 2009. D-Xylose degradation pathway in the halophilic archaeon *Haloferax volcanii*. J. Biol. Chem. 284: 27290–27303.

Johnsen, U., J.-M. Sutter, H. Zaiß and P. Schönheit. 2013. L-Arabinose degradation pathway in the haloarchaeon *Haloferax volcanii* involves a novel type of L-arabinose dehydrogenase. Extremophiles 17: 897–909.

Kaczowka, S.J. and J.A. Maupin-Furlow. 2003. Subunit topology of two 20S proteasomes from *Haloferax volcanii*. J. Bacteriol. 185: 165–174.

Kaminski, L., Z. Guan, S. Yurist-Doutsch and J. Eichler. 2013a. Two distinct N-glycosylation processes in the *Haloferax volcanii* S-layer glycoprotein upon changes in environmental salinity. mBio 4: e00716–13.

Kaminski, L., S. Naparstek, L. Kandiba, C. Cohen-Rosenzweig, A. Arbiv, Z. Konrad, et al. 2013b. Add salt, add sugar: N-glycosylation in *Haloferax volcanii*. Biochem. Soc. Trans. 41: 432–435.

Kaminski, L. and J. Eichler. 2014. *Haloferax volcanii* N-glycosylation: delineating the pathway of dTDP-rhamnose biosynthesis. PLoS One 9: e97441.

Kandiba, L., Z. Guan and J. Eichler. 2013. Lipid modification gives rise to two distinct *Haloferax volcanii* S-layer glycoprotein populations. Biochim. Biophys. Acta 1828: 938–943.

Kandiba, L. and J. Eichler. 2014. Archaeal S-layer glycoproteins: post-translational modification in the face of extremes. Frontiers Microbiol. 5: 661.

Kandiba, L., C.-W. Lin, M. Aebi, J. Eichler and Y. Guerardel. 2016. Structural characterizarion of the N-linked pentasaccharide decorating glycoproteins of the halophilic archaeon *Haloferax volcanii*. Glycobiology 27: 645–756.

Kapatai, G., A. Large, J.L.P. Benesch, C.V. Robinson, J.L. Carrascosa, J.M. Valpuesta, et al. 2006. All three chaperonin genes in the archaeon *Haloferax volcanii* are individually dispensable. Mol. Microbiol. 61: 1583–1597.

Kaplan, I.R. and A. Friedmann. 1970. Biological productivity in the Dead Sea. Part I. Microorganisms in the water column. Israel J. Chem. 8: 513–528.

Ken-Dror, S. and Y. Avi-Dor. 1985. Regulation of respiration by Na⁺ and K⁺ in the halotolerant bacterium, Ba₁. Arch. Biochem. Biophys. 243: 238–245.

Ken-Dror, S., J.K. Lanyi, B. Schobert, B. Silver and Y. Avi-Dor. 1986. An NADH: quinone oxidoreductase of the halotolerant bacterium Ba₁ is specifically dependent on sodium ions. Arch. Biochem. Biophys. 244: 766–772.

Kessel, M., E.L. Buhle, Jr., S. Cohen and U. Aebi. 1988a. The cell wall structure of a magnesium-dependent halobacterium, *Halobacterium volcanii* CD-2, from the Dead Sea. J. Ultrastr. Mol. Str. Res. 100: 94–106.

Kessel, M., I. Wildhaber, S. Cohen and W. Baumeister. 1988b. Three-dimensional structure of the regular surface glycoprotein layer of *Halobacterium volcanii* from the Dead Sea. EMBO J. 7: 1549–1554.

Kirk, R.G. and M. Ginzburg. 1972. Ultrastructure of two species of halobacterium. J. Ultrastr. Res. 41: 80–94.

Kis-Papo, T., I. Grishkan, A. Oren, S.P. Wasser and E. Nevo. 2001. Spatiotemporal diversity of filamentous fungi in the hypersaline Dead Sea. Mycol. Res. 105: 749–756.

Kis-Papo, T., A. Oren, S.P. Wasser and E. Nevo. 2003a. Survival of filamentous fungi in hypersaline Dead Sea water. Microb. Ecol. 45: 183–190.

Kis-Papo, T., V. Kirzhner, S.P. Wasser and E. Nevo. 2003b. Evolution of genomic diversity and sex at extreme environments: fungal life under hypersaline Dead Sea stress. Proc. Natl. Acad. Sci. USA 100: 14970–14975.

Knitsch, R., M. Schneefeld, K. Weitzel and F. Pfeifer. 2017. Mutations in the major gas vesicle protein GvpA and impacts on gas vesicle formation in *Haloferax volcanii*. Mol. Microbiol. 106: 530–542.

Konrad, Z. and J. Eichler. 2002. Protein glycosylation in *Haloferax volcanii*: partial characterization of a 98-kDa glycoprotein. FEMS Microbiol. Lett. 209: 197–202.

Kramer, P., K. Gäbel, F. Pfeiffer and J. Soppa. 2014. *Haloferax volcanii*, a prokaryotic species that does not use the Shine Dalgarno mechanism for translation initiation at the 5'-UTRs. PLoS One 9: e94979.

Lange, C., K. Zerulla, S. Breuert and J. Soppa. 2011. Gene conversion results in the equalization of genome copies in the polyploid haloarchaeon *Haloferax volcanii*. Mol. Microbiol. 80: 666–677.

Legerme, G., E. Yang, R.N. Esquivel, S. Kiljunen, H. Savilahti and M. Pohlschroder. 2016. Screening of *Haloferax volcanii* transposon library reveals novel motility and adhesion mutants. Life 6: 41.

Lin, Y.-C., H.-Y. Fu and C.-S. Yiang. 2010. Phototaxis of *Haloarcula marismortui* revealed through a novel microbial motion analysis algorithm. Photochem. Photobiol. 86: 1084–1090.

Lobasso, S., R. Vitale, P. Lopalco and A. Corcelli. 2015. *Haloferax volcanii*, as a novel tool for producing mammalian olfactory receptors embedded in archaeal lipid bilayer. Life 5: 770–782.

López-López, A., S. Benlloch, M. Bonfá, F. Rodríguez-Valera and A. Mira. 2007. Intragenomic 16S rDNA divergence in *Haloarcula marismortui* is an adaptation to different temperatures. J. Mol. Evol. 65: 687–696.

Mack, E.E., L. Mandelco, C.R. Woese and M.T. Madigan. 1993. *Rhodospirillum sodomense*, sp. nov., a Dead Sea *Rhodospirillum* species. Arch. Microbiol. 160: 363–371.

Madern, D., M. Camacho, A. Rodriguez-Arnedo, M.-J. Bonete and G. Zaccai. 2004. Salt-dependent studies of NADP-dependent isocitrate dehydrogenase from the halophilic archaeon *Haloferax volcanii*. Extremophiles 8: 377–384.

Magidovich, H., S. Yurist-Doutsch, Z. Konrad, V.V. Ventura, A. Dell, P.G. Hitchen, et al. 2010. AglP is a S-adenosyl-L-methionine-dependent methyltransferase that participates in the N-glycosylation pathway of *Haloferax volcanii*. Mol. Microbiol. 76: 190–199.

Maier, L.-K., B. Stoll, J. Brendel, S. Fischer, F. Pfeiffer, M. Dyall-Smith, et al. 2013. The ring of confidence: a haloarchaeal CRISPR/Cas system. Biochem. Soc. Trans. 41: 374–378.

Maier, L.-K., A.-E. Stachler, J. Brendel, B. Stoll, S. Fischer, K.A. Haas, et al. 2019. The nuts and bolts of the *Haloferax* CRISPR-Cas system 1-B. RNA Biol. 16: 469–480.

Maurer, S., K. Lundt and J. Soppa. 2017. Characterization of copy number control of two *Haloferax volcanii* replication origins using deletion mutants and haloarchaeal artificial chromosomes. J. Bacteriol. 200: e00517–17.

McGenity, T.J. and W.D. Grant. 1995. Transfer of *Halobacterium saccharovorum, Halobacterium sodomense, Halobacterium trapanicum* NRC 34041 and *Halobacterium lacusprofundi* to the genus *Halorubrum* gen. nov., as *Halorubrum saccharovorum* comb. nov., *Halorubrum sodomense* comb. nov., *Halorubrum trapanicum* comb. nov., and *Halorubrum lacusprofundi* comb. nov. Syst. Appl. Microbiol. 18: 237–243.

Megaw, J. and B.F. Gilmore. 2017. Archaeal persisters: persister cell formation as a stress response in *Haloferax volcanii*. Frontiers Microbiol. 8: 1589.

Meury, J. and M. Kohiyama. 1989. ATP is required for K⁺ active transport in the archaebacterium *Haloferax volcanii*. Arch. Microbiol. 151: 530–536.

Mevarech, M. and E. Neumann. 1977. Malate dehydrogenase isolated from extremely halophilic bacteria of the Dead Sea. 2. Effect of salt on the catalytic activity and structure. Biochemistry 16: 3786–3792.

Mevarech, M. and R. Werczberger. 1985. Genetic transfer in *Halobacterium volcanii*. J. Bacteriol. 162: 461–462.

Mevarech, M., F. Frolow, and L.M. Gloss. 2000. Halophilic enzymes: proteins with a grain of salt. Biophys. Chem. 86: 155–164.

Mullakhanbhai, M.F. and G.W. Francis. 1972. Bacterial lipids. 1. Lipid constituents of a moderately halophilic bacterium. Acta Chem. Scand. 26: 1399–1410.

Mullakhanbhai, M.F. and H. Larsen. 1975. *Halobacterium volcanii* spec. nov., a Dead Sea halobacterium with a moderate salt requirement. Arch. Microbiol. 104: 207–214.

Müller-Santos, M., E.M. de Souza, F. de O. Pedrosa, D.A. Mitchell, S. Longhi, F. Carrière, et al. 2009. First evidence for the salt-dependent folding and activity of an esterase from the halophilic archaea *Haloarcula marismortui*. Biochim. Biophys. Acta 1791: 719–729.

Mutnuri, S., N. Vasudevan, M. Kastner and H.J. Heipieper. 2005. Changes in fatty acid composition of *Chromohalobacter israelensis* with varying salt concentrations. Curr. Microbiol. 50: 151–154.

Mylvaganam, S. and P.P. Dennis. 1992. Sequence heterogeneity between the two genes encoding 16S rRNA from the halophilic archaebacterium *Haloarcula marismortui*. Genetics 130: 399–410.

Oremland, R.S., M.J. Herbel, J. Switzer Blum, S. Langley, T.J. Beveridge, P.M. Ajayan, et al. 2004. Structural and spectal features of selenium nanospheres produced by Se-respiring bacteria. Appl. Environ. Microbiol. 70: 52–60.

Oren, A. and M. Shilo. 1981. Bacteriorhodopsin in a bloom of halobacteria in the Dead Sea. Arch. Microbiol. 130: 185–187.

Oren, A. and M. Shilo. 1982. Population dynamics of *Dunaliella parva* in the Dead Sea. Limnol. Oceanogr. 27: 201–211.

Oren, A. 1983a. Population dynamics of halobacteria in the Dead Sea water column. Limnol. Oceanogr. 28: 1094–1103.

Oren, A. 1983b. *Halobacterium sodomense* sp. nov., a Dead Sea halobacterium with an extremely high magnesium requirement. Int. J. Syst. Bacteriol. 33: 381–386.

Oren, A. 1983c. *Clostridium lortetii* sp. nov., a halophilic obligatory anaerobic bacterium producing endospores with attached gas vacuoles. Arch. Microbiol. 136: 42–48.

Oren, A. 1983d. A thermophilic amyloglucosidase from *Halobacterium sodomense*, a halophilic bacterium from the Dead Sea. Curr. Microbiol. 8: 225–230.

Oren, A., W.G. Weisburg, M. Kessel and C.R. Woese. 1984. *Halobacteroides halobius* gen. nov., sp. nov., a moderately halophilic anaerobic bacterium from the bottom sediments of the Dead Sea. System. Appl. Microbiol. 5: 58–70.

Oren, A. 1986. Intracellular salt concentrations of the anaerobic halophilic eubacteria *Haloanaerobium praevalens* and *Halobacteroides halobius*. Can. J. Microbiol. 32: 4–9.

Oren, A., H. Pohla and E. Stackebrandt. 1987. Transfer of *Clostridium lortetii* to a new genus *Sporohalobacter* gen. nov. as *Sporohalobacter lortetii* comb. nov. and description of *Sporohalobacter marismortui* sp. nov. Syst. Appl. Microbiol. 9: 239–246.

Oren, A. 1989. Photosynthetic and heterotrophic benthic bacterial communities of a hypersaline sulfur spring on the shore of the Dead Sea (Hamei Mazor). pp. 64–76. *In*: Y. Cohen and E. Rosenberg [eds.]. Microbial Mats: Physiological Ecology of Benthic Microbial Communities. ASM Publications, Washington, D.C., USA.

Oren, A., M. Kessel and E. Stackebrandt. 1989. *Ectothiorhodospira marismortui* sp. nov., an obligately anaerobic, moderately halophilic purple sulfur bacterium from a hypersaline sulfur spring on the shore of the Dead Sea. Arch. Microbiol. 151: 524–529.

Oren, A., M. Ginzburg, B.Z. Ginzburg, L.I. Hochstein and B.E. Volcani. 1990. *Haloarcula marismortui* (Volcani) sp. nov., nom. rev., an extremely halophilic bacterium from the Dead Sea. Int. J. Syst. Bacteriol. 40: 209–210.

Oren, A., P. Gurevich and Y. Henis. 1991a. Reduction of nitro-substituted aromatic compounds by the halophilic anaerobic eubacteria *Haloanaerobium praevalens* and *Sporohalobacter marismortui*. Appl. Environ. Microbiol. 57: 3367–3370.

Oren, A., G. Simon and E.A. Galinski. 1991b. Intracellular salt and solute concentrations in *Ectothiorhodospira marismortui*: glycine betaine and Nα carbamoyl glutamineamide as osmotic solutes. Arch. Microbiol. 156: 350–355.

Oren, A. and P. Gurevich. 1993. Characterization of the dominant halophilic archaea in a bacterial bloom in the Dead Sea. FEMS Microbiol. Ecol. 12: 249–256.

Oren, A. and P. Gurevich. 1995. Dynamics of a bloom of halophilic archaea in the Dead Sea. Hydrobiologia 315: 149–158.

Oren, A., P. Gurevich, D.A. Anati, E. Barkan and B. Luz. 1995a. A bloom of *Dunaliella parva* in the Dead Sea in 1992: biological and biogeochemical aspects. Hydrobiologia 297: 173–185.

Oren, A., P. Gurevich, R.T. Gemmell and A. Teske. 1995b. *Halobaculum gomorrense* gen. nov., sp. nov., a novel extremely halophilic archaeon from the Dead Sea. Int. J. Syst. Bacteriol. 45: 747–754.

Oren, A. 2010. The dying Dead Sea: the microbiology of an increasingly extreme environment. Lakes & Reservoirs: Res. Manag. 15: 215–222.

Ortenberg, R., O. Rozenblatt-Rosen and M. Mevarech. 2000. The extremely halophilic archaeon *Haloferax volcanii* has two very different dihydrofolate reductases. Mol. Microbiol. 35: 1493–1505.

Ouellette, M., A.M. Makkay and R.T. Papke. 2013. Dihydroxyacetone metabolism in *Haloferax volcanii*. Frontiers Microbiol. 4: 376.

Paul, V.G. and M.R. Mormile. 2017. A case for the protection of saline and hypersaline environments: a microbiological perspective. FEMS Microbiol. Ecol. 93: fix091.

Peleg, E. and A. Tietz. 1971. Glycolipids of a halotolerant moderately halophilic bacterium. FEMS Lett. 15: 309–312.

Peleg, E. and A. Tietz. 1973. Phospholipids of a moderately halophilic bacterium. Isolation and identification of glucosylphosphatidylglycerol. Biochim. Biophys. Acta 306: 368–379.

Penczek, P., N. Ban, R.A. Grassucci, R.K Agrawal and J. Frank. 1999. *Haloarcula marismortui* 50S subunit – complementary of electron microscopy and X-ray crystallographic information. J. Struct. Biol. 128: 44–50.

Pickl, A., U. Johnsen and P. Schönheit. 2012. Fructose degradation in the haloarchaeon *Haloferax volcanii* involves a bacteria type phosphoenolpyruvate-dependent phosphotransferase system, fructose-1-phopshate kinase and Class II fructose-1,6-bisphosphate aldolase. J. Bacteriol. 194: 3088–3097.

Pickl, A., U. Johnsen, R.M. Archer and P. Schönheit. 2014. Identification and characterization of 2-keto-3-deoxygluconate kinase and 2-keto-3-deoxygalactonate kinase in the haloarchaeon *Haloferax volcanii*. FEMS Microbiol. Lett. 361: 76–83.

Pickl, A. and P. Schönheit. 2015. The oxidative pentose phosphate pathway in the haloarchaeon *Haloferax volcanii* involves a novel type of glucose-6-phosphate dehydrogenase – the archaeal Zwischenferment. FEBS Lett. 589: 1105–1111.

Pieper, U., G. Kapadia, M. Mevarech and O. Herzberg. 1998. Structural features of halophilicity derived from the crystal structure of dihydrofolate reductase from the Dead Sea halophilic archaeon, *Haloferax volcanii*. Structure 6: 75–88.

Prunetti, L., C.J. Reuter, N.L. Hepowit, Y. Wu, L. Barrueto, H.V. Miranda, et al. 2014. Structural and biochemical properties of an extreme 'salt-loving' proteasome activating nucleotidase from the archaeon *Haloferax volcanii*. Extremophiles 18: 283–293.

Pundak, S. and H. Eisenberg. 1981. Structure and activity of malate dehydrogenase from the extreme halophilic bacteria of the Dead Sea. I. Conformation and interaction with water and salt between 5 M and 1 M NaCl concentration. Eur. J. Biochem. 118: 463–470.

Qi, Q., Y. Ito, K. Yoshimatsu and T. Fujiwara. 2016. Transcriptional regulation of dimethyl sulfoxide respiration in a haloarchaeon, *Haloferax volcanii*. Extremophiles 20: 27–36.

Rafaeli-Eshkol, D. 1968. Studies on halotolerance in a moderately halophilic bacterium. Effect of growth conditions on salt resistance of the respiratory system. Biochem. J. 109: 679–685.

Rafaeli-Eshkol, D. and Y. Avi-Dor. 1968. Studies on halotolerance in a moderately halophilic bacterium. Effect of betaine in a moderately halophilic bacterium. Biochem. J. 109: 687–691.

Rainey, F.A., T.N. Zhilina, E.S. Boulygina, E. Stackebrandt, T.P. Tourova and G.A. Zavarzin. 1995. The taxonomic study of the fermentative halophilic bacteria: description of *Haloanaerobiales* ord. nov., *Halobacteroidaceae* fam. nov., *Orenia* gen. nov. and further taxonomic rearrangements at the genus and species level. Anaerobe 1: 185–199.

Rhodes, M.E., A. Oren and C.H. House. 2012. Dynamics and persistence of Dead Sea microbial populations as shown by high-throughput sequencing of rRNA. Appl. Environ. Microbiol. 78: 2489–2492.

Ring, G. and J. Eichler. 2004. Membrane binding of ribosomes occurs at SecYE-based sites in the Archaea *Haloferax volcanii*. J. Mol. Biol. 336: 997–1010.

Rodriguez-Oliveira, T., A.A. Souza, R. Kruger, B. Schuster, S.M. de Freitas and C.M. Kyaw. 2019. Environmental factors influence the *Haloferax volcanii* S-layer protein structure. PLoS One 24: e0216863.

Rose, R.W. and M. Pohlschröder. 2002. *In vivo* analysis of an essential archaeal signal recognition particle in its native host. J. Bacteriol. 184: 3260–3267.

Rosenshine, I., R. Tchelet and M. Mevarech. 1989. The mechanism of DNA transfer in the mating system of an archaebacterium. Science 245: 1387–1389.

Schroeder, S.J., G. Blaha, J. Tirado-Rives, T.A. Steitz and P.B. Moore. 2007. The structures of antibiotics bound to the *E* site region of the 50 S ribosomal subunit of *Haloarcula marismortui*: 13-deoxytedanolide and girodazole. J. Mol. Biol. 367: 1471–1479.

Serrano, J.A., M. Camacho and M.J. Bonete. 1998. Operation of glyoxylate cycle in halophilic archaea: presence of malate synthase and isocitrate lyase in *Haloferax volcanii*. FEMS Lett. 434: 13–16.

Seth-Pasricha, M., K.A. Bidle and K.D. Bidle. 2013. Specificity of archaeal caspase activity in the extreme halophile *Haloferax volcanii*. Environ. Microbiol. Rep. 5: 263–271.

Shalev, Y., I. Turgeman-Grott, A. Tamir, J. Eichler and U. Gophna. 2017. Cell surface glycosylation is required for efficient mating of *Haloferax volcanii*. Frontiers Microbiol. 8: 1253.

Sherwood, K.E., D.J. Cano and J.A. Maupin-Furlow. 2009. Glycerol-mediated repression of glucose metabolism and glycerol kinase as the sole route of glycerol catabolism in the haloarchaeon *Haloferax volcanii*. J. Bacteriol. 191: 4307–4315.

Shkedy-Vinkler, C. and Y. Avi-Dor. 1975. Betaine-induced stimulation of respiration at high osmolarities in a halotolerant bacterium. Biochem. J. 150: 219–226.

Shnaiderman, R. and Y. Avi-Dor. 1982. The uptake and extrusion of salts by the halotolerant bacterium, Ba_1. Arch. Biochem. Biophys. 213: 177–185.

Stachler, A.-E., I. Turgeman-Grott, E. Shtifman-Segal, T. Allers, A. Marchfelder and U. Gophna. 2017. High tolerance to self-targeting of the genome by the endogenous CRISR-Cas system in an archaeon. Nucl. Acids Res. 45: 5208–5216.

Stern, N. and A. Tietz. 1973a. Glycolipids of a halotolerant, moderately halophilic bacterium. I. The effect of growth medium and age of culture on lipid composition. Biochim. Biophys. Acta 296: 130–135.

Stern, N. and A. Tietz. 1973b. Glycolipids of a halotolerant, moderately halophilic bacterium. II. Biosynthesis of the glucuronosyldiglyceride by cell-free particles. Biochim. Biophys. Acta 296: 136–144.

Stern, N. and A. Tietz. 1978. Glycolipids of a halotolerant, moderately halophilic bacterium. Biosynthesis of the glucosylphosphatidylglycerol by cell-free particles. Biochim. Biophys. Acta 530: 357–366.

Sumper, M., E. Berg, R. Mengele and I. Strobel. 1990. Primary structure and glycosylation of the S-layer protein of *Haloferax volcanii*. J. Bacteriol. 172: 7111–7118.

Sussman, J.L., P. Zipori, M. Harel, A. Yonath and M.M. Werber. 1979. Preliminary X-ray diffraction studies on 2 Fe-ferredoxin from *Halobacterium* of the Dead Sea. J. Mol. Biol. 134: 375–377.

Sutter, J.-M., J.-B. Tästensen, U. Johnsen, J. Soppa and P. Schönheit. 2016. Key enzymes of the semiphosphorylative Entner-Doudoroff pathway in the haloarchaeon *Haloferax volcanii*: Characterization of glucose dehydrogenase, gluconate dehydratase, and 2-keto-3-deoxy-6-phosphogluconate aldolase. J. Bacteriol. 198: 2251–2262.

Sutter, J.-M., U. Johnsen and P. Schönheit. 2017. Characterization of a pentonolactonase involved in D-xylose and L-arabinose catabolism in the haloarchaeon *Haloferax volcanii*. FEMS Microbiol. Lett. 364: fnx140.

Switzer Blum, J., J.F. Stolz, A. Oren and R.S. Oremland. 2001. *Selenihalanaerobacter shriftii* gen. nov., sp. nov., a halophilic anaerobe from dead sea sediments that respires selenate. Arch. Microbiol. 175: 208–219.

Syutkin, A.S., M.G. Pyatibratov, O.V. Galzitskaya, F. Rodríguez-Valera and O.V. Fedorov. 2014. *Haloarcula marismortui* archaellin genes as ecoparalogs. Extremophiles 18: 341–349.

Syutkin, A.S., M. van Wolferen, A.K. Surin, S.V. Albers, M.G. Pyatibratov, O.V. Fedorov, et al. 2018. Salt-dependent regulation of archaellins in *Haloarcula marismortui*. Microbiology Open 2018: e718.

Tamir, A. and J. Eichler. 2017. N-glycosylation is important for proper *Haloferax volcanii* S-layer stability and function. Appl. Environ. Microbiol. 83: e03152–16.

Tchelet, R. and M. Mevarech. 1994. Interspecies genetic transfer in halophilic archaebacteria. Syst. Appl. Microbiol. 16: 578–581.

Tehei, M., B. Franzetti, K. Wood, F. Gabel, E. Fabiani, M. Jasnin, et al. 2007. Neutron scattering reveals extremely slow cell water in a Dead Sea organism. Proc. Natl. Acad. Sci. USA 104: 766–771.

Timpson, L.M., A.-K. Liliensiek, D. Alsafadi, J. Cassidy, M.A. Sharkey, S. Liddell, et al. 2013. A comparison of two novel alcohol dehydrogenase enzymes (ADH1 and ADH2) from the extreme halophile *Haloferax volcanii*. Appl. Microbiol. Biotechnol. 97: 195–203.

Torreblanca, M., F. Rodriguez-Valera, G. Juez, A. Ventosa, M. Kamekura and M. Kates. 1986. Classification of non-alkaliphilic halobacteria based on numerical taxonomy and polar lipid composition, and description of *Haloarcula* gen. nov. and *Haloferax* gen. nov. Syst. Appl. Microbiol. 8: 89–99.

Tozik, I., Q. Huang, C. Zwieb and J. Eichler. 2002. Reconstitution of the signal recognition particle of the halophilic archaeon *Haloferax volcanii*. Nucl. Acid. Res. 30: 4166–4175.

Trincone, A., B. Nicolaus, L. Lama, M. de Rosa, A. Gambacorta and W.D. Grant. 1990. The glycolipid of *Halobacterium sodomense*. J. Gen. Microbiol. 136: 2327–2331.

Tripepi, M., S. Imam and M. Pohlschröder. 2010. *Haloferax volcanii* flagella are required for motility but are not involved in PidD-dependent surface adhesion. J. Bacteriol. 192: 3093–3102.

Tripepi, M., J. You, S. Temel, Ö. Önder, D. Brisson and M. Pohlschröder. 2012. N-glycosylation of *Haloferax volcanii* flagellins requires known Agl proteins and is essential for biosynthesis of stable flagella. J. Bacteriol. 194: 4876–4887.

Tripepi, M., R.N. Esquivel, R. Wirth and M. Pohlschröder. 2013. *Haloferax volcanii* cells lacking the flagellin FlgA2 are hypermotile. Microbiology 159: 2249–2258.

Tästensen, J.-B. and P. Schönheit. 2018. Two distinct glyceraldehyde-3-phosphate dehydrogenases in glycolysis and gluconeogenesis in the archaeon *Haloferax volcanii*. FEBS Lett. 592: 1524–1534.

Tsai, F.-K., H.-Y. Fu, C.-S. Yang and L.-K. Chu. 2014. Photochemistry of a dual-bacteriorhodopsin system in *Haloarcula marismortui*: HmbR1 and HmbRII. J. Phys. Chem. B 118: 7290–7301.

Ventosa, A., M.C. Gutierrez, M.T. Garcia and F. Ruiz-Berraquero. 1989. Classification of "*Chromobacterium marismortui*" in a new genus, *Chromohalobacter* gen. nov., as *Chromohalobacter marismortui* comb. nov., nom. rev. Int. J. Syst. Bacteriol. 39: 382–386.

Ventosa, A., J.J. Nieto and A. Oren. 1998. Biology of moderately halophilic aerobic bacteria. Microbiol. Mol. Biol. Rev. 62: 504–544.

Volcani, B.E. 1944. The microorganisms of the Dead Sea. pp. 71–85. *In*: Papers Collected to Commemorate the 70th Anniversary of Dr. Chaim Weizmann, Collective Volume, Daniel Sieff Research Institute, Rehovoth.

Wei, X., Y. Jiang, X. Chen, Y. Jiang and H. Lai. 2015. *Amycolatopsis flava* sp. nov., a halophilic actinomycete isolated from Dead Sea. Antonie van Leeuwenhoek 108: 879–885.

Werber, M.M. and M. Mevarech. 1978a. Induction of a dissimilatory reduction pathway of nitrate in *Halobacterium* of the Dead Sea. A possible role for the 2 Fe-ferredoxin isolated from this organism. Arch. Biochem. Biophys. 186: 60–65.

Werber, M.M. and M. Mevarech. 1978b. Purification and characterization of a highly acidic 2Fe-ferredoxin from *Halobacterium* of the Dead Sea. Arch. Biochem. Biophys. 187: 447–456.

Wilkansky, B. 1936. Life in the Dead Sea. Nature 138: 467.

Yamada, Y., T. Fujiwara, T. Sato, N. Igarishi and N. Tanaka. 2002. The 2.0 Å crystal structure of catalase-peroxidase from *Haloarcula marismortui*. Nat. Struct. Biol. 9: 691–695.

Yoshimatsu, K., T. Iwasaki and T. Fujiwara. 2002. Sequence and electron paramagnetic resonance analyses of nitrate reductase NarGH from a denitrifying halophilic euryarchaeote *Haloarcula marismortui*. FEBS Lett. 516: 145–150.

Yurist-Doutsch, S., M. Abu-Qarn, F. Battaglia, H.R. Morris, P.G. Hitchen, A. Dell, et al. 2008. *aglF*, *aglG* and *aglI*, novel members of a gene island involved in the N-glycosylation of the *Haloferax volcanii* S-layer glycoprotein. Mol. Microbiol. 69: 1234–1245.

Zalazar, L., P. Pagola, M.V. Miró, M.S. Churio, M. Cerletti, C. Marínez, et al. 2018. Bacterioruberin extracts from a genetically modified hyperpigmented *Haloferax volcanii* strain: antioxidant activity and bioactive properties on sperm cells. J. Appl. Microbiol. 126: 796–810.

4

The Wonderful Halophiles: Microorganisms Producing Useful Compounds

Giuseppe Squillaci[1,2], Ismene Serino[1,3],
Francesco La Cara[1] and Alessandra Morana[1,*]

Introduction

The halophilic microorganisms are present in all three domains of life, namely Archaea, Bacteria, and Eukarya. They have the unique property of thriving in environments characterized by high salt concentration and are widespread in hypersaline lakes, solar salterns, saline soil, evaporation ponds, and marine environments (Ventosa et al. 1999, Deepa et al. 2015, Boyadzhieva et al. 2018).

Halophiles can be distinguished in moderately halophilic and extremely halophilic microorganisms. Moderate halophiles include both Gram-positive and Gram-negative microorganisms growing optimally in media with NaCl concentration ranging from 3% to 15%; however, they can also grow below and above these values (Ventosa et al. 1998). Extremely halophilic microorganisms show optimal growth in media containing from 15% to 30% NaCl, and the main representatives are halophilic aerobic Archaea (also known as Haloarchaea), although some extremely halophilic bacteria have been reported in this group (Grant et al. 2001).

As in such extreme environments, the salt concentration can reach saturation; halophiles succeed to survive by adopting several strategies: they accumulate in the cytoplasm a huge quantity of organic compounds that function as osmo-protectants or store in the cytoplasm potassium chloride offering an osmotic balance between cells and medium. Furthermore, they have also developed unique enzymes that are stable and active in the presence of salts (Lanyi 1974), and interesting biomolecules with protective functions as exopolysacchyarides (EPS), pigments, and polyhydroxyalkanoates (PHA) (Yin et al. 2015). Some halophilic microorganisms exhibit antimicrobial properties as they produce bacteriocin-like peptides, the halocins (Rodriguez-Valera et al. 1982). Consequently, the halophilic microorganisms can be regarded as amazing tools for producing compounds of significant importance for industries as they can found applications in many fields, such as pharmaceutical, food, and cosmetics. This review reports an overview of the most useful 'halophilic products' produced by Archaea, Bacteria, and Eukarya and their biotechnological potentials.

The World of Halophiles

In literature, there is a considerable amount of different definitions and different salt concentration limits for what a halophile is (Ventosa et al. 1998, Oren 2002a, Oren 2006). Here, the definition

[1] Research Institute on Terrestrial Ecosystems, National Research Council of Italy, Naples, Italy.
[2] Department of Precision Medicine, University of Campania "Luigi Vanvitelli", Naples, Italy.
[3] Department of Experimental Medicine, University of Campania "Luigi Vanvitelli", Naples, Italy.
* Corresponding author: alessandra.morana@cnr.it, alessandra.morana@iret.cnr.it

proposed by Oren in 2008 will be used: halophile means an organism capable of living and reproducing in environments characterized by high salt (NaCl) concentrations, 50 g/L (0.85 M NaCl) or higher, and tolerate at least 100 g/L of salt (1.7 M NaCl). On our planet, these particular conditions can be found in natural or artificial hypersaline water bodies, such as salt pans, hypersaline ponds, salt lakes, saline soils, and in salted foods.

Unlike halotolerant organisms, the high salt concentration is necessary for their vital functions; therefore, the halophiles represent a sub-category of the extremophilic microorganisms.

Despite their unusual propensity for saline environments, halophilic organisms can be traced in all three domains of life, Bacteria, Archaea, and Eukarya. Studies regarding the small subunit rRNA sequence revealed that halophiles are not neatly distributed in the tree of life (Madigan and Martinko 2006). Often genera and families of halophilic species can be found in mixed taxonomic groups including mesophiles. The only exceptions to this 'disorder' are the Archaea belonging to the *Halobacteriales* order and the aerobic fermentative bacteria belonging to the *Halanaerobiales* order. These orders include exclusively halophilic species.

Halophilic organisms evolved two different strategies to exclude NaCl out of the cytoplasm in order to avoid the salting-out phenomena consisting of proteins aggregation, and both strategies are ATP expensive. The first one consists in the accumulation of K^+ inside the cytoplasm and the extrusion of NaCl at the same time. This strategy requires many adaptations in the protein machinery of the cell. In fact, these proteins are effective in environments with high salt content, while they are ineffective or even undergo denaturation when the salt content decreases.

A second way used by halophilic microorganisms to contrast the extreme salinity of their environments consists in the cytoplasmic accumulation of organic 'compatible' solutes, also called osmolytes. Even in this strategy, NaCl has to be extruded outside the cell. The compatible solutes often consist of aminoacidic derivative substances, like glycine betaine, ectoine, and others. Unlike the KCl accumulation, the 'compatible solutes' strategy allows the organisms to be more adaptable to environmental NaCl concentration modifications. The protein machinery is not so different from the mesophilic enzymes, and only a few adaptations to this particular environment are needed.

The K^+ type osmotic control is peculiar of the extremely halophilic Archaea (*Halobacteriaceae*), of the moderately halophilic bacteria belonging to the *Halanaerobiales* (Oren 1999) and of the extremely halophilic bacterium *Salinibacter ruber* (Antón et al. 2002, Oren and Mana 2002b). The diffusion of this osmotic control in microorganisms so far from the evolutionary point of view, suggests the hypothesis of an independent co-evolution of the potassium cations intracytoplasmic accumulation ability, rather than for the hypothesis of a common genetic origin maintained only in these organisms. The K^+ type osmotic control may have requested for its evolution a greater quantity of mutations and gene adaptations than those necessary for the evolution of the osmotic control strategy through 'compatible' solutes. In fact, this second way is the most represented in the halophilic microbic world and has been adopted by a large number of microorganisms, belonging to Archaea, Bacteria, Yeast, filamentous Fungi, and Algae (Santos and da Costa 2002).

Prokaryotic Halophiles

Bacterial Halophiles

Halophilic bacteria can be found in the following phila: *Actinobacteria*, *Bacteroidetes*, *Cyanobacteria*, *Firmicutes*, *Proteobacteria*, and *Spirochaetes*. As above stated, extremely different species can be described, both for the range of tolerance to NaCl and the cytoplasmic adaptation strategies to low water activity in their environment. The most common strategy of cellular adaptation of halophilic bacteria to the saline environment is represented by the synthesis of compatible solutes (osmolytes). *Cyanobacteria*, for example, have been studied for their ability to synthesize different types of these molecules (Reed et al. 1984). They can synthesize glycine betaine, trehalose, glycosylglycerol, sucrose, glutamate, alone, or mixed together. The type of osmolyte that can be synthesized is related to the degree of tolerance of the microorganism to the salt. For instance, in the non-fermenting bacterial strains, the ability of glycine betaine synthesis has been associated with the extreme halophilicity (Trüper et al. 1991). The major compatible solutes and potential salt tolerance of these strains can be deduced from increasingly common genetic analyses (Klähn and Hagemann 2010).

Archaeal Halophiles

Most of the currently described halophiles are included in the Archaea domain. Within Archaea, the phylum Euryarchaeota is the largest, and the only one in which halophilic microorganisms have been counted. In this phylum some other extremophiles, such as aerobic and anaerobic hyperthermophilic microorganisms, have been registered. Euryarchaeota also contains anaerobic methanogens and marine microorganisms. The genera *Halomethanococcus* (Yu and Kawamura 1987), *Methanohalophilus*, and *Methanohalobium* (Zhilina and Zavarzin 1987) present halophilic character.

Halobacteria represent the most salt-requiring class inside Archaea. Studies on rRNA 16S, coupled with studies on lipidic environmental fraction, assessed that it encloses most of the prokaryotic population in hypersaline environments (Oren 2002c). Currently, the majority of halophilic Archaea is described within the *Halobacteriaceae* family. It contains microorganisms capable of living in different ecological niches, such as acidophilic, basophilic, neutrophilic, psychrophilic, aerobic, or anaerobic, characterized by different morphologies such as cocci, rods, pleomorphic or more unusual shapes like triangles (Oren 2014).

Eukaryotic Halophiles

Halophilic eukaryotes are exclusively represented by some Fungi and few algal species. Within Algae, the most studied is the green flagellated microalga *Dunaliella salina* (Schleper et al. 1995). It also represents the most abundant algal species in extreme saline environments, such as salt pans. Its adaptation and defense mechanisms against salt and oxidative stress, caused by the extreme concentration of salt and strong exposure to sunlight, include the peculiar ability to synthesize and retain glycerol within the cell as well as the strong production of the carotenoid beta-carotene useful to counteract oxidative reactions. The beta-carotene production by *Dunaliella salina* has been extensively used for the industrial production of this important vitamin-A precursor. It has application in cosmetic, pharmaceutic, and food industries and represents one of the success stories of the halophile biotechnology (Rüegg 1984).

Other halophilic algal species, showing polyextremophilic characteristics, are the psycrophilic green alga *Chlamydomonas nivalis* and the red thermophilic alga *Cyanidium caldarium*.

Among Fungi, the obligate requirement for salt is an exception. Many species that have been found capable of living in hypersaline environments, also thrive in common culture media lacking the addition of NaCl. For this reason, halophilic Fungi can be found not only in hypersaline environments but also in fresh or seawater. An exception is represented by *Wallemia ichthyophaga*, a basidiomycetous fungus that thrives in salt saturation conditions and requires at least 1.5 M NaCl for *in vitro* growth (Zalar et al. 2005).

In general, Fungi capable of living at salt concentrations above 1.7 M are considered halophilic Fungi (Butinar et al. 2005a). From a taxonomic point of view, only ten orders, among the 106 currently surveyed, include members capable of living in hypersaline environments. Among these, the most studied are *Capnodiales, Eurotiales, Dothideales* (*Ascomycota*) and *Wallemia* (*Basidiomycota*) (Al-Abri 2011, Butinar et al. 2005b).

Products From Halophiles

Exopolysaccharides

Exopolysaccharides (EPS) are high molecular weight polymers secreted by microorganisms into the surrounding environment. They can be found associated with the cell membrane (capsular EPS) or as dispersed slime with no association to any cell acting like an adhesin favoring interactions among microorganisms (Sutherland 1982). In this way, EPS create micro-environments where the transfer of genes and metabolites is habitual and provide a way for microorganisms to ensure their survival in nutrient-starved environments (Sutherland 2001). Another significant role concerns the protective function they provide against probable predators and low or high temperature and salinity (Nichols et al. 2005, Chug and Mathur 2013). Their composition greatly varies: EPS can be either homo- or heteropolysaccharides and may contain a number of different inorganic or organic substituents such as sulfate, phosphate, pyruvic acid, and acetic acid. They are gaining increasing interest because they can have a wide range of potential applications in many industrial fields. Thanks to their properties, these biopolymers can be used in cosmetic, food, pharmaceutical, and environmental sectors where emulsifying, viscosifying, and chelating capacities are required (Table 1) (Freitas et al. 2011, Barcelos et al. 2019). For example, the gellan gum, one of the most well-known microbial EPS, is used as a thickener and stabilizer in the food industry since 1990 after FDA approval for its high viscosity and thermal stability (Pszczola 1993). Its repeating unit consists of a tetrasaccharide with two D-glucose moieties, one L-rhamnose, and one D-glucuronic acid. The hyaluronic acid is a polysaccharide, discovered in 1934, and provided with high water retention ability and biocompatibility. Currently, it is produced by recombinant bacteria and it is mainly used in the cosmetic industry (Moscovici 2015). The hyaluronic acid a linear heteropolysaccharide made of units of D-glucuronic acid and D-N-acetyl-glucosamine.

TABLE 1: Most Known Polysaccharides

Product	Source	Composition	Properties and Applications	Reference
Alginate	Brown algae (*Phaeophyceae*)	D-mannuronate, L-guluronate	Thickening, gelling, and stabilizing agent in food and biomedicine	Quin et al. 2018 Lee and Mooney 2012
Gellan gum	*Pseudomonas elodea*	D-glucose, L-rhamnose, D-glucuronic acid	Thickener and stabilizer in the food industry, gelling agent in media for the growth of thermophiles	Banik et al. 2000 Lin and Casida 1984
Hyaluronic acid	Rooster comb *Streptococcus*	N-acetyl D-glucosamine, D-glucuronic acid	Moisturizing and viscoelastic agent in biomedicine and cosmetics	Sze et al. 2016 Fakhari and Berkland 2013
Levan	*Aerobacter levanicum*, *Erwinia herbicola*, etc.	D-Fructose	Thickener in the food industry, prebiotic, emollient, and anti-inflammatory in medicine	Srikanth et al. 2015
Mauran	*Halomonas maura*	D-glucose, D-mannose, D-galactose, D-glucuronic acid	Viscoelastic and thixotropic agent, chelator of heavy metals	Arias et al. 2003
Xanthan gum	*Xanthomonas campestris*	D-glucose, D-mannose, D-glucuronic acid	Gelling agent, emulsifier and stabilizer in both food, and non-food industries	Becker et al. 1998

The antioxidant power of the EPS has also been pointed out by several authors (Priyanka et al. 2014, Sun et al. 2015), and those having antioxidant activity could be used as food supplements or as protective agents in cosmetic and pharmaceutical fields. Microbial EPS are a promising material for drug delivery systems and tissue regeneration (Costa Rui et al. 2013). Furthermore, many of them have received great attention from the medical sector as anticancer, antiviral, and immune-stimulatory agents (Gugliandolo et al. 2013, Selim et al. 2018).

EPS are produced by halophiles belonging to the three domains of life (Table 2). Among the Eukaryotes, the microalga *Dunaliella salina* has the unique property of growing in a wide range of salt concentration from 0.05 to 0.5 M (NaCl saturation). This capacity seems to be linked to the number of EPS produced, which increases at increasing salt concentration: 56 mg EPS/L were obtained in 0.5 M NaCl-containing medium while 944 mg EPS/L were produced in 5.0 M NaCl. The EPS showed high emulsifying power (66.4%-85.8%) giving a stable emulsion with hexadecane (Mishra and Jha 2009).

The commonest EPS-producing halophilic microorganisms in the bacteria domain are those belonging to the genus *Halomonas*. In 1998, Béjar et al. described the production of a highly sulfatesulfated EPS from several strains of *Halomonas eurihalina*. A total of 18 strains isolated from saline soil in Alicante (Southern Spain) were grown in complex and minimal media. The maximum EPS production was reached in the complex media with yields ranging from 0.9 g/L (strains H214 and H217) to 1.6 g/L (strain H212). The high sulfate content makes this EPS interesting from an application point of view. The sulfated EPS resemble eukaryotic polymers, such as chondroitin sulfate and heparan sulfate, and hence they are of great potential interest for biomedical applications. They have a number of bioactive properties such as anticoagulant, anticancer, antiviral, and wound healing *in vitro* (Gross et al. 2005, Yamada and Sugahara 2008).

Halomonas sp. AAD6 (JCM 15723) strain, a halophilic bacterium isolated from soil samples in Turkey, produced a high amount of an exopolysaccharide composed of D-fructofuranosyl residues (levan) (Poli et al. 2009). The EPS yield reached 1.07 and 1.84 g/L when the microorganism was grown in flask and bioreactor, respectively. The numerous potential applications of levan in food, cosmetics, and pharmaceutical industries search for new levan-producer microorganisms of interest. It could be used as a sweetener, emulsifier, thickener, cryoprotector, and as a carrier for flavors. Moreover, it is used in medicine as a plasma substitute, and antihyperlipidemic and anticancer properties have been shown (Yoo et al. 2004, Ates and Oner 2018).

The anticancer activity is associated with many EPS biosynthesized by microalgae and prokaryotes, and the genus *Streptomyces* seems to produce EPS with promising anticancer activity. The moderate halophile *Streptomyces carpaticus*, isolated from marine sediments in Egypt, produced a highly sulfated (21.75%) EPS showing anticancer activity against breast (MCF-7) and colon (HCT-116) tumor cell lines (51.7% and 59.1% cytotoxicity, respectively). It was composed of many different sugars: D-glucose, D-xylose, D-galactose, D-mannose, D-fructose, and D-galacturonic acid in the ratio of 3:1:1:2:2:1 (Selim et al. 2018).

The antioxidant power is an additional feature of EPS that makes them highly attractive in several sectors as an active ingredient. In 2014, Sun et al. described a novel EPS with antioxidant activity isolated from the deep-sea bacterium *Zunongwangia profunda* SM-A87.

Its production was optimized through the development of economical medium containing 60.9% whey, 10 g/L soybean meal, and 2.9% NaCl; under fed-batch, fermentation conditions the yield reached 17.2 g/L compared to 8.9 g/L of the previous trials. The EPS exhibited a good ability to scavenge the following free radicals: $DPPH^{\bullet}$, $^{\bullet}OH$, and $O_2^{-\bullet}$. The antioxidant power was concentration-dependent, and at maximum value tested (10.0 mg/mL), the radical scavenging activity was 48.5, 58.7, and 27.2% for $DPPH^{\bullet}$, $^{\bullet}OH$, and $O_2^{-\bullet}$, respectively.

TABLE 2: EPS From Halophiles

	Source	EPS production (g/L)	Composition	MW (kDa)	Properties	Reference
EUKARYOTES	*Dunaliella salina*	0.056 g/L in 0.5 M NaCl; 0.944 g/L in 5 M NaCl	D-galactose, D-glucose, D-xylose, D-fructose	n.d.	Emulsifying activity	Mishra and Jha 2009
PROKARYOTES						
Bacteria	*Alteromonas hispanica* F32[T]	1.0 g/L	D-glucose, D-mannose, L-rhamnose, D-xylose	19,000.0	Emulsifying activity, Pb[++] binding capacity	Mata et al. 2008
	Halomonas sp. AAD6 (JCM 15723) strain	1.84 g/L	D-Fructose	1,000.0	Anti-cytotoxic agent	Poli et al. 2009
	Halomonas almeriensis strain M8[T]	1.7 g/L	D-mannose, D-glucose, L-rhamnose, sulfate groups; D-mannose, D-glucose, sulfate groups	6,300.0; 15.0	Pseudoplastic character, emulsifying activity, Cu[++], Pb[++], and Co[++] binding capacity	Llamas et al. 2012
	Halomonas anticariensis strain FP35[T]	0.296 g/L	D-glucose, D-mannose, D-galacturonic acid	20.0	Emulsifying activity, Cu[++], and Pb[++] binding capacity	Mata et al. 2006
	Halomonas anticariensis strain FP36	0.5 g/L	D-glucose, D-mannose, D-galacturonic acid	46.0	Emulsifying activity, Cu[++], and Pb[++] binding capacity	Mata et al. 2006
	Halomonas eurihalina	0.9-1.6 g/L	D-glucose, D-mannose, L-rhamnose, uronic acids, hexosamines, sulfate groups	n.d.	Gelification at low pH	Béjar et al. 1998
	Halomonas nitroreducens strain WB1	about 1.25 g/L	D-glucose, D-mannose, D-galactose; D-glucose, D-mannose, L-rhamnose, L-arabinose; D-glucose, D-mannose, D-galacturonic acid, D-galactose	5,200.0; 30.0; 1.3	Pseudoplastic character, emulsifying activity, Pb[++] binding capacity, antioxidant activity	Chikkanna et al. 2018
	Halomonas ventosae strain A112[T]	0.284 g/L	D-glucose, D-mannose, D-galactose	53.0	Emulsifying activity, Pb[++] binding capacity	Mata et al. 2006
	Halomonas ventosae strain A116	0.290 g/l	D-glucose, D-mannose, D-galactose	52.0	Emulsifying activity, Cu[++], and Pb[++] binding capacity	Mata et al. 2006

TABLE 2: Contd.....

TABLE 2: Contd.

Source	EPS production (g/L)	Composition	MW (kDa)	Properties	Reference
Idiomarina fontislapidosi F23[T]	1.45 g/L	EPS-1 D-glucose, D-mannose, D-galactose, traces of D-xylose EPS-2 D-glucose, D-mannose, D-galactose, traces of D-xylose	EPS-1 1500.0 EPS-2 15.0	Emulsifying activity, Cu^{++}, and Pb^{++} binding capacity Emulsifying activity, Cu^{++}, and Pb^{++} binding capacity	Mata et al. 2008
Idiomarina ramblicola R22[T]	1.50 g/L	EPS-1 D-glucose, D-mannose, L-rhamnose EPS-2 D-glucose, D-mannose, D-galacturonic acid, traces of D-xylose and L-rhamnose	EPS-1 550.0 EPS-2 20.0	Emulsifying activity, Cu^{++}, and Pb^{++} binding capacity Emulsifying activity, Cu^{++}, and Pb^{++} binding capacity	Mata et al. 2008
Polaribacter sp. SM1127	2.11 g/L	D-mannose, N-acetyl glucosamine, D-glucuronic acid, D-galactose, L-fucose, and traces of D-glucose and L-rhamnose	220.0	High viscosity, high tolerance to high salinity and a wide pH range, moisture-retention ability, antioxidant activity	Sun et al. 2015
Salipiger mucosus strain A3[T]	1.35 g/L	D-glucose, D-mannose, D-galactose, L-fucose	250.0	Pseudoplastic character, emulsifying activity	Llamas et al. 2010
Streptomyces carpaticus	7.45 g/L	D-glucose, D-xylose, D-galactose, D-mannose, D-galactouronic acid, D-fructose	118.0	Anticancer activity, antioxidant activity	Selim et al. 2018
Zunongwangia profunda SM-A87	17.2 g/L	D-glucose, D-mannose, D-galactose, L-fucose, D-xylose, D-glucuronic acid, unidentified sugar	3,760.0	Antioxidant activity, moisture-retention ability, high viscosity, good thermal and pH stability, salt resistance	Li et al. 2011 Sun et al. 2014

TABLE 2: Contd....

TABLE 2: Contd.

	Source	EPS production (g/L)	Composition	MW (kDa)	Properties	Reference
Archaea	*Haloarcula hispanica* ATCC33960	0.03 g/L	D-mannose, D-galactose, traces of D-glucose, sulfate groups	1,100.0	n.d.	Lü et al. 2017
	Haloarcula japonica strain T5	0.37 g/L	D-galactose, D-mannose, D-glucuronic acid	n.d.	n.d.	Nicolaus et al. 1999
	Haloarcula sp. strain T6	0.045 g/L	D-glucose, D-mannose, D-galactose	n.d.	n.d.	Nicolaus et al. 1999
	Haloarcula sp. strain T7	0.035 g/L	D-glucose, D-mannose, D-galactose	n.d.	n.d.	Nicolaus et al. 1999
	Haloferax mediterranei (ATCC 33500)	1.0 g/L	D-glucose, D-mannose, D-galactose, hexosamines, uronic acid, sulfate groups	< 100.0	High viscosity, tolerance to high salinity, temperature, and pH	Antón et al. 1988
	Haloterrigena turkmanica	0.21 g/L	D-glucose, D-galactose, D-glucosamine, D-galactosamine, D-glucuronic acid, sulfate groups	801.7 206.0	Antioxidant activity, moisture-absorption/retention ability, emulsifying activity	Squillaci et al. 2016

n.d.: not determined

The recently isolated *Halomonas nitroreducens* strain WB1 secreted an EPS with antioxidant activity comparable to that of the ascorbic acid, used as a comparison compound when assayed at concentrations between 0.1-5 mg/mL for hydroxyl and DPPH radical scavenging activities. The EPS was composed of three forms (A, B, and C) with very different molecular weights (Table 2). It showed a high binding affinity for lead in heavy metals mixture (182 mg/g EPS), and this capacity increased in the presence of lead alone (263 mg/g EPS). The emulsifying power was higher than Tween 80 and Triton X-100 with all hydrophobic compounds tested, and the highest and lowest emulsifying activities were shown with sunflower oil (85%) and hexadecane (56.3%), respectively (Chikkanna et al. 2018).

Marine bacteria, populating the Arctic and Antarctic areas, synthesize microbial EPS to survive the harsh environmental conditions. For example, the EPS from the Antarctic bacterium *Pseudoalteromonas arctica* KOPRI 21653 and the Antarctic fungus *Phoma herbarum* CCFEE 5080 have cryoprotective effects on the cells of these microorganisms (Selbmann et al. 2002, Kim and Yim 2007). Recently, Sun et al. (2015) characterized and established the biotechnological properties of a novel EPS synthesized by the Arctic marine bacterium *Polaribacter* sp. SM1127. It has an anionic nature with a high content of glucuronic acid (21.4%). It shows significant application properties, such as moisture-retention capacity and antioxidant power, that make the novel EPS interesting for applications as an active ingredient in the cosmetics field. The moisture-absorption/retention ability was compared to that of some commonly used moisturizing agents, namely hyaluronic acid, sodium alginate, and chitosan. The EPS exhibited a moisture-absorption ability intermediate among all the moisturizing agents tested but showed the highest moisture-retention ability. The antioxidant power was assayed as radical scavenging ability toward DPPH$^\bullet$, $^\bullet$OH, and $O_2^{-\bullet}$. The EPS (10 mg/mL) showed good activity with values of 55.4%, 52.1%, and 28.2%, respectively.

The majority of halophilic microorganisms that can synthesize EPS are Eubacteria, both moderate and extreme, but members belonging to extreme halophilic Archaea (Haloarchaea) also excrete them. Amounts are excreted by microorganisms of the halophilic genera *Haloarcula*, *Halococcus*, *Haloferax,* and *Halobacterium*. Antón et al. (1988) firstly reported on the production of EPS by a haloarchaeon. *Haloferax mediterranei* (ATCC 33500) was capable of producing an extracellular polymer that gave a typical mucous character to the colonies and was responsible for the appearance of a superficial layer in an unshaken liquid medium. Representatives of the genus *Haloarcula* were isolated from a Tunisian marine saltern close to Monasti by Nicolaus et al. (1999). In particular, the T5 isolate was classified as a new strain of *Haloarcula japonica* by DNA-DNA hybridization. The three described strains (T5, T6, and T7) biosynthesized EPS with chemical composition reported in Table 2. *Har. japonica* strain T5 gave the highest EPS yield with 0.37 g/L culture medium when glucose was used as a carbon source. Recently, sulfated EPS were isolated from the extreme halophiles *Haloterrigena turkmenica* and *Haloarcula hispanica* ATCC 33960, respectively (Squillaci et al. 2016, Lü et al. 2017). The ability of the haloarchaeon *Htg. turkmenica* to produce EPS was firstly assessed by observing the cells in unshaken cultures supplemented with glucose. The EPS was highly sulfated and composed of two main fractions of 801.7 and 206.0 kDa. It exhibited a significant emulsifying activity and a moisture-retention capacity higher than the hyaluronic acid used as a comparison. The biopolymer also showed antioxidant activity when assayed by using three different methods, namely, radical scavenging activity, total antioxidant capacity, and ferric reducing antioxidant power. The acidic EPS from *Har. Hispanica* ATCC 33960 was produced at a very low yield (0.03 mg/L). It showed a high percentage of sulfate groups (26%) responsible for the acidic character of the polymer, and its molecular weight was 1,100.0 kDa.

Polyhydroxyalkanoates

Polyhydroxyalkanoates (PHA) are very attractive macromolecules from a biotechnological point of view. They are renewable biopolymers synthesized by several species of Bacteria and Archaea and are accumulated inside the cells during the stationary phase of growth under peculiar growth conditions. Usually, they are produced when an excess of the carbon source occurs and when an essential nutrient such as nitrogen or phosphorous is depleted, thus serving as an internal reserve of carbon and energy that is depolymerized to hydroxybutyric acid and metabolized to acetoacetate and acetoacetyl-CoA when required (Poli et al. 2011). From a chemical point of view, PHA are polyesters, and the most extensively studied is composed of 3-hydroxybutyric acid (poly 3-hydroxybutyrate, PHB). These biopolymers can also be co-polymers of different organic acids with different carbon chain lengths, such as the co-polymer poly 3-hydroxybutyrate-co-3-hydroxyvalerate (PHBV), which shows better mechanical properties than PHB. According to the side chain length, PHA can be divided into short side chains and medium side chains PHA (ssc-PHA and msc-PHA, respectively).

PHA possesses interesting features similar to those of synthetic plastics, such as thermoplastic and elastomeric properties, but contrarily to conventional plastics, they are biodegradable. Due to their significant properties, studies about PHA production and applications are continuously increasing, and the research for microorganisms able to produce this very useful compound is very active.

Several halophilic bacteria can produce PHA, and the main producers belong to the genus *Halomonas* as shown in Table 3. The moderate halophilic bacterium *Halomonas boliviensis* LC1, isolated from a soil sample near the lake Laguna Colorada (Bolivia), produced PHB (58.8% of the cell dry weight) when grown in hydrolyzed starch as carbon source and under limited nitrogen amount (Quillaguamán et al. 2005). A higher PHB production was achieved in fed-batch culture using a defined growth medium containing glucose as a carbon source (81% of the cell dry weight) (Quillaguamán et al. 2008). PHA-producing halophilic bacteria, different from *Halomonas*, were isolated from a mangrove forest in Northern Vietnam. The five strains, QN271, QN187, ND240, ND199, and ND218, were found to be phylogenetically related to the proteobacterium *Yangia pacifica*, whereas the strains QN194, ND153, and ND97 belong to *Bacillus* species. These microorganisms were able to accumulate great amounts of PHA in the vacuoles inside the cells with values ranging from 24% to 65% for ND218 and ND153 strains, respectively, when glucose was added as the carbon source to the culture medium. PHA yield varied from 11% to 71% for QN195 and ND153, respectively, when a mixture of glucose and propionate was added to the medium. Interestingly, the addition of propionate changed the PHA composition from PHB to the co-polymer PHBV, with the only exception of the strain QN194, which continued to produce PHB even if in less amount (Van-Thuoc et al. 2012).

Fernandez-Castillo and co-workers (1986) were among the first to describe the production of PHA by microorganisms belonging to the Archaea domain. *Halobacterium mediterranei* and *Halobacterium volcanii* were able to accumulate PHB under a limited amount of nitrogen and a high amount of carbon source in the culture medium. Salt concentration was found to affect the biopolymer accumulation as well. PHB increased inside the cells when the salt concentration decreased. In particular, *Hfx. mediterranei* accumulated 38% and 19% PHB at 15% and 30% salt concentration, respectively, whereas *Haloferax volcanii* produced 7% PHB at 25% salt concentration. Among the extremely halophilic Archaea, *Hfx. mediterranei* is undoubtedly the most studied PHA producer. Lillo and Rodriguez-Valera (1990) demonstrated that phosphate limitation was essential for obtaining high accumulation of the biopolymer inside the cells, while nitrogen depletion gave the opposite result. Moreover, different carbon sources were tested in order to select the best one providing the highest PHA yield. Starch (2% w/v) was the best carbon source (6.48 g PHB/L) followed by glucose (2% w/v) with 4.16 g PHB/L. A further increase in glucose in the culture medium led to a decrease in PHB production (Table 3).

An interesting work of Cui et al. (2017) evaluates the effect of the NaCl concentration on EPS and PHA production. The authors demonstrated that it was possible to direct the biosynthesis toward the excreted polysaccharide or to the PHA by changing the salt concentration into the culture medium. The increase of NaCl favored PHA accumulation, reaching a maximum of 71% when the salt concentration was 250 g/L. Conversely, the production of EPS was 371.4 and 319.7 mg/g cell dry weight at 75 and 250 g/L, respectively. Don et al. (2006) reported on the ability of *Hfx. mediterranei* to synthesize PHBV (previously characterized as PHB). Using yeast extract (7.5 g/L) as nitrogen source and glucose (1 g/L) as a carbon source, the microorganism accumulated under fed-batch fermentation 48.6% PHBV inside the cells. The purification and fractionation of the biopolymer revealed that it was composed of two different forms: fraction P1, with a molecular weight of 569.5 kDa, represented the 93.4% of the total polymer and fraction P2, with a lower molecular weight (78.2 kDa), was the 6.6% of the polymer.

TABLE 3: PHA From Halophiles

	Source	PHA Production (% w/w of CDW)	Composition	Reference
PROKARYOTES				
Bacteria	*Halomonas boliviensis* LC1	81.0	PHB	Quillaguamán et al. 2008
	Halomonas elongata BK AG 18	21.4	PHB	Hertadi et al. 2017
	Halomonas halophile	82.0	PHB	Kucera et al. 2018
	Halomonas nitroreducens	33.0	PHB	Cervantes-Uc et al. 2014
	Halophile, strain ND97	[a]48.0/[b]53.0	[a]PHB/[b]PHBV	Van-Thuoc et al. 2012
	Halophile, strain ND153	[a]65.0/[b]71.0	[a]PHB/[b]PHBV	Van-Thuoc et al. 2012
	Halophile, strain ND199	[a]34.0/[b]12.0	[a]PHBV/[b]PHBV	Van-Thuoc et al. 2012
	Halophile, strain ND218	[a]24.0/[b]11.0	[a]PHBV/[b]PHBV	Van-Thuoc et al. 2012
	Halophile, strain ND240	[a]28.0/[b]12.0	[a]PHB/[b]PHBV	Van-Thuoc et al. 2012
	Halophile, strain QN187	[a]44.0/[b]27.0	[a]PHB/[b]PHBV	Van-Thuoc et al. 2012
	Halophile, strain QN194	[a]26.0/[b]11.0	[a]PHB/[b]PHB	Van-Thuoc et al. 2012
	Halophile, strain QN271	[a]48.0/[b]31.0	[a]PHB/[b]PHBV	Van-Thuoc et al. 2012
Archaea	*Haloarcula* sp. IRU1	63.0	PHB	Taran and Amirkhani 2010
	Haloarcula marismortui	21.0	PHB	Han et al. 2007
	Haloarcula marismortui	30.0	PHB	Pramanik et al. 2012
	Haloferax mediterranei	[c]6.48 (g/L)	PHB	Lillo and Rodriguez-Valera 1990
		[d]3.09 (g/L)	PHB	Lillo and Rodriguez-Valera 1990
		[e]4.16 (g/L)	PHB	Lillo and Rodriguez-Valera 1990
		[f]3.28 (g/L)	PHB	Lillo and Rodriguez-Valera 1990
		[g]3.52 (g/L)	PHB	Lillo and Rodriguez-Valera 1990
	Haloferax mediterranei	48.6	PHBV	Don et al. 2006
	Halogeometricum borinquense strain TN9	14.0	PHB	Salgaonkar et al. 2013
	Halogeometricum borinquense strain E3	73.5	PHBV	Salgaonkar and Bragança 2015
	Halopiger Aswanensis	53.0	PHB	Hezayen et al. 2000
	Haloterrigena turkmenica	n.d.	PHB	Squillaci et al. 2017a

CDW: cell dry weight; PHB: poly (3-hydroxybutyrate); PHBV: poly(3-hydroxybutyrate-co-3-hydroxyvalerate); [a]glucose as carbon source; [b]glucose/propionate as carbon source; [c]starch (2% w/v); [d]glucose (1% w/v); [e]glucose (2% w/v); [f]glucose (5% w/v); [g]glucose (10% w/v); n.d.: not determined

Salgaonkar and Braganca (2015) described the capacity of the extreme halophilic archaeon *Halogeometricum borinquense* strain E3, isolated from solar salterns of Marakkanam in Tamil Nadu (India), to synthesize PHA. This halophile accumulated a large quantity of the biopolymer inside the cells (73.51% of cell dry weight) when grown in the presence of glucose 2% (w/w) as the carbon source. ^{1}H NMR analysis established that the polymer accumulated was the co-polymer PHBV, constituted by 21.47% HV 3-hydroxyvalerate) units.

The extreme microorganism *Htg. turkmenica* is a halofilic archaeon isolated from a sulfate saline soil in Turkmenistan, requiring at least 2 M NaCl for growth. Recently, Squillaci et al. (2017a) demonstrated the capacity of *Htg. turkmenica* of accumulating PHB when grown in a nutrient deficient medium containing a reduced amount of yeast extract (1 g/L), glucose (10 g/L) as carbon source, and depletion of casamino acids. The presence of PHB inside the cells was assessed by optical microscopy with Nile Blue A, a dye that specifically bounds to poly-3-hydroxybutyrate (PHB). The biopolymer was accumulated in the cells characterized by a coccoid morphology, typical of stressed conditions, and mainly associated with the stationary phase of the microorganism.

In the last decades, PHA received great attention as potential material for medical uses, such as surgical or medical scaffolding material, due to their biocompatibility. The degradation leads to the production of 3-hydroxybutyrate, which is a natural metabolite known to be associated with ketone bodies formation. In addition, when PHB/PHBV-based materials are in contact with blood they do not affect the platelet response, while the sterilization process does not affect the biopolymer properties (Brigham and Sinskey 2012). Unfortunately, the diffusion of PHA uses is hampered by the high costs of production. In order to overcome this issue, solid and liquid agroindustrial wastes are considered as alternative and cheaper carbon sources. Here, a few examples are reported. *Hfx. Mediterranei* was grown in the presence of olive mill wastewaters (a residue generated from the olive oil production), and 25% pretreated vinasse (a residue from the ethanol industry), accumulating 43 and 70% PHBV inside the cells, respectively (Bhattacharyya et al. 2012, Alsafadi and Al-Mashaqbeh 2017). Sugarcane vinasse was used as a carbon source for the extremely halophilic archaeon *Haloarcula marismortui*. It produced 23% PHB when raw vinasse (10% w/w) was added to the culture medium. When the residue was subjected to pretreatment with activated charcoal, the biopolymer accumulation increased up to 30% (Pramanik et al. 2012).

Carotenoids

Carotenoids are pigments of natural origin found in a variety of microorganisms, plants, and animals. They originate from the biosynthetic pathway of terpenoids by a reaction between two geranylgeranyl diphosphate molecules that give a C_{40} skeleton from which other carotenoids are generated. They are usually used as colorants in the food industry to enhance the color in farmed salmons, such as astaxanthin and canthaxanthin (Kirti et al. 2014), or as nutraceuticals for pharmaceutical and cosmetic purposes (Anunciato and da Rocha Filho 2012). In addition, these pigments play key roles as antioxidant agents, light protectors, and cell membrane stabilizers. Several of them, like zeaxanthin and lutein, act as eye protectors against cataracts and macular degeneration (Jaswir et al. 2011). The high number of conjugated double bonds allows acting as a radical scavenger, thus protecting them against oxidative damages caused by harmful molecules as reactive oxygen and nitrogen species (Miller et al. 1996). One of the methods to produce carotenoids is chemical synthesis. Although this method gives highly pure molecules, the process is complex and time consuming for some of them. Furthermore, the increasing demand for natural compounds in substitution of the largely employed synthetic ones opens the way to the research of novel sources for the production of pigments. In this context, carotenoid-producing microorganisms represent a great opportunity to exploit. Among halophiles, *D. salina* represents the better known carotenoid-producing microorganism that is widely studied for beta-carotene natural production (Table 4). Moreover, it is the only halophilic organism at present used for the production of carotenoids in the industry. Additional pigments produced by *D. salina* are lycopene, phytoene, and some derivatives of bacterioruberin (Hosseini and Shariati 2009). Carotenoids usually consist of a C_{40} hydrocarbon backbone; however, the carotenoids produced by Haloarchaea are C_{50} carotenoid with bacterioruberin as the main representative (Figure 1).

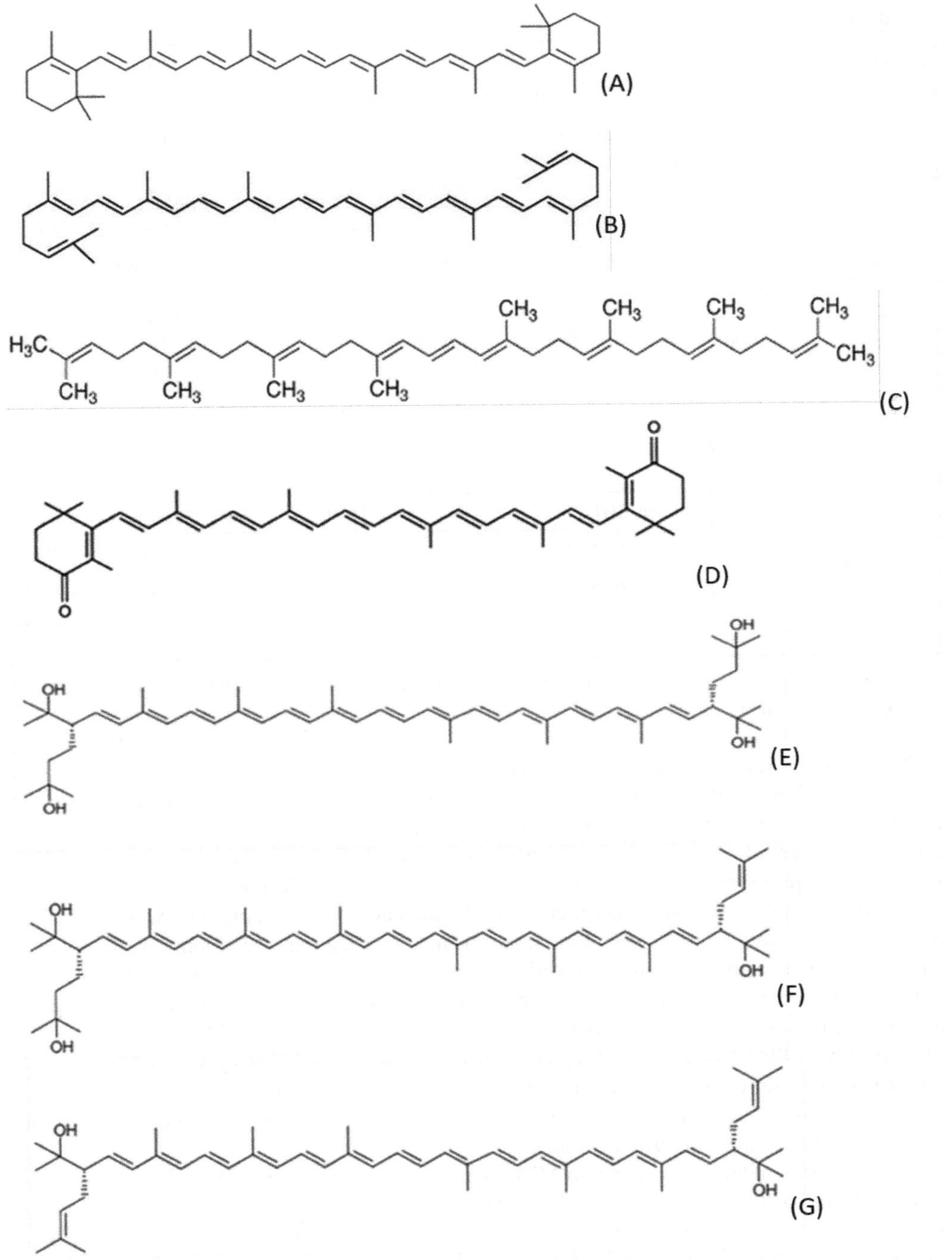

Figure 1. Chemical structures of beta-carotene (A), Lycopene (B), *trans*-Phytoene (C), Canthaxanthin (D), all-*trans*-bacterioruberin (E), all-*trans*-Monoanhydrobacterioruberin (F), all-*trans*-bisanhydrobacterioruberin (G).

TABLE 4: Carotenoids From Halophiles

	Source	Type	Reference
EUKARYOTES	*Dunaliella salina*	Lycopene, beta-carotene, phytoene, BR derivatives	Hosseini and Shariati 2009
PROKARYOTES			
Bacteria	*Aquisalibacillus elongates MB592*	BR derivatives	Fariq et al. 2019
	Halomonas aquamarina MB598	BR derivatives	Fariq et al. 2019
	Micrococcus yunnanensis strain AOY-1	sarcinaxanthin, sarcinaxanthin monoglucoside, sarcinaxanthin diglucoside	Osawa et al. 2010
	Salinicoccus sesuvii MB597	BR derivatives	Fariq et al. 2019
Archaea	*Haloarcula japonica*	BR, MABR, BABR	Yatsunami et al. 2014
	Halobacterium cutirubrum	BR, MABR, BABR	Kushwaha et al. 1975
	Halobacterium salinarium	BR, BABR, TABR	Mandelli et al. 2012
	Halococcus morrhuae	BR, BABR, TABR	Mandelli et al. 2012
	Haloferax mediterranei	BR, MABR, BABR	Fang et al. 2010
	Haloquadratum walsbyi	BR, beta-carotene, squalene, menaquinone MK-8	Lobasso et al. 2008
	Halorubrum sp. SH1	BR, BABR, TABR	De la Vega et al. 2016
	Halorubrum sp. TBZ126	BR, beta-carotene, lycopene	Naziri et al. 2014
	Haloterrigena turkmenica	BR, MABR, BABR, lycopene, phytoene, lycopersene, squalene, canthaxanthin, menaquinone MK-8	Squillaci et al. 2017b

BR: bacterioruberin; MABR: monoanhydrobacterioruberin; BABR: bisanhydrobacterioruberin; TABR, trisanhydrobacterioruberin

These microorganisms have colors varying from yellow-orange to pink-red as a consequence of the kind of pigments they synthesize. Halophilic carotenoid-producing bacteria were recently isolated from a soil sample in the hypersaline environment of Khewra Salt Range in Pakistan. The three strains were identified as *Aquisalibacillus elongatus* MB592, *Salinicoccus sesuvii* MB597, and *Halomonas aquamarina* MB598 (Fariq et al. 2019). Due to their capacity of growing optimally at 10% and 12% NaCl, they were classified as moderate halophiles. The pigment yield was comprised of 15.7 µg/mg. wet biomass (*H. aquamarina*) and 27.3 µg/mg. wet biomass (*S. sesuvii*). The characterization of the pigments allowed establishing that they were bacterioruberin derivatives, but the authors carried out no specific identification of the single molecules. The carotenoids from all three strains exhibited high antioxidant activity; in particular, 85% radical scavenging effect was measured by DPPH assay with 30 µg/mL. carotenoid extracts. Interestingly, the extracts showed noticeable antimicrobial effect against bacteria (*Enterococcus faecium, Enterococcusfaecalis, Bacillus cereus* and *Klebsiella pneumoniae* are some examples) and Fungi (*Aspergillus fumigatus, Aspergillus flavus, and Fusarium solani*).

The moderate halophilic bacterium *Micrococcus yunnanensis* strain AOY-1, a yellow bacterium isolated from hard coral growing on Akajima Island (Japan) by Osawa et al. (2010) was found to synthesize the rare C_{50} carotenoids sarcinaxanthin (S), sarcinaxanthin monoglucoside (SMG), and sarcinaxanthin diglucoside (SDG) provided with antioxidant activity. The $^1O_2$-quenching activity of the pigments (IC_{50}: 57, 54 and 74 µM for S, SMG, and SDG, respectively) was lower than the pigment astaxanthin (8.9 μM) but higher than beta-carotene (> 100 μM) chosen as reference antioxidant compounds.

The influence of the nutrients on the carotenoid production was studied in the halophilic archaeon *Hfx. mediterranei* by two-stage cultivation (Fang et al. 2010). The microorganism was grown in its usual *Halobacteria* medium in the first-stage cultivation. Then, the biomass was transferred in a new medium for the second-stage cultivation where the effects of sodium chloride and magnesium sulfate on the carotenoid production were studied. The highest yield was reached with 5% sodium chloride and 8% magnesium sulfate. Bacterioruberin, monoanhydrobacterioruberin, and bisanhydrobacterioruberin were the main pigments synthesized. As above stated, C_{50} carotenoids are the main pigments produced by Haloarchaea. In fact, those extracted from *Har. japonica* were composed of 68.1% bacterioruberin, 22.5% monoanhydrobacterioruberin, and 9.3% bisanhydrobacterioruberin. Additional carotenoids were isopentenyldehydrorhodopin, lycopene, and phytoene but in smaller quantities. The pigment yield was high in comparison to the usual amount produced by halophilic Archaea: 335 μg/g of dry biomass and the carotenoid extract exhibited an antioxidant power higher than beta-carotene (Yatsunami et al. 2014). Several *cis*-isomers of the bacterioruberin derivatives were identified in the Haloarchaea *Halococcus morrhuae* and *Halobacterium salinarium*, attesting that these pigments can have several forms due to the high numbers of conjugated double bonds. In addition to the usual all-*trans* form of bacterioruberin, bisanhydrobacterioruberin and trishanydrobacterioruberin, some of their *cis*-isomers were identified for a total of nine different compounds. The number of pigments produced corresponded to 89 μg/g of dry biomass and 45 μg/g of dry biomass for *Hcc. morrhuae* and *Hbt. Salinarium*, respectively. The antioxidant power of the archaeal carotenoids was compared to that of carotenoid extracted from the thermophilic carotenoid-producing bacterium *Thermus filiformis*. The Trolox Equivalent Antioxidant Capacity of the halophilic extracts was higher than that measured with the extract from the thermophilic bacterium which contained different kinds of pigments, such as zeaxanthin, thermozeaxanthin, and their derivatives (5.07 and 5.28 for the halophilic Archaea and 2.87 for *T. filiformis*, respectively). This diverse behavior seems to depend on the different chemical structures of the pigments, as interestingly reported in the work of Mandelli et al. (2012).

The extreme halophile *Htg. turkmenica* synthesized C_{50} carotenoids as main pigments, but C_{30}, C_{40}, and C_{51} carotenoids were also detected. Bacterioruberin, monoanhydrobacterioruberin, and bisanhydrobacterioruberin were detected together with the less represented lycopene, phytoene, lycopersene, canthaxanthin, menaquinone MK-8, and squalene. By means of HPLC/UV–Vis and APCI–ITMS analyses, a total of 30 carotenoids and seven geometric isomers for each bacterioruberin derivative were identified for the first time in a haloarchaeon. The total amount of carotenoids produced was 74.5 μg/g of dry biomass. They showed an antioxidant power, measured as Radical Scavenging Activity and Ferric Reducing Antioxidant Power, higher than alpha-tocopherol, butylhydroxytoluene, and ascorbic acid used as reference compounds (Squillaci et al. 2017b).

Several studies have evidenced that the cancer onset is inversely related to the carotenoids intake. Compounds such as alpha-carotene, beta-carotene, lutein, and lycopene are examples of molecules provided with anticancer activity (Nishino et al. 2000). Pigments from halophilic microorganisms have attracted the attention of researchers and companies in order to find novel anticancer agents as natural alternatives to the drugs currently used. Carotenoids from many halophiles have been tested with positive results against human cancer cells. The production of pigments from the haloarchaeon *Hfx. volcanii* was optimized by testing five different media. The best medium yielded 4 g/L carotenoids after extraction with *n*-hexane, and the resulting extract was tested against the human cancer cell line HepG2. After treating HepG2 cells with 25 μg, 50 μg, and 100 μg of carotenoids, the percentage of viability was 72.8, 60.8, and 46.5, respectively, indicating a dose-dependent cytotoxic effect (Sikkandar et al. 2013). A carotenoid extract from *Halobacterium halobium* was tested on HepG2 cells. Increasing concentrations of the extract led to a decrease of the cell viability up to 50% at 1.5 μM concentration (Abbes et al. 2013). Recently, the antihemolytic effect of carotenoids from the following halophilic Archaea: *Halogeometricum rufum, Halogeometricum limi, Haladaptatus litoreus, Haloplanus vescus, Halopelagius inordinatus, Halogranum rubrum,* and *Hfx. volcanii* has been described for the first time. They exhibited a significant protective effect against H_2O_2-induced hemolysis of mouse erythrocytes. The haemolysis inhibition rate ranged from 23.4% to 34.6% and was four to six times higher than beta-carotene. Furthermore, the carotenoid extracts from *Hgm. Limi* and *Hpn. Vescus* showed antiproliferative activity against HepG2 cells as well (Hou and Cui 2018).

Conclusions

Halophilic microorganisms possess high potential in the biotechnological field because they are able to produce several valuable compounds that can find utilization in various sectors. Although their products are considered interesting candidates for industrial applications, the yields are still insufficient for a wide utilization.

Studies aiming to obtain a higher production through optimization of the culture conditions and/ or downstream processing are required.

References

Abbes, M., H. Baati, S. Guermazi, C. Messina, A. Santulli, N. Gharsallah, et al. 2013. Biological properties of carotenoids extracted from *Halobacterium halobium* isolated from a Tunisian solar saltern. BMC Complement. Altern. Med. 13: 255.

Al-Abri, K. 2011. Use of Molecular Approaches to Study the Occurrence of Extremophiles, and Extremodures in Non-extreme Environments. Ph.D. Thesis, University of Sheffield, Sheffield, UK.

Alsafadi, D. and O. Al-Mashaqbeh. 2017. A one-stage cultivation process for the production of poly-3-(hydroxybutyrate-cohydroxyvalerate) from olive mill wastewater by *Haloferax mediterranei*. N. Biotechnol. 34: 47–53.

Antón, J., I. Meseguer and F. Rodríguez-Valera. 1988. Production of an extracellular polysaccharide by *Haloferax mediterranei*. Appl. Environ. Microbiol. 54: 2381–2386.

Antón, J., A. Oren, S. Benlloch, F. Rodriguez-Valera, R. Amann and R. Rosselló-Mora. 2002. *Salinibacter ruber* gen. nov., sp. nov., a novel, extremely halophilic member of the bacteria from saltern crystallizer ponds. Int. J. Syst. Evol. Microbiol. 52: 485–491.

Anunciato, T.P. and P.A. da Rocha Filho 2012. Carotenoids and polyphenols in nutricosmetics, nutraceuticals, and cosmeceuticals. J. Cosmet. Dermatol. 11: 51–54.

Arias, S., A. del Moral, M.R. Ferrer, R. Tallon, E. Quesada and V. Béjar. 2003. Mauran, an exopolysaccharide produced by the halophilic bacterium *Halomonas maura*, with a novel composition and interesting properties for biotechnology. Extremophiles. 7: 319–326.

Ateş, O. and E.T. Oner 2018. Microbial xanthan, levan, gellan and curdlan as food additives. pp. 149–174. *In*: V.K. Gupta, H. Treichel, V.O. Shapaval, L.A. de Oliveira and M.G. Tuohy [eds.]. Microbial Functional Foods and Nutraceuticals. John Wiley & Sons Ltd, West Sussex, UK.

Banik, R.M., B. Kanari and S.N. Upadhyay. 2000. Exopolysaccharide of the gellan family: prospects and potential. World J. Microbiol. Biotechnol. 16: 407–414.

Barcelos, M.C.S., K.A.C. Vespermann, F.M. Pelissari and G. Molina. 2019. Current status of biotechnological production and applications of microbial exopolysaccharides. Crit. Rev. Food Sci. Nutr. 11: 1–21.

Becker, A., F. Katzen, A. Pühler and L. Ielpi. 1998. Xanthan gum biosynthesis and application: a biochemical/genetic perspective. Appl. Microbiol. Biotechnol. 50: 145–152.

Bèjar, V., I. Llamas, C. Calvo and E. Quesada. 1998. Characterization of exopolysaccharides produced by 19 halophilic strains of the species *Halomonas eurihalina*. J. Biotechnol. 61: 135–141.

Bhattacharyya, A., A. Pramanik, S.K. Maji, S. Haldar, U.K. Mukhopadhyay and J. Mukherjee. 2012. Utilization of vinasse for production of poly-3-(hydroxybutyrate-co-hydroxyvalerate) by *Haloferax mediterranei*. AMB Express. 2: 34.

Boyadzhieva, I., I. Tomova, N. Radchenkova, M. Kambourova, A. Poli and E. Vasileva-Tonkova. 2018. Diversity of heterotrophic halophilic bacteria isolated from coastal solar salterns, Bulgaria and their ability to synthesize bioactive molecules with biotechnological impact. Microbiology 87: 519–528.

Brigham, C.J. and A.J. Sinskey 2012. Applications of polyhydroxyalkanoates in the medical industry. Int. J. Biotechnol. Wellness. Ind. 1: 53–60.

Butinar, L., P. Zalar, J.C. Frisvad and N. Gunde-Cimerman. 2005a. The genus Eurotium – members of indigenous fungal community in hypersaline waters of salterns. FEMS Microbiol. 51: 155–166.

Butinar, L., S. Sonjak, P. Zalar, A. Plemenitas and N. Gunde-Cimerman. 2005b. Melanized halophilic Fungi are eukaryotic members of microbial communities in hypersaline waters of solar salterns. Bot. Mar. 48: 73–79.

Cervantes-Uc, J.M., J. Catzin, I. Vargas, W. Herrera-Kao, F. Moguel, E. Ramirez, et al. 2014. Biosynthesis and characterization of polyhydroxyalkanoates produced by an extreme halophilic bacterium, *Halomonas nitroreducens*, isolated from hypersaline ponds. J. Appl. Microbiol. 117: 1056–1065.

Chikkanna, A., D. Ghosh and A. Kishore. 2018. Expression and characterization of a potential exopolysaccharide from a newly isolated halophilic thermotolerant bacteria *Halomonas nitroreducens* strain WB1. Peer J. 6: e4684.

Chug, R. and S. Mathur. 2013. Extracellular polymeric substances from cyanobacteria: characteristics, isolation and biotechnological applications: a review. I.J.A.S.E.A.T. 3: 49–53.

Costa Rui R., A.I. Neto, I. Calgeris, C.R. Correia, A.C.M. Pinho, J. Fonseca, et al. 2013. Adhesive nanostructured multilayer films using a bacterial exopolysaccharide for biomedical applications. J. Mater. Chem. B 1. 18: 2367–2374.

Cui, Y.W., X.Y. Gong, S. Yun-Peng and Z.D. Wang. 2017. Salinity effect on production of PHA and EPS by *Haloferax mediterranei*. RSC Adv. 7: 53587–53595.

Deepa, Y., S. Anuradha and N. Mathur. 2015. Halophiles: a review. Int. J. Curr. Microbiol. App. Sci. 4: 616–629.

De la Vega, M., A. Sayago, J. Ariza, A.G. Barneto and R. Leon. 2016. Characterization of a bacterioruberin-producing Haloarchaea isolated from the marshlands of the odiel river in the southwest of spain. Biotechnol. Prog. 32: 592–600.

Don, T.M., C.W. Chen and T.H. Chan. 2006. Preparation and characterization of poly (hydroxyalkanoate) from the fermentation of *Haloferax mediterranei*. J. Biomater. Sci. Polym. Ed. 17: 1425–1438.

Fakhari, A. and C. Berkland. 2013. Applications and emerging trends of hyaluronic acid in tissue engineering, as a dermal filler, and in osteoarthritis treatment. Acta Biomater. 9: 7081–7092.

Fang, C.J., K.L. Ku, M.H. Lee and N.W. Su. 2010. Influence of nutritive factors on C50 carotenoids production by *Haloferax mediterranei* ATCC 33500 with two-stage cultivation. Bioresour. Technol. 101: 6487–6493.

Fariq, A., A. Yasmin and J. Muhammad. 2019. Production, characterization, and antimicrobial activities of bio-pigments by *Aquisalibacillus elongatus* MB592, *Salinicoccus sesuvii* MB597, and *Halomonas aquamarina* MB598 isolated from Khewra Salt Range, Pakistan. Extremophiles 23: 435–449.

Fernandez-Castillo, R., F. Rodriguez-Valera, J. Gonzalez-Ramos and F. Ruiz-Berraquero. 1986. Accumulation of poly(beta-Hydroxybutyrate) by halobacteria. Appl. Environ. Microbiol. 51: 214–216.

Freitas, F., V.D. Alves and M.A. Reis. 2011. Advances in bacterial exopolysaccharides: from production to biotechnological applications. Trends Biotechnol. 29: 388–398.

Grant, W.D., M. Kamekura, T.J. Mc Genity and A. Ventosa. 2001. Class III. Halobacteria class nov. pp. 294–299. In: D.R. Boone, R.W. Castenholz and G.M. Garrity [eds.]. Bergey's Manual of Systematic Bacteriology, 2nd ed. Springer-Verlag, New York, NY, USA.

Gross, P.L., B.A. Esposito, J.J. Freedman, S. Hajmohammadi and N.W. Shworak. 2005. Anticoagulant heparan sulfate mediates antithrombin amelioration of leukocyte-endothelial interactions in sepsis. Blood 106: 3684.

Gugliandolo, C., A. Spanò, V. Lentini, A. Arena and T.L. Maugeri. 2013. Antiviral and immunomodulatory effects of a novel bacterial exopolysaccharide of shallow marine vent origin. J. Appl. Microbiol. 116: 1028–1034.

Han, J., Q. Lu, L. Zhou, J. Zhou and H. Xiang. 2007. Molecular characterization of the phaECHm genes, required for biosynthesis of poly(3-hydroxybutyrate) in the extremely halophilic archaeon *Haloarcula marismortui*. Appl. Environ. Microbiol. 73: 6058–6065.

Hertadi, R., K. Kurnia, W. Falahudin and M. Puspasari. 2017. Polyhydroxybutyrate (PHB) production by *Halomonas elongata* BK AG 18 indigenous from salty mud crater at central Java Indonesia. Malays. J. Microbiol. 13: 26–32.

Hezayen, F.F., B.H.A. Rehm, R. Eberhardt and A. Steinbüchel. 2000. Polymer production by two newly isolated extremely halophilic archaea: application of a novel corrosion-resistant bioreactor. Appl. Microbiol. Biotechnol. 54: 319–325.

Hosseini, T.A. and M. Shariati. 2009. *Dunaliella* Biotechnology: methods and applications. J. Appl. Microbiol. 107: 14–35.

Hou, J. and H.L. Cui. 2018. *In vitro* antioxidant, antihemolytic, and anticancer activity of the carotenoids from halophilic archaea. Curr. Microbiol. 75: 266–271.

Jaswir, I., D. Noviendri, R.F. Hasrini and F. Octavianti. 2011. Carotenoids: sources, medicinal properties and their application in food and nutraceutical industry. J. Med. Plant. Res. 5: 7119–7131.

Kim, S.J. and J.H. Yim. 2007. Cryoprotective properties of exopolysaccharide (P-21653) produced by the Antarctic bacterium, *Pseudoalteromonas arctica* KOPRI 21653. J. Microbiol. 45: 510–514.

Kirti, K., S. Amita, S. Priti, A.M. Kumar and S. Jyoti. 2014. Colorful world of microbes: carotenoids and their applications. Adv. Biol. 2014: 837891.

Klähn, S. and M. Hagemann. 2010. Compatible solute biosynthesis in cyanobacteria. Environ. Microbiol. 13: 551–562.

Kucera D, I. Pernicová, A. Kovalcik, M. Koller, L. Mullerova, P. Sedlacekand, et al. 2018. Characterization of the promising poly(3-hydroxybutyrate) producing halophilic bacterium *Halomonas halophila*. Bioresour. Technol. 256: 552–556.

Kushwaha, S.C., J.K. Kramer and M. Kates. 1975. Isolation and characterization of C50-carotenoid pigments and other polar isoprenoids from *Halobacterium Cutirubrum*. Biochim. Biophys. Acta. 398: 303–314.

Lanyi, J.K. 1974. Salt-dependent properties of proteins from extremely halophilic bacteria. Bacteriol. Rev. 38: 272–290.

Lee, K.Y. and D.J. Mooney. 2012. Alginate: properties and biomedical applications. Prog. Polym. Sci. 37: 106–126.

Li, H.P., W.G. Hou and Y.Z. Zhang. 2011. Rheological properties of aqueous solution of new exopolysaccharide secreted by a deep-sea mesophilic bacterium. Carbohyd. Polym. 84: 1117–1125.

Lillo, J.G. and F. Rodriguez-Valera. 1990. Effects of culture conditions on poly-(3-hydroxybutyric acid) production by *Haloferax mediterranei*. Appl. Environ. Microbiol. 56: 2517–2521.

Lin, C.C. and L.E. Casida. 1984. GELRITE as a gelling agent in media for the growth of thermophilic microorganisms. Appl. Environ. Microbiol. 47: 427–429.

Llamas, I., J.A. Mata, R. Tallon, P. Bressollier, M.C. Urdaci, E. Quesada, et al. 2010. Characterization of the exopolysacchararide produced by *Salipiger mucosus A3T*, a halophilic species belonging to the Alphaproteobacteria, isolated on the Spanish Mediterranean seaboard. Mar. Drugs 8: 2240–2251.

Llamas, I., H. Amjres, J.A. Mata, E. Quesada and V. Béjar. 2012. The potential biotechnological applications of the exopolysaccharide produced by the halophilic bacterium *Halomonas almeriensis*. Molecules 17: 7103–7120.

Lobasso, S., P. Lopalco, G. Mascolo and A. Corcelli. 2008. Lipids of the ultra-thin square halophilic archaeon *Haloquadratum walsbyi*. Archaea 2: 177–183.

Lü, Y., H. Lu, S. Wang, J. Han, H. Xiang and C Jin. 2017. An acidic exopolysaccharide from *Haloarcula hispanica* ATCC33960 and two genes responsible for its synthesis. Archaea. 2017: 5842958.

Madigan, M.T. and J.M. Martinko. 2006. Brock Biology of Microorganisms. 11th edition. Prentice Hall, Upper Saddle River, New Jersey, USA.

Mandelli, F., V.S. Miranda, E. Rodrigues and A.Z. Mercadante. 2012. Identification of carotenoids with high antioxidant capacity produced by extremophile microorganisms. World J. Microbiol. Biotechnol. 28: 1781–1790.

Mata, J.A., V. Béjar, I. Llamas, S. Arias, P. Bressollier, R. Tallon, et al. 2006. Exopolysaccharides produced by the recently described halophilic bacteria *Halomonas ventosae* and *Halomonas anticariensis*. Res. Microbiol. 157: 827–835.

Mata, J.A., V. Béjar, P. Bressollier, R. Tallon, M.C. Urdaci, E. Quesada, et al. 2008. Characterization of exopolysaccharides produced by three moderately halophilic bacteria belonging to the family *Alteromonadaceae*. J. Appl. Microbiol. 105: 521–528.

Miller, N.J., J. Sampson, L.P. Candeias, P.M. Bramley and C.A. Rice-Evans. 1996. Antioxidant activities of carotenoids and xanthophylls. FEBS Lett. 384: 240–242.

Mishra, A. and B. Jha. 2009. Isolation and characterization of extracellular polymeric substances from micro-algae *Dunaliella salina* under salt stress. Bioresour. Technol. 100: 3382–3386.

Moscovici, M. 2015. Present and future medical applications of microbial exopolysaccharides. Front. Microbiol. 6: 1012.

Naziri, D., M. Hamidi, S. Hassanzadeh, V. Tarhriz, B. Maleki Zanjani, H. Nazemyieh, et al. 2014. Analysis of carotenoid production by *Halorubrum* sp. TBZ126; an extremely halophilic archeon from urmia lake. Adv. Pharm. Bull. 4: 61–67.

Nichols, C.A., J. Guezennec and J.P. Bowman. 2005. Bacterial exopolysaccharides from extreme marine environments with special consideration of the Southern ocean, sea ice, and deep-sea hydrothermal vents: a review. Mar. Biotechnol. 7: 253–271.

Nicolaus, B., L. Lama, E. Esposito, M.C. Manca, R. Improta, M.R. Bellitti, et al. 1999. *Haloarcula* spp able to biosynthesize exo- and endopolymers. J. Ind. Microbiol. Biotechnol. 23: 489–496.

Nishino, H., H. Tokuda, M. Murakoshi, Y. Satomi, M. Masuda, M. Onozuka, et al. 2000. Cancer prevention by natural carotenoids. BioFactors. 13: 89–94.

Oren, A. 1999. Bioenergetic aspects of halophilism. Microbiol. Mol. Biol. Rev. 63: 334–348.

Oren, A. 2002a. Halophilic Microorganisms and their Environments. Kluwer Scientific Publishers, Dordrecht.

Oren, A. and L. Mana. 2002b. Amino acid composition of bulk protein and salt relationships of selected enzymes of *Salinibacter ruber*, an extremely halophilic bacterium. Extremophiles 6: 217–223.

Oren, A. 2002c. Molecular ecology of extremely halophilic Archaea and Bacteria. FEMS Microbiol. Ecol. 39: 1–7.

Oren, A. 2006. Life at high salt concentrations. pp. 263–282. *In*: M. Dworkin, S. Falkow, E. Rosenberg, K.H. Schleifer and E. Stackebrandt [eds.]. The Prokaryotes. A Handbook on the Biology of Bacteria. Ecophysiology and Biochemistry. Vol. 2. Springer, New York, NY, USA.

Oren, A. 2008. Microbial life at high salt concentrations: phylogenetic and metabolic diversity. Saline Systems. 4: 2.

Oren, A. 2014. Taxonomy of halophilic Archaea: current status and future challenges. Extremophiles 18: 825–834.

Osawa, A., Y. Ishii, N. Sasamura, M. Morita, H. Kasai, T. Maoka, et al. 2010. Characterization and antioxidative activities of rare C(50) carotenoids sarcinaxanthin, sarcinaxanthin monoglucoside, and sarcinaxanthin diglucoside obtained from *Micrococcus yunnanensis*. J. Oleo Sci. 59: 653–659.

Poli, A., H. Kazak, B. Gürleyendağ, G. Tommonaro, G. Pieretti, E.T. Öner, et al. 2009. High level synthesis of levan by a novel *Halomonas* species growing on defined media. Carbohydr. Polym. 78: 651–657.

Poli, A., P. Di Donato, G.R. Abbamondi and B. Nicolaus. 2011. Synthesis, production, and biotechnological applications of exopolysaccharides and polyhydroxyalkanoates by archaea. Archaea 2011: 693253.

Pramanik, A., A. Mitra, M. Arumugam, A. Bhattacharyya, S. Sadhukhan, A. Ray, et al. 2012. Utilization of vinasse for the production of polyhydroxybutyrate by *Haloarcula marismortui*. Folia Microbiol. (Praha). 57: 71–79.

Priyanka, P., A.B. Arun and P.D. Rekha. 2014. Sulfatedexopolysaccharide produced by Labrenzia sp. PRIM-30, characterization and prospective applications. Int. J. Biol. Macromol. 69: 290–295.

Pszczola, D.E. 1993. Gellan gum wins IFT's Food Technology Industrial Achievement Award. Food Technol. 47: 94–96.

Qin, Y., J. Jiang, L. Zhao, J. Zhang and F. Wang. 2018. Applications of alginate as a functional food Ingredient. Biopolymers for Food Design. 409–429.

Quillaguamán, J., S. Hashim, F. Bento, B. Mattiasson and R. Hatti-Kaul. 2005. Poly-(β-hydroxybutyrate) production by a moderate halophile, *Halomonas boliviensis* LC1 using starch hydrolysate as substrate. J. Appl. Microbiol. 99: 151–157.

Quillaguamán, J., T. Doan-Van, H. Guzmán, D. Guzmán, J. Martín, A. Everest, et al. 2008. Poly(3-hydroxybutyrate) production by *Halomonas boliviensis* in fed-batch culture. Appl. Microbiol. Biotechnol. 78: 227–232.

Reed, R.H., J.A. Chudek, R. Foster and W.D.P. Stewart. 1984. Osmotic adjustment in cyanobacteria from hypersaline environments. Arch. Microbiol. 138: 333–337.

Rodriguez-Valera, F., G. Juez and D.J. Kushner. 1982. Halocins: salt-dependent bacteriocins produced by extremelyhalophilic rods. Can. J. Microbiol. 28: 151–154.

Rüegg, R. 1984. Extraction Process for Beta-Carotene. U.S. Patent 4439629A.

Salgaonkar, B.B., K. Mani and J.M. Braganca. 2013. Accumulation of polyhydroxyalkanoates by halophilic archaea isolated from traditional solar salterns of India. Extremophiles. 17: 787–795.

Salgaonkar, B.B. and J.M. Bragança. 2015. Biosynthesis of poly(3-hydroxybutyrate-co-3-hydroxyvalerate) by *Halogeometricum borinquense* strain E3. Int. J. Biol. Macromol. 78: 339–346.

Santos, H. and M.S. da Costa. 2002. Compatible solutes of organisms that live in hot saline environments. Environ. Microbiol. 4: 501–509.

Schleper, C., G. Puhler, B. Kuhlmorgen and W. Zillig. 1995. Life at extremely low pH. Nature 375: 741–742.

Selbmann, L., S. Onofri, M. Fenice, F. Federici and M. Petruccioli. 2002. Production and structural characterization of the exopolysaccharide of the Antarctic fungus *Phomaherbarum* CCFEE 5080. Res. Microbiol. 153: 585–592.

Selim, M.S., S.K. Amer, S.S. Mohamed, M.M. Mounier and H.M. Rifaat. 2018. Production and characterisation of exopolysaccharide from *Streptomyces carpaticus* isolated from marine sediments in Egypt and its effect on breast and colon cell lines. J. Genet. Eng. Biotechnol. 16: 23–28.

Sikkandar, S., K. Murugan, S. Al-Sohaibani, F. Rayappan, A. Nair and F. Tilton. 2013. Halophilic bacteria-a potent source of carotenoids with antioxidant and anticancer potentials. J. Pure Appl. Microbio. 7: 2825–2830.

Squillaci, G., R. Finamore, P. Diana, O.F. Restaino, C. Schiraldi, S. Arbucci, et al. 2016. Production and properties of an exopolysaccharide synthesized by the extreme halophilic archaeon *Haloterrigena turkmenica*. Appl. Microbiol. Biotechnol. 100: 613–623.

Squillaci, G., R. Parrella, F. La Cara, S.M. Paixão, L. Alves and A. Morana. 2017a. Potential biotechnological applications of products from the Haloarchaeon *Haloterrigena turkmenica*. Adv. Appl. Sci. Res. 8: 42–48.

Squillaci, G., R. Parrella, V. Carbone, P. Minasi, F. La Cara and A. Morana. 2017b. Carotenoids from the extreme halophilic archaeon *Haloterrigena turkmenica*: identification and antioxidant activity. Extremophiles 21: 933–945.

Srikanth, R., C.H.S.S.S. Reddy, G. Siddartha, M.J. Ramaiah and K.B. Uppuluri. 2015. Review on production, characterization and applications of microbial levan. Carbohyd. Polym. 120: 102–114.

Sun, M.L., S.B. Liu, L.P. Qiao, X.L. Chen, X. Pang, M. Shi, et al. 2014. A novel exopolysaccharide from deep-sea bacterium *Zunongwangia profunda* SM-A87: low-cost fermentation, moisture retention, and antioxidant activities. Appl. Microbiol. Biotechnol. 98: 7437–7445.

Sun, M.L., F. Zhao, M. Shi, X.Y. Zhang, B.C. Zhou, Y.Z. Zhang, et al. 2015. Characterization and Biotechnological potential analysis of a new exopolysaccharide from the arctic marine bacterium *Polaribacter* sp. SM1127. Sci. Rep. 5: 18435.

Sutherland, I.W. 1982. Biosynthesis of microbial exopolysaccharides. Adv. Microb. Physiol. 23: 79–150.

Sutherland, I.W. 2001. Biofilm exopolysaccharides: a strong and sticky framework. Microbiology 147: 3–9.

Sze, J.H., J.C. Brownlie and C.A. Love. 2016. Biotechnological production of hyaluronic acid: a mini review. Biotech 6: 67.

Taran, M. and H. Amirkhani. 2010. Strategies of poly(3-hydroxybutyrate) synthesis by *Haloarcula* sp. IRU1 utilizing glucose as carbon source: optimization of culture conditions by Taguchi methodology. Int. J. Biol. Macromol. 47: 632–634.

Trüper, H.G., J. Severin, A. Wolhfarth, E. Müller and E.A. Galinksi. 1991. Halophily, taxonomy, phylogeny and nomenclature. pp. 3–7. *In*: F. Rodriguez-Valera [ed.]. General and Applied Aspects of Halophilism. Plenum Press, New York, NY, USA.

Van-Thuoc, D., T. Huu-Phong, N. Thi-Binh, N. Thi-Tho, D. Minh-Lam and J. Quillaguamàn. 2012. Polyester production by halophilic and halotolerant bacterial strains obtained from mangrove soil samples located in Northern Vietnam. Microbiology Open. 1: 395–406.

Ventosa, A., J.J. Nieto and A. Oren. 1998. Biology of moderately halophilic aerobic bacteria. Microbiol. Mol. Biol. Rev. 62: 504–544.

Ventosa, A., M.C. Gutierrez, M. Kamekura and M.L. Dyall-Smith. 1999. Proposal to transfer *Halococcus turkmenicus*, *Halobacterium trapanicum* JCM 9743 and strain GSL-11 to *Haloterrigena turkmenica* gen. nov., comb. nov. Int. J. Syst. Bacteriol. 49: 131–136.

Yamada, S. and K. Sugahara. 2008. Potential therapeutic application of chondroitin sulfate/dermatansulfate. Curr. Drug. Discov. Technol. 5: 289–301.

Yatsunami, R., A. Ando, Y. Yang, S. Takaichi, M. Kohno, Y. Matsumura, et al. 2014. Identification of carotenoids from the extremely halophilic archaeon *Haloarcula Japonica*. Front. Microbiol. 5: 100.

Yin, J., J.C. Chen, Q. Wu and G.Q. Chen. 2015. Halophiles, coming stars for industrial biotechnology. Biotechnol. Adv. 33: 1433–1442.

Yoo, S.H., E.J. Yoon, J. Cha and H.G. Lee. 2004. Antitumor activity of levan polysaccharides from selected microorganisms. Int. J. Biol. Macromol. 34: 37–41.

Yu, I. K. and F. Kawamura. 1987. Halomethanococcus doii gen. nov., sp. nov.: an obligately halophilic methanogenic bacterium from solar salt ponds. J. Gen. Appl. Microbiol. 33: 303–310.

Zalar, P., G. Sybren de Hoog, H.J. Schroers, J.M. Frank and N. Gunde-Cimerman. 2005. Taxonomy and phylogeny of the xerophilic genus Wallemia (Wallemiomycetes and Wallemiales, cl. et ord. nov.). Antonie van Leeuwenhoek. 87: 311–328.

Zhilina, T.N. and G.A. Zavarzin. 1987. Methanohalobium evestigatus, n. gen., n. sp. The extremely halophilic methanogenic Archaebacterium. Doklady Akademii Nauk SSSR 293: 464–468.

5

Exploring Microbial Diversity and Heavy Metals Resistance in an Active Volcanic Geothermal System

Urbieta María Sofía[1,*], Lima María Alejandra[1],
Massello Francisco[1], Lopez Bedogni Germán[1],
Giaveno María Alejandra[2] and Donati Edgardo[1]

Introduction – Geochemistry of Río Agri

Río Agrio is a naturally acidic, volcanic river that runs through the Copahue-Caviahue geothermal system, located on the Northwest corner of Neuquén province, in Patagonia, Argentina (Figure 1). The Copahue volcano (37°51′S, 71°10.2′W, 2965 m a.s.l.) is a predominantly andesitic stratovolcano that is still active and has frequent eruptive cycles (2018, 2015, 2014, 2013, and 2012). The crater of the Copahue volcano hosts an acidic lagoon and several acidic hot springs (some of them hidden below the glacier located in the peak of the mountain). The acidic crater lagoon has an erratic behavior associated with eruptive cycles as well as the rain and snow regimes; in fact, it had disappeared after the eruption in 2000 and it was only very recently that it became accessible again. The waters of the crater lagoon are characterized by their strong acidity (average pH between less than 1 and 1.7) and high contents of sulfate, chloride, and fluoride, although all the physicochemical parameters are subjected to change with the eruptive activity. The temperature also varies greatly with records from 0°C to near 60°C, measured close after the eruption of 2012 (Rodríguez et al. 2016). Independently from the crater lagoon, two streams originated in two different hot springs a few meters down the crater, merged downstream, and emerged at the eastern slope of the volcanic cone at approximately 2,740 m a.s.l to form the Upper Río Agrio (URA). The crater lagoon and hot springs that originate in Río Agrio are fed by different sources of the underlying Copahue volcano hydrothermal system, therefore they have differences in the chemical composition of the waters (Agusto and Varekamp 2016). The chemical composition of the URA is determined by the mixture of the emissions of the feeding acidic volcanic hot springs with glacial meltwater, the pH, ions (especially sulfate, chloride, and fluoride), and conductivity of the origin of the river, historically named as VA (for the Spanish 'Vertiente del Agrio'), are comparable with those of the crater lagoon.

The acidity of the Copahue volcano hydrothermal system comes from the condensation of magmatic gases, especially SO_2, H_2S, HCl, and HF, in waters from rain, snow, and melting of the local glacier. Equation 1 shows the aqueous dissolution and disproportionation of SO_2:

$$3SO_{2\,(g)} + 2H_2O \longrightarrow 2(HSO_4)^- + 2H^+ + S_{(s)} \qquad \text{Equation 1}$$

[1] CINDEFI (CCT La Plata-CONICET, UNLP), Facultad de Ciencias Exactas, Universidad Nacional de la Plata, La Plata, Argentina.

[2] PROBIEN-CONICET-UNCo, Departamento de Química, Facultad de Ingeniería, Universidad Nacional del Comahue, Neuquén, Argentina.

*Corresponding author: msurbieta@biol.unlp.edu.ar

The gas HCl also contributes to the high H^+ concentrations because of their almost complete dissociation in the water at low temperatures. Such hyperacid waters react with the surrounding rocks, leaching cations from minerals and partially dissolving the silica matrix on which it can also act HF (Rodríguez et al. 2016, Agusto et al. 2016). Thus, the waters of Río Agrio, especially at the origin, are also loaded with metals and heavy metals, predominantly iron.

After the origin the URA runs down the eastern slope of Copahue volcano hill for 13.5 Km until discharging in the Caviahue lake. Along its path, there are various waterfalls like Cascada del Gigante, Cascada de la Culebra, and Cabellera de la Virgen (abbreviated CAS from here on). In addition to that, the URA receives the input of at least three neutral water courses—Río Blanco, Río Rojo, and Arroyo Jara—that dilute the metals and heavy metals concentrations and thus lowering the conductivity, although they have very little effect on pH which remains highly acidic (average pH around 2.0-2.5) in all the course of the URA. After the hydrothermal origin, the average temperature of the Río Agrio is around 12°C.

Caviahue lake, located at 1,600 m a.s.l., has URA and the neutral Río Dulce as the main affluent and its only effluent is the Lower Río Agrio (LRA) (Figure 1). Despite having a glassier origin, the waters of the Caviahue lake are acidic (pH values around 3) due to the influence of the Copahue volcano hydrothermal system (Varekamp et al. 2009). On its west side, it hosts the Caviahue village, a small mainly touristic locality, that unfortunately pours its wastewaters into the lake and altering its natural conditions and, of course, its biodiversity.

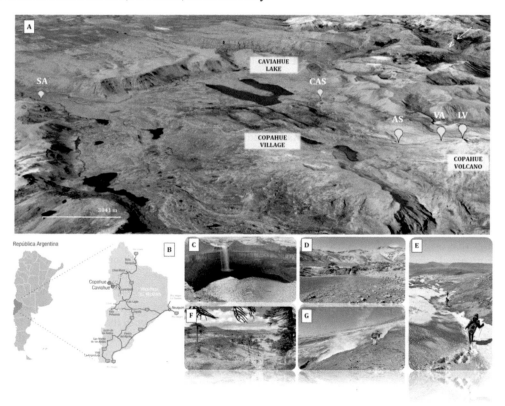

Figure 1. A. Satelite image of the Copahue-Caviahue region where the most relevant points are indicated. LV: Copahue volcano crater lagoon. VA: the area of the two geothermal origins of Río Agrio. AS: Upper Río Agrio after the input of snowmelt tributary courses. CAS: area of various waterfalls before URA discharges in the Caviahue lake. SA: big waterfall 'Salto del Agrio' in Lower Río Agrio. B. Location of Copahue-Caviahue region in Neuquén province and Argentina. C and D. Photographs of 'Salto del Agrio' after and before the waterfall. E, F, and G: different images of URA.

After emerging from the lake, the LRA is wider and less acidic than the URA (average pH 3.0-3.6); it runs for approximately 5 Km until a spectacular fall of 60 m named Salto del Agrio. At this point, the course and margins of the river are characterized by a strong orange color (Figure 1 C and D) caused by the precipitation of iron minerals due to the pH increase (Gammons et al. 2005) mainly caused by the irruption of neutral waters from local rivers and streams. Finally, the URA reaches neutrality at approximately 40 Km away from the Caviahue lake.

Table 1 presents some of the most relevant physicochemical characteristics of the most representative points of Río Agrio. In different sampling campaigns over twenty years, we have verified that parameters such as temperature and pH are almost constant through time, proving the acidic, moderate nature of temperature of Río Agrio, while others like iron concentration have suffered changes in scale but have always shown a dilution of more than one order of magnitude from the geothermal origin to Salto del Agrio in LRA.

TABLE 1: Most relevant physicochemical characteristics of representative points along Río Agrio: the geothermal origins (VA1 and VA2), the meeting point of the two geothermal origins (U2V), different points in Upper Río Agrio (AS1, AS2, and AS3), two different waterfalls of Upper Río Agrio (CAS1 and CAS2), the Caviahuelake (LC) and the big waterfall Salto del Agrio in Lower Río Agrio (SA)

Sampling Point	GPS Location	T (°C)	pH	Conductivity (µS/cm)	Fe (mg/L)	SO$_4^{2-}$ (mg/L)	Organic Matter (mg/L)
LV (crater lagoon)	37°51'21"S, 71°09'29"W	4.0	2.79	1,402	55	–	4.23
VA1	37°51'23"S, 71°09'04"W	53.0	1.0	47,200	540-998	2,841.01	8.84
VA2	37°51'19"S, 71°09'09"W	29.0	1.0	15,450	560-870	2,616.18	12.78
U2V	37°51'15"S, 71°08'45"W	9.0	1.5	10,690	470	2,467.10	–
AS1	37°50'59"S, 71°08'00"W	6.8	2.0	3,870	26-693	2,479.92	23.53
AS2	–	9.9	2.0	19,860	111	5,353.5	–
AS3	–	8.0	1.7	11,690	201	3,274.3	
CAS1	37°53'11"S, 71°04'15"W	18.0	2.0	3,560	32-241	1,046.60	13.22
CAS2	37°52'59"S, 71°04'00"W	15.9	2.0	3,290	30	1,120.00	–
LC (Caviahue lake)	37°53'14"S, 71°02'46"W	8.3	2.0	946	7	932.00	–
SA (Salto del Agrio)	37°48'43"S, 70°55'31"W	16.9	3.6	516	3-13	20.64	2.91

–: not determined

History of the Assessment of Prokaryotic Biodiversity of Río Agrio

Our group has been studying the prokaryotic biodiversity of the Copahue-Caviahue geothermal system for more than twenty years, and our works are among the first reports of the existence of extremophilic prokaryotic biodiversity in this unexplored, inhospitable habitat (Lavalle et al. 2005, Chiacchiarini et al. 2010). During this time, we have done several assessments using different approaches relying on cultivation and, in the last years, especially on molecular ecology techniques, from cloning and sequencing to massive next-generation sequencing. Besides, we have correlated biodiversity assessments with physicochemical data to schematize geochemical models of the main elements, such as C, S, and Fe (Urbieta et al. 2015a). At the very beginning, the driving force to search for microorganisms in Río Agrio was based on its physicochemical similarities with mining environments (marked acidity, moderate temperature, low content of organic matter, high content of sulfur compounds, and metals and heavy metals, especially iron) aiming to obtain native species of acidophilic, sulfur, or iron-oxidizing microorganisms with potential use in biomining operations. Thus, to fulfill this goal we made enrichment cultures using water and sediments from different parts of Río Agrio and cultivated them in acidic mineral media with iron and/or sulfur as an only energy source and gathered a collection of isolates associated with the genera and species *Acidithiobacillus ferrooxidans*, *At. thiooxidans*, *At. caldus*, *Leptospirillum*, and archaea *Sulfolobus* and *Acidianus* (Chiacchiarini et al. 2010). The different species found in these quite slanted cultivation assays encouraged us to go deeper into the biodiversity assessment by using molecular ecology techniques. The DGGE profile obtained using 16S rRNA universal primers to amplify the bacterial community of different points all through Río Agrio presented various bands that were almost in the same position in the origin of the river (VA1 and VA2), the merging of the two sources (U2V), various points in Upper Río Agrio (AS) including two of the waterfalls formed before the URA discharges in the Caviahue lake (CAS); data not shown due to poor image quality. It became the first evidence of the microbial diversity of this natural extreme environment and also showed that the community composition seemed to remain approximately constant as the river runs down the Copahue volcano hill; that is, as the physicochemical conditions become less extreme until the Caviahue lake (CL) is reached. Caviahue lake and many tributary courses that discharge in the LRA introduce significant changes in the physicochemical conditions, which have an impact on its microbial community as can be seen in Salto del Agrio (SA) (Urbieta et al. 2011).

At the same time, we assessed the bacterial and archaeal biodiversity of the most representative points of URA using cloning and Sanger sequencing of almost complete 16S rRNA genes complemented with catalyzed reporter deposition fluorescence *in situ* hybridization (CARD-FISH) using specific probes targeting different taxonomic levels. Such a combination of techniques allowed the semi-quantification of the most representative microbial species of the Río Agrio (Urbieta et al. 2012, Urbieta et al. 2015b). The results obtained, schematized in Figure 2, proved once again that the microbial community of the URA is formed by a relatively bounded number of species, and its structure is almost constant throughout its course. The Bacteria/Archaea ratio was maintained approximately constant with 40% of Archaea, represented almost exclusively by the extremely acidophilic iron oxidizer, euryarchaeota *Ferroplasma*. Regarding bacteria, there was also a marked preponderance of the same species from the origin to the last part of the URA; the autotrophic, acidophilic, iron, and sulfur oxidizer, gammaproteobacteria *Acidithiobacillus* as well as the presence (in much few proportions) of the autotrophic, acidophilic, and obligatory iron oxidizer *Leptospirillum* (Nitrospira)—especially near the origin where the ferrous iron concentration is higher—the acidophilic heterotroph *Acidiphilium*—a typical companion of the autotrophic acidophiles—and the mixotrophic iron and sulfur oxidizer *Sulfobacillus* (Firmicutes).

Finally, as the massive sequencing techniques became more accessible, we performed a deeper assessment of the whole microbial community of Río Agrio by Illumina sequencing of the V3-V4 hypervariable region of the 16S rRNA obtaining over 171000 gene fragments that were processed and taxonomically classified (Figure 3) (Lopez-Bedogni et al. 2020). In this analysis, we could include for the first time the acidic lagoon that sporadically forms in the crater of Copahue volcano (LV) as well as the geothermal origin of the river (VA), the middle part of URA (AS), one of the waterfalls before discharging in Caviahue lake (CAS), and the big waterfall 'Salto del Agrio' in LRA (SA) (the physicochemical parameters presented in Table 1).

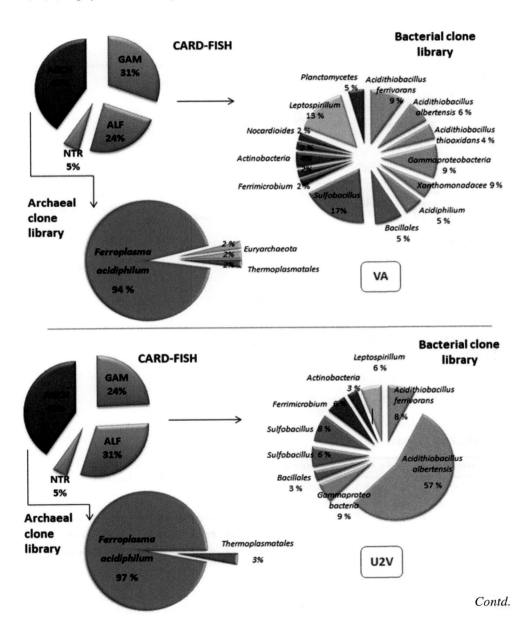

Contd.

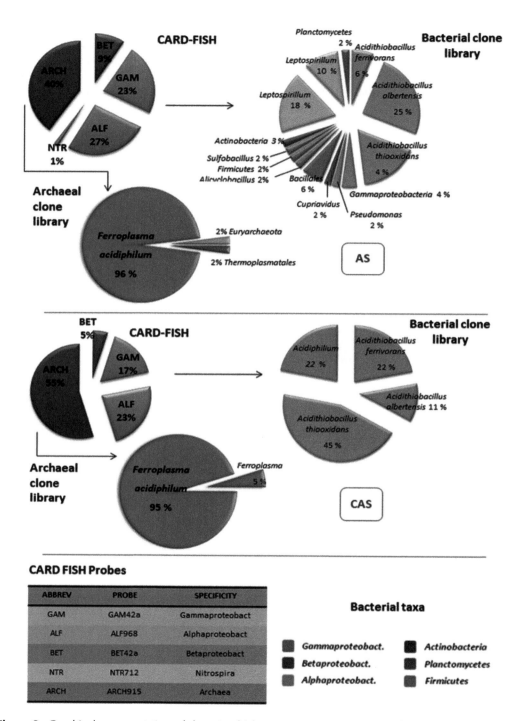

Figure 2. Graphical representation of the microbial community composition in four points of the Upper Río Agrio obtained by CARD-FISH hybridization and sequencing of the bacterial and archaeal 16S rRNA clone libraries. VA: the geothermal origin of the river, U2V: the meeting point of the two geothermal origins, AS: midpoint in URA after the discharge of tributary courses, CAS: waterfall in URA before discharging in Caviahue lake.

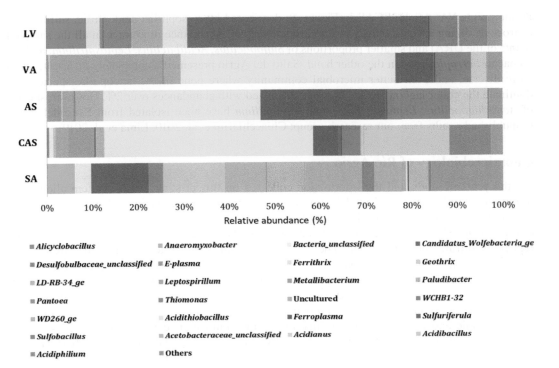

LV

VA

AS

CAS

SA

0%   10%   20%   30%   40%   50%   60%   70%   80%   90%   100%
Relative abundance (%)

■ *Alicyclobacillus*   ■ *Anaeromyxobacter*   *Bacteria_unclassified*   ■ *Candidatus_Wolfebacteria_ge*

■ *Desulfobulbaceae_unclassified* ■ *E-plasma*   *Ferrithrix*   *Geothrix*

■ *LD-RB-34_ge*   ■ *Leptospirillum*   ■ *Metallibacterium*   *Paludibacter*

■ *Pantoea*   ■ *Thiomonas*   *Uncultured*   ■ *WCHB1-32*

■ *WD260_ge*   *Acidithiobacillus*   ■ *Ferroplasma*   ■ *Sulfuriferula*

■ *Sulfobacillus*   ■ *Acetobacteraceae_unclassified* *Acidianus*   *Acidibacillus*

■ *Acidiphilium*   ■ Others

Figure 3. Graphical representation of the abundances of the microbial diversity found in Río Agrio and the Copahue volcano crater lagoon (genus level). Sequences with relative abundances that lower than 2% were grouped as 'Others'.

The high throughput metagenomic sequencing allowed us to analyze more than three orders of magnitude of sequences than the molecular ecology techniques used before. However, the results were the same; the URA is dominated by the acidophilic, mesophilic, iron, and/or sulfur oxidizers *Acidithiobacillus*, *Sulfobacillus*, and *Ferroplasma*. Surprisingly, when using this high throughput sequencing technology, sequences related to archaea were only found in URA (neither in the crater lagoon nor in Salto del Agrio) and their abundancy fell to less than 13%, even though the primers used were specifically checked to amplify the archaeal species previously detected in Río Agrio. In fact, all the archaeal sequences obtained were related to *Ferroplasma* or other euryarchaeal with a very high percentage of similitude with other clones found in previous studies of the river (Urbieta et al. 2011). The only exception was the appearance of *Acidianuscopahuensis*, a thermoacidophilic crenarchaeota apparently autochthonous of the Copahue-Caviahue system that was first isolated from the Copahue geothermal area (Giaveno et al. 2013) but never detected in the Río Agrio until the Illumina sequencing (Lopez Bedogni et al. 2020).

The canonical correlation of the biodiversity and physicochemical parameters of the points studied showed that the Copahue volcano-Río Agrio system could be divided into three different niches with different microbial communities: the crater lagoon, the URA, and Salto del Agrio (Lopez-Bedogni et al. 2020). The crater lagoon (LV) was dominated by *Sulfuriferula* (approximately 50%), a genus of neutrophilic mesophilic species able to oxidize a variety of sulfur compounds (Watanabe et al. 2016), that were not detected in any other point of the river. The rest of the microbial diversity of LV was completed by very low proportions of acidophilic sulfur-oxidizing species already found in Río Agrio, such as *Acidithiobacillus*, *Sulfobacillus*, and *Acidiphilium*, and others only found at this point, like *Thiomonas* and *Ferritrix*. This analysis confirmed that the microbial community of the URA is fairly constant from the geothermal origin to the discharge in the Caviahue lake (represented by the points VA, AS, and CAS in Figure 4), dominated by mesophilic, acidophilic, sulfur- and/ or iron- oxidizing species, chiefly *Acidithiobacillus* (50% abundance in average in all the sampled points of the URA) and smaller proportions of *Sulfobacillus, Acidibacillus, Leptospirillum*, and the archaeon *Ferroplasma*. On the other hand, Salto del Agrio presented a completely different, much more diverse, and even better microbial community where none of the acidophilic species that dominate the crater lagoon or the URA were detected with abundances over 2%. However, species of *Acidithiobacillus, Leptospirillum*, and *Acidiphilium* have been isolated from Salto del Agrio (unpublished results from our research group; Chiacchiarini et al. 2010, Lima et al. 2019).

Geomicrobiology of Río Agrio

All the years of study and the information collected on this pristine natural system allowed us to outline geomicrobiological models aiming to understand the interactions of the extremophilic microorganisms that develop all through the river with the physicochemical and environmental characteristic of the environment and their role in the cycles of the most relevant elements, especially sulfur, iron, and carbon. The main metabolism in the crater lake and the URA seems to be aerobic and chemoautotrophic at the expense of the oxidation of iron and chiefly sulfur compounds, which are highly abundant due to the particular geology of the place and the constant volcanic activity. The waters of both systems have a very low content of dissolved carbon, thus species such as *Sulfuriferula, Acidithiobacillus, Leptospirillum*, and *Ferroplasma*—which are litoautotrophic and some of them sensitive to organic compounds can develop properly—acting as primary producers by fixing atmospheric CO_2 and allowing the survival of *Acidibacillus, Acidiphilium*, and *Sulfobacillus*, all that are mixotrophic or heterotrophic species. Besides, the predominance of sulfur-oxidizing species helps to explain the maintenance of the low pH and the high concentration of sulfate measured in the URA (Table 1). In Salto del Agrio, if we had only considered the most abundant OTUs found in the metagenomics, we would have proposed a strictly heterotrophic metabolism, despite a very low organic carbon content (Figure 3, Table 1). However, as we have already mentioned, there are several examples of isolates of acidophilic iron and/or sulfur-oxidizing species, chiefly *Acidithiobacillus, Sulfobacillus, Leptospirillum*, and *Acidiphilium*. So, based on this, we performed a much thorough search on the high throughput sequencing results and found that all the same acidophilic, sulfur, and/or iron-oxidizing species that inhabit the rest of the Río Agrio were also in SA, only with much lower abundances. Thus, the chemoautotrophic, sulfur, and/or iron-oxidizing metabolisms are probably as important in Salto del Agrioas they are in the upper river. In the same way, the presence of iron oxidizers explains the orange color—which is quite noticeable in the images of Figure 1C and D and Figure 4—typical of ferrous iron precipitates, like jarosite, that is formed by the hydrolysis of ferrous iron when pH increases over 2 and there is enough sulfate; the two environmental conditions that are given in the proximities of Salto del Agrio. On the other hand, the metabolism of sulfur oxidizes justifies the persistence of a low pH so many kilometers away from the geothermal source, and after the input of many neutral tributary water courses.

Potential Biotechnological Applications of the Microbial Diversity of Río Agrio. A Practical Case: Selection of Heavy Metal and Arsenic Tolerant Microorganisms for Bioremediation

Extremophilic microorganisms are well known for their poly-resistance mechanisms and their flexibility under multiple stresses; these special features make them interesting for many biotechnological developments (Raddadi et al. 2015, Orellana et al. 2018). Considering the extreme conditions that prevail in the Caviahue-Copahue system and the extremophilic microbial diversity that there breeds, we assessed the heavy metal and arsenic tolerance of the community of Salto del Agrio in order to obtain different consortia of biotechnological potential. It is worth noting that the heavy metals content in Salto del Agrio is much lower than that of polluted sites, and even more, many of the metals and arsenic assayed could not be detected in the different environmental samples taken for study over the years.

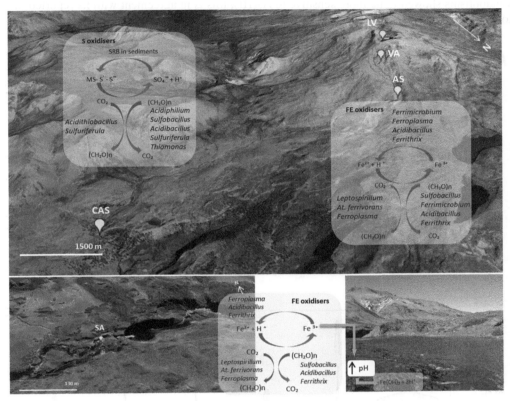

Figure 4. Schematic geomicrobiological model of the Río Agrio.

A sample consisting of sediment and water was enriched in different nutritional conditions to select microbial species with three different metabolisms: neutrophilic-organotrophic, neutrophilic-anaerobic, and acidophilic iron/sulfur-oxidizing microorganism. The cultures were later exposed to increasing concentrations of arsenic and different heavy metals (Cd, Co, Cu, Ni, and Zn) to obtain tolerant consortia.

The aerobic growth was performed in Luria Bertani (pH7) for organotrophs and Mackintosh (Mackintosh 1978) supplemented with $FeSO_4$ (11 g/L) and/or S (10 g/L) adjusted to pH 2.5 for acidophiles. Anaerobic microorganisms were cultivated in Postgate B (Postgate and Campbell 1966), adjusted to pH 6.8, in vials hermetically sealed with rubber and aluminum caps after purging with nitrogen for ten minutes. All cultures were incubated at 30°C and periodically monitored by optical density at 600 nm (organotrophic cultures), cell counts in a Neubauer chamber (acidophilic and anaerobic cultures), and iron titration with permanganate solution (iron-oxidizing acidophiles).

The interest in the chosen metabolisms relies on their potential uses in biotechnological processes; organotrophic consortia could be used for bioremediation in effluents with high organic load, such as residential wastewaters; anaerobic microorganisms, especially sulfate-reducing bacteria, can precipitate metals and metalloid cations and are useful in processes involving activated sludges, while acidophilic microorganisms are interesting for their biomining capacity such as the bioleaching of sulfide minerals.

Figure 5A represents the maximum concentration of heavy metals and arsenic tolerated by the selected consortia. Regarding heavy metals, acidophilic microorganisms were the most tolerant reaching concentration as high as 400 mM. On the other hand, organotrophs were the most tolerant of both arsenic species, being able to grow at 450 mM of As(V) and 20 mM of As(III). All the three metabolisms showed to be more sensitive to As(III); this is in agreement with its reported highest toxicity (Oremland and Stolz 2003).

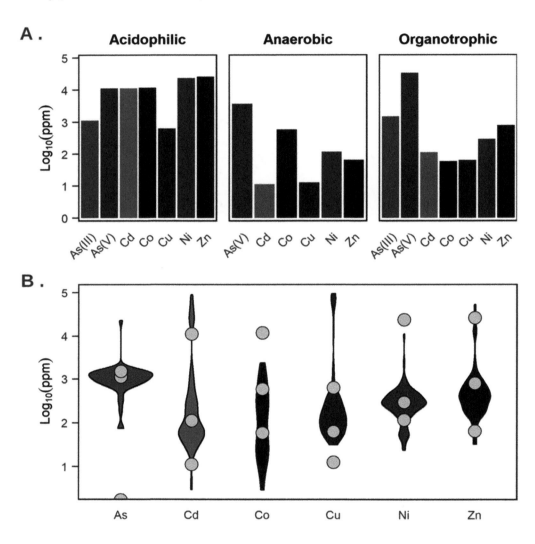

Figure 5: A. Maximum concentration of heavy metals and metalloids tolerated by the microbial consortia obtained from Salto del Agrio. B. Violin plot comparing Salto del Agrio consortia tolerance (yellow dots) and resistant microorganisms reported in the literature (violins). The color code of the metals is the same in both figures.

In order to evaluate the performance of the consortia obtained from Salto del Agrio, we compared their tolerance to heavy metals and metalloids with the values reported for many pure cultures and consortia in the literature (Mejias-Carpio et al. 2018) in a violin plot (Figure 5B). This analysis showed that most of the consortia obtained tolerated concentration in the order of those reported for resistant microorganisms. Even more, in the cases of Co, Ni, and Zn the concentrations tolerated by the consortia from Salto del Agrioare at least one order of magnitude higher than for others previously reported. Thus, such findings indicate that the microbial community of Salto del Agrio is probably naturally resistant, not only to extreme conditions like low pH but also to high concentrations of heavy metals and metalloids, even though they are not continuously exposed to them. Nevertheless, it is important to bear in mind that although some of these chemical species could not be detected in the different samples collected, the environmental conditions of the whole Copahue-Caviahue system, including Salto del Agrio, are influenced by the activity of the Copahue volcano; thus, they might have fluctuated significantly throughout time (Varekamp et al. 2009).

Regarding the actual biotechnological applications of the microbial species from Salto del Agrio, preliminary assays have shown that these consortia can diminish metallic and arsenic soluble concentration via bioprecipitation and adsorption processes. Moreover, an acidophilic consortium could successfully bioleach a zinc mineral recovering 50% of the metal in the presence of up to 200 ppm of arsenic (unpublished data).

Conclusions

In the more than twenty years of study of the Copahue-Caviahue geothermal system, we have gathered a huge amount of data regarding its geochemistry and its microbial diversity, even though we are sure that such a vast, almost pristine, extreme environment still has many hidden secrets. One of our main goals has always been the assessment of the microbial diversity of the Copahue volcano-Río Agrio system, and to achieve it, we have used different molecular ecology approaches as the technologies evolved and became more accessible. In this chapter, we collected and contrasted all that information which allowed us to confirm that independently of the methodology used, the conclusion always indicated that the microbial community of Río Agrio is formed by a bounded number of extremophilic species, chiefly acidophilic, litoautotrophic, iron and/or sulfur oxidizers and that this extreme microbial community remains almost unchanged as the river runs down the Copahue volcano hill from the geothermal origin near the crater. Besides, we have untangled some of the relations between the metabolisms of the members of the microbial community and the physicochemical characteristics of the environment, such as the importance of microbial sulfur oxidation on the maintenance of the acidic waters even after more than 30 Km of the river course, or the contribution of biological iron oxidation in the formation of the characteristic orange-red precipitate that color the LRA. Even though it is clear that the dominant species of Río Agrio are well known acidophilic bacteria and archaea, we have found many 16S rRNA sequences that could not be classified beyond the heist taxonomic levels, but they presented very high similitudes with other sequences retrieved from diverse acidic environments around the world, which might suggest the existence of a yet unexplored ubiquitous acidic microbial diversity.

A little-explored extreme environment with the characteristics of Río Agrio is unquestionably a source of microbial species with the potential to be used in a novel or enhanced biotechnological applications in many fields. Due to the history and expertise of our research group in biomining and bioremediation, we have assessed the tolerance to heavy metals and metalloids under different growth conditions of the microbial community of the river with surprisingly promising results; we hope to expand our knowledge on their applicability in other industrial areas in the near future.

Finally, our interest in the microbial biodiversity of the Copahue volcano-Río Agrio system is also motivated by our deep belief that the biodiversity of an environment is an intangible natural resource of a country that deserves to be acknowledged and protected, and the only way to fulfill that task properly is by knowing it first.

References

Agusto, M. and J.C. Varekamp. 2016. The Copahue volcanic-hydrothermal system and applications for volcanic surveillance. pp. 199–238. *In*: F. Tassi, O. Vaselli and A.T. Caselli [eds.]. Copahue Volcano. Springer, Heidelberg, Germany.

Chiacchiarini, P., L. Lavalle, M.A. Giaveno and E.R. Donati. 2010. First assessment of acidophilic microorganisms from geothermal Copahue-Caviahue system. Hydrometallurgy 104: 334–341.

Gammons, C.H., S.A. Wood, F. Pedrozo, J.C. Varekamp, B.J. Nelson, C.L. Shope, et al. 2005. Hydrogeochemistry and rare earth element behavior in a volcanically acidified watershed in Patagonia, Argentina. Chem. Geol. 222: 249–267.

Giaveno, M.A., M.S. Urbieta, J.R. Ulloa, E. González-Toril and E.R. Donati. 2013. Physiologic versatility and growth flexibility as the main characteristics of a novel thermoacidophilic *Acidianus* strain isolated from Copahue geothermal area in Argentina. Microb. Ecol. 65: 336–346.

Lavalle, L., P. Chiacchiarini, C. Pogliani and E.R. Donati. 2005. Isolation and characterization of acidophilic bacteria from Patagonia, Argentina. Process Biochem. 40: 1095–1099.

Lima, M.A., M.S. Urbieta and E. Donati. 2019. Arsenic-tolerant microbial consortia from sediments of Copahue geothermal system with potential applications in bioremediation. J. Basic Microbiol. 59(7): 680–691.

Lopez-Bedogni, G., F.L. Massello, M.A. Giaveno, E.R. Donati and M.S. Urbieta. 2020. A deeper look into the biodiversity of the extremely acidic Copahue Volcano-Río Agrio system in Neuquén, Argentina. Microorganisms 8: 58. doi:10.3390/microorganisms8010058.

Mackintosh, M.E. 1978. Nitrogen fixation by *Thiobacillusferrooxidans*. Microbiology 105: 215–218.

Mejias-Carpio, I.E., A. Ansari and D.F. Rodrigues. 2018. Relationship of biodiversity with heavy metal tolerance and sorption capacity: a meta-analysis approach. Environ. Sci. Technol. 52: 184–194.

Orellana, R., C. Macaya, G. Bravo, F. Dorochesi, A. Cumsille, R. Valencia, et al. 2018. Living at the frontiers of life: extremophiles in chile and their potential for bioremediation. Front. Microbiol. 9: 1–25.

Oremland, R.S and J.F. Stolz. 2003. The ecology of arsenic. Science 300: 939–944.

Postgate, J.R. and L.L. Campbell. 1966. Classification of *Desulfovibrio* species, the nonsporulating sulfate-reducing bacteria. Bacteriol. Rev. 30: 732–738.

Raddadi, N., A. Cherif, D. Daffonchio, M. Neifar and F. Fava. 2015. Biotechnological applications of extremophiles, extremozymes and extremolytes. Appl. Microbiol. Biotechnol. 99: 7907–7913.

Rodríguez, A., J.C. Varekamp, M.J. Van Bergen, T.J. Kading, P. Oonk, C.H. Gammons, et al. 2016. Acid rivers and lakes at Caviahue-Copahue volcano as potential terrestrial analogues for aqueous paleo-environments on mars. pp. 141–172. *In*: F. Tassi, O. Vaselli and A.T. Caselli [eds.]. Copahue Volcano. Springer, Heidelberg, Germany.

Urbieta, M.S., M.A. Giaveno, E. González-Toril and E.R. Donati. 2011. Molecular assessment of prokaryotes along Upper Río Agrio (Neuquén, Argentina). pp. 936–945. *In*: G. Qiu, T. Jiang, W. Qin, X. Liu, Y. Yang and H. Wang [eds.]. Biohydrometallurgy: Biotech Key to Unlock Mineral Resources Value. Proceedings of the 19th International Biohydrometallurgy Symposium Vol 2. Central South University Press, Changsha, China.

Urbieta, M.S., E.G. Toril, A. Aguilera, M.A. Giaveno and E.R. Donati. 2012. First prokaryotic biodiversity assessment using molecular techniques of an acidic river in Neuquén, Argentina. Microb. Ecol. 64: 91–104.

Urbieta, M.S., E. González-Toril, A. Aguilera-Bazán, M.A. Giaveno and E.R. Donati. 2015a. Comparison of the microbial communities of hot springs waters and the microbial biofilms in the acidic geothermal area of Copahue (Neuquén, Argentina). Extremophiles 19: 437–450.

Urbieta, M.S., G. Willis-Porati, A.B. Segretin, E. González-Toril, M.A. Giaveno and E.R. Donati. 2015b. Copahue geothermal system: a volcanic environment with rich extreme prokaryotic biodiversity. Microorganisms 3: 344–363.

Varekamp, J.C., A.P. Ouimette, S.W. Herman, K.S. Flynn, A. Bermudez and D. Delpino. 2009. Naturally acid waters from Copahue volcano, Argentina. App. Geochem. 24: 208–220.

Watanabe, T., H. Kojima and M. Fukui. 2016. *Sulfuriferulathiophila* sp nov., a chemolithoautotrophic sulfur-oxidizing bacterium, and correction of the name *Sulfuriferulaplum bophilus* Watanabe, Kojima and Fukui 2015 to *Sulfuriferulaplumbiphila* corrig. Int. J. Syst. Evol. Microbiol. 66: 2041–2045.

6

Metallophillic Bacteria and Bioremediation of Heavy Metals

Lily Shylla, Augustine Lamin Ka-ot, Macmillan Nongkhlaw and SR Joshi*

Introduction

Microbe-metal interactions have been around since life began to take shape on Earth. Microbes have learned to coexist with various forms of metals where some are beneficial for microbes and others have no known biological significance. However, essential and non-essential metals tend to have a toxic effect on the microorganisms if present at higher concentrations. Metal toxicity also depends on the bioavailability of metal. Bioavailability is the soluble fraction of metal that can enter into the cell cytoplasm; hence, larger the bioavailability, more will be the toxicity of that particular metal. Bioavailability depends on the speciation of that metal, which in turn is determined by the physicochemical state of that environment. Metals, like copper, zinc, iron, sodium, magnesium, manganese, nickel, potassium, calcium, molybdenum, and cobalt, serve as micronutrients where they play an important role in various cellular processes. These essential metals are involved in redox-processes, stabilizing biomolecules through ionic interaction as co-factors of various enzymes and regulation of osmotic pressure (Bruins et al. 2000). Microorganisms like bacteria have been found to exist in naturally occurring metal-rich habitat, like ore deposits, hydrothermal vent, and volcanic area, and metal-contaminated site, like mining sites, industrial effluents, agricultural, and waste disposal sites. Bacteria inhabiting these metal-rich environments have evolved with various metal tolerance mechanisms for their survival and have potential bio-resources for exploring metal tolerant bacteria that are useful for bioremediation processes.

Metals or heavy metals are found in soil or water due to natural phenomena like parent rock weathering, landslide, volcanic eruption, and anthropogenic activities (Garrett 2000). Except for a small fraction of metals that have arrived on Earth extraterrestrially, others have been here for more than 4.5 billion years as part of biogeochemical cycle interchanging the phase from rocks to sediments due to weathering condition and back to rocks and land due to tectonic processes (Dalziel et al. 1999, Fyfe 1998). However, from the advent of man, especially during the rise of the industrial revolution, an accumulation of metals in certain pockets of the Earth has been seen, and there is no equal redistribution of metals since then (Garrett 2000). Some of the anthropogenic activities that increase the accumulation of metals in one particular environment are waste disposal and incineration, atmospheric deposition, urban effluents, fertilizer, and traffic emission in addition to mining and smelting (Bilos et al. 2001, Koch 2001). The level of pollution due to anthropogenic

Department of Biotechnology and Bioinformatics, North Eastern Hill University, Mawlai, Umshing, Shillong-793022.
*Corresponding author: srjoshi2006@yahoo.co.in

activities was more than natural sources of heavy metal (Roozbahani et al. 2015). Due to anthropogenic activities, metals or heavy metals and toxic metalloids have accumulated in surface soils (Akoto et al. 2017), deposited sediments from run-off and industrial effluents (Moore et al. 2011), agricultural soil (Toth et al. 2016), water and soils (Adegoke et al. 2009), and underground water (Balli and Leghouchi 2018).

Metallophilic Bacteria

Bacteria are ubiquitous living creatures where their existence has been described everywhere, right from the glacier of high mountain ranges, deep seawater to the human guts. Bacteria were first discovered by Antony van Leeuwenhoek (1673) (Prescott 5th Edition, October 2002) and since then they are classified into diverse class and family depending on their morphology, physiology, and molecular traits. However, they have also been classified according to their ecological niches. Bacteria that exist in extreme habitats are called in general as extremophile (Gerday and Glansdorff 2007, Shrestha et al. 2018). Extremophiles that thrive in the acidic environment are called Acidophiles; alkaline environments are called Alkalinophiles; those that live under pressure are called Barophiles; those that thrive at elevated temperature are called Thermophiles, and those that thrive in metal-rich environments are called Metallophilic Bacteria (Bajpai 2014).

Physiology Response Toward Metal Stress

Heavy metals, irrespective of their biological significance, can enter the cytoplasm of microorganisms in two different ways. One is through constitutively expressed, fast, and non-specific channels, generally driven by chemiosmotic gradient across the cytoplasmic membrane of bacteria. The other is an ATP dependent specific metal uptake system, which is relatively slow, inducible, and expressed only during starvation or special metabolic situation (Brown et al. 1992).

With respect to the biological system, heavy metals have been classified into two distinct categories: (i) toxic metals with no known biological function, and (ii) metals which are important to the growth and function of living organisms but which are toxic when present in higher concentrations. A heavy metal that once enters the cytoplasm can interfere with various metabolic processes. For example, copper is an essential heavy metal commonly found in the environment but is toxic at high concentrations. Copper causes cell damage by interfering with the active sites of cellular enzymes and the peroxidation of membranes (Lee et al. 1990). Similarly, toxic effects of non-essential metal like lead (II) has been studied in *Escherichia coli* cells. Studies showed that due to the similarities between the ionic radii and charge/radius ratio of Pb(II) and Ca(II), Pb(II) can replace Ca(II) at the binding site of lipopolysaccharide of *E. coli* cell wall leading to the rupture of the cell surface (Peng et al. 2007).

Metal toxicity in microorganisms is generally a result of five mechanisms: (i) competition for metal-ligand binding, leading to the disruption of the biological function of the targeted molecules; (ii) generation of hazardous reactive oxygen species (ROS) as a by-product of the reaction of the metal ion with thiol compound (RSH) like glutathione, thiols, and oxyanions, such as Pinner-type reactions where Se and Te oxyanions (SeO_4^{2-}, SeO_3^{2-}, TeO_4^{2-}, and TeO_3^{2-}) reacts with thiols compounds; (iii) productions of ROS through Fenton-type reaction, which involves the reaction of extremely reactive transition metals, like Cu, Ni, and Fe, that can oxidize all biological macromolecules; (iv) disruption of transport of essential substrates by inhibiting membrane transport machinery by interfering with their binding sites or interrupting the membrane potential; (v) disruption of respiratory chain via thiol-disulfide oxidoreductase leading to the destruction of proton motive force (Harrison et al. 2007). The physiological requirement of the biological systems to survive the toxicity of heavy metals is optimized by balancing the resistance mechanism and the normal cellular metal metabolism, which permits the cell to accumulate metal for maintenance of metal-dependent activities while counteracting to excess metal concentrations (Brown et al. 1992).

Microorganisms like bacteria have evolved various direct and indirect mechanisms that enable them to coexist and adapt to the presence of heavy metals.

Expulsion of Metals by Permeability Barrier: Some bacteria prevent the entry of heavy metals by modification in the cell wall or outer membrane to protect metal sensitive internal components. Periplasmic proteins, specific to certain metals, have described in some species of microbes like Cu-binding protein (CopM) that is encoded by operon *CopMRS* in the cyanobacterium, *Synechocystis* sp. PCC6803. The presence of Cu(I) in the periplasm induces the expression of this operon, leading to the expulsion of Cu(I). CopM binds Cu from the periplasm and then export it outside the cell to avoid any intracellular accumulation of Cu. Similar proteins like CusF in *E. coli*, CopK in *C. metallidurans*, and CueP in *Salmonella typhimurium* involved in Cu resistance were found in these bacteria (Franke et al. 2003, Mergeay et al. 2003, Pontel and Soncini 2009). The production of exopolysaccharides (EPS) by microbes also plays an important role in the retention of metal ions outside the cell. The presence of many functional anionic ligands, like carboxyl, phosphoryl, sulfhydryl, phenolic, and hydroxyl group, function as metal-binding sites for physical sorption, complexation, and precipitation of heavy metal (More et al. 2014). Hence, EPS production was found to increase in some microorganisms when treated with copper and zinc. Other studies showed that there are competitive adsorption and preferential binding due to the metal ion acting as a complexation agent of protein and protein-like substance (Yang et al. 2015). Promising sorption capacities of EPS, produced by marine bacteria, have also been evaluated in other studies as well (Deschartre et al. 2015).

Bioprecipitation: Metals interact with phosphate, oxalate, glutathione, sulfur, and other types of metabolites forming insoluble complexes (metal-metabolite complex), which cannot pass through the cell membrane, hence avoiding the absorption and intracellular harm. Sulfate-reducing bacteria that generate hydrogen sulfide by dissimilatory sulfate reduction represent a good example of such a mechanism where metals such as Zn, Fe, and Cu formed insoluble precipitates and thus reduce the concentration of bioavailable metal. Another example is the production of hydrogen sulfide from thiosulfate by *Klebsiella planticola* under the anaerobic condition, which precipitate cadmium ions as insoluble sulfides (Sharma et al. 2000). Cadmium precipitation was observed for *Pseudomonas aeruginosa* strain under aerobic conditions (Wang et al. 2002) and *Vibrio harveyi* strain precipitated soluble divalent lead as a complex lead phosphate salt (Mire et al. 2004). Multi-resistant *Pseudomonas putida* S4 strain under carbon-limiting conditions formed insoluble precipitate composed of copper ions, hydroxyl, and phosphate residues (Saxena and Srivastava 1998).

Intracellular Sequestration of Metals: Some bacteria have the ability to interact with complex metals and accumulate them in the cytoplasm using complexants, such as metallothioneins and glutathione. Additionally, heavy metals such as Pb and Hg may get accumulated in polyphosphate inclusions (Guillian 2016). In *Cupriavidus metallidurans*, CH34 a regulatory protein PbrR induces the expression of a protein PbrD. PbrD is an intracellular Pb binding protein where the binding site is rich in cysteine residue and also have a large number of serine and proline residue that sequester Pb decreasing its toxic effects. Mer operon also have a similar mechanism that gets induced by the presence of Hg^{+2}, where merR is the activator of the expression, and merC is the protein involved in the transport and accumulation of Hg (Das et al. 2016). Bacteria such as *P. putida* strain have the ability of intracellular sequestration of zinc, cadmium, and copper ions with the help of cysteine-rich low molecular-weight proteins (Higham et al. 1986). *Pseudomonas diminuta* strain produced several low and high-molecular-weight silver-binding proteins (Ibrahim et al. 2001), and some marine gamma-proteobacteria produced cadmium-inducible low-molecular-weight proteins similar to phytochelatins (Ivanova et al. 2002). Similarly, *Rhizobium leguminosarum* can sequester cadmium ions by producing gluthathione (Lima et al. 2006).

Efflux by Active Transport: Active transport is a mechanism used by microorganisms to remove toxic metals from their cytoplasm. Metal transporters involved in this mechanism are encoded by chromosomes or plasmids (Lima et al. 2018). Most of the non-essential metals enter the cell through transport systems that are normally used to transport nutrients. However, these metals are rapidly exported; for example, cadmium, zinc, cobalt, nickel, and manganese enter the cells of *Ralstonia metallidurans* using systems of magnesium transport, chromate is transported inside the cell via sulfate transport system (Nies and Silver 1989, Cervantes and Gutierrez-Corona 1994), and for export metal ions from the cell, electrochemical gradient or ATP hydrolysis are being used (Nies 1995, Rensing et al. 1999). Efflux systems can be either be ATP dependent or independent but are highly specific for the cation or anion (Bruins et al. 2000). For example, Ni resistance founded in *Cupriavidus metallidurans* is mostly mediated by active transport where operon *cnrYHXCBAT*, which encoded the CnrCBA efflux pump that gets activated when Ni is present even in μM concentration. When periplasmic Ni binds to the membrane-bound protein, cnrX, the cnrCBA efflux pump gets transcript indirectly that expels Ni from the cell (Grass et al. 2000). Some bacteria can employ other mechanisms of heavy metal resistance combined with efflux systems. For example, *P. putida* S4 strain transports copper ions by ATPase efflux system from the cytoplasm with subsequent sequestration in the periplasm (Saxena et al. 2002). Another example of such a dualistic system is in arsenic resistance *ars* system, found both in gram-positive and gram-negative bacteria. *Ars* operon encodes ATPase pump ArsA/ArsB and ArsC reductase, where arsenate is first enzymatically reduced to arsenite by cytoplasmic ArsC arsenate reductase and then gets exported by efflux system through the plasma membrane (Mukhopadhyay et al. 2002).

Transformation and Detoxification: This mechanism involved processes such as oxidation or reduction, methylation, or volatilization. Detoxification of metal such As is one of the good examples, As(V) is less toxic and mobile compared to As (III), and some microorganisms use oxidation as a strategy for resistance. Protein such as *ArsC* in some microorganisms involved in the oxidation of As(III) to As(V) just for detoxification purposes only, and some chemolithoautotrops during fixation of CO_2 oxidase arsenite that couples reduction to arsenite oxidation with *aioAB* genes (Yamamura and Amachi 2014). Gene *arsM* which code for an arsenite S-adenosylmethionine methyltransferase transforms inorganic As to less toxic methylated form (Lima et al. 2018). Enzymatic reduction of mercury and chromium to their less toxic forms has been reported (Barkay et al. 2003, Viti et al. 2003). Biovolatilization occurs in microbes with the capability of removing Hg in a process mediated by a mercury ion reductase coded by the *merA* gene. This enzyme reduces Hg^{+2} to the volatile form Hg^0, which gets diffuses out of the cell (Lima et al. 2018).

Metallophilic Bacteria and Bioremediation

Heavy metals are naturally occurring elements, but their concentration in the environment is increasing day by day largely due to human activities (Camobreco et al. 1996, Nematian and Kazemeini 2013). There is a rapid acceleration of heavy metals pollution in the biosphere with the continuous spread of industrial activity, like smelting, mining, and synthetic compound creation. These have led to an increase in the amounts of heavy metals released into the water, soil, and atmosphere which poses major environmental problems (Gisbert et al. 2003, McConnell and Edwards 2008). Toxic heavy metals like Pb, Cr, Hg, Cu, Cd, Zn, and Cr have no biological role instead they are well-known for their toxicity, mutagenic, and carcinogenic impact on human beings and other living systems (Adriano 2001). Due to various reasons stated earlier, removal of toxic heavy metals from soil and water is of the utmost need of the hour. Furthermore, heavy metals unlike other environmental pollutants, are not chemically or biologically degradable (Malik et al. 2016). Treatment of metal-contaminated sites using physicochemical techniques has been effectively applied for the removal of toxic heavy metals. Conventional methods used for the removal of heavy metals in contaminated effluent include chemical precipitation, ion exchange, electrochemical treatment, evaporation, sorption, and reverse osmosis (Mustaphaa and Halimoon 2015). Similarly, physicochemical techniques like soil isolation, soil replacement, soil washing, electrokinetic remediation, vitrification, encapsulation, and immobilization techniques have been used for remediation of contaminated soil (Khalid et al. 2017). Physical and chemical methods have some unavoidable drawbacks as most of these techniques are inefficient when the heavy metals concentrations are less than 100 mg/L (Ahluwalia and Goyal 2007). Furthermore, most heavy metals get dissolved in wastewater facilely and cannot be separated by physical separation methods (Hussein et al. 2004). Physical and chemical techniques are usually associated with low capability, high-cost, the involvement of unwanted chemicals, and the release of enormous sludge to the environment as a by-product. Hence, the use of biological methods that are cost-effective and environmentally friendly has emerged as an alternative technique for the treatment of heavy metals pollution (Srivastava et al. 2008). Microorganisms thriving in heavy metal polluted environments develop resistance mechanisms to toxic metal ions. Microorganisms employ various methods in response to metal stress; these include biosorption to the cell walls, entrapment in extracellular capsules, complexation, precipitation, and oxidation-reduction reactions transport across the cell membrane (Ayangbenro and Babalola 2017). Metal-resistant bacteria also play an important role in the biogeochemical cycling of those metal ions and in cleaning up metal-contaminated sites (Nies 1999). The coal mine area in Bokaro, a 'Toxic Hotspot' in India, was investigated for the presence of bacteria that might possess uptake and efflux mechanisms, which enable these bacteria to exist in heavy metals contaminated environments. Metal-contaminated sites tend to harbor metal tolerant bacteria that can be exploited for bioremediation processes (Gandhi et al. 2015). Microbes-metals interactions and their potential to remove metals from contaminated sites represent a unique process and a potent bioremediator observed by the U.S. Geological Survey (USGS) (Mueller et al. 1989). As heavy metals are natural elements, degradation and metabolism are not possible, but microorganisms have evolved the ability to either transform the element to a less toxic form, bind metals intra- or extracellularly or actively transport metals out of the cell cytosol to prevent any kind of interactions in the host cell (Mowll and Gadd 1984, Hamlett et al. 1992, Wester 1992).

Removal of heavy metals by bioremediation has received great attention in recent years because it is eco-friendly and cost-effective as compared to the conventional chemical and physical methods. Moreover, bioremediation processes were found to be effective even when metal concentrations are low as had been reported by Blaylock et al. 1997. One study showed that 50-65% of the cost was saved when bioremediation was used for the treatment of one acre of Pb polluted soil as compared with the use of a conventional method such as excavation and landfill which produce significant amounts of toxic sludge (Blaylock et al. 1997). Bioremediation is a scientific origination which gives the probable application in industry effluent (Samantaray et al. 2014) and AMD sites (Natarajan 2008). A large number of bacterial isolates that inhabit metal-polluted soils have evolved various strategies to resist themselves against metal stress. Such metal-resistant bacteria isolates can be used as successful bioremediation agents (Ahemad 2012, Khan et al. 2009). Metal-resistant bacteria have been previously reported and isolated from metal-contaminated soils, waters, and sediments. Some of these isolates that have been identified and characterized belonged to *Bacillus* spp. (*Bacillus cereus, B. circulans, B. subtilis, B. firmus, B. licheniformis*), *Kocuria* sp., *Micrococcus* sp., *Pseudomonas* spp. (*P. aeruginosa*), *Cinetobacter* sp., *Sporosarcina* sp. (*S. saromensis*), *Staphylococcus* sp., *Acinetobacter* sp. (*A. haemolyticus*), *Streptomyces* sp., *Stenotrophomonas* sp., *Cellulosimicrobium* sp., *Methylobacterium* sp. (*M. organophilum*), *Gemella* sp., *Enterobacter* sp. (*Enterobacter cloacae*), *Flavobacterium* sp., *Arthrobacter* sp., *Klebsiella* sp. (*K. pneumonia*), *Vibrio* spp. (*V. parahaemolyticus, V. fluvialis*), *Alcaligenes* sp. (*A. faecalis*), *Brevibacterium* sp. (*B. iodinium*), and some sulfate-reducing bacteria like *D. desulfuricans* etc. can be used for bioremediation of heavy metals from the contaminated site (Igiri et al. 2018).

Bioremediation of Heavy Metals by Metallophilic Bacteria

Bioremediation involves the use of living organisms or their biomass as a tool to remove the pollutants or convert them into a non-toxic form. The use of microorganisms like bacteria for the removal of heavy metals has emerged as an environmentally friendly technology to solve the problem of heavy metal contamination. Bioremediation of heavy metal involves the use of the tolerant mechanism use by these microorganisms against heavy metal, hence the following methods have been exploited for the removal of heavy metals by bacteria.

Microbial Biosorption of Heavy Metals

Biosorption of heavy metals is one of the mechanisms that were exploited for bioremediation of heavy metal. Biosorption either involves the adsorption of metal on the cell surface or the intracellular absorption of metal. Generally, microorganisms accumulate heavy metals either actively (bioaccumulation) and/or passively (adsorption) (Hussein et al. 2001). In the adsorption process, the cellular structure of microbes can trap heavy metal ions and eventually sorb them onto the binding sites of the cell wall (Malik 2004). This mechanism is independent of the metabolic cycle, and both the kinetic equilibrium and constitution of the metal at the cellular surface play a key role in the amount of metal getting sorbed. Biosorption is a swift reaction that reaches equilibrium in few minutes (Ayangbenro and Babalola 2017), and there are many mechanisms that are involved, such as precipitation, the redox process, electrostatic interaction, ion exchange, and surface complexation (Yang et al. 2015). Biosorption either involves uptake by complexation of the surface part of cells or by dead biomass/living cells as passive onto the cell wall and other outer layers (Fomina and Gadd 2014). Extracellular polymeric substances are amongst the various reactive compounds associated with bacterial cell walls that play an important role in metal adsorption (Guiné et al. 2006). The bacterial cell wall is known to possess various functional groups, like phosphates, nitrates oxides, and sulfides, which act as nucleation centers for the adsorption of heavy metals. Metal cations form various coordination complexes with these functional groups, leading to the adsorption of heavy metal on the cell surface. Furthermore, the bacterial cell wall is rich in negatively charge phosphoric acid and carboxylic acid anions that serve as binding sites for metal cations. According to Wang et al. 2002, adsorption to the cell wall is the first line of defense against heavy metals, and this property has been exploited for the removal of heavy metals. While in bioaccumulation, the heavy metal ions move across the cell membrane into the cytoplasm through the cell metabolic cycle, and this process is dependent on a variety of mechanisms, such as physical, chemical, and biological. As it is a process taken up by the living cell and these factors include intracellular and extracellular processes, bioaccumulation plays a limited role (Fomina and Gadd 2014). In large scale applications, the bio-adsorption process seems to be more feasible than bioaccumulation as live microbes require the addition of nutrients to support their active uptake of heavy metals that in turn increases the chemical oxygen demand and biological oxygen demand in the waste. In addition to that, it is strenuous to maintain a viable number of microbes due to heavy metal stress and other environmental factors (Ajmal et al. 1996, Dilek et al. 1998). Bacteria can provide a large surface area that can quickly bind to metals (Haferburg and Kothe 2007). For example, one *Bacillus* sp. was able to remove 60% of Cu^+ within one minute of incubation and reached the equilibrium after ten minutes incubation. Microbial adsorption of heavy metals, like Fe(III) and Cr(VI), has been reported for *Streptococcus equisimalis* and likewise *Stenotrophomonas maltophilia*, and *Bacillus subtilis* have been found to show encouraging adsorption capacities for removal of Pb(II), Zn(II), and Ni(II) from industrial wastewater (Wierzba 2015). Biosorption also involves the accumulation of heavy metal in the cytoplasm of the cell. Bacteria are known to accumulate heavy metal in the cytoplasm with the help of glutathione and cysteine-rich biomolecules. Table 1 shows the various microbes that have been found to possess heavy metal biosorption potential. Biosorption presents a viable cost-effective and eco-friendly option for bioremediation of heavy metal and development of microbial biomass that can remove a variety of metal from contaminated effluents, which further enhances the effectiveness of biosorption. Furthermore, the development of recoverable biomass and metal recovery for reusable purposes will attract the use of this method for wider applications.

TABLE 1: List of Microbes Involved in Bioremediation

Biosorption	Metal	References
Pseudomonas fluorescens	Th, U	Tsezos and Volesky (1981)
Pseudomonas sp.	.U	Pons and Fuste (1993)
Ochrobactrum anthropi	Cr, Cd, Cu	Ozdemir et al. (2003)
Thiobacillus ferrooxidans	Zn, Cr	Baillet et al. (1998), Celaya et al. (2000)
Rhodococcus erythropolis	Cd, Zn	Plette et al. (1996)
Ocimum basilicium	Cr	Melo and D'Souza (2004)
Sphingomonas paucimobilis	Cd	Tangaromsuk et al. (2002)
Acinetobacter sp. *B9*	Ni	Bhattacharya and Gupta (2013)
Enterobacter cloacae	Co	Jafari et al. (2015)
Vibrio parahaemolyticus (PG02)	Hg	Jafari et al. (2015)
Stenotrophomonas sp.	Cr	Gunasundari and Muthukumar (2013)
Cellulosimicrobium sp. *(KX710177)*	Pb	Bharagava and Mishra (2018)
Bacillus firmus	Zn	Salehizadeh and Shojaosadati (2003)
Bioprecipitation		
Acinetobacter johnsonii	Lanthanum	Boswell et al. (1999, 2001)
Pseudomonas aeruginosa	Uranyl ions	Renninger et al. (2004)
Sporosarcina ginsengisoli	As	Achal et al. (2012a)
Exiguobacterium unda	Cd	Kumari et al. (2014)
Terrabacter tumescens	Cd	Li et al. (2013)
Kocuria flava	Cu	Achal et al. (2011)
Bacillus sp.	Cr(VI)	Achal et al. (2013)

Bioprecipitation of Metals

Soluble heavy metals present a more threat to the health of water bodies as it increases the bioavailability of these metals, which leads to a wider distribution and easy entry into living cells. Hence, one of the methods for removal of these heavy metals water bodies involves chemical precipitation, which requires the addition of chemicals that can precipitate the metals from the water bodies. This method was found to involve high operational cost and additional incorporation of chemicals into the water bodies could give rise to another environmental hazard. Bioprecipitation involves the use of microorganisms, especially bacteria to precipitate heavy metal from the aqueous phase. Bacteria are known to produce metabolites that react with metals to form metal-complexes that eventually precipitate. Some of the well-known mechanisms of bioprecipitation are the formation of metal sulfide by Sulfate Reducing Bacteria (SRB), carbonate precipitation, and phosphate precipitation (Postgate 1984, Macaskie et al. 1987, Kumari et al. 2016). SRB mediated bioprecipitation is one of the well-explored methods for the treatment of heavy metal contamination in an anaerobic environment. SRB have been used to treat acid mine, drainage, wastewater, and contaminated industrial soil (Bhagat et al. 2004, Costa et al. 2008, Torres et al. 2012, Zhao et al. 2018). Metal-phosphate bioprecipitation is an enzymatically driven reaction leading to the biomineralization of metals. Phosphate precipitation has been reported for the biomineralization of radionuclide like uranium (Mackaskie 1990). Bacteria like *Citrobacter* sp. use phosphatase enzyme to release inorganic phosphate into the environment which in turn reacts with uranium leading to the formation of insoluble uranium phosphate (Montgomery et al. 1995, Barkay and Schaeffer 2001). Microbial induced carbonate precipitation uses different kinds of metabolic pathways, such as ureolysis, denitrification, ammonification, photosynthesis, methane oxidation, sulfate reduction, and anaerobic sulfide oxidation (Phillips et al. 2013, Anbu et al. 2016, Zhu and Dittrich 2016). Carbonate bioprecipitation has been used as a soil improvement technique reported, and bacteria such as *Bacillus* sp., *Enterobacter cloacae*, *Sporosarcina* sp., and others have been studied for their application in immobilization of metal(loid)s by microbially-induced metal carbonate precipitation (Kumari et al. 2016, Whiffin et al. 2007).

Bioremediation by Microbial Transformation

Detoxification of heavy metals from toxic state to non-toxic state usually happened through a mechanism called valence transformation. This is true especially when metal toxicity is due to differences in the valence state of metals. For example, methyl mercury was converted to less toxic state Hg(II) by organomercurial lyase in mercury resistant bacteria, which is a hundredfold reduction in toxicity of methyl mercury (Wu et al. 2010). Another example is the reduction of chromium Cr(VI) to Cr(III) where Cr(III) is less toxic and mobile. Other mechanisms of metal detoxifications are metal-binding, volatilization, and vacuole compartmentalization. Binding of metals is facilitated by chelators like metallothein, phytochelatin, and metal-binding peptides. Volatilization happens with Se and Hg in which mercury resistant bacteria employed the MerA enzyme and reduced Hg(II) to volatile Hg(0) (Wu et al. 2010). *Bacillus pumilus*, *Bacillus* sp., *Alcaligenes faecalis*, *Brevibacterium iodinium*, and *Pseudomonas aeruginosa* employed several mechanisms of heavy metal detoxification namely through volatilization (for Hg) and putative entrapment in the extracellular polymeric substance (for Hg, Cd, and Pb) (Ramaiah and Vardanyan 2008).

Genetically Modified Microbes for Bioremediation

The uses of microorganisms have emerged as an effective supplementation to existing technologies to treat various forms of environmental pollution. Recent advances in biotechnology and genetic engineering have further improved the use of microorganisms for bioremediation. As mentioned earlier, bacteria have employed different mechanisms, such as biosorption, bioaccumulation, biotransformation, and biomineralization, for surviving in heavy metal-polluted habitats. These mechanisms can be exploited for bioremediation either *ex situ* or *in situ* (Gadd 2000, Lim et al. 2003, Malik 2004, Lin and Lin 2005). Bioremediation of heavy metals using indigenous microorganisms has been working in the removal of Cr, Pb, Ni, and Zn. However, most of the indigenous microorganisms have not shown effective results for certain groups of metals. For example, naturally occurring bacteria cannot remove heavy metals, such as Hg, from the environment, but a genetically engineered bacterium *Deinococcus geothemalis*, which express mer operon taken from *E. coli* have acquired Hg reduction properties, and modification of *Pseudomonas* strain with the pMR68 plasmid with novel genes (mer) made that strain resistant to mercury (Brim et al. 2003, Sone et al. 2013). Development of modern molecular biology provides the platform and opportunity to modify the genomic makeup of the microbes by inserting genes with a new function or manipulate the rate-limiting steps of intracellular or extracellular enzymes, which are involved in removing of metals or modification in the outer membrane proteins for improving metal-binding abilities that in turn could enhance the remediation process in comparison with the indigenous bacteria (Wang et al. 2014, Dixit et al. 2015, Das et al. 2016). For example, bioaccumulation of essential metals is a necessity for carrying out various cellular processes (such as signaling, enzyme catalysis, and stabilizing charges on biomolecules). Genetically engineered microorganisms expressing recombinant import-storage systems allow for superior uptake and sequestration of heavy metal ions. (Patrick et al. 2018).

Manipulation of bacterial systems for improving bioremediation of heavy metals includes the following strategies:

Modification a Single Gene or Operon: Microorganisms prevalent in the high heavy metals contaminated sites developed resistance toward heavy metal-mediated stress by modulating various genetic mechanisms. Thus, these strains are preferable for bioremediation application, and insertion of a single gene or gene cluster will enhance the bioremediation application as these strains already have the ability to adapt to the contaminated sites. For example, transformed *E. coli* JM109 shows enhanced expression of polyphosphate kinase (ppk) genes, and metallothionein (mtl) results in the accumulation of Hg more than 100 µM (Ruiz et al. 2011), *B. cereus* BW03 (pPW-05) that possess the mer operon-containing plasmid from a wild strain of *B. thuringiensis* PW-05 have both Hg volatilization as well as Hg biosorption, and this transgenic strains can be used for removal of mercury from the contaminated environments (Dash and Das 2015). Similarly, genetically modified *C. metallidurans* strain MSR33 shows tolerant to both methylmercury and inorganic mercury in addition to chromate and c copper (Rojas et al. 2011). Another example of the use of genetic engineering to enhance bioremediation of heavy metal is the expression of a phytochelatin analog and cyanobacterial metallothionein (MT) gene smtA in *Deinococcus radiodurans* R1 that leads to higher tolerance and accumulation to Cd^{2+} (Chaturvedi and Archana 2014). Insertion of desirous genes into compatible bacteria present naturally in the environment provides a better opportunity for remediation (Das et al. 2016).

Alteration of Existing Gene Sequences: Although the genetically engineered microorganisms provide a better approach for remediation, the expression of these genes in a foreign host can make these proteins unstable and may not give the desired outcome (Dixit et al. 2015, Kiyono and Pan-Hou 1999). To overcome this problem, other techniques can be employed where the intrinsic genes may be manipulated (Das et al. 2016). There are many studies where alteration/modification of the existing gene clusters of indigenous microbes for bioremediation application have been reported. Cadmium-resistant *Pseudomonas aeruginosa* isolated from industrial sludge when mutated by exposure to the dyes acridine orange and acriflavine increased the tolerance of these strains towards Cd^{2+} (Kermani et al. 2010). Similarly, overexpression of NfsA in *E. coli* increased chromate reduction, and *Pseudomonas putida* increased Cr^{6+} reduction with overproduction of ChrR (Ackerley et al. 2004, Gonzalez et al. 2005).

Pathway Switching: The construction, extension, and regulation of certain novel genetic mechanisms for bioremediation applications are involved in pathway switching. In order to achieve complete bioremediation for heavy metals, construction of a microbial consortium each with a specific path can be an effective measurement; however, there have been very limited studies carried out so far for the improvement of the bacterial strains for metal removal by the construction of novel consortium (Das et al. 2016). GMOs can also be employed for heavy metal removal by the pathway switching approach like in case of methanotrophic bacteria were mechanism can be developed for simultaneous uptake of Cu-Mb complexes rather than Cu dissociating from Mb prior to uptake (Balasubramaniam et al. 2011). Thus, having knowledge of pathways involved in metal transport machinery can facilitate the acquisition of metals from the environment and can play a key role in future research (Das et al. 2016).

Conclusions

Microbial bioremediation has been shown to be a feasible alternate technology for cleaning up heavy metal-contaminated environments. Furthermore, with the advancement of genetic engineering and a thorough understanding of the basic tolerance mechanism, these microbes can be engineered for better adaptability and a wider range of heavy metals removal capacities.

References

Achal, V., X. Pan and D. Zhang. 2011. Remediation of copper contaminated soil by Kocuriaflava CR1, based on microbially induced calcite precipitation. Ecol. Eng. 37: 1601–1605.

Achal, V., X. Pan, D. Zhang and Q.L. Fu. 2012. Biomineralization based remediation of As (III) contaminated soil by *Sporosarcinaginsengisoli*. J. Hazard. Mater. 201–202: 178–184.

Achal, V., X. Pan, D.J. Lee, D. Kumari and D. Zhang. 2013. Remediation of Cr (VI) from chromium slag by biocementation. Chemosphere 93: 1352–1358.

Ackerley, D.F., C.F. Gonzalez, M. Keyhan, R. Blake and A. Matin. 2004. Mechanism of chromate reduction by the *Escherichia coli* protein, NfsA, and the role of different chromate reductases in minimizing oxidative stress during chromate reduction. Environ. Microbiol. 6: 851–860.

Adegoke, J.A., W.B. Agbaje and O.O. Isaac. 2009. Evaluation of heavy metal status of water and soil at ikogosi warm spring, ondo state Nigeria. EJESM. 2(3): 88–93.

Adriano, D.C. 2001. Trace Elements in the Terrestrial Environment, 2nd edition. Springer-Verlag, New York.

Ahemad, M. 2012. Implications of bacterial resistance against heavy metals in bioremediation: a review. IIOABJ. 3: 39–46

Ahluwalia, S.S. and D. Goyal. 2007. Microbial and plant derived biomass for removal of heavy metals from wastewater. Bioresour. Technol. 98: 2243–2257.

Ajmal, M., A.K. Rafaqat and A.S. Bilquees. 1996. Studies on removal and recovery of Cr (VI) from electroplating wastes. Water Res. 30: 1478–1482.

Akoto O., N. Bortey–Sam, Y. Ikenaka, S.M.M. Nakayama, E. Baido, Y.B. Yohannes, et al. 2017. Contamination levels and sources of heavy metals and a metalloid in surface soils in the kumasi metropolis, Ghana. J. Health Pollut. 7(15): 28–39.

Anbu, P., C.H. Kang, Y.J. Shin and J.S. So. 2016. Formations of calcium carbonate minerals by bacteria and its multiple applications. Springerplus 5: 250.

Atkins, P.W. and L. Jones. 1997. Chemistry–Molecules, Matter and Change, 3rd Edition. New York. W.H. Freeman.

Ayangbenro, A.S. and O.O. Babalola. 2017. A new strategy for heavy metal polluted environments: a review of microbial biosorbents. Int. J. Environ. Res. Public Health 14(1): 94.

Azubuike, C.C., C.B. Chikere and G.C. Okpokwasili. 2016. Bioremediation techniques–classification based on site of application: principles, advantages, limitations and prospects. World J. Microbiol. Biotechnol. 32(11): 180.

Baillet, F., J.P. Magnin, A. Cheruy and P. Ozil. 1998. Chromium precipitation by the acidophilic bacterium *Thiobacillus ferrooxidans*. Biotechnol. Lett. 20(1): 95–99.

Bajpai, B. 2014. Extremophiles – a biotechnological perspective. Quest. 2(1): 14–19.

Balasubramanian, R., G.E. Kenney and A.C. Rosenzweig. 2011. Dual pathways for copper uptake by methanotrophic bacteria. J. Biol. Chem. 286: 37313–37319.

Balli, N. and E. leghouchi. 2018. Assessment of lead and cadmium in groundwater sources used for drinking purposes in Jijel (Northeastern Algeria). Global Nest J. 20(2): 417– 423.

Barkay, T., S.M. Miller and A.O. Summers. 2003. Bacterial mercury resistance from atoms to ecosystems. FEMS Microbiol. Rev. 27(2-3): 355–384.

Barr, D. 2002. Biological Methods for Assessment and Remediation of Contaminated Land: Case Studies. Construction Industry Research and Information Association, London.

Bhagat, M., J.E. Burgess, A.P.M. Antunes, C.G. Whiteley and J.R. Duncan. 2004. Precipitation of mixed metal residues from wastewater utilising biogenic sulphide. Miner. Eng.17(7–8): 925–932.

Bharagava, R.N. and S. Mishra. 2018. Hexavalent chromium reduction potential of *Cellulosimicrobium* sp. isolated from common effluent treatment plant of tannery industries. Ecotoxicol. Environ. Saf. 147: 102–109.

Bhattacharya, A. and A. Gupta. 2013. Evaluation of Acinetobacter sp. B9 for Cr (VI) resistance and detoxification with potential application in bioremediation of heavy–metals–rich industrial wastewater. Environ. Sci. Pollut. Res. 20(9): 6628–6637.

Bilos, C., J.C. Colombo, C.N. Skorupka and M.J. Rodriguez Presa. 2001. Sources, distribution and variability of airborne trace metals in La Plata City area, Argentina. Environ. Pollut. 111(1): 149–158.

Blaylock, M.J., D.E. Salt, S. Dushenkov, O.N. Zakharova, D. Christopher, C.D. Gussman, et al. 1997. Enhanced accumulation of Pb in Indian mustard by soil–applied chelating agents. Environ. Sci. technol. 31(3): 860–865.

Boswell, C.D., R.E. Dick and L.E. Macaskie. 1999. The effect of heavy metals and other environmental conditions on the anaerobic phosphate metabolism of *Acinetobacterjohnsonii*. Microbiology 145: 1711–1720.

Boswell, C.D., R.E. Dick, H. Eccles and L.E. Macaskie. 2001. Phosphate uptake and release by Acinetobacterjohnsonii in continuous culture and coupling of phosphate release to heavy metal accumulation. J. Ind. Microbiol. Biotechnol. 26: 333–340.

Brim, H., A. Venkateshwaran, H.M. Kostandarithes, J.K. Fredrickson and M.J. Daly. 2003. Engineering *Deinococcusgeothermalis* for bioremediation of high temperature radioactive waste environments. Appl. Environ. Microbiol. 69: 4575–4582.

Brown, N.L., D.A. Rouch and B.T.O. Lee. 1992. Copper resistance determinants in bacteria. Plasmid 27: 41–51.

Bruins M.R., S. Kapil and F.W. Oehme. 2000. Microbial resistance to metals in the environment. Ecotoxicol. Environ. Saf. 45: 198–207.

Camobreco, V.J., B.K. Richards, T.S. Steenhuis, J.H. Peverly and M.B. McBride. 1996. Movement of heavy metals through undisturbed and homogenized soil columns. Soil Sci. 161: 740–750.

Celaya, R.J., J.A. Noriega, J.H. Yeomans, L.J. Ortega and A. Ruiz-Manri-quex. 2000. Biosorption of Zn(II) by *Thiobacillus ferrooxidans*. Bioprocess. Eng. 22: 539–542.

Cervantes, C. and F. Gutierrz-Corona. 1994. Copper resistance mechanisms in bacteria and fungi. FEMS Microbiol. Rev. 14(2): 121–138.

Chaturvedi, R. and G. Archana. 2014. Cytosolic expression of synthetic phytochelatin and bacterial metallothionein genes in *Deinococcusradiodurans* R1 for enhanced tolerance and bioaccumulation of cadmium. Biometals 27(3): 471–482.

Choudhury, R. and S. Srivastava. 2001. Zinc resistance mechanisms in bacteria. Curr. Sci. 81(7): 768–775.

Costa, M., M. Martins, C. Jesus and J. Duarte. 2008. Treatment of acid mine drainage by sulphate–reducing bacteria using low cost matrices. Water Air Soil Pollut. 189(1): 149–162.

Dalziel, I.W.D. 1999. Vestiges of a beginning and the prospect of an end. pp. 119–155. *In*: G.Y. Craig and J.H. Hull [eds.]. James Hutton–Past and Future, Geological Society of London Special Publications.

Das, S., H.R. Dash and J. Chakraborty. 2016. Genetic basis and importance of metal resistant genes in bacteria for bioremediation of contaminated environments with toxic metal pollutants. Appl. Microbiol. Biotechnol. 100: 2967–2984.

Dash, H.R. and S. Das. 2015. Enhanced bioremediation of inorganic mercury through simultaneous volatilization and biosorption by transgenic marine bacterium *Bacillus cereus* BW–03(pPW–05). Int. Biodeter. Biodegr. 103: 179–185.

de Oliveira Martins, P.S., N.F. de Almeida and S.G. Leite. 2008. Application of a bacterial extracellular polymeric substance in heavy metal adsorption in a co–contaminated aqueous system. Braz. J. Microbiol. 39(4): 780–786.

Deschartre, M., F. Ghillebaert, J. Guezennec and C. Simon–Colin. 2015. Study of biosorption of copper and silver by marine bacterial exopolysaccharides. WIT Trans. Ecol. Environ. 196: 549–559.

Dilek, F.B., C.F. Gokcay and U. Yetis. 1998. Combined effects of Ni(II) and Cr(VI) on activated sludge. Water Res. 32: 303–312.

Dixit, R., D. Malaviya, K. Pandiyan, U.B. Singh, A. Sahu, R. Shukla, et al. 2015. Bioremediation of heavy metals from soil and aquatic environment: an overview of principles and criteria of fundamental processes. Sustainability 7: 2189–2212.

Duffus, J.H. 2002. Heavy metals—a meaningless term? Pure Appl. Chem. 74(5): 793–807.

Fergusson, J.E. 1990. The Heavy Elements: Chemistry, Environmental Impact and Health Effects. Oxford, Pergamon, Press.

Fomina, M. and G.M. Gadd. 2014. Biosorption: current perspectives on concept, definition and application. Bioresour. Technol. 160: 3–14.

Franke, S., G. Grass, C. Rensing and D.H. Nies. 2003. Molecular analysis of copper–transporting efflux system CusCFBA of *Escherichia coli*. J. Bacteriol. 185: 3804–3812.

Fyfe, W.S. 1998. Toward 2050; the past is not the key to the future; challenges for the science of geochemistry. J. Environ. Geol. 33: 92–95.

Gadd, G.M. 2000. Bioremedial potential of microbial mechanisms of metal mobilization and immobilization. Curr. Opin. Biotechnol. 11: 271–279.

Gadd, G.M. 2001. Accumulation and transformation of metals by microorganisms. pp. 225–264. *In*: H.J. Rehm, G. Reed, A. Puhler and P. Stadler [eds.]. Biotechnology Set, 2nd ed.; John Wiley and Sons Inc., New York, USA.

Gandhi, V.P., A. Priya, S. Priya, V. Daiya, J. Kesari, K. Prakash, et al. 2015. Isolation and molecular characterization of bacteria to heavy metals isolated from soil samples in Bokaro coal mines, India. Pollution 1: 287–295.

Garrett, R.G. 2000. Natural sources of metals to the Environment. Hum. Ecol. Risk Assess 6(6): 945–963.

Gerday, C. and N. Glansdorff. 2007. Physiology and Biochemistry of Extremophiles. ASM. pp. 429.

Gisbert, G., R. Ros, A.D Haro, D.J. Walker, M.P. Bernal, R. Serrano, et al. 2003. A plant genetically modified that accumulates Pb is especially promising for phytoremediation. Biochem. Biophy. Res. Com. 303: 440–445.

Gonzalez, C.F., D.F. Ackerley, S.V. Lynch and A. Matin. 2005. ChrR, a soluble quinonereductase of Pseudomonas putida that defends against H_2O_2. J. Biol. Chem. 280: 22590–22595.

Goyal, N., S.C. Jain and U.C. Banerjee. 2003. Comparative studies on the microbial adsorption of heavy metals. Adv. Environ. Res. 7(2): 311–319.

Grass, G., C. Grose and D.H. Nies. 2000. Regulation of the cnr cobalt and nickel resistance determinant from *Ralstonia* sp. strain CH34. J. Bacteriol. 182: 1390–1398.

Guillian, D.C. 2016. Metal resistance systems in cultivated bacteria: are they found in complex communities? Curr. Opin. Biotechno. 38: 123–130.

Guiné, V., L. Spadini, G. Sarret, M. Muris, C. Delolme, J.P. Gaudet, et al. 2006. Zinc sorption to three gram-negative bacteria: Combined titration, modeling and EXAFS study. Environ. Sci. Technol. 40: 1806–1813.

Gunasundari, D. and K. Muthukumar. 2013. Simultaneous Cr(VI) reduction and phenol degradation using Stenotrophomonas sp. isolated from tannery effluent contaminated soil. Environ. Sci. Pollu. Res. 20(9): 6563–6573.

Haferburg, G. and E. Kothe. 2007. Microbes and metals: interactions in the environment. J. Basic Microbiol. 47(6): 453–467.

Hamlett, N.V., E.C. Landale, B.H. Davis and A.O. Summers. 1992. Roles of the Tn21 merT, merP and merC gene products in mercury resistance and mercury binding. J. Bacteriol. 174: 6377– 6385.

Harrison, J.J., H. Ceri and R.J. Turner. 2007. Multimetal resistance and tolerance in microbial biofilms. Nat. Rev. Microbiol. 5: 928–938.

Higham, D.P., P.J. Sadler and M.D. Scawen. 1986. Cadmium–binding proteins in Pseudomonas putida: pseudothioneins. Environmental Health Perspect. 65(3): 5–11.

Hussein, H., R. Krull, S.I. Abou El–Ela and D.C. Hempel. 2001. Interaction of the different heavy metal ions with immobilized bacterial culture degrading xenobiotic wastewater compounds. *In*: Proceedings of the Second International Water Association World Water Conference, Berlin, Germany. 15–19.

Hussein, H., S. Farag and H. Moawad. 2004. Isolation and characterization of Pseudomonas resistant to heavy metals contaminants. Arab. J. Biotehnol. 7: 13–22.

Ibrahim, Z., W.A. Ahmad and A.B. Baba. 2001. Bioaccumulation of silver and the isolation of metal–binding protein from P. diminuta. Braz. Arch. Biol. Technol. 44(3): 223–225.

Igiri, B.E., S.I.R. Okoduwa, G.O. Idoko, E.P. Akabuogu, A.O. Adeyi and E.K. Ejiogu. 2018. Toxicity and bioremediation of heavy metals contaminated ecosystem from tannery wastewater: a review. J. Toxicol. 16: 16.

Ince Y.E. 2003. Metal tolerance and biosorption capacity of Bacillus circulans strain EB1. Res. Microbiol. 154: 409–415.

Issazadeh. K., N. Jahanpour, F. Pourghorbanali, G. Raeisi and J. Faekhondeh. 2013. Heavy metals resistance by bacterial strains. Schol. Res. Lib. 4(2): 60–63.

Ivanova, E.P., V.V. Kurilenko, A.V. Kurilenko, N.M. Gorshkova, F.N. Shubin, D.V. Nicolau, et al. 2002. Tolerance to cadmium of free–living and associated with marine animals and eelgrass marine gamma-proteobacteria. Curr. Microbiol. 44(4): 357–362.

Jafari, S.A., S. Cheraghi, M. Mirbakhsh, R. Mirza and A. Maryamabadi. 2015. Employing response surface methodology for optimization of mercury bioremediation by Vibrio parahaemolyticus PG02 in coastal sediments of Bushehr, Iran. Clean – Soil, Air, Water. 43(1): 118–126.

Karigar, C.S. and S.S. Rao. 2011. Role of microbial enzymes in the bioremediation of pollutants: a review. Res. Enzyme. 11: 1–11.

Kermani, A.J.N., M.F. Ghasemi, A. Khosravan, A. Farahmand and M.R. Shakibaie. 2010. Cadmium bioremediation bymetal–resistant mutated bacteria isolated from active sludge of industrial effluent. J. Environ. Health Sci. Eng. 7(4): 279–286.

Khalid, S., M. Shahid, N.K. Niazi, B. Murtaza, I. Bibi and C. Dumat. 2017. A comparison of technologies for remediation of heavy metal contaminated soils. J. Geochem. Explor. 182: 247–268. doi: 10.1016/j.gexplo.2016.11.02.

Khan, M.S., A. Zaidi, P.A. Wani and M. Oves. 2009. Role of plant growth promoting rhizobacteria in the remediation of metal contaminated soils: a review. pp. 319–350. *In*: E. Lichtfouse [ed.]. Organic Farming, Pest Control and Remediation of Soil Pollutants, Sustainable Agriculture Reviews 1, Springer Science and Business Media B.V.

Kiliç, N.K., G. Kürkçü, D. Kumruoglu and G. Dönmez. 2015. EPS production and bioremoval of heavy metals by mixed and pure bacterial cultures isolated from Ankara stream. Water Sci. Technol. 72: 1488–1494.

Kiyono, M. and H. Pan–Hou. 1999. The merG gene product is involved in phenylmercury resistance in Pseudomonas strain K–62. J. Bacteriol. 181: 726–730.

Koch, M. and W. Rotard. 2001. On the contribution of background sources to the heavy metal content of municipal sewage sludge. Water Sci. and Technol. 43(2): 67–74.

Kumari, D., X. Pan, D.J. Lee and V. Achal. 2014. Immobilization of cadmium in soil by microbially induced carbonate precipitation with Exiguobacteriumundae at low temperature. Int. Biodeterior. Biodegradation. 94: 98–102.

Kumari, D., X. Qian, X. Pan, A. Varenyam, Q. Li and G.M. Geoffrey. 2016. Microbially–induced carbonate precipitation for immobilization of toxic metals. Adv. Appl. Microbiol. 94: 79–108.

Kure, J.T., M. Gana, A. Emmanuel, R.M. Isah and C.C. Ukubuiwe. 2018. Bacteria associated with heavy metal remediation: a review. Int. J. Appl. Biol. Res. 9(1): 134–148.

Lee B.T.O., N.L. Brown, S. Rogers, A. Bergemann, J. Camakaris and D.A. Rouch. 1990. Bacterial response to copper in the environment: copper resistance in *Escherichia coli* as a model system. pp. 625–632. *In*: J.A.C. Broekaert, Ş. Güçer and F. Adams [eds.]. Metal Speciation in the Environment. NATO ASI Series (Series G: Ecological Sciences). Springer, Berlin, Heidelberg.

Li, M., X. Cheng and H. Guo. 2013. Heavy metal removal by biomineralization of urease producing bacteria isolated from soil. Int. Biodeterior. Biodegradation. 76: 81–85.

Lim, P.E., K.Y. Mak, N. Mohamed and A.M. Noor. 2003. Removal and speciation of heavy metals along the treatment path of wastewater in subsurface–flow constructed wetlands. Water Sci. Technol. 48: 307–313.

Lima, A.I.G., S.C. Corticeiro and E.M.A.P. Figueira. 2006. Glutathione–mediated cadmium sequestration in Rhizobium leguminosarum. Enzyme Microb. Technol. 39(4): 763–769.

Lima, M.A., M.S. Urbieta and E.R. Donati. 2018. Microbial communities and the interaction with heavy metals and metalloids: impact and adaptation. pp. 5–14. *In*: E.R. Donati [ed.]. Heavy Metals in the Environment, CRC Press, Boca Raton.

Lin, C.C. and H.L. Lin. 2005. Remediation of soil contaminated with the heavy metal (Cd^{2+}). J. Hazard. Mater. 122: 7–15.

Macaskie, L.E., A.C.R. Dean, A.K. Cheetham, R.J.B. Jakeman and A.J. Skarnulis. 1987. Cadmium accumulation by a citrobacter sp.: the chemical nature of the accumulated metal precipitate and its location on the bacterial cells. J. Gen. Microbiol. 133(3): 539–544.

Macaskie, L.E. 1990. An immobilized cell bioprocess for the removal of heavy metals from aqueous flows. J. Chem. Technol. Biotechnol. 49: 357–379.

Malik, A. 2004. Metal bioremediation through growing cells. Environ. Int. 30: 261–278.

Malik, S., S.A.L. Andrade, M.H. Mirjalili, R.R.J. Arroo, M. Bonfill and P. Mazzafera. 2016. Biotechnological Approaches for bioremediation: *in vitro* hairy root culture. *In*: Jha S. [eds]. Transgenesis and Secondary Metabolism. Reference Series in Phytochemistry. Springer, Cham. doi: https://doi.org/10.1007/978-3-319-27490-4_28-1.

McConnell, J.R. and R. Edwards. 2008. Coal burning leaves toxic heavy metal legacy in the Arctic. Proc. Natl. Acad. Sci. U.S.A. 105: 12140–12144.

Melo, J.S. and S.F. D'Souza. 2004. Removal of chromium by mucilaginous seeds of *Ocimum basilicum*. Bioresour. Technol. 92: 151–155.

Mergeay, M., S. Monchy, T. Vallaeys, V. Auquier, A. Benotmane, P. Bertin, et al. 2003. Ralstonia Metallidurans, a bacterium specifically adapted to toxic metals: towards a catalogue of metals–responsive genes. FEMS Microbiol. Rev. 27: 385–410.

Mire, C.E., J.A. Tourjee, W.F. O'Brien, K.V. Ramanujachary and G.B. Hecht. 2004. Lead precipitation by Vibrio harveyi: evidence for novel quorum–sensing interactions. Appl. Environ. Microbiol. 70(2): 855–864.

Monachese, M., J.P. Burton and J. Reid. 2012. Bioremediation and tolerance of humans to heavy metals through microbial processes: a potential role for probiotics? Appl. Environ. Microbiol. 78: 6397–6404.

Montgomery, D.M., A.C.R. Dean, P. Wiffen and L.E. Macaskie. 1995. Phosphatase production and activity in Citrobacterfreundii and a naturally occurring, heavy–metal–accumulating Citrobacter sp. Microbiol. 141(10): 2433–2441.

Moore, F., A. Attar and F. Rastmanesh. 2011. Anthropogenic sources of heavy metals in deposited sediments from runoff and industrial effluents, Shiraz, SW Iran. 2nd International conference on environmental science and technology IPCBEE. Singapore. IACSIT Press. 6: 215–219.

More, T.T., J.S.S. Yadav, S. Yan, R.D. Tyagi and R.Y. Surampalli. 2014. Extracellular polymeric substances of bacteria and their potential environmental applications. J. Environ. Manage. 144: 1–25.

Mowll, J.L. and G.M. Gadd 1984. Cadmium uptake by Aureobasidiumpullulans. J. Gen. Microbiol. 130: 279–284.

Mueller, J.G., P.J. Chapman and P.H. Pritchard. 1989. Creosote–contaminated sites. Their potential for bioremediation. Environ. Sci. Technol. 23: 1197–1201.

Mukhopadhyay, R., B.P. Rosen, L.T. Phung and S. Silver. 2002. Microbial arsenic: from geocycles to genes and enzymes. FEMS Micobiol. Rev. 26(3): 311–325.

Mustaphaa, M.U. and N. Halimoon. 2015. Screening and isolation of heavy metal tolerant bacteria in industrial effluent. Proc. Environ. Sci. 30: 33–37.

Natarajan, K.A. 2008. Microbial aspects of acid mine drainage and its bioremediation. T. Nonferr. Metals SOC. 18(6): 1352–1360.

Naz, N., H.K. Young, N. Ahmed and G.M. Gadd. 2005. Cadmium Accumulation and DNA homology with metal resistance genes in sulfate–reducing bacteria. Appl. Environ. Microbiol. 71(8): 4610–4618.

Nematian, M.A. and F. Kazemeini. 2013. Accumulation of Pb, Zn, Cu and Fe in plants and hyperaccumulator choice in galali iron mine area, Iran. Int. J. Agric. Crop Sci. 5(4): 426–432.

Nies, D.H. and S. Silver. 1989. Metal ion uptake by a plasmid-free metal-sensitive *Alcaligenes eutrophus* strain. J. Bacteriol. 171(7): 4073–4075.

Nies, D.H. 1995. The cobalt, zinc, and cadmium efflux system CzcABC from *Alcaligenes eutrophus* functions as a cationproton antiporter in *Escherichia coli*. J. Bacteriol. 177(10): 2707–2712.

Nies, D.H. 1999. Microbial heavy-metal resistance. Appl. Microbiol. Biotechnol. 51: 730–750.

Ozdemir, G., T. Ozturk, N. Ceyhan, R. Isler and T. Cosar. 2003. Heavy metal biosorption by biomass of *Ochrobactrum anthropi* producing exopolysaccharide in activated sludge. Bioresour. Technol. 90: 71–74.

Patrick, D., R.K. Mahadevan and A.F. Yakunin. 2018. Heavy metal removal by bioaccumulation using genetically engineered microorganisms. Front. Bioeng. Biotechnol. 6: 1–20.

Peng, L., R. Lifang and X. Hongyu, L. Xi and Z. Chaocan. 2007. Study on the toxic effect of lead (II) ion on *Escherichia coli*. Biol. Trace Elem. Res. 115(2): 195–202.

Perpetuo, E.A., C.B. Souza and C.A.O. Nascimento. 2011. Engineering bacteria for bioremediation. pp. 605–632. *In*: A. Carpi [ed.]. Progress in Molecular and Environmental Bioengineering-From Analysis and Modelling to Technology Applications. InTech Open Access Publisher, Rijeka, Croatia.

Phillips, A.J., R. Gerlach, E. Lauchnor, A.C. Mitchell, A.B. Cunningham and L. Spangler. 2013. Engineered applications of ureolyticbiomineralization: a review. Biofouling 29: 715–733. doi:10.1080/08927014.2013.796550.

Philp J. and R. Atlas. 2005. Bioremediation of contaminated soils and aquifers. pp. 139–236. *In*: R. Atlas, J. Philip [ed.]. Bioremediation. ASM Press, Washington, DC.

Plette, A.C.C., M.F. Benedetti and W.H. Riemsdijk. 1996. Competitive binding of protons, calcium, cadmium, and zinc to isolated cell walls of Gram-positive bacterium. Environ. Sci. Technol. 30: 1902–1910.

Pons, M.P. and C.M. Fuste. 1993. Uranium uptake by immobilized cells of Pseudomonas strain EPS 5028. Appl. Microbiol. Biot. 39: 661–665.

Pontel, L.B. and F.C. Soncini. 2009. Alternative periplasmic copper–resistance mechanisms in Gram-negative bacteria. Mol. Microbiol. 73: 212–225.

Postgate, J.R. 1984. The Sulphate-Reducing Bacteria. Cambridge University Press.

Prescott, L.M. and J.P. Harley. 2002. Laboratory Exercise in Microbiology. 5th Edition. New York. McGraw–Hill Companies Inc.

Ramaiah, J.D.N and L. Vardanyan. 2008. Detoxification of toxic heavy metals by marine bacteria highly resistant to mercury. Mar. Biotechnol. 10(4): 471–477.

Renninger, N., R. Knopp, H. Nitsche, D.S. Clark, D. Jay and J.D. Keasling. 2004. Uranyl precipitation by Pseudomonas aeruginosa via controlled polyphosphate metabolism. Appl. Environ. Microbiol. 70: 7404–7412.

Rensing, C., M. Ghosh and B. Rosen. 1999. Families of soft-metal-ion-transporting ATPases. J. Bacteriol. 181(9): 5891–5897.

Rojas, L.A., C. Yáñez, M. González, S, Lobos, K. Smalla and M. Seeger. 2011. Characterization of the metabolically modified heavy metal–resistant Cupriavidusmetallidurans strain MSR33 generated for mercury bioremediation. PLoS one 6(3): e17555.

Roozbahani, M.M., S. Sobhanardakani, H. Karimi and R. Sorooshnia. 2015. Natural and anthropogenic source of heavy metals pollution in the soil samples of an industrial complex; a case study. I.J. Toxicol. 9(29):1336–1341.

Ruiz, O.N, D. Alvarez, G. Gonzalez-Ruiz and C. Torres. 2011. Characterization of mercury bioremediation by transgenic bacteria expressing metallothionein and polyphosphate kinase. BMC Biotechnol. 11: 82.

Salehizadeh, H. and S.A. Shojaosadati. 2003. Removal of metal ions from aqueous solution by polysaccharide produced from Bacillus firmus. Water Res. 37(17): 4231–4235.

Samantaray, D.V., S. Mohapatra and B.B. Mishra. 2014. Microbial bioremediation of industrial effluents. pp. 325–339. *In*: S. Das [ed.]. Microbial Biodegradation and Bioremediation. Elsevier.

Samantha, A., P. Bera, M. Khatun, C. Sinha, P. Pal, A. Lalee, et al. 2012. An investigation on heavy metal tolerance and antibiotic resistance properties of bacterial strain Bacillus sp. isolated from municipal waste. J. Microbiol. Biotechnol. Res. 2(1): 178–189.

Saxena, D. and S. Srivastava. 1998. Carbon source–starvation–induced precipitation of copper by Pseudomonas putida strain S4. World J. Microbiol. Biotechnol. 14(6): 921–923.

Saxena, D., N. Joshi and S. Srivastava. 2002. Mechanism of copper resistance in a copper mine isolate *Pseudomonas putida* strain S4. Curr. Microbiol. 45(6): 410–414.

Sharma, P.K., D.L. Balkwill, A. Frenkel and M.A. Vairavamurthy. 2000. A new Klebsiellaplanticola strain (Cd–1) grows anaerobically at high cadmium concentrations and precipitates cadmium sulphide. Appl. Environ. Microbiol. 66(7): 3083–3087.

Shrestha, N., G. Chilkoor, B. Vemuri, N. Rathinam, R.K. Sani and V. Gadhamshetty. 2018. Extremophiles for microbial-electrochemistry applications: a critical review. Bioresour. Technol. 255: 318–330.

Sone, Y., Y. Mochizuki, K. Koizawa, R. Nakamura, H. Pan–Hou, T. Itoh, et al. 2013. Mercurial resistance determinants in Pseudomonas strain K–62 plasmid pMR68. AMB Express, 3, Article 41.

Sprocati, A.R., C. Alisi, L. Segre, F. Tasso, M. Galletti and C. Cremisini. 2006. Investigating heavy metal resistance, bioaccumulation and metabolic profile of a metallophile microbial consortium native to an abandoned mine. Sci. Total Environ. 366(2–3): 649–658.

Srivastava V.C., D. MallI and M. MishraI. 2008. Removal of cadmium (II) and zinc (II) metal ions from binary aqueous solution by rice husk ash. Colloids Surf. A Physicochem. Eng. Asp. 312(2): 172–184.

Tangaromsuk, J., P. Pokethitiyook, M. Kruatrachue and E.S. Upatham. 2002. Cadmium biosorption by *Sphingomonas paucimobilis* biomass. Bioresour. Technol. 85: 103–105.

Tsezos, M. and B. Volesky. 1981. Biosorption of uranium and thorium. Biotechnol. Bioeng. 23: 583–604.

Torres, L.G., R.B. Lopez and M. Beltran. 2012. Removal of As, Cd, Cu, Ni, Pb and Zn from a highly contaminated industrial soil using surfactant enhanced soil washing. Phys. Chem. Earth. 37–39: 30–36.

Toth, G., T. Hermann, M.R. Da Silva and L. Montanarella. 2016. Heavy metals in agricultural soils of the European union with implications for food safety. Environ. Int. 88: 299–309.

Valls, M. and V. de Lorenzo. 2002. Exploiting the genetic and biochemical capacities of bacteria for the remediation of heavy metal pollution. FEMS Microbiol. Rev. 26: 327–338.

Verma, J.P. and D.K. Jaiswal. 2016. Book review: advances in biodegradation and bioremediation of industrial waste. Front Microbiol. 6: 1555.

Vidali, M. 2001. Bioremediation: an overview. Pure Appl. Chem. 73(7): 1163–1172.

Viti C., A. Pace and L. Giovannetti. 2003. Characterization of Cr(VI)-resistant bacteria isolated from chromium-contaminated soil by tannery activity. Curr. Microbiol. 46(1): 1–5.

Wang, C.L., S.C. Ozuna, D.S. Clark and J.D. Keasling. 2002. A deep-sea hydrothermal vent isolate, Pseudomonas aeruginosa CW961, requires thiosulfate for Cd^{2+} tolerance and precipitation. Biotechnol. Lett. 24(8): 637–641.

Wang, J.H., D. Zhu, S.S. Zhang and J.F Guan. 2014. Construction and application of genetic engineering bacteria in contaminated environment bioremediation. Adv. Mat. Res. 955–959: 1935–1938. https://doi.org/10.4028/www.scientific.net/amr.955-959.1935.

Wester, R.C., H.I. Maibach, L. Sedik, J. Melendres, S. Di Zio and M. Wade. 1992. *In vitro* percutaneous absorption of cadmium from water and soil into human skin. Fund. Appl. Toxicol. 19: 1–5.

Whiffin VS., L.A. Van Paassen and M.P. Harkes. 2007. Microbial carbonate precipitation as a soil improvement technique. Geomicrobiol. J. 24: 417–423.

Wierzba, S. 2018. Biosorption of lead(II), zinc(II) and nickel(II) from industrial wastewater by *Stenotrophomonas maltophilia* and *Bacillus subtilis*. Pol. J. Chem. Technol. 34(1): 33–38.

Wong, S.J. and E.R. Rene. 2017. Bioprecipitation–a promising technique for heavy metal removal and recovery from contaminated wastewater streams. MOJ Civ. Eng. 2(6): 191–193.

Wu, G., H. Kang, X. Zhang, H. Shao, L. Chu and C. Ruan. 2010. A critical review on the bio–removal of hazardous heavy metals from contaminated soils: Issues, progress, eco–environmental concerns and opportunities. J. Hazard. Mater. 174: 1–8.

Yamamura, S. and S. Amachi. 2014. Microbiology of inorganic arsenic: from metabolism to bioremediation. J. Biosci. Bioeng. 118: 1–9.

Yang, J., W. Wei, S. Pi, F. Ma, A. Li, D. Wu, et al. 2015. Competitive adsorption of heavy metals by extracellular polymeric substances extracted from Klebsiella sp. J1. Bioresour. Technol. 196: 533–539.

Yang, T., M.L. Chen and J.H. Wang. 2015. Genetic and chemical modification of cells for selective separation and analysis of heavy metals of biological or environmental significance. TrAC-Trend. Anal. Chem. 66: 90–102.

Zhao, Y., Z. Fu, X. Chen and G. Zhang. 2018. Bioremediation process and bioremoval mechanism of heavy metal ions in acidic mine drainage. Chem. Res. Chin. Univ. 34(1): 33–38.

Zhu, T. and D. Maria. 2016. Carbonate precipitation through microbial activities in natural environment, and their potential in biotechnology: a review. Front. Bioeng. Biotechnol. 4(4): 1–21.

7

New Trends in Antarctic Bioprospecting: The Case of Cold-Adapted Bacteria

Carmen Rizzo[1], Maria Papale[2] and Angelina Lo Giudice[2,*]

Introduction

The bioprospecting research area is rapidly developing with the acquisition of an ever-increasing number of approaches, both in terms of scientific rationales and hypotheses, and tools and methodologies that are being used. In particular, bioprospecting needs more than ever to draw on new and unexplored sources, but at the same time, to focus on sources with a high potential of isolation of new compounds with novel and more specific functionalities. In this context, the polar environments, lands of extremes, and ground of richness and biodiversity, harsh environments that conceal fullness of life have established themselves as a perfect study basin in the eyes of bioprospectors. Scientific knowledge of Poles is very scant in comparison with other areas in the world, and a lot of aspects and sites are still to be explored and potentially exploited. The significant degree of uncertainty about what lies to be discovered beyond the austerity of the ice makes them particularly compelling for researchers.

The paucity of in-depth knowledge on the polar biota and on the biodiversity ranges that are only at the beginning of their discovery, the pristine aspect of these environments, and in the meanwhile the numberless genetic, physiological, and metabolic specializations that their inhabitants have developed are some of the key points that have made these areas so exciting to researchers. Overall, the environments considered as extreme in different forms are interesting because they are a potential resource for highly specific bioproducts, employable in numerous commercial and biotechnological fields (Cavicchioli et al. 2011; Cowan et al. 2015).

Antarctica assumes a great scientific value because, despite an apparent homogeneity and poverty of biological resources, it offers a great variety of habitats and micro-habitats, all characterized by extreme and peculiar conditions. The great attention aroused in bioprospectors lies in the fact that a difficult environment necessarily stimulates the development of communities of organisms which, individually from microscale to macroscale, or in relation to each other in the case of symbiotic relationships possess uncommon adaptation strategies and consequent specified enzymes that are not detected in other environments. The strong interest in searching and detecting new bioprocesses and/or bioproducts has shifted the focus on the microbial world with the support of a series of advantages over the macroorganisms, such as a rapid growth rate (and as a consequence, high yields of obtained biomass and time reduction), replication possibility, easier chance of genetic manipulation, and valuable alternative to limit the oversampling of higher organisms specimens. Microbial biosynthesis processes thus become a cost-effective and environmentally friendly alternative to produce biomolecules of technological interest (Raddadi et al. 2015).

[1] Stazione Zoologica Anton Dohrn, Messina (Italy).
[2] Institute of Polar Sciences (CNR-ISP), National Research Council, Messina (Italy).
Institute of Polar Sciences (CNR-ISP), National Research Council, Spianata San Raineri 86, 98122 Messina (Italy).
*Corresponding author: angelina.logiudice@cnr.it

Many researchers have found in the extremophiles the key for bioprospecting investigations in Antarctica as a main concrete vehicle for scientific purposes and commercial application of genetic resources (Watt and Dean 2000, De Pascale et al. 2012). The paucity of data about Antarctic organisms provides a concrete opportunity for researchers to discover new microbial extremophiles with possible biotechnological exploitation. In consideration of the number and the variety of habitats that delineate the Antarctic ecosystem, several basins could be considered as an optimal source for different microorganisms, and some attempts could be listed among the recent efforts. Both abiotic and biotic matrices have been used with a preliminary approach for isolation of cold-adapted bacteria with different ecological valence and consequent biotechnological applications.

Cold-Adapted Bacteria

A precise definition of the terms used for a microorganism living in the cold is far to be established, and the debate on the topic remains always open. Conventional classification of cold-adapted bacteria includes psychrophiles (cold-loving) and psychrotrophs (cold-tolerant or psychrotolerant) on the base of the optimum temperature range for growth (Morita 1975). Psychrophiles prefer a growth temperature at 15°C or below, but they do not proliferate above 20°C. As a consequence, strict psychrophiles could be found only in permanently cold habitats. Otherwise, psychrotolerants exhibit a wide temperature range and grow fastest above 20°C and are widespread also in environments experiencing thermal fluctuations. Feller and Gerday (2003) suggested the terms eurypsychrophile (i.e., facultative psychrophiles or psychrotrophs) and stenopsychrophile (i.e., true or obligate psychrophiles) for microorganisms showing a wide and narrow tolerance to temperature, respectively. According to Cavicchioli (2016), temperature-dependent growth rate as a metric for cold adaptation could provide severe misconceptions about how well adapted microorganisms are to their native cold habitats (Mocali et al. 2017). Therefore, the author suggested the term psychrophile as the more appropriate and sufficient for describing microbes indigenous to cold environments (thus, contrasting with the terms psychrotolerant, psychrotroph, and facultative psychrophile that defy logic).

In order to overcome the diversified nomenclature system attributed to cold lifestyle, here the term 'cold-adapted bacteria' will be used to collectively designate bacteria autochthonous to cold environments.

The Cold Lifestyle in Antarctic Microbial Habitats

In the beginning, the polar environments appeared monochromatic and monotonous so that it was believed they were simply desolate and lifeless lands. In reality, over the years, a great diversity of environments has emerged with unique and peculiar features. Low temperature environments are affected by several stressful factors that could act individually or synergistically, i.e., dryness, nutrient accessibility, osmotic stress, high salt concentration, low biochemical activity, adverse solar radiation, and a highly variable photoperiod (with alternance of light and dark for long period) (Lo Giudice and Rizzo 2018).

In Antarctica, the incredibly hostile conditions often exclude all but microbial life, and the biomass of certain microorganisms (also in the absence of strong competition and grazing pressure) can become dominant (Vincent 1988). What in Antarctic areas should stem and limit life stimulates it insistently by leading Antarctic inhabitants to develop and adopt a uniquely adaptive and survival strategies, minimizing the impact of such severe and sometimes permanent, environmental conditions (D'Amico et al. 2006). Recent advances suggested that the cold-adapted lifestyle most likely derives from the synergistic interaction between genomic and metabolic features and not merely from a unique set of genes (Tribelli and López 2018). Some examples are given below.

Low Temperature

About 85% of the biosphere is characterized by an enduring exposition to temperatures below 5°C all along the year and 14% of the total biosphere is represented by polar regions (including the Arctic and Antarctica). The summer period (December-February) in Antarctic coastal environments is featured by temperatures close to freezing (or often a bit positive in the northern part of the Antarctic Peninsula), whereas during winter areas mean temperatures range between −10°C and −30°C. Higher elevation, higher latitude, and greater distance from the ocean make the high interior plateau much colder (around −20°C and below −60°C in summer and winter, respectively). As an example, the Vostok station holds the record for the lowest ever temperature recorded at the surface of the Earth (−89.2°C). Lakewater temperature beneath the ice in winter lies near 0°C, whereas air temperature drops below −15°C. The marine Antarctic environments are more stable and the temperature does not drop below −1.86°C.

Low temperature thus becomes the main restrictive environmental factor in Antarctica, exerting a fundamental control on the typical traits of cold-adapted bacteria (Cavicchioli 2016). Traditionally, cold adaptation involves a number of structural and physiological modifications of enzymes and membranes, expression of cold shock proteins, an adaptation of translation and transcription processes, and the presence of cryoprotectants (Maccario et al. 2015). Cold-adapted bacteria actuate uncommon metabolic mechanisms for a cold lifestyle. Recently, comparative genomic analyzes, carried out on the huge amount of data available on the genomes of cold-adapted bacteria and their mesophilic counterparts, have allowed obtaining significant deepening on the metabolic versatilities of cold-adapted living forms across different temperature ranges or ecological niches (Tribelli and López 2018). The application of other 'multi-omics' approaches (e.g., metatranscriptomics, metaproteomics, metabolomics, and metagenomics) has been usefully used in the recent past to acquire novel information related to the functionality of microbial genotypic and metabolic traits in response to cold temperatures. Most phenotypic modifications seem to be permanent and genetically-driven and not merely derived from short-term acclimatization (Chintalapati et al. 2004).

The dynamic interface between the external and internal sides of the cell is represented by cell envelopes. Among them, the cell membrane and its structural modifications at low temperatures have been intensively explored. Membrane rigidity, as the main effect of the cold, strongly interferes with some vital processes operated by membrane proteins (such as respiration and trans-membrane transports). The maintenance of membrane fluidity at low temperatures is therefore guaranteed by the presence of high amounts of fatty acids (polyunsaturated and methyl branched) and carotenoids pigments in the membrane bilayer. The membrane fluidity could be modulated by cold-adapted bacteria based on the fatty acid chain length, *cis-trans* fatty acids ratio, and type of carotenoids. For instance, the inactivation of fatty acid desaturases was reported as membrane fluidity preserving factor against temperature lowering in *Planococcus halocryophilus* from Antarctic permafrost (Mykytczuk et al. 2013, 2016). Pigmented bacteria often predominate in certain Antarctic habitats. Higher production of polar carotenoids (instead of non-polar carotenoids), which stabilize the membrane structure under cold growth conditions, has been demonstrated (e.g., in the case of *Sphingobacterium antarcticus* and *Micrococcus roseus* from Antarctic soil) (Chattopadhyay et al. 1997, Jagannadham et al. 2000, Dieser et al. 2010).

Cell envelopes different from membranes have gained less attention. Genomic studies on *Pseudoalteromonas haloplanktis* from Antarctic seawater have demonstrated the occurrence of a high percentage of cell envelope genes as a specific adjustment to manage the cold (Médigue et al. 2005). As it was reported by Benforte et al. (2018) for *Pseudomonas extremaustralis*, lipopolysaccharides within the outer membrane of Gram-negative bacteria play a fundamental role in the active growth of cold-adapted bacteria at low temperature. In mutant strains, the deficiency in the gene (wapH) encoding a core LPS glycosyltransferase resulted in a lower flexibility and higher turgor pressure of the cell envelope, cell permeability, and surface area to volume ratio (Benforte et al. 2018). In Gram-positive bacteria, the cell wall and the inner membrane both play a fundamental role by protecting the cell against ice-caused disruption and/or osmotic pressure under cold conditions. A genetic control mechanism was evidenced in *Planococcus halocryophilus*, whose expression of genes encoding for enzymes involved in calcium carbonate mineralization raises at subzero temperatures by enhancing calcium carbonate precipitation (Mykytczuk et al. 2016).

However, cold-adapted bacteria have to adopt strategies useful to maintain an efficient enzyme-catalyzed reaction rate at low temperatures that inhibit enzyme-mediated primary cell processes, such as transcription and translation. Cold-adapted enzymes are characterized by structural adjustments aimed at obtaining a less compact three-dimensional structure and the destabilization of the active site (Feller and Gerday 2003). Generally, proline and arginine residues, stabilizing the protein structure, are reduced in number, whereas glycine residues, which essentially have no side chains, are present in abundance as they allow localized chain mobility. As a consequence, the catalytic center becomes more flexible and more accessible to ligands at freezing temperatures, favoring the release of the reaction products (Feller and Gerday 2003).

The number of proteins and solutes may minimize or eliminate ice dangerous effects (e.g., cellular damage and osmotic imbalance) during the freezing process. In cold-adapted bacteria, cold shock proteins (CSPs), cold-acclimation proteins (CAPs), anti-freeze proteins (AFPs), and antinucleating proteins (ANPs) play key functions in protecting the cell structure(s) from the formation of ice crystals, in preventing the ice penetration into cells, and in stabilizing the mRNAs (thus acting also at the translation or transcriptional level) (Margesin et al. 2007).

Among substances involved in cold adaptation and cryoprotection, there are a number of compatible solutes (such as glycine, betaine, proline, sucrose, and mannitol) that accumulate in the cytoplasm to lower its freezing point and avoid desiccation (De Maayer et al. 2014) and trehalose, which protects cells undergoing shock exposure to high and low temperatures and/or osmotic stress (Phadtare and Inouye 2008). In particular, trehalose stabilizes the cell membrane and enables in removing free radicals and preventing protein denaturation (Fendrihan and Negoita 2012). Extracellular polymeric substances (EPSs) are complex compounds with high molecular weight, mostly constituted by high carbohydrates content with the presence of several organic and inorganic substituents (e.g., sulfate, phosphate, acetic acid, and acetylate). In addition to a number of ecological roles (trapping water, nutrients, and metals; cellular aggregation; biofilm formation), EPSs act through cryoscopic lowering with the defensive role of extracellular enzymes against cold denaturation and autolysis. Several Antarctic isolates, from sea-ice, seawater, and sponges, have been reported as producers of EPSs with cryoprotective function (Mancuso Nichols et al. 2005a, Caruso et al. 2018a, 2018b, 2019). Finally, certain cold-adapted bacteria produce polyhydroxyalkanoates (PHAs), which can reduce oxidative stress at low temperatures, maintaining the redox state (Ayub et al. 2009). The accumulation of polyhydroxybutyrate (a short-chain length PHA) in *Pseudomonas extremaustralis* is involved in biofilm formation and motility at low temperatures, favoring surface colonization under cold conditions (Tribelli and López 2018).

UV Radiation

In Antarctica, the surface of glaciers, ice sheets, snow, soil, and rock (as well as lake water and marine environments) experience a prolonged and intense input of UV radiation during summer. UV-mediated cellular stress can cause an increase in reactive oxygen species (ROS) with subsequent dangerous effects on genetic material, lipids, and proteins (Correa-Llantén 2012). The Antarctic microbial communities living in such habitats have developed a number of successful defense mechanisms (e.g., UV-stress avoidance, antioxidant response, and active repair systems) to survive such constant exposure to UV radiation (Dieser et al. 2010). In particular, Antarctic bacterial isolates recovered from ice cores, glaciers, or marine surface waters have been often reported as harboring increased synthesis (as secondary metabolites) of photoprotective pigments, such as carotenoids (with light adsorption between 400 and 500 nm), suggesting that pigmentation plays a role in adaptation to cold environments (as a combined response to low temperature and UV radiation). The production of carotenoids pigments by Antarctic bacteria may vary in abundance depending on light exposure and dissolved oxygen content (antioxidant defense against oxidative stress) (Dieser et al. 2010, Correa-Llantén et al. 2012, Reis-Mansur et al. 2019).

High Salinity

In a number of microbial habitats, freezing and evaporation processes lead to a high concentration of dissolved ions. In the marine environment, during the sea-ice formation, the salt accumulates into droplets called brines. These latter are typically expelled back into the ocean, leading the near-surface water salinity to raise. However, the sea-ice microstructure is permeated by brine, different cavities containing concentrated seawater-derived brine (Butler et al. 2016). The typical Antarctic continental basins in which hypersaline brines occurred include permafrost, and more recently, also glaciers, and below ice-sealed and subglacial lakes have been reported (Papale et al. 2019). Their high salt content (which lowers the freezing point of water) is responsible for the maintenance of their unfrozen conditions several degrees below 0°C, thus providing liquid but an extremely cold environment (up to –30°C) for microbial growth during winter (Vincent 1988). Hypersaline environments are less susceptible to freezing also at the diurnal scale. Finally, continental Antarctica hosts also a number of hypersaline soils (e.g., in the McMurdo Dry Valleys), whose salt content extends the period of summer thaw (Levy et al. 2012).

Bacteria inhabiting such saline Antarctic habitats experience wide fluctuation in salt content, also over relatively short periods, but damages related to cellular ice crystal formation are reduced (Vincent 1988). Among the main negative effects of high salinity at the cellular level, there are dehydration and water activity changes. The high osmolarity of the surroundings leads to a rapid loss of intracellular water with consequences on enzyme activity and other cellular functions. To achieve the osmotic equilibrium (with the cytoplasm at least isoosmotic with the extracellular environments), microbes mainly adopt two different strategies called the 'salt-in' strategy and the 'compatible-solute' strategy. The former consists of the accumulation of equimolar concentrations of inorganic ions in the cytoplasm, while the second consists of the storage of highly organic compatible solutes (Mei et al. 2017).

Low Water Content

The terrestrial environment in Antarctica is considered as a special kind of desertic place, in which the precipitation rate is very low (less than 250 mm water equivalents; Nylen et al. 2004), the humidity rate is poor, but in contrast with the common image of a desert, the temperatures are strictly cold. Moreover, despite what one can expect, not all the continent is covered by snow and ice with some inner regions characterized by large uncovered rocks. These areas, permanently free of ice formation, amount for a total of 0.32-04% of the total surface area in the Antarctic and are mainly constituted by sand material (Convey et al. 2008). Despite different kind of soils occurred in the continent, which could be defined as habitats, a common structure correlates them that generally comprises of a surface layer formed by stones and boulders exposed to the atmospheric agents, and an active unconsolidated layer extended up to one meter of depth (Bockheim 2015b). These abiotic component of the Antarctic ecosystem are poor in water, nutrients, and organic matter (Kennedy 1993, Cary et al. 2010, Mergelov et al. 2015, Zazovskaya et al. 2015), but at the same time, they are strongly affected by external and unexpected events which could promote the nutrient intake sustaining the living communities within the soils (Cowan and Tow 2004, McKnight et al. 2004, Nelson et al. 2008, Cary et al. 2010, Tiao et al. 2012, Bockheim 2015a), such as wind erosion, meltwater runoff, and animal carcasses occurrence.

Relevant Molecules in Biotechnology

The possible synthesis of still unknown natural biomolecules, as a consequence of the adopted complex strategies to survive under harsh environmental conditions, makes cold-adapted bacteria interesting by a biotechnological point of view. The set of natural molecules deriving from cold-adapted bacteria is really wide. Each kind of bioactive molecule could dispatch one or more functions that have a precise ecological role, which is dependent on the surrounding environmental conditions. Indeed, each organism is stimulated to produce a particular polymer, metabolite, and similar compounds as a reaction to an extrinsic condition, which could represent a stress source. Antimicrobial compounds (ACs), extracellular polymeric substances (EPSs), and biosurfactants (BSs) are among the scarcely explored molecules from Antarctic bacteria.

If we can speak of an engine that has set in motion bioprospecting studies, it should be identified in the search for molecules with antimicrobial activity. The urgency to discover new natural substances with antagonistic effects toward bacterial pathogens, viruses, or other etiological agents arises from the need to replace commonly used antibiotics and to front the antibiotic resistance developed during the last decades. After an initial phase in which terrestrial sources started to be explored in this field, the attention began to be addressed to marine environments and specifically to marine microorganisms. In addition to the high potential of new molecule detection and isolation, more investigations are necessary to understand the mechanisms at the base of the antagonistic effect and to compare it with those of common antibiotics in order to prospect an effective alternative (Rizzo and Lo Giudice 2019). The biotechnological challenge is to obtain a compound that can control and regulate the multi-resistant bacteria associated with human health or to fish species bred in aquaculture. Antimicrobial activity from Antarctic culturable bacteria has been studied by a number of researchers with promising results (Moncheva et al. 2002, Nedialkova and Naidenova 2004, O'Brien et al. 2004, De Souza et al. 2006, Lo Giudice et al. 2007, Biondi et al. 2008, Shekh et al. 2011, Solecka et al. 2012, Dong et al. 2013, Encheva et al. 2013, Asencio et al. 2014).

If some molecules are produced to front the menace represented by the competition with other organisms, others are produced to face different conditions such as contamination by organic and inorganic pollutants (e.g. hydrocarbons and heavy metals or temperatures below zero accompanied by ice formation with consequent possible damages to the cellular structure). The cellular uptake of insoluble compounds dissolved in the water column or stored into the soils and sediments could be enhanced by the production of molecules with amphipathic molecular structure and peculiar properties of emulsification and interface action. The production of BSs has been investigated especially with bioremediation purposes, but often it meets the demand of a lot of industrial processes, commercial applications, and medical challenges. The existence of bacteria able to produce these tensioactive agents in cold environments has been documented by several authors (Perfumo et al. 2019 and therein references), who reported also Antarctic BS producers and highlighted the important fallouts of discovering natural product stable at low temperatures and possible acting as pollutant removers. The BSs potentialities have been largely studied starting from contaminated matrices (Rizzo and Lo Giudice 2018) in order to increase the possibility to isolate potential producers. However, they have been extensively reported for temperate environments (Rizzo et al. 2014, 2015), but no enough efforts have been made in polar ones, and results are limited to the use of abiotic sources for the isolation (Yakimov et al. 1999, Pepi et al. 2005, Pini et al. 2007, Jadhav et al. 2013, Malavenda et al. 2015, Parhi et al. 2016).

EPSs have some common characteristics with BSs including properties like emulsifying and chelating functions. They present a high molecular weight and a strong polysaccharidic composition, generally accompanied by functional groups that confer them peculiar functions as well as sulfates, phosphates, or uronic acids. The most-reported genera of cold-adapted EPS producers include *Pseudoalteromonas* and *Halomonas* (Corsaro et al. 2004, Mancuso Nichols et al. 2004, 2005b, Kim and Yim 2007). Also in the case of EPSs, most of investigations have been focused on the use of abiotic sources, such as seawater (Corsaro et al. 2004, Mancuso Nichols et al. 2004, 2005b, Caruso et al. 2018b) and sediments (Kim and Yim 2007, Carriòn et al. 2015), while the literature on the isolation of EPS producers from Antarctic biotic sources is scant (Caruso et al. 2018a).

Despite a general overview of the most relevant molecules that have been provided here, it is also important to note that there are very often no clear boundaries between these kinds of molecules, which could have similar chemical structures as well as peculiar functions. For example, as we will point out in the following sections, EPSs are proven to be optimal cryoprotective agents, but they also have a good emulsifying activity toward hydrocarbon like BSs have. Moreover, often the molecules with antibacterial activities reported in some works are classified as BSs or EPSs.

Antarctic Matrices for Bioprospectors

In order to provide a contribution that would underline aspects not yet considered on the topic in a more ecological interpretation, here we provide an analysis of different matrices used for bioprospecting purposes, classified in abiotic and biotic, and in the function of the different extreme conditions evidenced in the previous section.

Terrestrial Environments

Despite the harsh conditions and the large exposition to extreme factors, organic carbon and microbial assemblages have been detected in Antarctic soils (Cowan et al. 2014), and indeed the abiotic factors are largely responsible for soil-associated bacterial communities structure modeling (Cary et al. 2010, Tytgat et al. 2016). Lambrechts et al. (2019) highlighted that a limited number of available sites have been effectively used as a source of isolation for cultivable bacteria with greater attention focused on some of them, such as the King George Island (Antarctic Peninsula) and Victoria Land (continental Antarctica). Interestingly, most studies were oriented not merely to detect the bacterial diversity but to determine bacterial functions, especially in terms of antibacterial activities (O'Brien et al. 2004, Lo Giudice et al. 2007, Gesheva and Vasileva-Tonkova 2012, Lee et al. 2012, Asencio et al. 2014, Chea et al. 2015, Tomova et al. 2015, Lavin et al. 2016) and biodegradative potential (Arenas et al. 2014, Gran-Scheuch et al. 2017, Lamilla et al. 2018).

Among a total number of more than 200 isolates from King George Island soils, Asencio et al. (2014) investigated the inhibitory activity of a violet Gram-negative *Janthinobacterium* sp. and found an ethanolic extract active against the beta-lactamase-producing strain of *Serratia marcescens* and *Escherichia coli* and the carbapenemase-producing strains of *Acinetobacter baumannii* and *Pseudomonas aeruginosa* with concentrations ranging from 0.5 and 16 µg ml^{-1}. Volcanic soils from Deception Island, a volcanic island located at the west of the Antarctic Peninsula, was proven to be an optimal source for the discovery of new ACs, which as much led to the isolation of bioactive strains (*Gordonia terrae*, *Leifsonia soli*, and *Terrabacter lapilli*) with antimicrobial activity against *Salmonella* species (Cleah et al. 2015). Gesheva and Vasileva-Tonkova (2012) underlined the influence of culturing conditions on the production of antimicrobial activity exhibited by an Antarctic soil-associated *Nocardioides* strain A-1. Indeed, different carbon sources favored the production of secondary metabolites active against different bacterial pathogens (i.e., *Bacillus subtilis* ATCC 6633, *Micrococcus* sp., and *Xanthomonas oryzae*) and at a different measure.

The use of Antarctic soils for bacterial isolation could be a promising tool for bioprospecting in terms of high specificity in shaping the bacterial communities. Indeed, it was reported that the harsh conditions recognizable in the Antarctic soils tend to favor the establishment of microbial communities characterized by Firmicutes (rich in spore-forming members) as the dominant phylogenetic group (Forsyth and Logan 2000, Allan et al. 2005, Rodríguez-Díaz et al. 2005, Saul et al. 2005, Singh et al. 2007). These fraction of cold-adapted bacteria is particularly resistant to the harsh conditions encountered in the soils, thanks to the spore-forming ability that helps them to preserve all functionalities from the extreme exposition to disturbing factors. In addition to that, they are known as able producers of cold enzymes and bioproducts used in several industrial fields (Huston 2008), thus particularly useful for the discovery of new cold-adapted/active enzymes and other biomolecules. As a support of this assumption, recent investigations reported a strict correlation between soil physicochemical properties and phylogenetic distribution of bacteria, and inhibitory activity against *Staphylococcus aureus* and *Candida albicans* was evidenced by spore-forming bacterial species (Lee et al. 2012).

Another interesting phylogenetic group, extremely endemic and widely distributed in Antarctic soils, is represented by Actinobacteria (Wawrik et al. 2007, Aislabie et al. 2008), whose bioprospecting potential was investigated by Lee et al. (2012). The antibacterial and antifungal activity of 39 strains isolated from Barrientos Island, an area occupied by several breeders as penguins, giant petrels, gulls, and skuas, was investigated. Authors reported inhibition against the growth of different strains of *Candida albicans, Staphylococcus aureus*, and *Pseudomonas aeruginosa*, and support the concreteness of the profitable use of soils for bioprospecting purposes. Similarly, Lavin et al. (2016) investigated an Antarctic *Streptomyces* strain isolated from soils and detected a genic pool encoding ACs with specific activity inhibiting the growth of seven Gram-negative and eight Gram-positive food-borne pathogens. *Streptomyces, Nocardia*, and *Geodermatophilus* spp. were isolated from Haswell Island (Gesheva and Negoita 2011) and proven to synthesize a series of multispectrum compounds from extracellular enzymes to antifungal and antibacterial substances. The soils of Deception and Galindez Islands were also used as a source of isolation of bacteria with antibacterial activity by detecting 25 strains, which were able to inhibit the growth of *Bacillus subtilis, B. cereus, A. johnsonii, Escherichia coli, Micrococcus luteus, S. lutea*, and *Pseudomonas aeruginosa* in addition to several yeast species. Among all isolates, a *Pseudomonas* strain presented the broadest activity spectrum against all bacterial and yeast targets (Tomova et al. 2015).

The ACs applications include many commercial fields as well as food preservation. In this regard, growth-inhibitor producers, mainly affiliated to *Arthrobacter, Planococcus, Pedobacter*, and *Pseudomonas*, were obtained from Antarctic soil samples and proven to be active against food-borne microorganisms at low temperatures (O'Brien et al. 2004, Wong et al. 2011).

The intrinsic characteristic of a solid and relatively static matrix gives to soils a highly welcoming nature for contaminants particularly recalcitrant, such as hydrocarbons and heavy metals. A delicate balance of attachment/detachment of pollutants to the granulometric components of the soils makes it difficult for their mobilization and therefore their disposal. This condition favors the establishment of highly competitive bacterial communities, specialized in the biodegradation processes or contaminants chelation. Often, some of the adopted strategies to face a pollution event or a chronic condition of contamination include the production of peculiar metabolites, such as BSs or bioemulsifiers (BEs) and amphipatic agents acting at the interface between phases with different degrees of polarity.

Perfumo et al. (2018) reported that microbial BS production could be considered a common feature among cold-adapted microorganisms, especially in those associated with polar soils and sediments for an amount of 50% of positive strains on the total screened. Despite that Antarctic environments are still considered quite pristine and uncontaminated; some authors demonstrated the occurrence of BSs and BEs bacterial producers associated with soils of different origin. A *Streptomyces* strain, resulting from a screening procedure for BS production carried out on a total of 59 strains isolated from five different locations at the South Shetland Islands, was detected as bioemulsifier-producer in the presence of different carbon sources (Lamilla et al. 2018). The best performance in terms of BS production was obtained after culturing the isolate in a mineral medium supplemented with hexadecane, suggesting the strong potential in the bioremediation field. Similarly, 17 strains isolated from three Antarctic sites (Casey Station, Dewart Island, and Terra Nova Bay) were screened for their hydrocarbon-degrading capacity, cell hydrophobicity, and glycolipid production (Vasileva-Tonkova and Gesheva 2004). The interesting findings highlighted a widespread degradation ability among the tested strains and a broader ability to produce glycolipids, not only in the presence of hydrocarbons but also in the presence of soluble carbon sources by suggesting a possible dual role in the hydrocarbon uptake and adhesion/de-adhesion processes. Moreover, the authors suggested a model consortium including most promising isolates, mainly affiliated to coryneform and nocardioform bacteria for the treatment of polluted soils. Gesheva et al. (2010) achieved the same conclusion while studying a *Rhodococcus* strain isolated from soil in Casey Station with hydrocarbon-degrading capacity and ability to produce a rhamnose-containing glycolipid by using both glucose or kerosene as carbon source. The compound is one example of multiactive agents with optimal surface tension reducing function, inhibition growth of *B. subtilis* ATCC6633, and hemolytic activity.

In support of the assumption according to which Antarctic sources are particularly promising for isolation of new bacterial producers, less common taxonomic groups also have been detected as producers of glycolipids at low temperature as in the case of *Pantoea* sp. (*Enterobacteriaceae* family) isolated from ornithogenic soil in the Frazier Islands, Antarctica (Vasileva-Tonkova et al. 2007).

Snow samples have been also used for the isolation of antimicrobial producers, even if a small amount of reports is available and limited only to *Janthinobacterium lividum* (Baricz et al. 2018). The strain allowed to isolate an extract corresponding to 30 µg of violacein and a ratio violacein:deoxyviolacein of 10:1, which was tested against a wide range of multidrug-resistant bacterial isolates and showed bactericidal and bacteriostatic effects. Among Antarctic terrestrial habitat, permafrost has gained a lot of interest during last decades as a model study of bacterial adaptation as it was recognized for high physiological flexibility of such bacterial communities, which were able to survive to the succession of geological events by facing freeze/thawing cycles, rigid temperature, radiation exposure, and exsiccation (Steven et al. 2006, Wagner 2008). The promising perspectives offered by the study of permafrost-associated bacteria have been evidenced by La Ferla et al. (2017), who detected interesting metabolic pathways and enzymatic activity rates. Despite this, such environment results are still poorly explored in the BS bioprospecting field, and to the best of our knowledge, no works have reported the production of bioactive molecules from bacterial populations inhabiting it.

Marine Environment

The marine environment is a very dynamic system in which each external parameter could change also in a limited spatial range from the water column to the sediment, from the surface to the bottom, in the dependence of the depth or hydrodynamism, the kind of sediment, of the local geological conformation and so on. The marine ecosystem, different from the terrestrial one, is more variegated and contains a wide kind of habitats and micro-habitats. All these features acquire peculiar shadow in the Antarctic environment. Seawater is characterized by the presence of dissolved salts so that the freezing point is lowered at –1.9°C, and the organic matter concentration is very low, especially in the oceans. For this reason, as suggested by Perfumo and coauthors (2018), microbial assemblages are particularly abundant in areas with sources of organic material, such as undissolved material or unsoluble liquids. Polluted areas, despite what is expected, could be also particularly populated and sometimes the presence of contaminants in the water column or at sediment level could stimulate the development and establishment of specialized microorganisms, which can degrade pollutants through the production of special molecules that are so useful for bioprospecting fields.

Polluted marine environments, especially hydrocarbon polluted marine areas, have been considered for a long time as the optimal source of isolation of bacterial BS producers. Indeed, BS production represents one of the main strategies used by bacteria to facilitate the cellular uptake of insoluble compounds, such as hydrocarburic substrates. In the Antarctic environment, the biodegradation processes are complicated by the harsh conditions, in particular by the ice formation and the rigid temperature. Several bacterial taxonomic groups have been detected as BS producers in the polar oceans, namely *Pseudomonas, Pseudoalteromonas, Marinomonas, Halomonas, Rhodococcus*, and *Cobetia* representatives (Ibacache-Quiroga et al. 2013, Cai et al. 2014, Konishi et al. 2014). Interestingly, different marine matrices have been used as a source of isolation of BS producers, and several approaches have been adopted. Seawater samples are the most used source of isolation for BS bacterial producers, which lead sometimes to the detection of uncommon taxonomic groups. *Oceanobacillus* spp. BS producers we isolated by Jadhav et al. (2013) and Parhi et al. (2016), who demonstrated the biosurfactant production also in the presence of alternative carbon sources with optimal properties of stability at harsh temperature and pH conditions. Additional scientific contributes to the support of the use of seawater samples. In particular, noteworthy are the works of Yakimov et al. (1999), who described the production of a cell-bound BS by the best-studied bacterial isolate in the context of hydrocarbon degradation, namely *Alcanivorax borkumensis* strain (from Terra Nova Bay, Ross Sea) and Pini et al. (2007) who reported BS-mediated hydrocarbon uptake by *Rhodococcus* sp. isolate. Abraham et al. (1998) finely analyzed the chemical nature of BS produced by *Alcanivorax borkumensis* strain by indicating an anionic glucose lipidic structure with a tetrameric oxyacyl side chain and identified ten derivatives of these lipids, whose structure was elucidated by MSMS technique. Interestingly, the authors reported that the relative abundance of different lipids identified is variable in relation to the strain species.

An interesting contribution providing a useful BS application was reported by Ibacache-Quiroga et al. (2013), who defined a *Cobetia* strain deriving from Montemar seawater samples (Chile) as a producer of a BS constituted by hydroxy fatty acids and extracellular lipidic structures and observed an antagonistic activity exhibited by the BS with the quorum sensing communication system of *Aeromonas salmonicida* sub sp. *salmonicida*. In line with this finding, the authors proposed the use of BS as a strategy for the control of fish pathogen infection in fish aquaculture.

In addition to the most expected contaminants retrievable in seawater, the impact of bacterial biofilms is increasing and becoming dangerous for human health and industrial activities. This kind of impact has to be focused too with the discovery of natural molecules, which interfere with the biofilm formation processes. Papa et al. (2013) recently demonstrated the production of an anti-biofilm product of saccharidic composition by a *Pseudoalteromonas* strain, active against different staphylococci. Similarly, Leyton and coauthors (2015) evaluated the *Flavobacterium psychrophilum* 19749 biofilm inhibition exhibited by Antarctic bacteria bioproducts by underlying the importance of such findings in the aquaculture managing of fish infections.

Malavenda et al. (2015) experimented on microcosms assessed with Antarctic sediments enriched with different hydrocarburic carbon sources (crude oil or commercial diesel oil) and used them to isolate hydrocarbon-degrading bacteria. Interesting results were obtained from the investigation with 17 Antarctic strains, which were able to produce BSs as proved by emulsification rates and surface tension reduction ability. The strains were mainly affiliated not only to genera already known as BS producers, like *Rhodococcus* and *Pseudomonas*, but also to bacterial genera that are scarcely reported for this topic from cold environments, such as *Pseudoalteromonas* and *Idiomarina*. Strains within the former genus can produce EPSs in polar environments as well (Bozal et al. 1994, Mancuso Nichols et al. 2004, 2005b, Casillo et al. 2018).

BS producers have been successfully isolated also from seawater/ice interface systems in which ice and water component seasonally alternate and occasionally coexist by threatening the survival of microbial communities because of the crystal formation with damages for the cells, the limited circulation, and diffusion of nutrients or physical injuries. These events stimulate bacteria to produce molecules, namely BSs and EPSs, which help them to face such harmful conditions. The success of the use of seawater/ice systems as isolation sources has been proven by several reports (Pepi et al. 2015, Mancuso Nichols et al. 2004, 2005a, b). An *Halomonas* sp. strain was isolated by Pepi et al. (2005) from the Terra Nova Bay (Ross Sea) and was able to produce an emulsifier agent of 18 kDa weight and a composition including a mixture of different fatty acids and sugars and also active at low temperature. The same isolation matrix (seawater/ice) was used by Mancuso Nichols and coauthors (2004) to isolate bacterial strains and to produce EPSs. They were affiliated to *Pseudoalteromonas, Shewanella, Polaribacter*, and *Flavobacterium* (Mancuso Nichols et al. 2004, 2005b). The authors demonstrated the stimulating effect of cold temperature on EPS production but also the strong influence of the carbon source on the chemical structure of the produced compound.

The severe influence of concomitant surrounding conditions on biosynthesis processes in bacterial communities inhabiting seawater habitats was recently demonstrated by Caruso et al.

(2018b) for a *Pseudoalteromonas* isolate from Terra Nova Bay (Lo Giudice et al. 2012). Optimal EPS production was detected at 4°C, pH 7, a sucrose concentration of 2% (w/v), and 3% NaCl. In this study, the stimulating effect of inorganic pollutants, i.e., heavy metals, was also assessed. Another EPS-producing *Pseudoalteromonas* strain was also obtained from Antarctic seawater. The composition of produced EPS was strongly affected by temperature incubation (Mancuso Nichols et al. 2004).

Antarctic sediments were also exploited for the isolation of EPS producers. Such molecules, as BSs, possess an important ecological role for bacteria by facilitating them to front the cryodamages and also the accumulation of pollutants, which could occur in sediments with greater impact than in the dynamic seawater. Kim and Yim (2007) and Carrión et al. (2015) isolated different EPS producers from sediments of the South Shetland Islands and King George Island, respectively, evidencing promising cryoprotective properties.

In the context of antimicrobial activity, three rhamnolipids have been reported as produced by *Pseudomonas* sp. BTN1 isolated from Antarctic sediments in the Ross Sea and resulted in being interestingly active against human pathogens belonging to the *Burkholderia cepacia* complex (Tedesco et al. 2016). Antifungal activity was evidenced for the strain *Bacillus* sp. Pc3 deriving from Antarctic seawater, whose complete genome sequence was also provided by contributing to the enhancement of knowledge about the metabolic pathway of antimicrobial compounds in this strain (Guo et al. 2015).

Lake and Ponds

The Antarctic lake system includes about 400 lakes, some of which are hydrologically connected, and for which a microbial life beneath the ice was proven (Smith et al. 2009, Fricker et al. 2011, Ross et al. 2011, Wright and Siegert 2012). The Antarctic lake microbial biomass was estimated at ~10^{29} cells (Priscu et al. 2008). However, its bioprospecting potential has not been still well investigated. Antarctic lakes could assume uncommon conditions that make them complex habitats, which lead them to include a lot of peculiar micro-habitats that are not only represented by water or sediments as in the case of other freshwater basins. As an example, Antarctic brines are really interesting liquid water pockets within the permafrost matrix, which are also stable at subzero temperatures. The conditions, which lead to the formation and stabilization of such small ecosystems, force the dominance of bacterial communities that are extremely specialized and can survive under unusual osmotic pressure and salinity oscillations (Price 2000, Mader et al. 2006). To the best of our knowledge, such matrices have been not explored for bioprospecting field, but we have recently analyzed the physiological patterns of cultivable bacterial fractions from two different Antarctic lakes and detected promising properties of resistance to heavy metals and hydrocarbon degradation activity (*unpublished data*).

To date, few researchers have explored lakes and ponds for bioprospecting. Mojib et al. (2010) described the antimycobacterial activity of pigments from a *Janthinobacterium* sp. and a *Flavobacterium* sp. isolated from freshwater lakes in Schirmacher Oasis, East Antarctica. Coronel-León et al. (2015) reported about a strain of *Bacillus licheniformis* isolated from sand samples in Kroner lake (Deception Island), and its ability to produce BSs. Interestingly, the obtained BS showed homologies in chemical structure with the known lichensyn groups, exhibited optimal surface tension reduction properties (28.5 mN/m), and showed stability under a wide pH range, high temperatures, and different salt concentrations.

Biotic Matrices

Taking into account that in the bioprospecting field, the main focus should be on the discovery of new producing species or new biomolecule structures and no useful sources should be neglected. As pointed out by Lo Giudice and Rizzo (2018) in a recent review, the abiotic sources have been extensively exploited in comparison with the biotic ones. The latter, as living organisms used for the isolation of bacterial producers, have been scarcely considered in most common and accessible environments, and this is even more true for Antarctic matrices. Especially in the marine environment, marine invertebrates represent real microbial habitats, hosting numerous and abundant bacterial populations. Such assemblages are finely regulated by complex interactions for the maintenance of the correct equilibrium among the bacterial components between themselves and between the associated bacteria and their host—which often needs the production of complex molecules— mostly involved in defense strategies. Biotic matrices have gained the interest of researchers since the last decades and are considered promising sources in temperate conditions. But in the case of Antarctica, so unique and peculiar, such living components acquire an added scientific value in the function of their high under-exploration level.

Sponges are the most exploited Antarctic marine invertebrates, even if the number of scientific contributes is yet scant. Only recently, it was scientifically proven that the real responsible bioactive molecule production is the abundant bacterial communities associated with sponges and not the sponge themselves (Hill 2004, Wakimoto et al. 2014). These benthic filter-feeders gained a lot of attention for the high potential to isolate bacterial producers of molecules with antimicrobial properties. The first report was provided by Jayatilake et al. (1996), who isolated a *Pseudomonas aeruginosa* strain associated with the Antarctic sponge *Isodictya setifera*, producer of natural molecules active against Gram-positive microorganisms. The strong potential of Antarctic sponge-associated bacteria for the ACs production was reported by Mangano et al. (2009), who tested the antagonistic interactions among the bacterial communities associated with the two sponges species i.e., *Anoxycalyx joubini* and *Lissodendoryx nobilis*. Bacterial isolates from three Antarctic sponge species (i.e., *Haliclonissa verrucosa*, *Anoxycalyx joubini*, and *Lissodendoryx nobilis*), collected from Terra Nova Bay (Ross Sea), were screened for growth inhibition ability against bacteria involved in the cystic fibrosis disease (Papaleo et al. 2012), showing the production of volatile organic compounds effective against *Burkholderia cepacia*.

To the best of our knowledge, no investigations are available about BS producers among bacteria associated with Antarctic marine invertebrates, but one important contribution was provided by Caruso et al. (2018a) about EPS production. The biotechnological aspect was strongly underlined by authors, who found EPS producers among the cultivable bacterial fraction associated with the Antarctic sponges *Tedania charcoti*, *Hemigellius pilosus,* and *Haliclonissa verrucosa*. The strains were taxonomically affiliated to *Winogradskyella* spp., *Colwellia* sp., and *Shewanella* sp. The EPSs were proven to be excellent emulsifying and cryoprotecting agents alongside heavy metal chelating molecules.

Within biotic matrices, benthic Antarctica microbial mats were also used as a source of bacterial ACs. In addition to Cyanobacteria members (Asthana et al. 2009), Gammaproteobacteria and Betaproteobacteria members also were reported, namely *Psychrobacter* sp., *Shewanella* sp., and *Janthinobacterium* sp. (Taton et al. 2006, Rojas et al. 2009).

Useful Peculiarities of Bioactive Molecules of Antarctic Origin

Bioprospecting research offers surely many promising perspectives, but in this particular topic some points have been addressed for a concrete and safe joining of discoveries. In all the processes in which one product is synthesized by a living organism, and it represents a candidate as an alternative to chemical counterparts, two aspects should be focused: cost reduction and time optimization. Moreover, a natural molecule can be considered as a valuable substitute to a synthetic one if it exhibits peculiar, more specific, or better properties in order to have more than one reason to prefer it over the substances in use. Perfumo et al. (2018) underlined an aspect in relation to BSs from cold environments that could be extended to all molecules deriving from Antarctica. Beyond the eco-friendly nature of BSs, EPSs, or ACs—which is a common feature to the same molecules from temperate environments—the specificity of action, the stability at a wide range of pH, salinity, and temperature have to be considered. As Perfumo stated, the eco-friendly being of such molecules is strongly related to an assessment of energy saving. The cost reduction for production procedure should be simplified in the case of molecules that could be produced at low temperatures and that are also effective below 0°C. Among the main useful compounds deriving from Antarctica, surely glycoprotein with anti-freeze function, isolated from Antarctic fish, are the most interesting from a commercial and applicative point of view (Cheng and Cheng 1999). Indeed, these molecules could be employed in several industrial processes—or the extension of frozen food shell-life—in the preservation of tissues in medical fields, in the fish production in aquaculture, or in the increase of freeze tolerance for commercial plants (Lohan and Johnston 2003).

The strong limitation could be represented by the time of production as cold-adapted bacteria need time to achieve a consistent biomass value, thus producing satisfactory biomolecule yields. For this reason, a study on the enhancement and improvement of optimal production conditions have to be performed, and deep comprehension of bacterial needs, responses, and behavior is necessary. In any case, cold-adapted bacteria have been reported as able to produce amounts of bioactive molecules comparable to those reported for temperate bacterial producers in terms of both costs and amounts. In addition to this, the wide range of tolerance of temperature values, showed by some extremophiles, does not prevent the possibility to use higher temperature of incubation to optimize the whole production process.

Some other advantages are correlated to the intrinsic nature of compounds—as in the case of the BS MELs produced by an Antarctic yeast (Morita et al. 2013)—with specific properties of anti-agglomerants for a lot of expensive processes acting at ice-water interfaces, such as ice storage, refrigeration, air conditioning, and industrial cooling (Perfumo et al. 2018), with considerable economic advantages. Despite these relevant and cost- and time-saving findings are related to yeast, the same opportunity could be found within the bacterial Antarctic communities.

Finally, by taking into account that the Antarctic environments are characterized by harsh conditions, also in terms of carbon source and nutrient availability, it is expected that cold-adapted bacteria could exhibit optimal bioactive molecules production performances without the need of complex (and expensive) carbon source or nutrient supply. They should be able to achieve optimal productivity also by furnishing them low-cost alternative in line with data reported by Jadhav et al. (2013) and Parhi et al. (2016) for BS production.

Ethical Issues of Antarctic Bioprospecting

With the recent fast development of new innovative techniques of investigation, the researchers' approach toward the bioprospecting field has changed, and the efforts are focused on the discovery of new commercially exploitable natural compounds that have increased significantly. However, in the case of Antarctic resources, such an issue implies numerous troubles especially when the investigations are moved not only by a scientific intention but with the main aim of a possible commercial application. Despite all the environmental resources that have to be protected, in the case of Antarctic ones a particular safeguard has been adopted in the defense of the high scientific value globally attributed to them. The exploitation of resources in Antarctica is regulated by some official references, which mainly include the Antarctic Treaty and the Madrid Protocol. However, there are neither precise rules regarding the collection and treatment of samples nor a clear distinction between the different matrices (water, soil, sediments, and living organisms). In the case of bioprospecting, the more involved matrices are all the categories of living organisms that could be used as a direct or indirect source of novel bioactive compounds. The Antarctic Treaty establishes peaceful and scientific purposes of research in the Antarctic continent and promotes the cooperation between countries involved. International cooperation must be pursued in terms of transparency and sharing of scientific programs, global availability of research station facilities, scientific staff, and obtained data. The second important reference in this context is the Madrid Protocol, which represents an additional safeguard action toward Antarctica as a natural reserve in all its ecosystems and forbids any non-scientific activity concerning mineral resources. Therefore, in summary, in Antarctica, the principles of freedom of research, peace, commercial exploitation, and environmental protection should coexist together.

The first discussion about bioprospecting dates back to 1999 within the Treaty System and was further supervised at several meetings during the following years (Committee for Environmental Protection [CEP], Scientific Committee on Antarctic Research [SCAR], and the Antarctic Treaty Consultative Meeting [ATCM]) (Lohan and Johnston 2005). The theme of Antarctic bioprospecting is very intricate and difficult to treat and interpret, compounded by the fact that the term itself is still under discussion and acquires many means. In some cases, it is defined only as research for genetic material and chemical compounds with bioactivity useful to mankind. Another more wide definition included in bioprospecting terms the study of biological and genetic resources, the discovery of a product, the marketing of it, and the protection through intellectual property (Goyes 2013). Indeed, behind the pure scientific vision of the activities contemplated in bioprospecting matter, also exist activities mainly oriented to the commercial application. Among the main application with possible business returns in Antarctica, surely the anti-freeze proteins represent one of the most concrete and immediate one, but generally, the focus is on the discovery of new compounds with extraordinary properties of efficiency, stability, and resistance at low temperature. Another term usually associated with Antarctic bioprospecting is biopiracy, intended as any appropriation of natural products, genetic codes, and knowledge associated with them.

The first requirement that emerges in the approach to bioprospecting in Antarctica lies in the need for a redefinition of the activities connected to it and for new and clearer regulation. What differentiates Antarctic bioprospecting from the ones carried out in other terrestrial areas is that in addition to the objectives normally pursued, there are specific ethical, environmental, and legal implications for its territory imposed by the Antarctic Treaty. The correlation with the concept of biopiracy could be considered excludable without an indigenous population on the continent or sovereign able to collect any possible direct benefit. The other concerns are ethical and involve questions regarding the possession of natural biological/genetic resources or their possible recognition as the property of mankind, the intellectual property deriving by the use of particular resources, and the potential impact of biotechnological related activities. If according to someone the Antarctic area has to be considered a global commons and therefore also its resources could be used for global benefits, others categorically reject their exploitation. Although it is still to be clarified in detail if the possible impact deriving from bioprospecting activities in Antarctica is negligible, greater, or lesser compared to other research activities; to face this uncertainty, a precautionary approach is probably the most desirable. According to Farrel and Duncan (2005), bioprospecting has to be considered a scientific exploration activity aimed at discovering something useful from nature, which may contribute to fundamental science. The impact of all activities concerning it is correlated also to how they are carried out. In other terms, a sustainable form of bioprospecting should be globally assessed by regulating the removal of biological material without threatening its presence in Antarctica and by correct harvesting of biological information.

Conclusions

In conclusion, Antarctic bioprospecting is carried on for several years and continues to be an intriguing research area because of its still unexplored land and promising perspectives. Some authors delineated four phases of Antarctic bioprospecting by biological resources, generally resumed in a step of discovery (involving collection, screening, and describing), a step of research (with isolation and purification of useful products), a step of manufacturing (large scale production and testing activities), and a final phase of marketing (with registration requirements and commercialization).

A list of main Antarctic matrices and sites exploited for bioprospecting purposes until now is presented in Table 1, showing how most contributions were focused on the same areas.

TABLE 1: List of Main Antarctic Sites Exploited for Bioprospecting Purposes

	Matrix	Mol.*	Site
	Antarctic Bioprospecting		
	Soil		
	Asencio et al. 2014	AC	King George Island
	Cheah et al. 2015	AC	Deception Island
	Gesheva and Vasileva-Tonkova 2012	AC	Casey Station, Wilkes Land
	Lavin et al. 2016	AC	King George Island
	Lee et al. 2012	AC	Barrientos Island
	Lo Giudice et al. 2007	AC	Terra Nova Bay
Terrestrial Habitats	O'Brien et al. 2004	AC	–
	Tomova et al. 2015	AC	Deception and Galindez Islands
	Wong et al. 2011	AC	King George Island
	Vollù et al. 2014	AC	King George Island
	Gesheva and Negoita 2011	BS	Haswell Island
	Gesheva et al. 2010	BS	Casey Station, Wilkes Land
	Lamilla et al. 2018	BS	Shetland Islands
	Vasileva-Tonkova and Gesheva 2004	BS	Casey Station, Dewart Island, Terra Nova Bay
	Vasileva-Tonkova and Gesheva 2007	EPS	Frazier Islands
	Snow		
	Baricz et al. 2018	AC	King George Island
	Sediments		
	Abraham et al. 1998	BS	North Sea
	Malavenda et al. 2015	BS	South Shetlands Islands, Byers Peninsula
	Yakimov et al. 1998	BS	North Sea
	Carriòn et al. 2015	EPS	King George Island
	Kim et al. 2007	EPS	Shetland Islands
Marine Habitats	*Seawater*		
	Guo et al. 2015	AC	–
	Leyton et al. 2015	AC	Rey Jorge Island
	Tedesco et al. 2016	AC	Terra Nova Bay
	Papa et al. 2013	AC	Antarctic Station Dumont d'Urville
	Ibacache-Quiroga et al. 2013	BS	Montemar, Chile
	Jadhav et al. 2013	BS	–
	Parhi et al. 2016	BS	–
	Pepi et al. 2005	BS	Terra Nova Bay

TABLE 1: Contd....

TABLE 1: Contd.

	Matrix	Mol.*	Site
	Antarctic Bioprospecting		
	Seawater		
Marine Habitats	Pini et al. 2007	BS	Terra Nova Bay
	Yakimov et al. 1999	BS	North Sea
	Caruso et al. 2017	EPS	Terra Nova Bay
	Corsaro et al. 2004	EPS	Antarctic Station Dumont d'Urville
	Mancuso Nichols et al. 2004, 2005	EPS	Southern Ocean
Living Organisms	*Biotic*		
	Asthana et al. 2009	AC	–
	Jayatilake et al. 1996	AC	–
	Mangano et al. 2008	AC	Terra Nova Bay
	Papaleo et al. 2012	AC	Terra Nova Bay
	Rojas et al. 2009	AC	Larsemann Hills, Vestfold Hills, McMurdo Dry Valleys
	Taton et al. 2006	AC	East Antarctica**
	Caruso et al. 2018	EPS	Terra Nova Bay
	Lake and ponds		
	Mojib et al. 2010	AC	Schirmacher Oasis, Dronning Maud Land
	Coronel-León et al. 2015	BS	South Shetlands Islands
	Morita et al. 2013	BS	Lake Vanda

* Molecules

** Larsemann Hills, Bølingen Islands, Vestfold Hills, Rauer Islands, McMurdo Dry Valleys

As far as we know, several Antarctic resources have been exploited with particular regard for soils in the case of terrestrial ones and sediments in the case of the marine environment (Figure 1). The living organisms are the less used matrix, despite interesting results have been obtained especially from sponge samples. In this context, the use of associated or free-living microorganisms could be considered a good solver tool, which would make research operations more fluid and rapid and overcome the problem of oversampling communities of higher marine organisms. With regard to the kind of biomolecules, surely antibiotics have been the most investigated, while fewer reports are available for BSs and EPSs.

The development of new innovative techniques have given a further boost to Antarctic bioprospecting and should be enhanced. Such technologies are to be considered a fundamental tool for bioprospectors as key for a safe investigation with minimum impact on this sensitive environment. In any case, the investigations carried out until now are inconsistent to provide a critical and exhaustive drawing of the topic. The adherence to the Antarctic Treaty should be maintained to ensure the preservation of the Antarctic environment in an attempt to pursue common benefits across all the nations involved and a sustainable administration and conservation of resources. In other words, a sustainable model of bioprospecting has to be globally adopted, while keeping faith with the principles of sharing, respect for nature, and scientific interest.

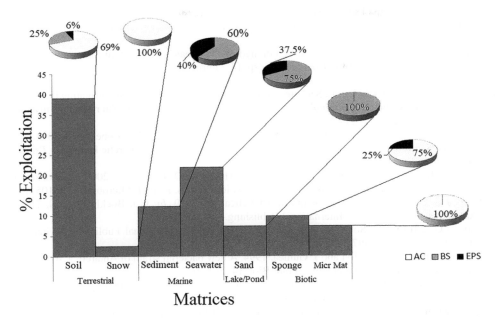

Figure 1. Exploitation rate of abiotic and biotic Antarctic matrices for the isolation of antimicrobial compound (AC), biosurfactant (BS) and extracellular polymeric substance (EPS) bacterial producers.

Once this is assumed, the question of the lack of clarity in the rules is the second important point to be taken into consideration. The main issues are the absence of clear regulation with particular regard for the biological and genetic resources, which affect in a significant way the stakeholders use them. Moreover, the inadequacy and unavailability of protocols for scientists that inhibit their efficiency and representatives of industry are thwarted by the ambiguity of ownership of samples. Finally, governments have to overcome the negotiation troubles regarding the correct sharing of benefits.

Acknowledgments

This chapter reports also data obtained within the Research Project PNRA16_00020 granted by the PNRA (Programma Nazionale di Ricerche in Antartide), Italian Ministry of Education and Research.

References

Abraham, W.-R., H. Meyer and M. Yakimov. 1998. Novel glycine containing glucolipids from the alkane using bacterium *Alcanivorax borkumensis*. Biochim. Biophys. Acta 1393: 57–62.

Aislabie, J.M., S. Jordan and G.M. Barker. 2008. Relation between soil classification and bacterial diversity in soils of the Ross Sea region, Antarctica. Geoderma 144: 9–20.

Allan, R.N., L. Lebbe, J. Heyrman, P. De Vos, C.J. Buchanan and N.A. Logan. 2005. *Brevibacillus levickii* sp. nov. and *Aneurinibacillus terranovensis* sp. nov., two novel thermoacidophiles isolated from geothermal soils of northern Victoria Land, Antarctica. Int. J. Syst. Evol. Microbiol. 55: 1039–1050.

Arenas, F.A., B. Pugin, N.A. Henriquez, M.A. Arenas-Salinas, W.A. Diaz-Vasquez, M.F. Pozo, et al. 2014. Isolation, identification and characterization of highly tellurite-resistant, tellurite-reducing bacteria from Antarctica. Polar Sci. 8: 40–52.

Asencio, G., P. Lavin, K. Alegria, M. Dominguez, H. Bello, G. Gonzalez-Rocha, et al. 2014. Antibacterial activity of the Antarctic bacterium *Janthinobacterium* sp. SMN 33.6 against multi-resistant Gram-negative bacteria. Electron. J. Biotechnol. 17: 1–5.

Asthana, R.K., M.K. Deepali, A. Tripathi, A.P. Srivastava, S.P. Singh, G. Singh, et al. 2009. Isolation and identification of a new antibacterial entity from the Antarctic cyanobacterium *Nostoc* CCC 537. J. Appl. Phycol. 21: 81.

Ayub, N.D., P.M. Tribelli and N. Lopez. 2009. Polyhydroxyalkanoates are essential for maintenance of redox state in the Antarctic bacterium *Pseudomonas* sp. 14-3 during low temperature adaptation. Extremophiles 13: 59–66.

Baricz, A., A. Teban, C.M. Chiriac, E. Szekeres, A. Farkas, M. Nica, et al. 2018. Investigating the potential use of an Antarctic variant of *Janthinobacterium lividum* for tackling antimicrobial resistance in a one health approach. Sci. Rep. 8: 15272.

Benforte, F.C., M.A. Colonnella, M.M. Ricardi, E.C.S. Venero, L. Lizarraga, N.I. López, et al. 2018. Novel role of the LPS core glycosyltransferase WapH for cold adaptation in the Antarctic bacterium *Pseudomonas extremaustralis*. PLOS ONE 13: e0192559.

Biondi, N., M.R. Tredici, A. Taton, A. Wilmotte, D.A. Hodgson, D. Losi, et al. 2008. Cyanobacteria from benthic mats of Antarctic lakes as a source of new bioactivities. J. Appl. Microbiol. 105: 105–115.

Bockheim, J.G. 2015a. Soil-forming factors in Antarctica. pp. 5–20. *In:* G.J. Bockheim [ed.]. The Soils of Antarctica. Cham: Springer International Publishing.

Bockheim, J.G. 2015b. The Soils of Antarctica. Basel: Springer International Publishing Switzerland. doi: 10.1007/978-3-319-05497-1.

Bozal, N., A. Manresa, J. Castellvi and J. Guinea. 1994. A new bacterial strain of Antarctica, Alteromonas sp., that produces a heteropolymer slime. Polar Biol. 14: 561–567.

Butler, B.M., S. Papadimitriou, A. Santoro and H. Kennedy. 2016. Mirabilite solubility in equilibrium sea ice brines. Geochim. Cosmochim. Acta 82: 40–54.

Cai, Q., B. Zhang, B. Chen, Z. Zhu, W. Lin and T. Cao. 2014. Screening of biosurfactant producers from petroleum hydrocarbon contaminated sources in cold marine environments. Mar. Pollut. Bull. 86: 402–410.

Carriòn, O., L. Delgado and E. Mercade. 2015. New emulsifying and cryoprotective exopolysaccharides from Antarctic *Pseudomonas* sp. ID1. Carbohydr. Polym. 117: 1028–1034.

Caruso, C., C. Rizzo, S. Mangano, A. Poli, P. Di Donato, I. Finore, et al. 2018a. Production and biotechnological potentialities of extracellular polymeric substances from sponge-associated Antarctic bacteria. Appl. Environ. Microbiol. 84: e01624–17.

Caruso, C., C. Rizzo, S. Mangano, A. Poli, P. Di Donato, B. Nicolaus, et al. 2018b. Extracellular polymeric substances with metal adsorption capacity produced by *Pseudoalteromonas* sp. MER144 from Antarctic seawater. Environ. Sci. Poll. Res. 25: 4667–4677.

Caruso, C., C. Rizzo, S. Mangano, A. Poli, P. Di Donato, B. Nicolaus, et al. 2019. Isolation, characterization, and optimization of extracellular polymeric substances produced by a cold-adapted *Marinobacter* isolate from Antarctic seawater. Antarc. Sci. 31: 69–79.

Cary, S.C., I.R. Mcdonald, J.E. Barrett and D.A. Cowan. 2010. On the rocks: the microbiology of Antarctic dry valley soils. Nat. Rev. Microbiol. 8: 129–138.

Casillo, A., R. Lanzetta, M. Parrilli and M.M. Corsaro. 2018. Exopolysaccharides from marine and marine extremophilic bacteria: structures, properties, ecological roles and applications. Mar. Drugs. 16: 69.

Cavicchioli, R., T. Charlton, H. Ertan, S.M. Omar, K.S. Siddiqui and T.J. Williams. 2011. Biotechnological uses of enzymes from psychrophiles. Microb. Biotechnol. 4: 449–460.

Cavicchioli, R. 2016. On the concept of a psychrophile. ISME J. 10: 793–795.

Chattopadhyay, M.K., M.V. Jagannadham, M. Vairamani and S. Shivaji. 1997. Carotenoid pigments of an Antarctic psychrotrophic bacterium *Micrococcus roseus*: temperature dependent biosynthesis, structure and interaction with synthetic membranes. Biochem. Biophys. Res. Comm. 239: 85–90.

Cheah, Y.K., L.H. Lee, C.C.Y. Chieng and V.-L.C.M. Wong. 2015. Isolation, identification and screening of actinobacteria in volcanic soil of Deception Island (the Antarctic) for antimicrobial metabolites. Pol. Polar Res. 36(1): 67–78.

Cheng, C.C. and L. Cheng. 1999. Evolution of an antifreeze glycoprotein. Nature 443–444.

Chintalapati, S., M.D. Kiran and S. Shivaji. 2004. Role of membrane lipid fatty acids in cold adaptation. Cell. Mol. Biol. 50: 631–642.

Convey, P., J.A.E. Gibson, C.D. Hillenbrand, D.A. Hodgson, P.J.A. Pugh, J.L. Smellie, et al. 2008. Antarctic terrestrial life – challenging the history of the frozen continent? Biol. Rev. 83: 103–117.

Coronel-León J., G. de Grau, A. Grau-Campistany, M. Farfan, F. Rabanal, A. Manresa, et al. 2015. Biosurfactant production by AL 1.1, a *Bacillus licheniformis* strain isolated from Antarctica: production, chemical characterization and properties. Ann. Microbiol. 65: 2065–2078.

Correa-Llantén, D.N., M.J. Amenábar and J.M. Blamey. 2012. Antioxidant capacity of novel pigments from an Antarctic bacterium. J. Microbiol. 50: 374–379.

Corsaro, M.M., R. Lanzetta, E. Parrilli, M. Parrilli, M.L. Tutino and S. Ammarino. 2004. Influence of growth temperature on lipid and phosphate contents of surface polysaccharides from the Antarctic bacterium *Pseudoalteromonas haloplanktis* TAC 125. J. Bacteriol. 186: 29–34.

Cowan, D.A. and L.A. Tow. 2004. Endangered Antarctic environments. Annu. Rev. Microbiol. 58: 649–690.

Cowan, D.A., T.P. Makhalanyane, P.G. Dennis and D.W. Hopkins. 2014. Microbial ecology and biogeochemistry of continental Antarctic soils. Front. Microbiol. 5: 154.

Cowan, D.A., J.B. Ramond, T.P. Makhalanyane and P. De Maayer. 2015. Metagenomics of extreme environments. Curr. Opin. Microbiol. 25: 97–102.

De Maayer, P., D. Anderson, G. Cary and D.A. Cowan. 2014. Some like it cold: understanding the survival strategies of psychrophiles. EMBO Rep. 15: 508–517.

De Pascale, D., C. De Santi, J. Fu and B. Landfald. 2012. The microbial diversity of polar environments is a fertile ground for bioprospecting. Mar. Genomics. 8: 15–22.

De Souza, M.-J., S. Nair, P.A. Loka Bharathi and D. Chandramohan. 2006. Metal and antibiotic-resistance in psychrotrophic bacterial from Antarctic marine waters. Ecotoxicology 15: 379–384.

Dieser, M., M. Greenwood and C.M. Foreman. 2010. Carotenoid pigmentation in Antarctic heterotrophic bacteria as a strategy to withstand environmental stresses. Arctic, Antarctic and Alpine Res. 42: 396–405.

Dong, N., Z. Di, Y. Yu, M. Yuan, X. Zhang and H. Li. 2013. Extracellular enzyme activity and antimicrobial activity of culturable bacteria isolated from soil of Grove Mountains, East Antarctica. Acta Microbiol. Sin. 53: 1295–1306.

D'Amico, S., T. Collins, J.C. Marx, G. Feller and C. Gerday. 2006. Psychrophilic microorganisms: challenges for life. EMBO Rep. 7(4): 385–389.

Encheva, M., N. Zaharieva, A. Kenarova, N. Chipev, V. Chipeva, P. Hristova, et al. 2013. Abundance and activity of soil actinomycetes from livingston Island, Antarctica. Bulgarian J. Agric. Sci. 19(2): 68–71.

Farrel, R. and S. Duncan. 2005. Uniqueness of Antarctica and potential for commercial success. Paper presented at the bioprospecting in Antarctica workshop. *In*: A.D. Hemmings and M. Rogan-Finnemore [eds.]. University of Canterbury Private Bag 4800 Christchurch, New Zealand.

Feller, G. and C. Gerday. 2003. Psychrophilic enzymes: hot topics in cold adaptation. Nat. Rev. Microbiol. 1: 200–208.

Fendrihan, S. and T.G. Negoita. 2012. Psychrophilic microorganisms as important source for biotechnological processes. pp. 133–172. *In*: H. Stan-Lotter and S. Fendrihan [eds.]. Adaption of Microbial Life to Environmental Extremes. Springer-Verlag, Vienna.

Forsyth, G. and N.A. Logan. 2000. Isolation of *Bacillus thuringiensis* from northern Victoria Land, Antarctica. Lett. Appl. Microbiol. 30: 263–266.

Fricker, H.A., R. Powell, J. Priscu, S. Tulaczyk, S. Anandakrishnan, B. Christner, et al. 2011. Siple coast subglacial aquatic environments: the whillans ice stream subglacial access research drilling project. Geophys. Monogr. Ser. 194: 199–219.

Gesheva, V., E. Stackebrandt and E. Vasileva-Tonkova. 2010. Biosurfactant production by halotolerant *Rhodococcus fascians* from Casey Station, Wilkes Land, Antarctica. Curr. Microbiol. 61: 112–117.

Gesheva, V. and T. Negoita. 2011. Psychrotrophic microorganism communities in soils of Haswell Island, Antarctica, and their biosynthetic potential. Polar Biol. 35: 291–297.

Gesheva, V. and E. Vasileva-Tonkova. 2012. Production of enzymes and antimicrobial compounds by halophilic Antarctic *Nocardioides* sp. grown on different carbon sources. World J. Microbiol. Biotechnol. 28(5): 2069–2076.

Goyes, D.R. 2013. General introduction: Determinants as social phenomena to take into consideration when constructing public policy for bioprospecting in Colombia. pp. 19–42. *In*: L.M. Melgarejo and Pèrex C.T. [eds.]. Determinantes Cientìficas, Econòmicas y Socio-ambientales de la Bioprospecciòn en Colombia. Bogtà: Universidad Nacional de Colombia.

Gran-Scheuch, A., E. Fuentes, D.M. Bravo, J.C. Jiménez and J.M. Pérez-Donoso. 2017. Isolation and characterization of phenanthrene degrading bacteria from diesel fuel-contaminated Antarctic soils. Front. Microbiol. 8: 1634.

Guo, W., P. Cui and X. Chen. 2015. Complete genome of *Bacillus* sp. Pc3 isolated from the Antarctic seawater with antimicrobial activity. Mar. Genomics 20: 1–2.

Hill, R.T. 2004. Microbes from marine sponges: a treasure trove of biodiversity for natural products discovery. pp. 177–190. *In:* A.T. Bull [ed.]. Microbial Diversity and Bioprospecting. ASM Press: Washington, DC, USA.

Huston, A.L. 2008. Biotechnological aspects of cold-adapted enzymes. pp. 347–363. *In:* R. Margesin, F. Schinner, J.C. Marx and C. Gerday [eds.]. Psychrophiles: from Biodiversity to Biotechnology. Springer, Berlin.

Ibacache-Quiroga, C., J. Ojeda, G. Espinoza-Vergara, P. Olivero, M. Cuellar and M.A. Dinamarca. 2013. The hydrocarbon-degrading marine bacterium *Cobetia* sp. strain MM1IDA2H-1 produces a biosurfactant that interferes with quorum sensing of fish pathogens by signal hijacking. Microb. Biotechnol. 6: 394–405.

Jadhav, V., A. Yadav, Y. Shouche, S. Aphale, A. Moghe, S. Pillai, et al. 2013. Studies on biosurfactant from *Oceanobacillus* sp. BRI 10 isolated from Antarctic sea water. Desalination 318: 64–71.

Jagannadham, M.V., M.K. Chattopadhyay, C. Subbalakshmi, M. Vairamani, K. Narayanan, C.M. Rao, et al. 2000. Carotenoids of an Antarctic psychrotolerant bacterium, *Sphingobacterium antarcticus*, and a mesophilic bacterium, *Sphingobacterium multivorum*. Arch. Microbiol. 173: 418–424.

Jayatilake, G.S., M.P. Thornton, A.C. Leonard, J.E. Grimwade and B.J. Baker. 1996. Metabolites from an Antarctic sponge-associated bacterium, *Pseudomonas aeruginosa*. J. Nat. Prod. 59: 293–296.

Kennedy, A.D. 1993. Water as a limiting factor in the Antarctic terrestrial environment – a biogeographical synthesis. Arctic Alpine Res. 25: 308–315.

Kim, S.K. and J.H. Yim. 2007. Cryoprotective properties of exopolysaccharide (P-21653) produced by the Antarctic bacterium, *Pseudoalteromonas arctica* KOPRI 21653. J. Microbiol. 45: 510–514.

Konishi, M., S. Nishi, T. Fukuoka, D. Kitamoto, T.O. Watsuji, Y. Nagano, et al. 2014. Deep-sea *Rhodococcus* sp. BS-15, lacking the phytopathogenic fas genes, produces a novel glucotriose lipid biosurfactant. Mar. Biotechnol. 6: 484–493.

La Ferla, R., M. Azzaro, L. Michaud, G. Caruso, A. Lo Giudice, R. Paranhos, et al. 2017. Prokaryotic abundance and activity in permafrost of the Northern Victoria Land and Upper Victoria Valley (Antarctica). Microb. Ecol. 74: 402.

Lambrechts, S., A. Willems and G. Tahon. 2019. Uncovering the uncultivated majority in Antarctic soils: toward a synergistic approach. Front. Microbiol. 10: 242.

Lamilla C., D. Braga, R. Castro, C. Guimarães, L.V.A. de Castilho, D.M.G. Freire, et al. 2018. *Streptomyces luridus* So3.2 from Antarctic soil as a novel producer of compounds with bioemulsification potential. PLoS ONE 13(4): e0196054.

Lavin, P.L., S.T. Yong, C.M.V.L. Wong and M. De Stefano. 2016. Isolation and characterization of Antarctic psychrotroph *Streptomyces* sp. strain INACH3013. Antarc. Sci. 28: 433–442.

Lee, L.H., Y.K. Cheah, S.M. Sidik, N.S.A. Mutalib, Y.L. Tang, H.P. Lin, et al. 2012. Molecular characterization of Antarctic actinobacteria and screening for antimicrobial metabolite production. World J. Microbiol. Biotechnol. 28: 2125–2137.

Lee, Y.M., K. GoHeung, Y-J. Jung, C-D. Choe, J.H. Jim, H.K. Lee, et al. 2012. Polar and Alpine Microbial Collection (PAMC): a culture collection dedicated to polar and alpine microorganisms. Polar Biol. 35: 1433–1438.

Levy, J.S., A.G. Fountain, K.A. Welch and W.B. Lyons. 2012. Hypersaline "wet patches" in Taylor Valley, Antarctica, Geophys. Res. Lett. 39: L05402.

Leyton, A., H. Urrutia, J.M. Vidal, M. de la Fuente, M. Alarcón, G. Aroca, et al. 2015. Inhibitory activity of Antarctic bacteria *Pseudomonas* sp. M19B on the biofilm formation of *Flavobacterium psychrophilum* 19749. Rev. Biol. Mar. Oceanog. 50(2): 375–381.

Lo Giudice, A., V. Bruni and L. Michaud. 2007. Characterization of Antarctic psychrotrophic bacteria with antibacterial activities against terrestrial microorganisms. J. Basic Microbiol. 47: 496–505.

Lo Giudice, A., C. Caruso, S. Mangano, V. Bruni, M. De Domenico and L. Michaud. 2012. Marine bacterioplankton diversity and community composition in an Antarctic coastal environment. Microb. Ecol. 63(1): 210–223.

Lo Giudice, A. and C. Rizzo. 2018. Bacteria associated with marine benthic invertebrates from polar environments: Unexplored Frontiers for Biodiscovery? Diversity 10: 80.

Lohan, D. and S. Johnston. 2003. The International Regime for Bioprospecting Existing Pocilices and Emerging issues for Antarctica. United Nations University Institute for Advanced Studies Report, Tokyo, Japan.

Maccario, L., L. Sanguino, T.M. Vogel and C. Larose. 2015. Snow and ice ecosystems: not so extreme. Res. Microbiol. 166: 782–795.

Mader, H.M., M.E. Pettitt, J.L. Wadham, E.W. Wolff and R.J. Parkes. 2006. Subsurface ice as a microbial habitat. Geology 34: 169–172.

Malavenda R., C. Rizzo, L. Michaud, B. Gerçe, V. Bruni, C. Syldatk, et al. 2015. Biosurfactant production by Arctic and Antarctic bacteria growing on hydrocarbons. Polar Biol. 38: 1565–1574.

Mancuso Nichols, C.A., S. Garron, J.P. Bowman, G. Raguénès and J. Guèzennec. 2004. Production of exopolysaccharides by Antarctic marine bacterial isolates. J. Appl. Microbiol. 96: 1057–1066.

Mancuso Nichols, C.M., J.P. Bowman and J. Guézennec. 2005a. Effects of incubation temperature on growth and production of exopolysaccharides by an Antarctic sea ice bacterium grown in batch culture. Appl. Environ. Microbiol. 71: 3519–3523.

Mancuso Nichols, C.M., S.G. Lardiere, J.P. Bowman, P.D. Nichols, J.A.E. Gibson and J. Guézennec. 2005b. Chemical characterization of exopolysaccharides from Antarctic marine bacteria. Microb. Ecol. 49: 578–589.

Mangano, S., L. Michaud, C. Caruso, M. Brilli, V. Bruni, R. Fani, et al. 2009. Antagonistic interactions among psychrotrophic cultivable bacteria isolated from Antarctic sponges: a preliminary analysis. Res. Microbiol. 160: 27–37.

Margesin, R. 2007. Alpine microorganism: useful tools for low-temperature bioremediation. J. Microbiol. 45: 281–285.

McKnight, D.M., R.L. Runkel, C.M. Tate, J.H. Duff and D.L. Moorhead. 2004. Inorganic N and P dynamics of Antarctic glacial meltwater streams as controlled by hyporheic exchange and benthic autotrophic communities. J. N. Am. Benthol. Soc. 23: 171–188.

Médigue, C., E. Krin, G. Pascal, V. Barbe, A. Bernsel, P.N. Bertin, et al. 2005. Coping with cold: the genome of the versatile marine Antarctica bacterium *Pseudoalteromonas haloplanktis* TAC125. Genome Res. 15: 1325–1335.

Mei, Y., H. Liu, S. Zhang, M. Yang, C. Hu, J. Zhang, et al. 2017. Effects of salinity on the cellular physiological responses of *Natrinema* sp. J7-2. PLoS One 12: e0184974.

Mergelov, N.S., D.E. Konyushkov, A.V. Lupachev and S.V. Goryachkin. 2015. Soils of macrobertson land. pp. 65–86. *In*: G.J. Bockheim [ed.]. The Soils of Antarctica. Cham. Springer International Publishing doi: 10.1007/978-3-319-05497-1_5.

Mocali, S., C. Chiellini, A. Fabiani, S. Decuzzi, D. de Pascale, E. Parrilli, et al. 2017. Ecology of cold environments: new insights of bacterial metabolic adaptation through an integrated genomic-phenomic approach. Sci. Rep. 7: 839.

Mojib, N., R. Philpott, J.P. Huang, M. Niederweis and A.K. Bej. 2010. Antimycobacterial activity *in vitro* of pigments isolated from Antarctic bacteria. Anton. van Leeuw. 98: 531–540.

Moncheva, P., S. Tishkov, N. Dimitrova, V. Chipeva, S. Antonova-Nikolova and N. Bogatzevska 2002. Characteristics of soil actinomycetes from Antarctica. J. Culture Collections 3: 3–14.

Morita, R.Y. 1975. Psychrophilic bacteria. Bacteriol Rev. 39(2): 144–167.

Morita, T., T. Fukuoka, T. Imura and D. Kitamoto. 2013. Production of mannosylerythritol lipids and their application in cosmetics. Appl. Microbiol. Biotechnol. 97: 4691–4700.

Mykytczuk, N.C.S., S.J. Foote, C.R. Omelon, G. Southam, C.W. Greer and L.G. Whyte. 2013. Bacterial growth at –15°C: molecular insights from the permafrost bacterium *Planococcus halocryophilus* Or1. ISME J. 7: 1211–1226.

Mykytczuk, N.C.S., J.R. Lawrence, C.R. Omelon, G. Southam and L.G. Whyte. 2016. Microscopic characterization of the bacterial cell envelope of *Planococcus halocryophilus* Or1 during subzero growth at –15°C. Polar Biol. 39: 701–712.

Nedialkova, D. and M. Naidenova. 2005. Screening the antimicrobial activity of Actinomycetes strains isolated from Antarctica. J. Culture Collections 4: 29–35.

Nelson, A.E., J.L. Smellie, M. Williams and S. Moreton. 2008. Age, geographical distribution and taphonomy of an unusual occurrence of mummified crabeater seals on James Ross Island, Antarctic Peninsula. Antarctic Sci. 20: 485–493.

Nylen, T.H., A.G. Fountain and P. Doran. 2004. Climatology of katabatic winds in the McMurdo Dry Valleys, southern Victoria Land, Antarctica. J. Geoph. Res. 109: D03114.

O' Brien, A., R. Sharp, N.J. Russell and S. Roller. 2004. Antarctic bacteria inhibit growth of food-borne microorganisms at low temperatures. FEMS Microbiol. Ecol. 48: 157–167.

Papa, R., E. Parrilli, F. Sannino, G. Barbato, M.L. Tutino, M. Artini, et al. 2013. Anti-biofilm activity of the Antarctic marine bacterium *Pseudoalteromonas haloplanktis* TAC125. Res. Microbiol. 164: 450–456.

Papale, M., A. Lo Giudice, A. Conte, C. Rizzo, C. Rappazzo, G. Maimone, et al. 2019. Microbial assemblages in pressurized Antarctic brine pockets (Tarn Flat, Northern Victoria Land): a hotspot of biodiversity and activity. Microorganisms 7: 333.

Papaleo M.C., M. Fondi, I. Maida, E. Perrin, A. Lo Giudice, L. Michaud, et al. 2012. Sponge-associated microbial Antarctic communities exhibiting antimicrobial activity against *Burkholderia* cepacia complex bacteria. Biotechnol. Adv. 30: 272–293.

Parhi P., V.V. Jadhav and R. Bhadekar. 2016. Increase in production of biosurfactant from *Oceanobacillus* sp. BRI 10 using low cost substrates. Songklanakarin J. Sci. Technol. 38(2): 207–211.

Pepi, M., A. Cesaro, G. Liut and F. Baldi. 2005. An Antarctic psychrotrophic bacterium *Halomonas* sp. ANT-3b, growing on n-hexadecane, produces a new emulsyfying glycolipid. Microb. Ecol. 53: 157–166.

Perfumo A., I.M. Banat and R. Marchant. 2019. Going green and cold: biosurfactants from low-temperature environments to biotechnology applications. Trends Biotechnol. 36: 3.

Phadtare, S. and M. Inouye. 2008. Cold shock proteins. pp. 191–210. *In*: R. Margesin, F. Schinner, J.C. Marx and G. Gerday [eds.]. Psychrophiles: From Biodiversity to Biotechnology. Springer, Berlin, Heidelberg.

Pini, F., C. Grossi, S. Nereo, L. Michaud, A. Lo Giudice, V. Bruni, et al. 2007. Molecular and physiological characterisation of psychrotrophic hydrocarbon-degrading bacteria isolated from Terra Nova Bay (Antarctica). Eur. J. Soil Biol. 43: 368–379.

Price, P.B. 2000. A habitat for psychrophiles in deep Antarctic ice. PNAS 97: 1247–1251.

Priscu, J.C., S. Tulaczyk, M. Studinger, M. Kennicutt, B.C. Christner and C.M. Foreman. 2008. Antarctic subglacial water: origin, evolution and ecology. pp. 119–135. *In*: W.F. Vincent and J. Laybourn-Parry [eds.]. Polar Lakes and Rivers: Limnology of Antarctic and Antarctic Aquatic Ecosystems. Oxford: Oxford University Press.

Raddadi, N., A. Cherif, D. Daffonchio, M. Neifar and F. Fava. 2015. Biotechnological applications of extremophiles, extremozymes and extremolytes. Appl. Microbiol. Biotechnol. 99: 7907–7913.

Reis-Mansur, M.C.P.P., J.S. Cardoso-Rurr, J.V.M.A. Silva, G.R. de Souza, V. da Silva Cardoso, F.R. Passos Mansoldo, et al. 2019. Carotenoids from UV-resistant Antarctic *Microbacterium* sp. LEMMJ01. Sci. Rep. 9: 9554.

Rizzo, C., L. Michaud, C. Syldatk, R. Hausmann, E. De Domenico and A. Lo Giudice. 2014. Influence of salinity and temperature on the activity of biosurfactants by polychaete-associated isolates. Env. Sci. Poll. Res. 21: 2988–3004.

Rizzo, C., L. Michaud, M. Graziano, E. De Domenico, C. Syldatk, R. Hausmann, et al. 2015. Biosurfactant activity, heavy metal tolerance and characterization of *Joostella* strain A8 from the Mediterranean polychaete *Megalomma claparedei* (Gravier, 1906). Ecotoxicology 24: 1294–1304.

Rizzo, C. and A. Lo Giudice. 2018. Marine invertebrates: underexplored sources of bacteria producing biologically active molecules. Diversity 10: 52.

Rodríguez-Díaz, M., L. Lebbe, B. Rodelas, J. Heyrman, P. De Vos and N.A. Logan. 2005. *Paenibacillus wynnii* sp. nov., a novel species harbouring the nifH gene, isolated from Alexander Island, Antarctica. Int. J. Syst. Evol. Microbiol. 55: 2093–2099.

Rojas, J.L., J. Martín, J.R. Tormo, F. Vicente, M. Brunati, I. Ciciliato, et al. 2009. Bacterial diversity from benthic mats of Antarctic lakes as a source of new bioactive metabolites. Mar. Genom. 2(1): 33–41.

Ross, N., M. Siegert, A. Rivera, M. Bentley, D. Blake, L. Capper, et al. 2011. Ellsworth Subglacial Lake, West Antarctica: a review of its history and recent field campaigns. pp. 221–233. *In*: M.J. Siegert and M.C. Kennicutt [eds.]. Antarctic Subglacial Aquatic Environments. Washington, DC: American Geophysical Union.

Saul, D.J., J.M. Aislabie, C.E. Brown, L. Harris and J.M. Foght. 2005. Hydrocarbon contamination changes the bacterial diversity of soil from around Scott Base, Antarctica. FEMS Microbiol. Ecol. 53: 141–155.

Shekh, R.M., P. Singh, S.M. Singh and U. Roy. 2011. Antifungal activity of Arctic and Antarctic bacteria isolates. Polar Biol. 34: 139–143.

Singh, B.K., S. Munro, J.M. Potts and P. Millard. 2007. Influence of grass species in soil type on rhizosphere microbial community structure in grassland soils. Appl. Soil Ecol. 36: 147–155.

Smith, B.E., H.A. Fricker, I.R. Joughin and S. Tulaczyk. 2009. An inventory of active subglacial lakes in Antarctica detected by ICESat (2003–2008). J. Glaciol. 55: 573–595.

Solecka, J., J. Zajko, M. Postek and A. Rajnisz. 2012. Biologically active secondary metabolites from Actinomycetes. Cent. Eur. J. Biol. 7: 373–390.

Steven, B., R. Léveillé, W.H. Pollard and Whyte L.G. 2006. Microbial ecology and biodiversity in permafrost. Extremophiles 10: 259–267.

Taton, A., S. Grubisic, D. Ertz, D.A. Hodgson, R. Piccardi, N. Biondi, et al. 2006. Polyphasic study of Antarctic cyanobacterial strains. J. Phycol. 42: 1257–1270.

Tedesco, P., I. Maida, F. Palma Esposito, E. Tortorella, K. Subko, C.C. Ezeofor, et al. 2016. Antimicrobial activity of monoramnholipids produced by bacterial strains isolated from the Ross Sea (Antarctica). Mar. Drugs 14: 83.

Tiao, G., C.K. Lee, I.R. Mcdonald, D.A. Cowan and S.C. Cary. 2012. Rapid microbial response to the presence of an ancient relic in the Antarctic dry valleys. Nat. Commun. 3: 660.

Tomova, I., M. Stoilova-Disheva, I. Lazarkevich and E. Vasileva-Tonkova. 2015. Antimicrobial activity and resistance to heavy metals and antibiotics of heterotrophic bacteria isolated from sediment and soil samples collected from two Antarctic islands. Front. Life Sci. 8: 348–357.

Tribelli, P.M. and N.I. López. 2018. Reporting key features in cold-adapted bacteria. Life 8(1): 8.

Tytgat, B., E. Verleyen, M. Sweetlove, S. D'hondt, P. Clercx, E. Van Ranst, et al. 2016. Bacterial community composition in relation to bedrock type and macrobiota in soils from the Sor Rondane Mountains, East Antarctica. FEMS Microbiol. Ecol. 92: 9.

Vasileva-Tonkova, E. and V. Gesheva. 2004. Potential for biodegradation of hydrocarbons by microorganisms isolated from antarctic soils. Z. Naturforsch. 59c: 140–145.

Vasileva-Tonkova, E. and V. Gesheva. 2007. Biosurfactant production by Antarctic facultative anaerobe *Pantoea* sp. during growth on hydrocarbons. Curr. Microbiol. 54: 136–141.

Vincent, W.F. 1988. Microbial Ecosystems of Antarctica. Cambridge University Press, Cambridge (UK).

Wagner, D. 2008. Microbial communities and processes in Arctic permafrost environments. pp. 133–154. *In*: P. Dion and C. Shekhar Nautiyal [eds]. Soil Biology Microbiology of Extreme soils. Springer-Verlag, Berlin Heidelberg.

Wakimoto, T., Y. Egami, Y. Nakashima, Y. Wakimoto, T. Mori, T. Awakawa, et al. 2014. Calyculin biogenesis from a pyrophosphate protoxin produced by a sponge symbiont. Nat. Chem. Biol. 10: 648–655.

Watt, W.B. and A.M. Dean. 2000. Molecular-functional studies of adaptive genetic variation in prokaryotes and eukaryotes. Annu. Rev. Genet. 34: 593–622.

Wawrik, B., D. Kutliev, U.A. Abdivasievna, J.J. Kukor, G.J. Zylstra and L. Kerkhof. 2007. Biogeography of actinomycete communities and type II polyketide synthase genes in soils collected in New Jersey and Central Asia. Appl. Environ. Microbiol. 73: 2982–2989.

Wong, C.M.V.L., H.K. Tam, S.A. Alias, M. González, G. González-Rocha and M. Domínguez-Yévenes. *Pseudomonas* and *Pedobacter* isolates from King George Island inhibited the growth of foodborne pathogens. Pol. Polar Res. 32: 3–14.

Wright, A. and M. Siegert. 2012. A fourth inventory of Antarctic subglacial lakes. Antarct. Sci. 24: 659–664.

Yakimov, M.M., L. Giuliano, V. Bruni, S. Scarfi and Golyshin P.N. 1999. Characterization of Antarctic hydrocarbon-degrading bacteria capable of producing bioemulsifiers. New Microbiol. 22: 249–259.

Zazovskaya, E., D. Fedorov-Davydov, T. Alekseeva and Dergacheva M. 2015. Soils of queen maud land. pp. 21–44. *In*: J.G. Bockheim [ed.]. The Soils of Antarctica. Basel: Springer International Publishing. doi: 10.1007/978-3-319-05497-1_3.

8

Yeasts on King George Island (Maritime Antarctica): Biodiversity and Biotechnological Applications

Silvana Vero[1,*], Gabriela Garmendia[1], Ivana Cavello[2],
Florencia Ruscasso[2], Angie Alvarez[1], Eloisa Arrarte[1],
Mariana Gonda[1], Adalgisa Martínez-Silveira[1],
Sebastián Cavalitto[2] and Michael Wisniewski[3]

Introduction

Antarctica has been classified into three biogeographic zones, mainly based on climate conditions. These regions are designated as sub, maritime, and continental Antarctica (Terauds et al. 2012). King George Island is the biggest island in the South Shetlands archipelago, which is situated 120 km North of the Antarctica Peninsula in maritime Antarctica (Korczak-Abshire and Angiel 2011). Extreme conditions for life, such as low temperatures, high solar radiation, aridness, and low nutrient availability, characterize the climate of this island, although they are not as extreme as in continental Antarctica. However, microorganisms, including Bacteria, Archea, and Eukarya domains, have been found as common inhabitants of soils, continental and marine waters, subglacial ice, and even rocks in these biogeographic zones (Martinez et al. 2016, González-Rocha et al. 2017, Pershina et al. 2018). In particular, the presence of psychrophilic and psychrotolerant yeasts and their role as organic matter decomposers have been well documented (Carrasco et al. 2012, Martinez et al. 2016).

Biodiversity

Most studies about yeast diversity in cold environments, including King George Island, have been based on culture methods (Cary et al. 2010, de Pascale et al. 2012, Gugliandolo et al. 2016). In all studies, yeasts belonging to the Basidiomycota have been recognized as predominant over ascomycetous yeasts (Buzzini et al. 2012, Carrasco et al. 2012, Rovati et al. 2013, Martinez et al. 2016, Brandao et al. 2017). The prevalence of Basidiomycetous yeasts in Antarctic habitats has been primarily attributed to their ability to produce polysaccharide capsules that confer desiccation

[1] Laboratory of Biotechnology, Microbiology Area, Department of Bioscience, Faculty of Chemistry, Universidad de la República, Gral. Flores 2124, Montevideo, Uruguay.

[2] Research and Development Center for Industrial Fermentations, CINDEFI (CONICET-La Plata, UNLP). Calle 47 y 115. (B1900ASH), La Plata, Argentina.

[3] Appalachian Fruit Research Station, Agricultural Research Service, United States Department of Agriculture, Wiltshire Road, Kearneysville, WV 25443, USA.

*Corresponding author: svero@fq.edu.uy

resistance and to the higher proportion of unsaturated fatty acids in their cellular membranes, which helps to maintain membrane fluidity at low temperatures. Moreover, in some cases, the ability to grow at subzero temperatures has also been associated with the presence of antifreeze proteins (Connell et al. 2008, Buzzini et al. 2012, Tsuji et al. 2013, Baeza et al. 2017). *Rhodotorula* and *Cryptococcus* have been reported as the predominant genera among basidiomycetous psychrotolerant yeasts isolated from King George Island (Rovati et al. 2013, Martinez et al. 2016). However, yeasts from other genera, such as *Cystofilobasidium, Bullera, Dioszegia, Sporidiobolus, Filobasidium, Vishniacozyma, Leucosporidium*, and *Leucosporidiella*, have also been frequently isolated (Carrasco et al. 2012, Duarte et al. 2013, Rovati et al. 2013, Martinez et al. 2016, Brandao et al. 2017). *Leucosporidium* spp., which includes psychrotolerant and psychrophilic species (Laich et al. 2014), was reported as the most dominant genus in soil samples collected from Admiralty Bay (King George Island) (Wentzel et al. 2019), while *Sporidiobolus salmonicolor* was the most ubiquitous species found by Carrasco et al. (2012) in different terrestrial habitats of the island. Phsycrophilic, basidiomycetous yeasts belonging to the genera *Mrakia, Glaciozyma*, and *Phenolipheria* have been frequently isolated from soils and waters of King George Island (Turchetti et al. 2011, Carrasco et al. 2012, Martinez et al. 2016) and also in other extremely cold regions. *Mrakia* was reported as the predominant genus in lake sediments and soil surrounding lakes in Antarctica, such as the Skarvsnes area (Tsuji et al. 2013). This genus was also reported as predominant in the Vestfold Hills area of Davis Base in Antarctica (Thomas-Hall et al. 2010). *Mrakia* spp. was also found in the Arctic, Siberia, Alaska, and Patagonia (Poliakova et al. 2001, Pathan et al. 2010, de Garcia et al. 2012, Singh and Singh 2012). Species in this genus are obligate psychrophiles that cannot grow over 20°C (except *M. curviuscula*) (Kurtzman et al. 2011). The ability of *Mrakia* spp. to grow at subzero temperatures has been well documented. Tsuji et al. (2013) and Tsuji (2016) indicated that this ability could be related to the high proportion of unsaturated fatty acids in cellular membranes and the intracellular accumulation of compounds, such as aromatic amino acids and polyamines, rather than the production of extracellular polysaccharides, or the accumulation of trehalose or glycerol as cryoprotectants.

Many genera of ascomycetous yeasts have also been reported from King George Island. Among them, *Metschnikowia* has been reported in several studies as the predominant ascomycetous genus when studying culturable yeast biodiversity from lakes and marine environments (Vaca Cerezo et al. 2013, Brandao et al. 2017, Wentzel et al. 2019). Other genera, including *Meyerozyma, Candida*, and *Debaryomyces*, have also been frequently found in Antarctic habitats (Carrasco et al. 2012, Duarte et al. 2013, Martinez et al. 2016). In particular, *D. hansenii* was the most abundant and ubiquitous species found by Martinez et al. (2016) in ice-free areas in King George Island.

Studies about yeast biodiversity in cold environments have mainly explored culturable yeast communities. According to Hawksworth (2001), only a small proportion (around 17%) of fungi could be detected using current culture methodologies. The use of metagenomic approaches has confirmed this assumption and revealed significantly higher levels of diversity. Baeza et al. (2017) utilized high-throughput sequencing to examine the fungal diversity in soils from different locations in Antarctica, including King George Island. They obtained sequences corresponding to 87 genera and 123 putative species of fungi. Among the fungal sequences, 37 corresponded to genera not previously reported from Antarctica by culture methods. Notably, the predominant yeast genera identified were *Xanthophyllomyces* and *Malassezia*, which had not been reported in previous studies of these soils using culture methods of isolation. This finding underscores the need to perform more comprehensive studies of yeast biodiversity in these zones. Such studies would provide information that could help to understand nutrient cycles and identify greater potential biotechnological applications of the microorgaisms present in these environments. This information would also be essential for the development of new culture methods that could be used to isolate specific yeast genera or species that have not been previously recovered from these zones.

Biotechnological Applications

Production of Cold-Active Enzymes

Cold-active enzymes are characterized by their high specific activity at moderate or even low temperature and their thermolability (Feller et al. 1996). Their use provides energy savings to the processes in which they are involved since high temperatures are not needed to achieve high yields. Such enzymes can be used for catalytic reactions involving heat-sensitive substrates or products, and their use in processes occurring at low temperatures greatly reduces the chances of contamination with mesophilic microorganisms (Javed and Qazi 2016). In general, these enzymes can be efficiently inactivated by moderate heat input after the catalytic process has finished (Martorell et al. 2019). This characteristic is very important in processes involving sequential catalytic steps that require the inactivation of each enzyme after it has performed its function. The use of thermolabile enzymes in such processes enables heat inactivation at temperatures that do not cause undesirable changes in substrates or products and avoids the incorporation of chemical extraction steps (Cavicchioli et al. 2011).

The characteristics of cold-active enzymes make them a valuable input resource for different biotechnological applications in a wide variety of industries, including the production of food, beverage, and household products (Sarmiento et al. 2015).

Numerous reports have been published on the identification of about cold-adapted, extracellular hydrolytic enzymes produced by Antarctic yeasts and potential technological applications. Most have focused on yeasts isolated in the sub-Antarctic region and in particular from King George Island (Carrasco et al. 2012, Duarte et al. 2013, Rovati et al. 2013, Martinez et al. 2016, Cavello et al. 2017). Cold-adapted, hydrolytic enzymes, including cellulases, xylanases, pectinases, chitinases, lipases, proteases, amylases and glucoamylases, b-galactosidases, phytases, inulinases, and invertases have all been identified (Margesin and Miteva 2011, Duarte et al. 2018).

Yeast producing esterases and, in particular, lipases have been frequently found in Antarctic habitats (Carrasco et al. 2012, Martinez et al. 2016). Vaz et al. (2011) reported that esterase activity was present in 76% of the yeast isolates obtained from different sites in Antarctica, including King George Island. Duarte et al. (2013) found that more than 46% of the yeast isolates obtained from different Antarctic habitats were able to produce lipase. Esterases and lipases are useful in many industrial applications. They have been reported to play a role in hydrocarbon degradation in bioremediation processes in polluted, cold environments (Martorell et al. 2019). These enzymes also have high biotechnological potential in food and feed processing and also as catalysts in many organic synthesis reactions utilized in the pharmaceutical and cosmetic industries. Lipases A and B from *Pseudozyma* (*Candida*) *antarctica* are the most studied lipases produced by Antarctic yeasts and have been patented and are commercially available (Shivaji and Prasad 2009).

The production of cold-adapted pectinases by Antarctic yeasts has also been reported (Vaz et al. 2011, Carrasco et al. 2012, Martinez et al. 2016). Pectinolytic enzymes are important enzymes in the fruit processing industry where they are used to hydrolyze pectic substances. Based on their specific activity, they can be divided into depolymerizing (polymethyl galacturonases, pectin lyases, polygalacturonases, and pectate lyases) and deesterifying (pectinesterases) enzymes (Tapre and Jai 2014, Cavello et al. 2017). Endo-polygalacturonase and pectate lyase activities were demonstrated when the yeast, *Guehomyces pullulans* 8E, isolated by Martinez et al. (2016) from King George Island, was cultivated in pectin-containing fruit wastes (Cavello et al. 2017). The authors also reported that the enzymatic extract produced an 80% clarification of apple juice at 20°C at an enzyme concentration of 4 U/ml, demonstrating great potential for its application in the food industry.

Many cold-adapted enzymatic activities have been reported for isolates of *G. pullulans* obtained from cold climates. Ligninase and manganese peroxidase activity, involved in the degradation of lignin-containing wastes, were reported by Sláviková et al. (2002). Nakagawa et al. (2006) reported the production of cold-active β-galactosidase by strains of *G. pullulans*. Duarte et al. (2013) reported that *G. pullulans* isolates from different Antarctic zones were able to produce lipases, xylanases, and proteases. Martínez et al. (2016) also demonstrated that, in addition to pectinases, *G. pullulans* 8E could produce proteases, esterases, amylases, and inulinases that were functionally active at cold temperatures.

Many yeast isolates from Antarctica produce more than one type of extracellular enzyme. In their survey of King George Island, Carrasco et al. (2012) reported that most of the yeast isolates identified as *Leuconeurospora* spp. could produce cellulase, esterase, lipase, protease, pectinase, and chitinase, while those identified as *Dioszegia fristingensis* exhibited amylolytic, cellulolytic, pectinolytic, and xylanolytic activity.

A review of the studies on yeasts from Antarctic and sub-Antarctic regions indicates that all of the yeast should be considered a source of cold-adapted enzymes with potential use in different industrial processes. Further studies on the production and purification of these enzymes and the exploration of new catalytic activities should be encouraged.

Potential Use in Bioremediation

The high diversity of enzymatic activities exhibited by cold-adapted yeast have also been explored for their use in bioremediation strategies. Bioremediation refers to the use of microorganisms, plants, or enzymes to detoxify contaminants in polluted habitats (de Lorenzo 2008).

Currently, environmental pollution is one of the most critical problems in the world. The need for remediation strategies, including bioremediation processes, is an urgent need (Vidali 2001). In this regard, microorganisms could serve an important role in transforming, detoxifying or even mineralizing pollutants to innocuous compounds (Abatenh et al. 2017).

Several environmental contaminants can be fully degraded or transformed in less toxic or even harmless derivatives by microbial action, restoring the value and utility of contaminated sites. In other cases, chemicals can be intracellularly accumulated or adsorbed to microbial surfaces, thus requiring the need to dispose of the bioaccumulated pollutant. As bioremediation requires minimum technology, this approach is more viable economically than chemical or physical methods. It is also non-invasive and can be often conducted on-site and has high public acceptance (Boopathy 2000, Dzionek et al. 2016).

In the last two decades the use of extremophilic microorganisms, including psychrophilic or psychrotrophic yeasts, in bioremediation processes has been actively explored (Margesin 2000, Peeples 2014, Raddadi et al. 2016, Giovanella et al. 2019). Fernández et al. (2017) examined the utilization of some organic pollutants (phenol, methanol, and n-hexadecane) as carbon sources by Antarctic yeasts and also investigated their tolerance to heavy metal ions (Cu(II), Cd(II), and Cr(VI)). Among 128 yeast isolates, 17 were able to use methanol as a carbon source, while phenol and n-hexadecane were assimilated by 32% and 78% of the isolates, respectively. A high percentage of yeasts with tolerance to heavy metals was also demonstrated, notably, 19% of the total isolates were able to grow in the presence of heavy metals and were able to use phenol as a carbon source. Thus, these isolates have great potential to be used in the bioremediation of effluents from petroleum refineries.

Martorell et al. (2017) studied the ability of yeast isolates from King George Island to degrade n-alkanes (nC11, nC12, nC13, and nC14) and gasoil in liquid culture at 15°C. Only *Pichia caribbica* 171, was able to grow in those conditions, indicating that the degradation of hydrocarbons by yeasts is not an extensive or common characteristic. Zhang et al. (2019) reported the ability of a strain of a novel species, *Exophiala macquariensis* strain CZ06, to use toluene as a carbon and energy source at 10°C. This yeast was isolated from soils from a sub-Antarctic island (Macquarie Island) that was contaminated with petroleum hydrocarbons. Its use in bioremediation of polluted soils under cold conditions was presented as a promising approach.

Other authors have also studied the production of extracellular hydrolytic enzymes by Antarctic yeasts to select optimal strains for use in bioremediation processes (Martínez et al. 2016, Wentzel et al. 2018). Tsuji et al. (2013) reported the biological decomposition of milk fat under low temperature by the Antarctic yeast *Mrakia blollopis* SK-4. The production of an extracellular lipase was considered as the major enzyme involved in the process.

Microbial degradation of azo dyes, pollutants present in the effluents of textile dyeing factories, is also well documented. Textile effluents are commonly characterized by a high concentration of azo dyes which have a negative impact on the aquatic ecosystems, mainly due to their inhibition of photosynthetic activity. The complete degradation of azodyes by yeasts in processes, involving reductive and oxidative reactions mediated by different enzymes, has been demonstrated. In general, the first step involves a reduction mediated by azoreductases that catalyze the cleavage of azo dyes into aromatic amines, which are then further degraded. Ligninolytic enzymes, such as laccase, manganese-dependent peroxidase, and lignin peroxidase, can also be involved in the oxidative degradation of dyes (Jafaria et al. 2014, Marco et al. 2005, Kurade et al. 2016). Rovati et al. (2013) reported the ability of some Antarctic yeasts to decolorize solid media supplemented with dyes (Reactive Blue 221, Reactive Red 141, Reactive Black 5, and Reactive Yellow 84) commonly present in textile effluents. Thirty-three percent of the evaluated yeast strains produced significant decolorization of the media, indicating their potential application for there mediation of textile effluents. The potential application of Antarctic yeasts for the treatment of textile effluents was also examined by Ruscasso et al. (2017). The authors demonstrated that biodegradation of Reactive Black 5 and Reactive Orange 6 dyes by the Antarctic yeast *Candida sake* 41E was achieved after 24 and 48 hours with 98% and 94% decolorization, respectively (Figure 1). The toxicity of the degradation products was also evaluated on seeds of *Lactuca sativa* L with promising results (Ruscasso et al. 2019a). The Antarctic yeast, *Debaryomyces hansenii* F39A, was also demonstrated to be useful for the removal of Reactive Red 141 and Reactive Blue 19 by a process of biosorption. In this process, the dye moleules are bound to molecules present on yeast cell walls by electrostatic attractions, dispersive interactions, hydrophobic attraction, hydrogen bonding interactions, and physical adsorption. Bioaccumulation was also demonstrated when *Debaryomyces hansenii* F39A was grown in the liquid containing Reactive Red 141 (RR-141), Reactive Green 19 (RG-19), Reactive Violet 5 (RV-5), Reactive Orange 16 (RO-16), and Reactive Blue 19 (RB-19). More than 73% of the dyes were removed after 48 hours of culture (Ruscasso et al. 2019b). The use of actively-growing yeasts as bioadsorbents may represent a reliable alternative option to the use of activated carbon. However, more importantly, this type of bioremediation process still entails the need for the subsequent disposal of the contaminated biomass.

On the basis of the current literature, it can be concluded that the use of psychrotrophic or psychrophilic yeasts in different bioremediation strategies in cold areas may represent a promising alternative. Further studies are necessary, however, to develop and optimize the bioremediation processes.

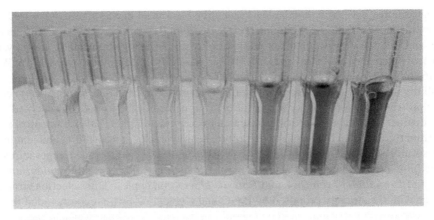

(A)

(B)

Figure 1. Degradation of azo dyes by the Antarctic yeast, *Candida sake* 41E (A) Supernatants from *Candida sake* 41E cultures growing in the presence of the azo-dye, Black Reagent 5, taken at different times between 0 and 30 hours after inoculation. Biodegradation of the dye can be observed from the right (time = 0 h) to the left (time = 30 h) of the picture. (B) Biodegradation of Reactive Orange 16 by *Candida sake* 41E growing in a bioreactor supplemented with 100 mg l^{-1} of dye.

Antarctic Yeast as Biocontrol Agents During the Postharvest Cold Storage of Fruit

Due to their ability to grow in extreme cold and stressful environments, Antarctic yeasts have also been studied as biocontrol agents of postharvest diseases on cold-stored fruit. Harvested fruit is often stored in refrigerated chambers to lower their metabolism and to delay decay, thus making high-quality fruit available for sale for an extended period (Usall et al. 2016). Optimal storage temperatures are dependent on the fruit being conserved. Apples, pears, and grapes are typically stored at 0 or 1°C, while in the case of other fruits, such as citrus and banana, higher temperatures are required (Gross et al. 2016). Despite the use of low temperatures, several pathogenic fungi can develop on cold-stored fruit resulting in significant economic losses. For decades, the application of synthetic fungicides prior to fruit storage has been the main strategy used to reduce fungal rot. Due to safety concerns and the demands of consumers to reduce exposure to chemical residues on produce, alternative approaches have been explored (Spadaro and Droby 2016).

In particular, the utilization of yeasts as postharvest biological control agents has received considerable attention as a reliable alternative for the control of the major postharvest pathogens of fruit. Many studies have been conducted to identify biocontrol agents that can be used to prevent the development of *Penicillium digitatum* and *Penicillium italicum* on citrus, *Penicillium expansum* on apple, and *Botrytis cinerea* on different types of fruit, including apple and strawberry, as well as other postharvest pathogens (Droby et al. 2009, Liu et al. 2013, Di Francesco et al. 2016, Wisniewski et al. 2016).

The ability to grow at cold storage temperature is an important attribute to consider when selecting a yeast to protect fruit stored under refrigeration. In this regard, the use of cold-adapted yeasts may present a distinct advantage. In previous research by Vero et al. (2009, 2011), biocontrol agents were isolated from the superficial microflora present on healthy apples stored at 0°C and lemons stored at 10°C for several months. Selected yeasts were well adapted to cold storage temperatures and most isolates exhibited good biocontrol activity against *Penicillium italicum* and *Penicillium digitatum* on citrus and *Penicillium expansum* and *Botrytis cinerea* on apples. This selection strategy was in accordance with the strategy proposed by Baker and Cook (1974) who recommended that potential antagonists against a particular pathogen should be isolated from places where the disease could be expected but did not occur (Sangorrin et al. 2014). Similar results have been reported by Lutz et al. 2012 and Robiglio et al. 2011 who found that microbiota associated with fruit stored in cold chambers showed better biocontrol performance against postharvest pathogens than mesophilic yeasts. The inclusion of the cold adaptation as a selective characteristic in the first step of identifying a suitable biocontrol agent reduces the number of isolates to be tested and increases the success of the selection process (Vero et al. 2013).

The mechanisms by which antagonists exert their activity against pathogens have not yet been fully elucidated. Many modes of action, including antibiosis, competition for nutrients and space, parasitism, and the induction of host resistance have been proposed as playing a role in the biocontrol activity of yeast antagonists against postharvest pathogens (Sharma et al. 2009, Liu et al. 2013, Di Francesco et al. 2016, Spadaro and Droby 2016). Competition for space and nutrients is considered the primary way in which yeasts suppress the development of necrotrophic pathogens (Droby et al. 2009). For this reason, the selection of a yeast strain as a biocontrol agent to be applied on cold-stored fruit should be based on its ability to grow rapidly and efficiently and to remain viable under the prevailing storage conditions. In addition to this, low temperature antagonists need to overcome many other abiotic stresses to be successful (Sui et al. 2015). Antarctic yeasts are adapted to several stresses, including low temperatures, low water availability, osmotic stress, desiccation, and low nutrient availability (Ruisi et al. 2007). Notably, these stresses are similar to those that biocontrol agents experience when applied to fruit stored at low temperatures (Wang et al. 2010, Liu et al. 2011).

Given this background, Vero et al. (2013) evaluated the potential of cold-adapted yeasts from Antarctica to control postharvest diseases of apples. Five yeast isolates obtained from soils samples near the Uruguayan research station (Artigas Antarctic Scientific Station) located on King George Island, Antarctica (Lat 62°11′04S; Long 58°54″W) were demonstrated to be good biocontrol agents against *Penicillium expansum* and *Botrytis cinerea* on 'Red Delicious' and 'Pink Lady' apples stored at 1.0 ± 0.5°C. Yeast isolates were obtained after enrichment in apple juice at low temperature followed by isolation on Apple Juice Agar. All of the obtained isolates exhibited a certain level of biocontrol activity, thus demonstrating that this isolation strategy represents an efficient approach for identifying and obtaining potential biocontrol agents to reduce fungal rot on cold-stored apples. The selected yeasts were identified as *Cryptococcus terricola*, *Cryptococcus gastricus*, *Rhodotorula mucilaginosa*, *Rhodotorula laryngis*, and *Leucosporidium scottii*. Reductions of at least 85% in the incidence of both pathogens were achieved when the strain identified as *L. scottii* was used as a biocontrol agent on cold-stored apples (Figure 2).

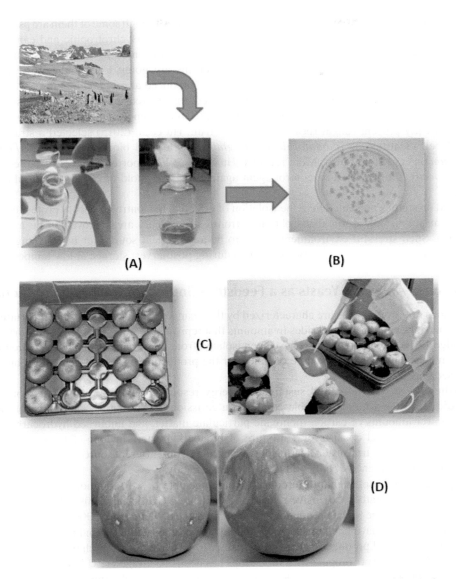

Figure 2. Strategy for obtaining biocontrol yeasts to prevent blue mold decay on cold-stored apples. (A) Enrichment of cold-adapted yeasts that could grow in apple juice (B) Isolation of yeasts on Apple Juice Agar. Broth and agar plates were incubated at low temperatures. (C) Biological control assay on artificially-wounded fruit. Wounds were inoculated with each yeast isolates and the pathogen. Control fruit was inoculated only with the pathogen. Control and treated fruit were stored at 0°C. (D) Treated and control fruit after three months at 0°C. Rot appeared in the control fruit (right) while wounds in treated fruit (left) remained free of decay.

In further studies, biocontrol agents of *P. expansum* on cold-stored apples were selected from among a group of non-pectinolytic yeast isolates obtained from different locations on King George Island. One of them, *Candida sake* 41E, was able to produce volatile antifungal compounds that inhibited the development of the pathogen in wounded fruit stored at 0-1°C (Arrarte et al. 2017). Recently, Alvarez et al. (2019) demonstrated the ability of this strain to protect apple wounds against *P. expansum* at low and room temperatures and to degrade patulin, a mycotoxin produced by the fungal pathogen under the studied conditions.

Yeasts from Antarctica can be endemic or cosmopolitan. Although some of them are psychrophilic species, most of the isolates found in King George Island are psychrotrophic and can grow over a wide temperature range with an optimal growth temperature near 25-28°C. This characteristic represents a biotechnological advantage if yeast strains are going to be industrially produced. Large-scale volumes of a psychrotrophic yeast can be produced at room temperature while in case of a psychrophilic yeast refrigeration would be needed, thus increasing production costs.

Psychrotrophic yeasts isolated from cold areas different from Antarctica have also been used to successfully prevent fungal rot on refrigerated fruit. Hu et al. (2015) selected a strain isolated from Tibet, identified as *Rhodotorula mucilaginosa*, as a good biocontrol agent of *Penicillium expansum* on cold-stored pear. More recently, Hu et al. (2017) reported on a cold-adapted strain of *Cryptococcus laurentii* as a biocontrol agent against the gray mold of cherry tomato during cold storage.

Based on studies presented here the isolation of effective biocontrol agents to protect cold-stored fruit from among microorganisms obtained from extreme cold areas proved to be a successful approach. Such stressful environments seem to be an excellent source for novel antagonists that optimize the prevention of decay of fresh fruits during postharvest storage at low temperatures.

Oleaginous Antarctic Yeasts as a Feedstock Source for Biodiesel Production

Oleaginous microorganisms are characterized by their capacity to accumulate, under certain culture conditions, intracellular triglycerides in amounts that represent more than 20% of their dry biomass (Ratledge 1979). Oleaginous yeasts have been isolated from many habitats including extremely cold environments. Their use as a source of lipids for the production of second-generation biodiesel is discussed here.

Biodiesel is a renewable and sustainable energy resource with several benefits compared to petro-diesel. It has a more favorable combustion emission profile with no net emission of sulfur oxides, lower amounts of carbon monoxide particulate matter, and unburned hydrocarbons (Kaya et al. 2018). Biodiesel consists of a mixture of monoalkyl esters of long-chain fatty acids. It is typically produced by transesterification of triglycerides from different sources with alcohol (mainly methanol) in the presence of either a base or acid as a catalyst.

At present, biodiesel is mainly produced from vegetable oils. Production costs are quite high and so biodiesel is rather expensive related to fossil fuels (Pan et al. 2009). The overall cost of biodiesel is mainly due to the production of the oil feedstock (Meng et al. 2009, Vicente et al. 2009). Moreover, concerns have been raised about the diversion of agricultural land and the production of food crops for biofuel production (Kraisintu et al. 2010). Therefore, research on the development of new biofuels that do not compete with food crops has been encouraged (EU Directive 2015). In this regard, microbial oils produced by oleaginous microorganisms, including bacteria, yeasts, molds, and algae, represent promising potential sources for biodiesel production. In the case of oleaginous yeasts, numerous studies have reported that lipid accumulation can range from 20-60% of dry biomass under certain culture conditions (Sitepu et al. 2014). Intracellular lipids in oleaginous yeasts are mainly represented by triacylglycerides stored in lipid bodies (Shapaval et al. 2019).

Biodiesel, regardless of its origin, must meet certain requirements to be used as a fuel. International standards, such as ASTM 6751 (USA) and EN 14214 (Europe), have been set up to monitor the quality parameters of biodiesel (Botella et al. 2014), some of which depend on the fatty acid profile of the alkylesters that constitute the fuel. In the case of yeast, the fatty acid composition of intracellular triglycerides is very similar to that of vegetable oils, so they are considered as a potential alternative feedstock source for biodiesel production (Mukhtar et al. 2018). Studies on the use of oleaginous yeast to produce biodiesel (Sitepu et al. 2014) are numerous; however, most are based on mesophilic yeast and studies involving cold-adapted yeasts are still uncommon.

Pereyra et al. (2014) conducted the initial studies on yeast isolates from King George Island for their ability to accumulate intracellular, saponifiable lipids in a nitrogen-limited medium. Two isolates, *Leucosporidium scotti* At17 and *Rhosotorula mucilaginosa* At7, were identified as oleaginous yeasts. *L. scotti* At 17 accumulated saponifiable lipids representing 47% of its dry biomass with oleic acid as the main fatty acid produced. The cetane number of the mixture of fatty acids from saponifiable lipids was 55.6, which fulfilled international standards for biodiesel (Patel et al. 2017).

In further research, Martínez et al. (2016) evaluated the ability of 68 isolates from King George Island to accumulate intracellular saponifiable lipids. Psychrophilic and psychrotolerant yeasts were included in the study. Almost half of the isolates were identified as oleaginous yeasts and the identified yeasts were *Cryptococcus aerius, Cryptococcus glivescens, Candida glaebosa, Leucosporidiella fragaria, Guehomyces pullulans,* and *Rhodotorula laryngis*. All of the oleaginous isolates were classified as psychrotolerant based on their optimal growth temperature. No psychrophilic oleaginous yeasts were identified. This study was the first report of *C. glaebosa* and *L. fragaria* as oleaginous yeasts. All the isolates belonging to those species accumulated saponifiable lipids in amounts that represent more than 20% of their dry biomass.

In the same year, Viñarta et al. (2016) evaluated the ability of 17 yeast strains from King George Island, identified as *Rhodotorula* spp., to accumulate intracellular lipids. More than 70% of the strains were characterized as oleaginous microorganisms. Six strains, belonging to three species (*Rhodotorula glutinis, Rhodotorula glacialis,* and *Rhodotorula laryngis*), exhibited the highest lipid accumulation. In all the above-mentioned studies, fatty acids composition from intracellular yeast lipids were similar to that from vegetable oils, indicating that lipids accumulated by oleaginous Antarctic yeasts can be considered an alternative feedstock source for biodiesel production.

Antarctic yeasts have developed adaptive responses in order to survive in extreme climate conditions. One of the most studied adaptations is a shift in the lipid composition of the cell, including those in cell membranes and lipid bodies. In general, a decrease in temperature favors the presence of unsaturated fatty acids (Rossi et al. 2009). *Rhodotorula glacialis* DBVPG 4785, an oleaginous psychrophilic yeast, was demonstrated to grow and accumulate lipids at temperatures between –3 to 20°C and produce lipids with a higher degree of unsaturation at lower temperatures (Amaretti et al. 2009). In most cases, cold-adapted yeasts cultured at lower temperatures would produce a feedstock source that would result in biodiesel with a lower cold filter plugging point, which is the lowest temperature at which biodiesel easily passes through a standardized filtration device in a specific time (Patel et al. 2017). One of the major concerns about biodiesel is its unfavorable cold flow properties since it tends to gel at higher temperatures relative to conventional diesel. The temperature at which biodiesel freezes depends on the type of oil used to make it. Thus, oils accumulated by psychrophilic or psychotropic oleaginous yeasts cultured at low temperatures appear to represent an excellent feedstock source for the production of biodiesel suitable for use in cold climates.

Concluding Remarks

The diversity of culturable cold-adapted yeasts from ice-free areas in King George Island has been extensively explored. Studies based on metagenomic approaches that reveal higher diversity are uncommon. Metagenomic studies are needed to increase our knowledge of yeast biodiversity in these cold regions and to understand their role in nutrient cycles. Such information would also be fundamental for the development of culture media and methods to isolate specific yeast genera or species that have not been recovered under standard conditions.

Many biotechnological applications involving the use of cold-adapted yeasts have been explored. Among them, their use in bioremediation strategies at cold temperatures appears to have great potential and should be further explored. Their ability to produce extracellular hydrolytic enzymes with high levels of functional activity at low temperatures represents a valuable resource for a wide range of industrial processes, including the production of novel chemicals, pharmaceuticals, food and feed products, and detergents. Further studies on the production and purification of cold-adapted enzymes and the exploration of new catalytic activities should be encouraged. Cold-adapted yeasts have been successfully used as biocontrol agents of postharvest diseases on cold-stored apples and pears due to their ability to grow at low temperatures. Their use to protect fruits and vegetables from postharvest decay under refrigerated conditions appears to be a promising alternative to the use of synthetic fungicides. The existence of oleaginous yeasts in Antarctica has been demonstrated but not extensively studied. Their potential as a source for the production of second-generation biodiesel for use in cold climates should be further explored.

Acknowledgments

The authors thank the Uruguayan Antarctic Institute for the logistic support during the stay in the Antarctic Base Artigas. This work was partially supported by PEDECIBA (Programa de Desarrollo de las CienciasBásicas), CSIC (Comisión Sectorial de Investigación Científica), and ANII (AgenciaNacional de Investigación e Innovación).

References

Abatenh, E., B. Gizaw, Z. Tsegaye and M. Wassie. 2017. Application of microorganisms in bioremediation-review. J. Environ. Microbiol. 1(1): 2–9.

Alvarez, A., R. Gelezoglo, G. Garmendia, M.L. González, A.P. Magnoli, E. Arrarte, et al. 2019. Role of Antarctic yeast in biocontrol of *Penicillium expansum* and patulin reduction of apples. Environ. Sustain. 2(3): 277–283.

Amaretti A., S. Raimondi, M. Sala, L. Roncaglia, M. De Lucia and A. Leonardi. 2010. Single cell oils of the cold-adapted oleaginous yeast *Rhodotorula glacialis* DBVPG 4785. Microb. Cell Fact. 9(73): 2–6.

Arrarte, E., G. Garmendia, C. Rossini, M. Wisniewski and S. Vero. 2017. Volatile organic compounds produced by Antarctic strains of *Candida sake* play a role in the control of postharvest pathogens of apples. Biol. Control. 109: 14–20.

Baeza, M., S. Barahona, J. Alcaíno and V. Cifuentes. 2017. Amplicon-metagenomic analysis of fungi from antarctic terrestrial habitats. Front Microbiol. 8: 2235–2246.

Baker, K.F. and R.J. Cook. 1974. Biological Control of Plant Pathogens. W.H. Freeman. San Francisco.

Boopathy, R. 2000. Factors limiting bioremediation technologies. Bioresour Technol. 74: 63–67.

Botella, L., F. Bimbela, L. Martín, J. Arauzo and J.L. Sánchez. 2014. Oxidation stability of biodiesel fuels and blends using the Rancimat and PetroOXY methods. Effect of 4-allyl-2, 6-dimethoxyphenol and catechol as biodiesel additives on oxidation stability. Front Chem. 2(43): 1–9.

Brandao, L.R., A.B.M. Vaz, L.C. Espírito Santo, R.S. Pimenta, P.B. Morais, D. Libkind, et al. 2017. Diversity and biogeographical patterns of yeast communities in Antarctic, Patagonian and tropical lakes. Fungal Ecol. 28: 33–43.

Buzzini, P., E. Branda, M. Goretti and B. Turchetti. 2012. Psychrophilic yeasts from worldwide glacial habitats: Diversity, adaptation strategies and biotechnological potential. FEMS Microbiol. Ecol. 82: 217–241.

Carrasco, M., J.M. Rozas, S. Barahona, J. Alcaíno, V. Cifuentes and M. Baeza. 2012. Diversity and extracellular enzymatic activities of yeasts isolated from King George Island, the sub-Antarctic region. BMC Microbiol. 12: 251–259.

Cary, S.C., I.R. McDonald, J.E. Barrett and D.A. Cowan. 2010. On the rocks: the microbiology of Antarctic Dry Valley soils. Nat. Rev. Microbial. 8: 129–138.

Cavello, I., A. Albanesi, D. Fratebianchi, G. Garmedia, S. Vero and S. Cavalitto. 2017. Pectinolytic yeasts from cold environments: novel findings of *Guehomyces pullulans*, *Cystofilobasidium infirmominiatum* and *Cryptococcus adeliensis* producing pectinases. Extremophiles 21: 319–329.

Cavicchioli, R., T. Charlton, H. Ertan, S. Mohd Omar, K.S. Siddiqui and T.J. Williams. 2011. Biotechnological uses of enzymes from psychrophiles. Microb. Biotechnol. 4(4): 449–460.

Connell, L., R. Redman, S. Craig, G. Scorzetti, M. Iszard and R. Rodriguez. 2008. Diversity of soil yeasts isolated from South Victoria Land, Antarctica. Microb. Ecol. 56: 448–459.

de Garcia, V., S. Brizzio and M. van Broock. 2012. Yeasts from glacial ice of Patagonian Andes, Argentina. FEMS Microbiol. Ecol. 82(2): 540–550.

de Lorenzo, V. 2008. Systems biology approaches to bioremediation. Curr. Opin. Biotechnol. 19(6): 579–589.

de Pascale, D., C. De Santi, J. Fu and B. Landfald. 2012. The microbial diversity of Polar environments is a fertile ground for bioprospecting. Mar. Genomics 8: 15–22.

Di Francesco, A., L. Ugolini, L. Lazzeri and M. Mari. 2015. Production of volatile organic compounds by *Aureobasidiumpullulans* as a potential mechanism of action against postharvest fruit pathogens. Biol. Control. 81: 8–14.

Droby, S., M. Wisniewski, D. Macarisin and C. Wilson. 2009. Twenty years of postharvest biocontrol research: is it time for a new paradigm? Postharvest Biol. Technol. 52: 137–145.

Duarte, A.W.F., I. Dayo-Owoyemi, F.S. Nobre, F.C. Pagnocca, L.C.S. Chaud, A. Pessoa, et al. 2013. Taxonomic assessment and enzymes production by yeasts isolated from marine and terrestrial Antarctic samples. Extremophiles 17: 1023–1035.

Duarte, A.W.F., J.A. dos Santos, M.V. Vianna, J.M.F. Vieira, V.H. Mallagutti, F.J. Inforsato, et al. 2018. Cold-adapted enzymes produced by fungi from terrestrial and marine Antarctic environments. Crit. Rev. Biotechnol. 38: 600–619.

Dzionek, A., D. Wojcieszyńska and U. Guzik. 2016. Natural carriers in bioremediation: a review. Electron. J. Biotechnol. 23: 28–36.

EU Directive (2015) 2015/1513 of the European Parliament and of the Council of 9 September 2015, amending directive 98/70/EC relating to the quality of petrol and diesel fuels and amending directive 2009/28/EC on the promotion of the use of energy from renewable sources. Off. J. Eur. Union. L. 239: 1–29.

Feller, G., E. Narinx, J.L. Arpigny, M. Aittaleb, E. Baise, S. Genicot, et al. 1996. Enzymes from psychrophilic organisms. FEMS Microbiol. Rev. 18: 189–202.

Fernández P.M., M.M. Martorell, M.G. Blaser, L.A.M. Ruberto, L.I.C. de Figueroa and W.P. Mac Cormack. 2017. Phenol degradation and heavy metal tolerance of Antarctic yeasts. Extremophiles 21: 445–457.

Giovanella, P., G.A.L. Vieira, I.V. Ramos Otero, E. PaisPellizzer, B.J. Fontes and L.D. Sette. 2019. Metal and organic pollutants bioremediation by extremophile microorganisms. J. Hazard. Mater. 121024.

González-Rocha, G., G. Muñoz-Cartes, C.B. Canales-Aguirre, C.A. Lima, M. Domínguez-Yévenes, H. Bello-Toledo, et al. 2017. Diversity structure of culturable bacteria isolated from the Fildes Peninsula (King George Island, Antarctica): a phylogenetic analysis perspective. Plos One 12(6): e0179390.

Gross, K.C., C.Y. Wang and M.E. Saltveit (eds.). 2016. The Commercial Storage of Fruits, Vegetables, and Florist and Nursery Stocks. United States Department of Agriculture, Agricultural Research Service.

Gugliandolo, C., L. Michaud, A. Lo Giudice, V. Lentini, C. Rochera and A. Camacho. 2016. Prokaryotic community in lacustrine sediments of byers peninsula (Livingston Island, Maritime Antarctica). Microb. Ecol. 71: 387–400.

Hawksworth, D. 2001. The magnitude of fungal diversity: the 1.5 million species estimate revisited. Mycol. Res. 105: 1422–1432.

Hu, H., F. Yan, C. Wilson, Q. Shen and X. Zheng X. 2015. The ability of a cold-adapted *Rhodotorula mucilaginosa* strain from Tibet to control blue mold in pear fruit. Antonie Van Leeuwenhoek 108(6): 1391–1404.

Hu, H., M.E. Wisniewski, A. Abdelfattah and X. Zheng. 2017. Biocontrol activity of a cold-adapted yeast from Tibet against gray mold in cherry tomato and its action mechanism. Extremophiles 21: 789–803.

Jafari N., M.R. Soudi and R. Kasra-Kermanshahi. 2014. Biodegradation perspectives of azo dyes by yeasts. Microbiology 83(5): 484–497.

Javed, A. and J.I. Qazi. 2016. Psychrophilic microbial enzymes implications in coming biotechnological processes. Am. Sci. Res. J. Eng. Technol. Sci. 23(1): 103–120.

Kaya, T., O. Kutlar and O. Taskiran. 2018. Evaluation of the effects of biodiesel on emissions and performance by comparing the results of the New European drive cycle and worldwide harmonized light vehicles test cycle. Energies 11(10): 2814–2827.

Korczak-Abshire, M. and P.J. Angiel. 2011. Records of white-rumped sandpiper (Calidrisfuscicollis) on the South Shetland Islands. Polar Rec. 47(242): 262–267.

Kraisintu, P., Y. Wichien and S. Limtong. 2010. Selection and optimization for lipid production of a newly isolated oleaginous yeast, rhodosporidiumtoruloides DMKU3-TK16. Kasetsart J. (Nat. Sci.) 44: 436–445.

Kurade M.B., T.R. Waghmode, R.V. Khandare, B. Jeon and S. Govindwar. 2016. Biodegradation and detoxification of textile dye Disperse Red 54 by Brevibacillus laterosporus and determination of its metabolic fate. J. Biosci. Bioeng. 121: 442–449.

Kurtzman, C.P., J.W. Fell and T. Boekhout. 2011. The Yeasts: A Taxonomic Study. Burlington: Elsevier Science.

Laich, F., R. Chavez and I. Vaca. 2014. *Leucosporidiumescuderoi* f.a., spnov., a basidiomycetous yeast associated with an Antarctic marine sponge. Antonie Van Leeuwenhoek 105(3): 593–601.

Liu, J., M. Wisniewski, S. Droby, S. Tian, V. Hershkovitz and T. Tworkoski. 2011. Effect of heat shock treatment on stress tolerance and biocontrol efficacy of *Metschnikowiafructicola*. FEMS Microbiol. Ecol. 76(1): 145–155.

Liu, J., M. Wisniewski, T. Artlip, Y. Sui, S. Droby and J. Norelli. 2013. The potential role of PR-8 gene of apple fruit in the mode of action of the yeast antagonist, *Candida oleophila*, in postharvest biocontrol of Botrytis cinerea. Postharvest Biol. Technol. 85: 203–209.

Lutz, M.C., C.A. Lopes, M.C. Sosa and M.P. Sangorrín. 2012. A new improved strategy for the selection of cold-adapted antagonist yeasts to control postharvest pear diseases. Biocontrol. Sci. Technol. 22: 1465–1483.

Marco S.L, C. Amaral, A. Sampaio, J. Peres and A. Díaz. 2005. Biodegradation of the diazo dye reactive black 5 by the wild isolate of *Candida oleophila*. Enzyme and Microb. Technol. 39: 51–55.

Margesin, R. 2000. Potential of cold-adapted microorganisms for bioremediation of oil-polluted Alpine soils. Int. Biodeter. Biodegr. 46(1): 3–10.

Margesin, R. and V. Miteva. 2011. Diversity and ecology of psychrophilic microorganisms. Res. Microbiol. 162(3): 346–361.

Martínez, A., I. Cavello, G. Garmendia, C. Rufo, S. Cavalitto and S. Vero. 2016. Yeasts from sub-Antarctic region: biodiversity, enzymatic activities and their potential as oleaginous microorganisms. Extremophiles 20: 759–769.

Martorell, M., L. Ruberto, P. Fernández, L. De Figueroa and W. Mac Cormack. 2019. Biodiversity and enzymes bioprospection of Antarctic filamentous fungi. Antarct Sci. 31(1): 3–12.

Martorell, M.M., P.M. Fernández, L. Adolfo, L.I. Castellanos de Figueroa and W.P. Mac Cormack. 2017. Bioprospection of cold-adapted yeasts with biotechnological potential from Antarctica. J. Basic Microbiol. 57(6): 504–516.

Meng, X., X. Yang, X. Xu, L. Zhang, Q. Nie and M. Xian. 2009. Biodiesel production from oleaginous microorganisms. Renew Energy 34: 1–5.

Mukhtar, H., S.M. Suliman, A. Shabbir, M.W. Mumtaz, U. Rashid and S.A. Rahimuddin. 2018. Evaluating the potential of oleaginous yeasts as feedstock for biodiesel production. Protein Pept. Lett. 25(2): 195–201.

Nakagawa, T., R. Ikehata, M. Uchino, T. Miyaji, K. Takano and N. Tomizuka. 2006. Cold-active acid β-galactosidase activity of isolated psychrophilic-basidiomycetous yeast Guehomyces pullulans. Microbiol. Res. 161(1): 75–79.

Pan, L., D. Yang, L. Shao, W. Li, C. Wei, G. Chen and Z. Liang. 2009. Isolation of the oleaginous yeasts from the soil and studies of their lipid-producing capacities. Food Technol. and Biotechnol. 47(2): 215–220.

Patel, A., N. Arora, J. Mehtani, V. Pruthi and P.A. Pruthi. 2017. Assessment of fuel properties on the basis of fatty acid profiles of oleaginous yeast for potential biodiesel production. Renew. Sust. Energ. Rev. 77: 604–616.

Pathan, A.A., B. Bhadra, Z. Begum and S. Shivaji. 2010. Diversity of yeasts from puddles in the vicinity of midre lovénbreen glacier, arctic and bioprospecting for enzymes and fatty acids. Curr. Microbiol. 60(4): 307–314.

Peeples, T.L. 2014. Bioremediation using extremophiles. pp. 251–268. *In*: Surajit Das [ed.]. Microbial Biodegradation and Bioremediation. Elsevier, Rourkela Odisha, India.

Pereyra, V., A. Martinez, C. Rufo and S. Vero. 2014. Oleaginous yeasts form Uruguay and Antarctica as renewable raw material for biodiesel production. Am J. Biosci. 2(6): 251–257.

Pershina, E.V., E.A. Ivanova, E.V. Abakumov and E.E. Andronov. 2018. The impacts of deglaciation and human activity on the taxonomic structure of prokaryotic communities in Antarctic soils on King George Island. Antarct Sci. 30(5): 278–288.

Poliakova, A.V., I.Y. Chernov and N.S. Panikov. 2001. Yeast biodiversity in hydromorphic soils with reference to grass-sphagnum swamp in Western Siberia and the hammocky tundra region (Barrow, Alaska) Microbiology 70: 617–622.

Raddadi, N., A. Cherif, D. Daffonchio, M. Neifar and F. Fava 2015. Biotechnological applications of extremophiles, extremozymes and extremolytes. Appl. Microbiol. Biotechnol. 99(19): 7907–7913.

Ratledge, C. 1979. Resources conservation by novel biological processes. Part I: Grow fats from wastes. Chem Soc. Rev. 8(2): 283–296.

Robiglio, A., M.C. Sosa, M.C. Lutz, C.A. Lopes, M.P. Sangorrín. 2011. Yeast biocontrol of fungal spoilage of pears stored at low temperature. Int. J. Food. Microbiol. 147(3): 211–216.

Rossi, M., P. Buzzini, L. Cordisco, A. Amaretti, M. Sala, S. Raimondi, et al. 2009. Growth, lipid accumulation, and fatty acid composition in obligate psychrophilic, facultative psychrophilic, and mesophilic yeasts. FEMS Microbiol. Ecol. 69: 363–372.

Rovati, J.I., H.F. Pajot, L. Ruberto, W. Mac Cormack and L.I.C. Figueroa. 2013. Polyphenolic substrates and dyes degradation by yeasts from 25 de Mayo/King George Island (Antarctica). Yeast 30: 459–470.

Ruisi S., D. Barreca, L. Selbmann, L. Zucconi and S. Onofri. 2007. Fungi in Antarctica. Rev. Environ. Sci. Bio./ Technol. 6: 127–141.

Ruscasso, F., G. Garmendia, S. Vero, S. Cavalitto and I. Cavello. 2017. Aerobic degradation of reactive orange 16 by an antarctic yeast. In: Guaiquil, I., Leppe, M., Rojas, P., yR. Canales, Eds. Visiones de Ciencia Antártica, Libro de Resúmenes, IX Congreso Latinoamericano de Ciencias Antártica, Punta Arenas-Chile.

Ruscasso F., B. Bezus, G. Curutchet, I. Cavello, et al. 2019a. Reducción de la toxicidad de un efluente simulado de la industria textil. In: XV Congreso Argentino de Microbiología (CAM 2019). Buenos Aires, Argentina.

Ruscasso, F., B. Bezus, G. Garmendia, S. Vero, G. Curutchet, I. Cavello, et al. 2019b. Estudio de la bioacumulación de colorantes textiles por parte de la levadura antártica *Debaryomyceshansenii* F39A. *In*: XV Congreso Argentino de Microbiología (CAM 2019). Buenos Aires, Argentina.

Sangorrin, M.P., C.A. Lopes, S. Vero and M. Wisniewski. 2014. Cold adapted yeasts as biocontrol agents: Biodiversity, adaptation strategies and biocontrol potential. pp. 441–464. *In*: P. Buzzini and R. Margesin [eds.]. Cold-adapted Yeasts. Springer, Berlin, Heidelberg.

Sarmiento, F., R. Peralta and J.M. Blamey. 2015. Cold and hot extremozymes: Industrial relevance and current trends. Front Bioeng. Biotechnol. 3(148): 1–15.

Shapaval, V., J. Brandenburg, J. Blomqvist, V. Tafintseva, V. Passoth, M. Sandgren, et al. 2019. Biochemical profiling, prediction of total lipid content and fatty acid profile in oleaginous yeasts by FTIR spectroscopy. Biotechnol. Biofuels. 12(1): 140.

Sharma, R., D. Singh and R. Singh. 2009. Biological control of postharvest diseases of fruits and vegetables by microbial antagonists: a review. Biol. Control. 50: 205–221. 10.1016/j.biocontrol.2009.05.001.

Shivaji, S. and G.S. Prasad. 2009. Antarctic yeasts: biodiversity and potential applications. pp. 3–18. *In*: T. Satyanarayana and G. Kunze [eds.]. Yeast Biotechnology: Diversity and Applications. Springer, Dordrecht.

Singh. P. and S.M. Singh. 2012. Characterization of yeast and filamentous fungi isolated from cryoconite holes of Svalbard, Arctic. Polar Biol. 35: 575–583.

Sitepu, I.R., L.A. Garay, R. Sestric, D. Levin, D.E. Block, J.B. German, et al. 2014. Oleaginous yeasts for biodiesel: current and future trends in biology and production. Biotechnol. Adv. 32: 1336–1360.

Sláviková, E., B. Kosíková and M. Mikulasova. 2002. Biotransformation of waste lignin products by the soil-inhabiting yeast Trichosporon pullulans. Can. J. Microbiol. 48: 200–203.

Spadaro, D. and S. Droby. 2016. Development of biocontrol products for postharvest diseases of fruit: the importance of elucidating the mechanisms of action of yeast antagonists. Trends Food Sci. Tech. 47: 39–49.

Sui, Y., M. Wisniewski, M., S. Droby and J. Liu. 2015. Responses of yeast biocontrol agents to environmental stress. Appl. Environ. Microbiol. 81(9): 2968–2975.

Tapre, A.R. and R.K. Jai. 2014. Pectinases, enzymes for fruit processing industry. Int. Food Res. J. 21: 447–453.

Terauds, A., S.L. Chown, F. Morgan, H.J. Peat, D.J. Watts, H. Keys, et al. 2012. Conservation biogeography of the Antarctic. Divers Distrib. 18: 726–741.

Thomas-Hall, S.R., B. Turchetti, P. Buzzini, E. Branda, T. Boekhout, B. Theelen, et al. 2010. Cold-adapted yeasts from Antarctica and the Italian Alps-description of three novel species: *Mrakia robertii* sp. nov., *Mrakia blollopis* sp. nov. and *Mrakiella niccombsii* sp. nov. Extremophiles 14(1): 47–59.

Tsuji, M., S. Fujiu, N. Xiao, Y. Hanada, S. Kudoh, H. Kondo, et al. 2013. Cold adaptation of fungi obtained from soil and lake sediment in the Skarvsnes ice-free area, Antarctica. FEMS Microbiol. Lett. 346(2): 121–130.

Tsuji, M. 2016. Cold-stress responses in the Antarctic basidiomycetous yeast *Mrakiablollopis*. R. Soc. Open. Sci. 3(7): 1–13.

Turchetti, B., S.R. Thomas-Hall, L.B. Connell, E. Branda, P. Buzzini, B. Theelen, et al. 2011. Psychrophilic yeasts from Antarctica and European glaciers: description of *Glaciozyma* gen. nov., *Glaciozyma martinii* sp. nov. and *Glaciozyma watsonii* sp. nov. Extremophiles 15(5): 573–586.

Usall, J., A. Ippolito, M. Sisquella and F. Neri. 2016. Physical treatments to control postharvest diseases of fresh fruits and vegetables. Postharvest Biol. Technol. 122: 30–40.

Vaca Cerezo, I., C. Faúndez, F. Maza, B. Paillavil, V. Hernández, F. Acosta, et al. 2013. Cultivable psychrotolerant yeasts associated with Antarctic marine sponges. World J. Microbiol. Biotechnol. 29: 183–189.

Vaz, A.B.M., L.H. Rosa, M.L.A. Vieira, V. de Garcia, L.R. Brandão, L. Teixeira, et al. 2011. The diversity, extracellular enzymatic activities and photoprotective compounds of yeasts isolated in antarctica. Braz. J. Microbiol. 42: 937–947.

Vero, S., G. Garmendia, M.B. González González, F. Garat and M. Wisniewski. 2009. Aureobasidiumpullulans as a biocontrol agent of postharvest pathogens of apples in Uruguay. Biocontrol. Sci. Techn. 10: 1033–1049.

Vero, S., G. Garmendia, F. Garat, I. de Aurrecochea and M. Wisniewski. 2011. *Cystofilobasidiumin firmominiatum*as biocontrol agent of postharvest disease on apples and citrus. Acta Hortic. 905: 169–180.

Vero, S., G. Garmendia, M.B. González, O. Bentancur and M. Wisniewski. 2013. Evaluation of yeast obtained from Antarctic soils samples as biocontrol agents for the management of postharvest diseases of Apple (*Malusdomestica*). Fems. Yeast Res. 13: 189–199.

Vicente, G., L. Bautista, R. Rodríguez, F. Gutiérrez, I. Sadaba, M.R. Ruiz-Vazquez, et al. 2009. Biodiesel production from biomass of an oleaginous fungus. Biochemical Eng. J. 48: 22–27.

Vidali, M. 2001. Bioremediation: an overview. Pure Appl. Chem. 73(7): 1163–1172.

Viñarta, S.C., M.V. Angelicola, J.M. Barros, P.M. Fernández, W. MacCormak, M.J. Aybar, et al. 2016. Oleaginous yeasts from Antarctica: screening and preliminary approach on lipid accumulation. J.Basic Microbiol. 56(12): 1360–1368.

Wang, Y., Z. Xu, P. Zhu, Y. Liu, Z. Zhang, Y. Mastuda, et al. 2010. Postharvest biological control of melon pathogens using *Bacillus subtilis* EXWB1. J. Plant Pathol. 92: 645–652.

Wentzel L.C.P., F.J. Inforsato and Q.V. Montoya. 2018. Fungi from Admiralty Bay (King George Island, Antarctica) Soils and Marine Sediments. Microb. Ecol. 77: 12–24. doi: https://doi.org/10.1007/s00248-018-1217-x.

Wentzel, L.C.P., F.J. Inforsato, Q.V. Montoya, B.G. Rossin, N.R. Nascimento, A. Rodrigues, et al. 2019. Fungi from admiralty bay (King George Island, Antarctica) soils and marine sediments. Microbial Ecology 77(1): 12–24.

Wisniewski, M., S. Droby, J. Norelli, J. Liu and L. Schena. 2016. Alternative management technologies for postharvest disease control: the journey from simplicity to complexity. Postharvest Biol. Technol. 122: 3–10.

Zhang C., N. Sirijovski, L. Adler and B.C. Ferrari. 2019. *Exophiala macquariensis* sp. nov., a cold adapted black yeast species recovered from a hydrocarbon contaminated sub-Antarctic soil. Fungal Biol. 123: 151–158.

9

Bacterial Communities Associated With Macroalgae In Antarctica

Sergio Leiva

Introduction: Antarctic Macroalgae

The Antarctic continent is of exceptional biological interest, possessing a rich and highly endemic terrestrial and marine biodiversity. Despite the harsh abiotic environment, Antarctica contains a rich biodiversity of macroalgae and a high degree of endemism. Although a more precise estimation of the species richness of Antarctic macroalgae needs improvement in the collection techniques—the use of combined taxonomical methods and more systematic collection efforts in East Antarctica (Pellizzari et al. 2017)—the recent taxonomic surveys have identified 124 species of Antarctic macroalgae of which 35% are endemic with the highest levels in the Heterokontophyta and the Rhodophyta (Wiencke et al. 2014). In the western Antarctic Peninsula, macroalgae often reach biomass levels of 5-10 wet kg m^{-2}, which is comparable to temperate kelp forests (Wiencke and Amsler 2012). Antarctic macroalgae are unique in several aspects of their biology and ecology and have developed remarkable ecophysiological adaptations to deal with strong seasonal changes in ice cover and amount of daylight. They are extremely shade-adapted organisms, capable of efficient photosynthesis under weak light conditions in winter and well-adapted to cope with very high irradiances in summer without photoinhibition (Gómez et al. 2009, Wiencke and Amsler 2012). Antarctic macroalgae are strongly adapted to the low seawater temperatures of the Southern Ocean. For example, the endemic species *Himantothallus grandifolius* (Phaeophyta) is a shade-adapted species with a wide range of vertical distribution (Huovinen and Gómez 2013) being reported from 10 to 90 m depth in Admiralty Bay, King George Island, South Shetland Islands (maritime Antarctica) (Zielinski 1990); this species grows only up to 5°C and have upper survival temperatures (USTs) of 11-13°C (Wiencke et al. 2014). Macroalgae from the Arctic are comparatively less adapted to low temperatures (Wiencke and Amsler 2012). Endemic brown and red macroalgae exhibit maximum photosynthetic rates in culture at 0°C, which is comparable to the values reported for cold temperate species measured at 10 to 15°C (Wiencke et al. 1993). Reproductive cells of Antarctic macroalgal species are also low light-adapted and susceptible to UV radiation, although propagules possess effective DNA repair mechanisms against UV damage (Roleda et al. 2007). It seems that UV radiation along with ice scouring, temperature, and morpho-functional characteristics of macroalgae are important factors influencing their zonation pattern along the Antarctic coast (Huovinen and Gómez 2013, Gómez et al. 2019). Macroalgae are also key habitat-structuring agents and an important food source for benthic grazers in Antarctic ecosystems, providing shelter to diverse

Instituto de Bioquímica & Microbiología, Facultad de Ciencias, Universidad Austral de Chile, Casilla 567, Valdivia, Chile.
E-mail: sleiva@uach.cl

assemblages of marine invertebrates and fishes, particularly amphipods and gastropods (Iken et al. 1998, Amsler et al. 2014).

Bacterial Communities on Antarctic Macroalgae

The surface of macroalgae is a protected and nutrient-rich habitat that harbours a dense, complex, and highly dynamic microbial community, which plays an important role in the physiology, ecology, and evolution of the host (Goecke et al. 2010, Hollants et al. 2013). Studies in tropical and temperate waters have shown that marine macroalgae associate with microbial populations that differ significantly from those occurring in the surrounding seawater (Burke et al. 2011, Lachnit et al. 2011) and in which bacteria are by far the dominant colonisers (Wahl et al. 2012). Using manipulative experiments, Chen and Parfrey (2018) showed that macroalgae surfaces are highly selective and exert a strong influence on the microbiome structure of the surrounding water column but a much weaker influence on the surface microbiota of neighbouring macroalgae. Biological factors such as host health condition, disease, seaweed growth cycle, age of the algal tissue, grazing pressure, and competitive interactions between bacteria on the algal surface are important in shaping microbial community composition (Rao et al. 2006, Bengtsson et al. 2010, Fernandes et al. 2012). The composition and structure of the seaweed microbiome are also influenced by abiotic factors, such as irradiance, nutrient concentration, pollution, salinity, and temperature, as well as site-specific environmental conditions, and even habitat modification caused by coastal urbanisation (Hengst et al. 2010, Wahl et al. 2010, Stratil et al. 2013, Stratil et al. 2014, Brodie et al. 2016, Marzinelli et al. 2018, Florez et al. 2019). The composition of the seaweed-associated microbiota is spatially and temporally dynamic and influenced by changing environmental conditions and host physiology (Lachnit et al. 2011, Serebryakova et al. 2018).

It is increasingly argued that the macroalgal hosts and its associated microbiota form a complex ecological unit termed holobiont in which microbes play crucial roles in host functioning (van der Loos et al. 2019). However, the ecological relevance of most associated bacteria remains undetermined, but it has been postulated a mutualistic relationship in which the microbes play roles related to host's growth and normal development and morphogenesis, metabolism regulation, nutrient supply, prevention of extensive biofouling or infection by microbial pathogens, and adaptation and physiological response of seaweed to environmental change and thus possibly acting as the drivers of speciation; in turn, the algal host may provide nutrients and physical protection to the bacteria (Goecke et al. 2010, Singh and Reddy 2014, Dittami et al. 2016). Members of the *Cytophaga-Flavobacterium-Bacteroides* group and the *Roseaobacter* clade are particularly important in the life cycle of *Ulva* species by releasing infochemicals that enhance the zoospore settlement and algal morphogenesis (Wichard 2015). Because of the association of the microbiome with macroalgal health and function, the modulation of some beneficial microbial members provides an attractive strategy for promoting macroalgal health and performance, which can lead to the improvement of production in macroalgae aquaculture (Singh and Reddy 2016).

Few studies to date have explored the diversity and ecological role of microorganisms associated with the surface of Antarctic macroalgae. Recent culture-based studies have shed light on the diversity and bioactive potential of the fungal community living in association with Antarctic macroalgae, concluding that it is composed by endemic species, such as *Metschnikowia australis* and *Antarctomyces psychrotrophicus* as well as cold-adapted cosmopolitan species of *Penicillium*, *Cryptococcus*, and *Rhodotorula* (Ogaki et al. 2019). Godinho et al. (2013) studied the distribution and diversity of fungi associated with Antarctic brown, green, and red macroalgae and found that the fungal community is diverse and complex with only a few dominant species (*Geomyces* sp., *Penicillium* sp., and *Metschnikowia australis*) as well as a large number of rare and/or endemic taxa. In a later study, these investigators reported a diverse fungal community associated with the algae *Monostroma hariotti* (green alga) and *Pyropia endiviifolia* (red alga), recording 48 taxa within the phyla Ascomycota, Basidiomycota, and Zygomycota, although they found low similarities among the fungal assemblages of these two macroalgae (Furbino et al. 2014). In turn, Duarte et al. (2016) found a diverse yeast community from macroalgae collected in the South Shetland Islands, Antarctica, which was dominated by Basidiomycota, with a total of 24 distinct taxa and potential new species isolated from *Adenocystis utricularis* (brown alga), *Palmaria decipiens* (red alga) and *Himantothallus grandifolius*.

Antarctic biota and flora are emerging as important sources of bacterial biodiversity with enormous biotechnological potential. Recent culture-independent and traditional culture-based studies have greatly increased our understanding of the microbiology of Antarctic marine invertebrates (Herrera et al. 2017, Lo Giudice and Rizzo 2018, Silva et al. 2018, Clarke et al. 2019). However, very little is known about the natural distribution, diversity and ecological functions of bacteria associated with marine Antarctic macroalgae. The few studies, so far performed, have shown that the epiphytic culturable bacterial community is dominated by Gram-negative bacteria and include representatives of γ-Proteobacteria (e.g., *Colwellia, Pseudoaltermonas, Pseudomonas*), α-Proteobacteria (*Sulfitobacter*), and Bacteroidetes (e.g., *Cellulophaga, Flavobacterium, Winogradskyella*). Figure 1 shows some macroalgal species occurring in King George Island and some of the bacterial species isolated as epibionts. The list of Gram-negative bacteria recovered from Antarctic macroalgae is summarised in Table 1. In two studies conducted in King George Island, Tropeano et al. (2012, 2013) explored the enzymatic capabilities of bacteria isolated from different types of marine samples (fishes, invertebrates, macroalgae, seawater, sediments), obtaining 277 isolates of which 19 were recovered from the surface of mostly brown macroalga (*Adenocystis utricularis, Ascoseira mirabilis, Desmarestia anceps*, and *Phaeurus antarcticus*). The epiphytic isolates were affiliated to the genera *Flavobacterium, Olleya, Colwellia, Pseudoalteromonas*, and *Psychromonas*. Alvarado and Leiva (2017) and Sánchez Hinojosa et al. (2018) studied the diversity of agar-degrading bacteria associated with the surface of intertidal (*Adenocystis utricularis, Iridaea cordata, Monostroma hariotti*) and subtidal macroalgae (*Himantothallus grandifolius, Plocamium cartilagineum, Pantoneura plocamioides*) of King George Island. Both works obtained a total of 13 phylotypes of Gammaproteobacteria and Flavobacteriia, with isolates of *Cellulophaga algicola* and *Pseudoalteromonas* sp. being the most prevalent agarolytic bacteria. Barreto et al. (2017) recovered 33 isolates from the surface of four marine macroalgae (red and brown) of Admiralty Bay, of which nearly 80% were Gram-negative bacteria. These authors identified 11 genera of Gram-negative bacteria with Gammaproteobacteria as the most abundant and diverse phylum.

Pseudoalteromonas represents one of the most common genera isolated from marine macroalgae (brown, green, and red) of different latitudes (Hollants et al. 2013). The genus *Pseudoalteromonas* is considered one of the most important model systems for studying microbial adaptation to extreme environments, and it has developed unique cold adaptation strategies (Parrilli et al. 2019). This genus has been widely reported to be isolated from Antarctica, with novel species discovered from this continent and many other isolates recovered from various marine sources, such as invertebrates, sea-ice, seawater, and sediments (Bozal et al. 1997, Bowman 1998, Tropeano et al. 2013, González-Aravena et al. 2016, Lo Giudice and Rizzo 2018); in addition to this, isolates of seven species of *Pseudoalteromonas* have been reported from nine Antarctic macroalgal species (Tropeano et al. 2012, Tropeano et al. 2013, Barreto et al. 2017, Sánchez Hinojosa et al. 2018) (Table 1).

Cellulophaga is another genus that has been reported from a wide range of samples in Antarctica, including seaweeds. *Cellulophaga* spp. can hydrolyse algal phycocolloids, display gliding motility and are strongly proteolytic and algicidal (Mayali and Azam 2004, Krieg et al. 2010). Antarctic *Cellulophaga* isolates have been recovered from bivalve mollusks (Tropeano et al. 2012), marine sediments (Tropeano et al. 2013), sea-ice diatoms (Bowman 2000), seawater (Bianchi et al. 2014), sponges (Silva et al. 2018), and water from an ice-covered lake (Stingl et al. 2008). Strains of *Cellulophaga* have been isolated from non-Antarctic macroalgal species (Kientz et al. 2013, Lafleur et al. 2015), and Alvarado and Leiva (2017) reported a wide diversity of colony morphotypes of *Cellulophaga algicola* strains isolated from the Antarctic macroalgae *Adenocystis utricularis, Iridaea cordata, Monostroma hariotii*, and *Pantoneura plocamioides*.

TABLE 1: List of Taxa of Gram-Negative Bacteria Isolated From Antarctic Marine Macroalgae

Phylum	Species	Origin	Reference
Bacteroidetes	*Algibacter lectus*	IC	(Alvarado and Leiva 2017)
	Cellulophaga algicola	AU, IC, MH, PP	(Alvarado and Leiva 2017; Sánchez Hinojosa et al. 2018)
	Flavobacterium sp.	UA	(Ferrés et al. 2015)
	Flavobacterium faecale	PO	(Lavín et al. 2016)
	Flavobacterium frigidarium	DC, PA, UA	(Tropeano et al. 2013)
	Maribacter arcticus	AM	(Barreto et al. 2017)
	Lacinutrix himadriensis	HG	(Sánchez Hinojosa et al. 2018)
	Olleya namhaensis	DC, HG	(Sánchez Hinojosa et al. 2018; Tropeano et al. 2013)
	Polaribacter sp.	PD	(Barreto et al. 2017)
	Putridiphycobacter roseus	UA	(Wang et al. 2019)
	Winogradskyella thalassocola	DA	(Barreto et al. 2017)
	Winogradskyella undariae	HG	(Sánchez Hinojosa et al. 2018)
	Zobellia laminariae	AU	(Alvarado and Leiva, 2017)
Proteobacteria	*Cobetia* sp.	PD	(Barreto et al. 2017)
	Colwellia sp.	PD	(Barreto et al. 2017)
	Colwellia aestuarii	AM, PA	(Tropeano et al. 2013)
	Colwellia sediminilitoris	PC	(Sánchez Hinojosa et al. 2018)
	Halomonas sp.	DA	(Barreto et al. 2017)
	Marinomonas sp.	HG	(Barreto et al. 2017)
	Paraglaciecola psychrophila	PC	(Sánchez Hinojosa et al. 2018)
	Pseudoalteromonas sp.	AM, DA, HG	(Barreto et al. 2017)
	Pseudoalteromonas arctica	AM, AU, HG, PC, PP, UA	(Sánchez Hinojosa et al. 2018; Tropeano et al. 2013)
	Pseudoalteromonas distincta	PC, PP	(Sánchez Hinojosa et al. 2018)
	Pseudoalteromonas haloplanktis	AM, DA, PD	(Barreto et al. 2017)
	Pseudoalteromonas nigrifaciens	DC, HG, UA	(Sánchez Hinojosa et al. 2018; Tropeano et al. 2013)
	Pseudoalteromonas paragorgicola	AU	(Tropeano et al. 2012)
	Pseudoalteromonas prydzensis	PP	(Sánchez Hinojosa et al. 2018)
	Pseudoalteromonas translucida	DC, PA, PC, UA	(Sánchez Hinojosa et al. 2018; Tropeano et al. 2013)
	Pseudomonas sp.	HG	(Barreto et al. 2017)
	Pseudomonas putida	PD	(Melo et al. 2009)
	Pseudomonas savastanoi	PD	(Melo et al. 2009)
	Psychrobacter sp.	AM, DA, UA	(Barreto et al. 2017; Ferrés et al. 2015)
	Psychrobacter glacincola	AM	(Barreto et al. 2017)
	Psychromonas arctica	AU	(Tropeano et al. 2012)
	Sulfitobacter sp.	HG	(Barreto et al. 2017)

The two letters of the origin indicate the algae from which the taxon was isolated. AM *Ascoseira mirabilis,* AU *Adenocystis utricularis,* DA *Desmarestia antarctica,* DC *Desmarestia anceps,* HG *Himantothallus grandifolius,* IC *Iridaea cordata,* MH *Monostroma hariotii,* PA *Phaeurus antarcticus,* PC *Plocamium cartilagineum,* PD *Palmaria decipiens,* PO *Porphyra* sp., PP *Pantoneura plocamioides,* UA unidentified alga.

TABLE 2: List of Taxa of Gram-Positive Bacteria Isolated From Antarctic Marine Macroalgae

Phylum	Species	Origin	Reference
Actinobacteria	*Aeromicrobium alkaliterrae*	IC	(Alvarado et al. 2018)
	Aeromicrobium ginsengisoli	AU	(Alvarado et al. 2018)
	Agrococcus baldri	MH	(Leiva et al. 2015)
	Amycolatopsis antarctica	AU	(Wang et al. 2018)
	Arthrobacter agilis	AU, MH	(Alvarado et al. 2018, Leiva et al. 2015)
	Arthrobacter flavus	AU	(Leiva et al. 2015)
	Arthrobacter psychrochitiniphilus	UA	(Vila et al. 2019)
	Arthrobacter rhombi	PE	(Leiva et al. 2015)
	Brachybacterium paraconglomeratum	AU, IC	(Alvarado et al. 2018, Leiva et al. 2015)
	Brachybacterium rhamnosum	MH	(Leiva et al. 2015)
	Citricoccus zhacaiensis	IC	(Leiva et al. 2015)
	Janibacter anophelis	IC, MH	(Alvarado et al. 2018)
	Kocuria palustris	AU, PC	(Leiva et al. 2015)
	Labedella endophytica	AU	(Alvarado et al. 2018)
	Microbacterium oxydans	AU	(Leiva et al. 2015)
	Microbacterium hatanonis	AU	(Alvarado et al. 2018)
	Micrococcus luteus	AU, IC	(Alvarado et al. 2018)
	Pseudarthrobacter oxydans	AU	(Alvarado et al. 2018)
	Pseudonocardia adelaidensis	MH	(Alvarado et al. 2018)
	Rhodococcus cerastii	AU, MH	(Alvarado et al. 2018, Leiva et al. 2015)
	Rhodococcus fascians	MH	(Alvarado et al. 2018)
	Salinibacterium sp.	MH	(Leiva et al. 2015)
	Salinibacterium amurskyense	IC	(Leiva et al. 2015)
	Sanguibacter inulinus	AU	(Leiva et al. 2015)
	Streptomyces brevispora	MH	(Alvarado et al. 2018)
	Streptomyces pratensis	IC, MH	(Alvarado et al. 2018)
	Tessaracoccus flavescens	AU	(Alvarado et al. 2018)
Firmicutes	*Aerococcus viridans*	PD	(Barreto et al. 2017)
	Bacillus sp.	UA	(Lavin et al. 2013)
	Planococcus sp.	AM	(Barreto et al. 2017)
	Planomicrobium flavidum	IC	(Alvarado et al. 2018)
	Staphylococcus arlettae	IC	(Leiva et al. 2015)
	Staphylococcus cohnii subsp. *cohnii*	AU	(Leiva et al. 2015)
	Staphylococcus haemolyticus	AU	(Alvarado et al. 2018)
	Staphylococcus sciuri	PH	(Leiva et al. 2015)

The two letters of the origin indicate the algae from which the taxon was isolated. AM *Ascoseira mirabilis*, AU *Adenocystis utricularis*, IC *Iridaea cordata*, MH *Monostroma hariotii*, PC *Plocamium cartilagineum*, PD *Palmaria decipiens*, PE *Pyropia endiviifolia*, PH *Phycodrys antarctica*, UA unidentified alga.

Culture-based and culture-independent studies show that Antarctic marine habitats such as invertebrates, sediments, and surface seawater harbour a rich community of Gram-positive bacteria (Webster and Bourne 2007, Papaleo et al. 2012, Li et al. 2016, Wang et al. 2017). Research on non-Antarctic seaweeds shows that Actinobacteria and Firmicutes are prevalent colonisers of algal surfaces (Salaün et al. 2010, Goecke et al. 2013, Singh and Reddy 2014). However, the information regarding the association between Antarctic macroalgae and Gram-positive bacteria is scarce. Recently, it has been reported an unprecedented diversity of pigmented, Gram-positive epiphytic bacteria isolated from different macroalgae species of King George Island (Leiva et al. 2015). Based on 16S rRNA gene sequence analysis, 18 different bacterial phylotypes were recorded, which were clustered into 11 genera of *Actinobacteria* and one genus of the *Firmicutes*, with isolates exhibiting different shades of yellow, red, orange, and amber. In this study, most of the pigmented isolates were recovered from algae collected in the intertidal zone, which suggests that carotenoid-type compounds may help epiphytic bacteria tolerate the harsh and changing abiotic environment in this zone during low tide (high solar radiation, freeze-thaw cycles) and in a region that is prominently affected by the ozone layer depletion (Singh et al. 2018). In another previous study, Vazquez and Mac Cormack (2002) also isolated a high number of pigmented bacteria (mostly yellow and orange coloured, non-spore forming, Gram-negative bacilli) from both marine and non-marine sources in King George Island, which was associated to the high temperatures and levels of UV radiation during the Austral summer. In another recent study, Alvarado et al. (2018) investigated the culturable Gram-positive bacteria associated with the surface of three intertidal, co-occurring Antarctic macroalgae (*Adenocystis utricularis*, *Iridaea cordata*, and *Monostroma hariotii*). It was found a diverse assemblage of Gram-positive bacteria, which were affiliated to 17 genera of Actinobacteria and two of Firmicutes. The authors concluded that the culturable Gram-positive bacterial community exhibits little overlap at the species level among the three co-occurring macroalgae species, suggesting that Antarctic macroalgae host species-specific Gram-positive epiphytes. Barreto et al. (2017) recovered seven Gram-positive isolates from the surface of the endemic Antarctic macroalgae *Ascoseira mirabilis* and *Palmaria decipiens*, which were identified as members of *Planococcus* sp. and *Aerococcus* sp., respectively (Table 2).

Many new bacterial species, genera, and orders have been identified from marine algae, mostly from macroalgal sources, although most of the new bacterial taxa have been described from a few widespread macroalgal genera such as *Fucus*, *Porphyra*, and *Ulva* (Goecke et al. 2013). Several new bacterial species have been isolated from Antarctic algal sources, such as from microalgae and sea-ice algal assemblages (Bowman 2000, Bowman and Nichols 2005), and recently, new bacterial taxa have been reported from macroalgal samples collected in King George Island. Wang et al. (2019) described a novel species of a new genus in the phylum Bacteroidetes (*Putridiphycobacter roseus*), which was cultivated from an unidentified rotten seaweed. In turn, Wang et al. (2018) discovered a new species of the actinobacterial genus *Amycolatopsis* (*Amycolatopsis antarctica*) from sporophytes of *Adenocystis utricularis* collected in King George Island (Figure 1). *A. antarctica* is the only species of *Amycolatopsis* isolated from a marine macroalga, which suggest that Antarctic macroalgae are an interesting environment for the discovery of novel prokaryotic taxa.

Microbial Decomposition of Antarctic Macroalgae

It has been estimated that macroalgae cover about 30% of the seabed surface in maritime Antarctica and produce ca. 74,000 tons of wet biomass around Admiralty Bay (Zielinski 1990), thus constituting an important source of organic matter and nutrients in this area (Nedzarek and Rakusa-Suszczewski 2004) that can explain the high relative abundance of heterotrophic microorganisms found in sediments (Franco et al. 2017). For instance, the brown alga *Himantothallus grandifolius* is reported as one of the most common species in King George Island. According to Quartino et al. (2001), in Potter Cove (King George Island), this species exhibited one of the highest biomasses (306.77 ± 591.85 g dry weight m^{-2}), and along with the brown algal species *Desmarestia menziesii* and *Desmarestia anceps* account for almost 80% of the biomass in summer (Quartino and Boraso de Zaixso 2008).

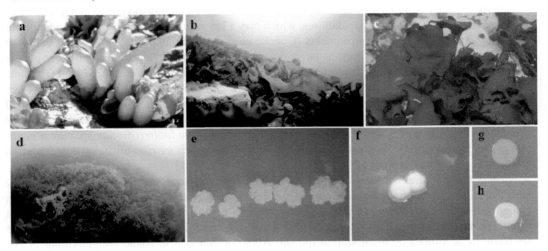

Figure 1. Antarctic macroalgae and associated-epiphytic bacteria. (a) *Adenocytis utricularis*, (b) *Himantothallus grandifolius*, (c) *Iridaea cordata*, (d) *Plocamium cartilagineum*, (e) colonies of *Amycolatopsis antarctica* sp. nov. isolated from *A. utricularis*, (f) colonies of *Pseudonocardia* sp. isolated from *Monostroma hariotii*, (g) colony of *Salinibacterium amurskyense* isolated from *I. cordata*, (h) colony of *Agrococcus baldri* isolated from *M. hariotii*. Photo credits: a, e, f, g, h, Sergio Leiva; b, c, d, Ignacio Garrido, Universidad Austral de Chile.

Microbial communities are important components of the microbial food loop, and the bacterial community plays a pivotal role in the formation of organic carbon in Antarctica. According to Zdanowski (1998), the bacterial biomass in Admiralty Bay is equal or exceed that of the estimated standing stock of krill or other major living components. Similarly, bacteria account for more than one-third of the microbial community biomass in Great Wall Cove and Ardley Cove, King George Island (Tian et al. 2015).

Although the initial steps in the macroalgal decomposition process involve the rapid leaching of soluble substances in a cellular autolytic process, microorganisms play key roles as decomposers of algal detritus (Arroyo and Bonsdorff 2016). Algae-degradation studies and SEM microscopy have revealed great numbers of bacteria colonising cell debris of the Antarctic brown alga *Desmarestia anceps* and that viable heterotrophic bacteria increase their numbers over time during algal degradation, suggesting that bacteria play an important role in the decomposition of algae and on nutrient cycling dynamics in Antarctic marine ecosystems (Quartino et al. 2015). Kinetic studies on the bacterial degradation of Antarctic macroalgae show that the bacterial colonisation rates of macroalgae debris are comparable to those reported for temperate regions and that some of the bacterial enzymes involved in the degradation of algal polymers are optimally adapted to the cold Antarctic environment (Reichardt and Dieckmann 1985). Although the heterotrophic bacterial fraction is significantly active in Antarctic waters (Gillespie et al. 1976), the effectiveness of the degradation of the algal tissue also depends on the prior mechanical breakdown of the macroalgal thalli by sea-ice abrasion and stones (Quartino et al. 2015) and grazer activities (Rieper-Kirchner 1990, Catenazzi and Donnelly 2007). The degradation rate of algal material is related to its nutritional quality (%C, %N, C/N ratio), palatability, the content of fibrous material, concentration of deterrent secondary metabolites, and is largely dependent on the extent of microbial colonisation (Krumhansl and Scheibling 2012). In a recent mesocosm study in the subtidal area of Potter Cove, Braeckman et al. (2019) concluded that the phytodetritus from the herbivore-palatable red alga *Palmaria decipiens* was recycled at a higher rate than that of the less palatable *Desmarestia anceps*. They also found that the early stages of the degradation pathway of *Palmaria decipiens* are more herbivore driven, with bacteria playing a stronger role at later stages, whereas the early degradation of *Desmarestia anceps* was more microbial driven.

Epiphytic bacteria of marine macroalgae produce cell wall degrading enzymes such as agarases, carrageenases, alginases, and fucoidanases, which play a key role in biotransformation and nutrient recycling in coastal marine environments (Goecke et al. 2010, Egan et al. 2014). Algal cell wall degrading bacteria are mostly Gram-negative organisms, including phylogenetically diverse genera like *Alteromonas, Cytophaga, Flavobacterium, Pseudoalteromonas, Vibrio*, and *Zobellia* (Hollants et al. 2013), although Gram-positive bacteria such as *Bacillus* sp. has also been reported as key degraders of macroalgae (Zhao and Ruan 2011). Complex metabolic interrelations are established among members of the host-associated microbiota during the degradation of macroalgae as it has been shown for the brown alga *Fucus evanescens* in which *Pseudoalteromonas* sp. play a major role in the degradation of algal polysaccharides, while *Halomonas marina* utilises the degradation products of the former (Ivanova et al. 2002).

Most of these studies have been performed with macroalgae from warmer geographical regions, but to date, very few studies have explored the cell wall degrading capacity of microorganisms associated with Antarctic macroalgae. Heterotrophic organisms, such as fungi, may have a significant effect on the decomposition of algal tissue in Antarctica. It has been found that yeast are abundant in decomposing giant kelp at the sub-Antarctic Macquarie Island (Raghukumar 2017). Loque et al. (2010) suggest that the psychrophilic filamentous fungus *Geomyces pannorum* may play an important role in the decomposition and nutrient cycling of the brown algae *Adenocystis utricularis* and *Desmarestia anceps* in Admiralty Bay. The study of Furbino et al. (2018) highlighted that Antarctic macroalgae are a source of algicolous fungi with the capacity to degrade algal complex polysaccharides. Among the few studies of this type on Antarctic bacteria, Lavin et al. (2013, 2016) reported the characterisation of isolates of *Bacillus* sp. and *Flavobacterium* sp. retrieved from decomposing macroalgae in King George Island, which exhibited agarase and alginate-lyase activities. Working with subtidal macroalgae from King George Island, Sánchez Hinojosa et al. (2018) found a diverse agar-degrading, culturable epiphytic bacterial community, which exhibited hydrolytic activity at low temperatures. Alvarado and Leiva (2017) also reported a diverse community of agar-degrading bacteria isolated from the surface of intertidal macroalgae collected in King George Island, comprising representatives of Gram-negative and Gram-positive bacteria (*Algibacter, Arthrobacter, Brachybacterium, Cellulophaga, Citricoccus, Labedella, Microbacterium, Micrococcus, Salinibacterium, Sanguibacter*, and *Zobellia*). In this study, a substantial fraction of the agarolytic isolates were yellow- and orange pigmented bacteria. Interestingly, microcosm studies of microbial degradation of North Sea macroalgae showed that yellow-pigmented bacteria are primary agents in the macroalgal decomposition (Rieper-Kirchner 1990). All these studies suggest that the Antarctic macroalgal epibiota harbour diverse culturable bacteria capable of degrading and metabolising algal polysaccharides, a function that is central to the carbon cycling in the Antarctic ocean. Finally, Herrera et al. (2017) reported a diverse hydrolytic enzyme-producing microbiota associated with the gut of the Antarctic oligochaete *Grania* sp., which suggests that the gut microbiome of marine invertebrates also contribute to the degradation of macroalgal compounds and nutrient recycling in Antarctica.

Biotechnological Potential of Bacteria Associated With Antarctic Macroalgae

The antimicrobial resistance is a global problem and one of the biggest public health threats in the present century. However, the development of new antibiotics has declined dramatically with only two new classes reaching the market in the last 40 years (Rahimi 2019). Thus, there is an urgent need for innovative antimicrobial research and development of new drugs. In this sense, extreme underexplored environments, such as those present in polar regions, are a promising source of novel antimicrobials. Polar microorganisms are exposed to distinct environmental pressures and have evolved complex and unique metabolic pathways by which they produce unusual secondary metabolites, quite different from many of the classes of antimicrobials currently in use derived from their counterparts from warmer environments (Borchert et al. 2017). In recent years, the screening of Antarctic microorganisms for pharmacologically active compounds has received increasing interest, and a number of bacteria and fungi have been reported to have antimicrobial activity, although in the majority of the cases the active principle responsible for the activity has not been identified yet (Yarzábal 2016). Furthermore, although it is increasingly clear that macroalgae-associated microbes are a major resource for the discovery of novel secondary metabolites with unprecedented chemical structures and a broad range of promising pharmaceutical properties (Singh et al. 2015, Kizhakkekalam and Chakraborty 2019), so far only a few studies have investigated the antimicrobial potential of Antarctic macroalgae-associated microbes. For instance, the works of Godinho et al. (2013) and Furbino et al. (2014) have shown that crude extracts of fungi associated with endemic macroalgae display high and selective antifungal, antiviral, and trypanocidal activities. Alvarado et al. (2018) reported the isolation of different Actinobacteria with antimicrobial activity from Antarctic macroalgae. Fourteen Actinobacteria phylogenetically affiliated to the genera *Agrococcus, Arthrobacter, Micrococcus, Pseudarthrobacter, Pseudonocardia, Sanguibacter, Streptomyces,* and *Tessaracoccus* were active against at least one pathogenic microorganism. Interestingly, isolates of the potential new species of *Pseudonocardia* and *Tessaracoccus* exhibited activity against *Candida albicans,* which suggests that these potentially novel Actinobacteria could represent a source of new antifungal compounds.

Marine pigmented microbes, and bacteria in particular, are recognised as an important source of antimicrobial molecules (Soliev et al. 2011, Ramesh et al. 2019) and bioactive carotenoid pigments (Torregrosa-Crespo et al. 2018). Pigmented bacteria with antimicrobial activity have been isolated from Antarctic samples, such as freshwater lakes (Mojib et al. 2010), marine macroalgae (Leiva et al. 2015), and soil (Asencio et al. 2014). Leiva et al. (2015) reported a diverse community of culturable pigmented Gram-positive bacteria (mostly *Actinobacteria*) associated with the surface of marine Antarctic macroalgae with 16% of the isolates exhibiting antimicrobial activity against a set of macroalgae-associated bacteria. The antagonistic organisms were phylogenetically-related to *Agrococcus baldri, Brachybacterium rhamnosum, Citricoccus zhacaiensis,* and *Kocuria palustris*; these isolates were recovered from the Antarctic macroalgae *Adenocystis utricularis, Iridaea cordata,* and *Monostroma hariotii*. Vila et al. (2019) characterised pigmented bacteria from various types of samples collected from Fildes Peninsula, King George Island, reporting the isolation of a yellow-pigmented *Arthrobacter psychrochitiniphilus* from dry seaweed and pigmented isolates of *Salinibacterium amurskyense* and *Planococcus halocryophilus* from algal mats, being the last two novel sources of C50 carotenoids. Isolates of *S. amurskyense* and *Planococcus* sp. also have been recovered from Antarctic macroalgae (Leiva et al. 2015, Barreto et al. 2017). The C50 carotenoids are synthesised by a restricted number of bacterial species, mainly Actinobacteria, and these compounds exhibit strong antioxidative and reactive oxygen-quenching activities, and effective UV- and visible light filtering properties, which make these pigments very attractive for pharmaceutical and cosmetic applications, such as additives in sunscreen products (Henke et al. 2017). Thus, Antarctic macroalgae are a potential source of epiphytic bacteria producing these rare photoprotective pigments.

Cold-adapted enzymes produced by microorganisms thriving in permanently cold environments possess unique catalytic properties, which make them highly attractive for a variety of industrial applications (Barroca et al. 2017). Different studies have shown that marine and non-marine Antarctic environments represent an important reservoir of cold-adapted, polysaccharide-degrading marine bacteria (Tropeano et al. 2012, Lavin et al. 2013, Tropeano et al. 2013, Ferrés et al. 2015, Lavín et al. 2016, Lamilla et al. 2017). Although it is well established that epibiotic bacteria of marine macroalgae are a good source of carbohydrate-active enzymes (Martin et al. 2014), little is known about the polysaccharide-degrading enzymes produced by bacteria living on the surface of Antarctic marine macroalgae. The studies of Tropeano et al. (2012, 2013) shed some light on the potential of Antarctic macroalgae-associated bacteria as a source of polysaccharide-active enzymes. They investigated the capacity of epiphytic bacteria to degrade complex polysaccharides (agar, carboxymethyl-cellulose, cellulose, pectin, starch, and xylan) and found that isolates of *Pseudoalteromonas* sp. exhibited the broadest range of hydrolytic activities. Ferrés et al. (2015) isolated a high number of cellulose-degrading bacteria from unidentified brown and red seaweeds from Fildes Peninsula; among the cellulolytic bacteria, isolates of *Psychrobacter* sp. also showed lipolytic and ligninolytic activity.

Macroalgae are the most abundant source of commercially important polysaccharides, such as agar, alginate, carrageenan, fucoidan, and ulvan. Marine algae-degrading enzymes (agarases, carrageenanases, alginate lyases, and fucoidanases) are powerful tools for hydrolysing these polysaccharides into functional oligosaccharides with excellent solubility, bioavailability, and a wide range of biological activities, which make them highly attractive for applications in the food biotechnology and healthcare industries (Zhu et al. 2019). Sánchez Hinojosa et al. (2018) reported a diverse, agar-degrading bacterial community associated with the surface of three subtidal Antarctic macroalgae. The phylogenetic analysis performed by 16S rRNA gene sequencing affiliated the agarolytic isolates to the genera *Cellulophaga*, *Colwellia*, *Lacinutrix*, *Olleya*, *Paraglaciecola*, *Pseudoalteromonas* and *Winogradskyella*, with all agarolytic isolates displaying the hydrolytic activity at low temperature on agar plates containing agar as the sole carbon source. Additionally, the authors presented the first report on the purification of an agarase from an epiphytic Antarctic bacteria. They purified a low temperature-active agarase from the culture supernatant of a potential new species of the genus *Olleya*. The purified enzyme showed activity at 4°C, retaining >50% of its maximum activity at this temperature. In turn, Lavin et al. (2013) isolated a strain closely related to *Bacillus cereus* from an Antarctic decaying macroalga, which exhibited the highest agarase activity at 4°C. Taken together, these studies suggest that endemic Antarctic macroalgae can be a valuable source of culturable epiphytic bacteria producing low temperature-active polysaccharidases with enormous potential in biotechnological innovations.

Conclusions

The study of macroalgae-associated bacteria in Antarctica is still very much in its infancy. Data available so far come from studies focused on the epibacterial community of a few macroalgal holobionts occurring in the northern parts of the Antarctic Peninsula, but the bacterial microbiota of most of the endemic macroalgae and the exceptional diversity and bioprospecting potential of the microbiome of seaweeds occurring in the south-western Antarctic Peninsula and East Antarctica remain unexplored. Thus, microbiological bioprospecting efforts in unexplored Antarctic areas certainly merit to be taken into consideration in future research programs. Moreover, no information is available about bacterial endophytes associated with Antarctic macroalgae. Investigations are also needed to understand the complexity of the host-microbiome interactions and their responses to host factors and changing environmental conditions, which may provide interesting models of algal-bacteria interactions under the harsh and variable Antarctic environment. Additionally, more research is needed to characterise the microbial communities associated with decaying macroalgae and thus expand our understanding of the decomposition process of algal biomass in Antarctic ecosystems.

Bacteria associated with Antarctic macroalgae are an untapped source of potential new bioactive compounds and cold-adapted enzymes. Further research utilising innovative cultivation methods and sequence-based and functional metagenomics to the macroalgal holobiont will significantly advance our understanding of the unknown genetic diversity of the uncultured bacterial world associated with Antarctic seaweeds, which will undoubtedly contribute to accelerating the identification of novel bioactive molecules and industrially relevant catalysts.

Conflict of Interests

The author declares no conflict of interest.

Acknowledgements

This work was supported by a grant of Instituto Antartico Chileno (INACH, Grant RT_06-13).

References

Alvarado, P., Y. Huang, J. Wang, I. Garrido and S. Leiva. 2018. Phylogeny and bioactivity of epiphytic Gram-positive bacteria isolated from three co-occurring antarctic macroalgae. Antonie van Leeuwenhoek 111: 1543–1555. doi: 10.1007/s10482-018-1044-6.

Alvarado, R. and S. Leiva. 2017. Agar-degrading bacteria isolated from Antarctic macroalgae. Folia Microbiol. 62: 409–416. doi: 10.1007/s12223-017-0511-1.

Amsler, C.D., J.B. McClintock and B.J. Baker. 2014. Chemical mediation of mutualistic interactions between macroalgae and mesograzers structure unique coastal communities along the western Antarctic Peninsula. J. Phycol. 50: 1–10. doi: 10.1111/jpy.12137.

Arroyo, N.L. and E. Bonsdorff. 2016. The role of drifting algae for marine biodiversity. pp. 100–129. In: E. Ólafsson [ed.]. Marine Macrophytes as Foundation Species. CRC Press, Boca Raton.

Asencio, G., P. Lavin, K. Alegría, M. Domínguez, H. Bello, G. González-Rocha et al. 2014. Antibacterial activity of the Antarctic bacterium Janthinobacterium sp: SMN 33.6 against multi-resistant Gram-negative bacteria. Electron. J. Biotechnol. 17: 1–1.

Barreto, C., R. Duarte, D. Lima, E. de Oliveira Filho, T. Absher and V. Pellizari. 2017. The cultivable microbiota associated with four antarctic macroalgae. Open Access J. Microbiol. Biotechnol. 2: 000122.

Barroca, M., G. Santos, C. Gerday and T. Collins. 2017. Biotechnological aspects of cold-active enzymes. pp. 461–475. In: R. Margesin [ed.]. Psychrophiles: From Biodiversity to Biotechnology. Springer International Publishing, Cham.

Bengtsson, M.M., K. Sjøtun and L. Øvreås. 2010. Seasonal dynamics of bacterial biofilms on the kelp Laminaria hyperborea. Aquat. Microb. Ecol. 60: 71–83. doi: 10.3354/ame01409.

Bianchi, A.C., L. Olazábal, A. Torre and L. Loperena. 2014. Antarctic microorganisms as source of the omega-3 polyunsaturated fatty acids. World J. Microbiol. Biotechnol. 30: 1869–1878. doi: 10.1007/s11274-014-1607-2.

Borchert, E., S.A. Jackson, F. O'Gara and A.D.W. Dobson. 2017. Psychrophiles as a source of novel antimicrobials. pp. 527–540. In: R. Margesin [ed.]. Psychrophiles: From Biodiversity to Biotechnology. Springer International Publishing, Cham.

Bowman, J.P. 1998. Pseudoalteromonas prydzensis sp. nov., a psychrotrophic, halotolerant bacterium from Antarctic sea ice. Int. J. Syst. Evol. Microbiol. 48: 1037–1041. doi: 10.1099/00207713-48-3-1037.

Bowman, J.P. 2000. Description of Cellulophaga algicola sp. nov., isolated from the surfaces of Antarctic algae, and reclassification of Cytophaga uliginosa (ZoBell and Upham 1944) Reichenbach 1989 as Cellulophaga uliginosa comb. nov. Int. J. Syst. Evol. Microbiol. 50: 1861–1868. doi: 10.1099/00207713-50-5-1861.

Bowman, J.P. and D.S. Nichols. 2005. Novel members of the family Flavobacteriaceae from Antarctic maritime habitats including Subsaximicrobium wynnwilliamsii gen. nov., sp. nov., Subsaximicrobium saxinquilinus sp. nov., Subsaxibacter broadyi gen. nov., sp. nov., Lacinutrix copepodicola gen. nov., sp. nov., and novel species of the genera Bizionia, Gelidibacter and Gillisia. Int. J. Syst. Evol. Microbiol. 55: 1471–1486. doi: 10.1099/ijs.0.63527-0.

Bozal, N., E. Tudela, R. Rosselló-Mora, J. Lalucat and J. Guinea. 1997. Pseudoalteromonas antarctica sp. nov., isolated from an Antarctic coastal environment. Int. J. Syst. Evol. Microbiol. 47: 345–351. doi: 10.1099/00207713-47-2-345.

Braeckman, U., F. Pasotti, S. Vázquez, K. Zacher, R. Hoffmann, M. Elvert, et al. 2019. Degradation of macroalgal detritus in shallow coastal Antarctic sediments. Limnol. Oceanogr. 64: 1423–1441. doi: 10.1002/lno.11125.

Brodie, J., C. Williamson, G.L. Barker, R.H. Walker, A. Briscoe and M. Yallop. 2016. Characterising the microbiome of Corallina officinalis, a dominant calcified intertidal red alga. FEMS Microbiol. Ecol. 92: fiw110. doi: 10.1093/femsec/fiw110.

Burke, C., T. Thomas, M. Lewis, P. Steinberg and S. Kjelleberg. 2011. Composition, uniqueness and variability of the epiphytic bacterial community of the green alga Ulva australis. ISME J. 5: 590–600. doi: 10.1038/ismej.2010.164.

Catenazzi, A. and M.A. Donnelly. 2007. Role of supratidal invertebrates in the decomposition of beach-cast green algae *Ulva* sp. Mar. Ecol. Prog. Ser. 349: 33–42. doi: 10.3354/meps07106.

Chen, M.Y. and L.W. Parfrey. 2018. Incubation with macroalgae induces large shifts in water column microbiota, but minor changes to the epibiota of co-occurring macroalgae. Mol. Ecol. 27: 1966–1979. doi: 10.1111/mec.14548.

Clarke, L.J., L. Suter, R. King, A. Bissett and B.E. Deagle. 2019. Antarctic krill are reservoirs for distinct Southern Ocean microbial communities. Front. Microbiol. 9: 3226. doi: 10.3389/fmicb.2018.03226.

Dittami, S.M., L. Duboscq-Bidot, M. Perennou, A. Gobet, E. Corre, C. Boyen, et al. 2016. Host–microbe interactions as a driver of acclimation to salinity gradients in brown algal cultures. ISME J. 10: 51–63. doi: 10.1038/ismej.2015.104.

Duarte, A.W.F., M.R.Z. Passarini, T.P. Delforno, F.M. Pellizzari, C.V.Z. Cipro, R.C. Montone, et al. 2016. Yeasts from macroalgae and lichens that inhabit the South Shetland Islands, Antarctica. Environ. Microbiol. Rep. 8: 874–885. doi: 10.1111/1758-2229.12452.

Egan, S., N.D. Fernandes, V. Kumar, M. Gardiner and T. Thomas. 2014. Bacterial pathogens, virulence mechanism and host defence in marine macroalgae. Environ. Microbiol. 16: 925–938. doi:10.1111/1462-2920.12288.

Fernandes, N., P. Steinberg, D. Rusch, S. Kjelleberg and T. Thomas. 2012. Community structure and functional gene profile of bacteria on healthy and diseased thalli of the red seaweed *Delisea pulchra*. PLoS One 7: e50854. doi: 10.1371/journal.pone.0050854.

Ferrés, I., V. Amarelle, F. Noya and E. Fabiano. 2015. Identification of Antarctic culturable bacteria able to produce diverse enzymes of potential biotechnological interest. Adv. Polar Sci. 26: 71–79. doi: 10.13679/j.advps.2015.1.00071.

Florez, J.Z., C. Camus, M.B. Hengst, F. Marchant and A.H. Buschmann. 2019. Structure of the epiphytic bacterial communities of *Macrocystis pyrifera* in localities with contrasting nitrogen concentrations and temperature. Algal Res. 44: 101706. doi: 10.1016/j.algal.2019.101706.

Franco, D.C., C.N. Signori, R.T.D. Duarte, C.R. Nakayama, L.S. Campos and V.H. Pellizari. 2017. High prevalence of Gammaproteobacteria in the sediments of Admiralty Bay and North Bransfield Basin, Northwestern Antarctic Peninsula. Front. Microbiol. 8: 153. doi: 10.3389/fmicb.2017.00153.

Furbino, L., V. Godinho, I. Santiago, F. Pellizzari, T. Alves, C. Zani, et al. 2014. Diversity patterns, ecology and biological activities of fungal communities associated with the endemic macroalgae across the Antarctic peninsula. Microb. Ecol. 67: 775–787. doi: 10.1007/s00248-014-0374-9.

Furbino, L.E., F.M. Pellizzari, P.C. Neto, C.A. Rosa and L.H. Rosa. 2018. Isolation of fungi associated with macroalgae from maritime Antarctica and their production of agarolytic and carrageenolytic activities. Polar Biol. 41: 527–535. doi: 10.1007/s00300-017-2213-1.

Gillespie, P.A., R.Y. Morita and L.P. Jones. 1976. The heterotrophic activity for amino acids, glucose and acetate in Antarctic waters. J. Oceanogr. Soc. Japan 32: 74–82. doi: 10.1007/bf02107374.

Godinho, V.M., L.E. Furbino, I.F. Santiago, F.M. Pellizzari, N.S. Yokoya, D. Pupo, et al. 2013. Diversity and bioprospecting of fungal communities associated with endemic and cold-adapted macroalgae in Antarctica. ISME J. 7: 1434–1451.

Goecke, F., A. Labes, J. Wiese and J.F. Imhoff. 2010. Chemical interactions between marine macroalgae and bacteria. Mar. Ecol. Prog. Ser. 409: 267–299. doi: 10.3354/meps08607.

Goecke, F., V. Thiel, J. Wiese, A. Labes and J.F. Imhoff. 2013. Algae as an important environment for bacteria – phylogenetic relationships among new bacterial species isolated from algae. Phycologia 52: 14–24. doi: 10.2216/12-24.1.

Gómez, I., A. Wulff, Y. Roleda Michael, P. Huovinen, U. Karsten, M.L. Quartino, et al. 2009. Light and temperature demands of marine benthic microalgae and seaweeds in polar regions. Bot. Mar. 52: 593–608. doi: 10.1515/BOT.2009.073.

Gómez, I., N.P. Navarro and P. Huovinen. 2019. Bio-optical and physiological patterns in Antarctic seaweeds: a functional trait based approach to characterize vertical zonation. Prog. Oceanogr. 174: 17–27. doi: 10.1016/j.pocean.2018.03.013.

González-Aravena, M., R. Urtubia, K. Del Campo, P. Lavín, C.M.V.L. Wong, C.A. Cárdenas, et al. 2016. Antibiotic and metal resistance of cultivable bacteria in the Antarctic sea urchin. Antarct. Sci. 28: 261–268. doi: 10.1017/S0954102016000109.

Hengst, M.B., S. Andrade, B. González and J.A. Correa. 2010. Changes in epiphytic bacterial communities of intertidal seaweeds modulated by host, temporality, and copper enrichment. Microb. Ecol. 60: 282–290. doi: 10.1007/s00248-010-9647-0.

Henke, N.A., P. Peters-Wendisch, V.F. Wendisch and S.A. Heider. 2017. C50 carotenoids: occurrence, biosynthesis, glycosylation, and metabolic engineering for their overproduction. pp. 107–126. *In*: O.V. Singh [ed.]. Bio-pigmentation and Biotechnological Implementations.

Herrera, L.M., C.X. García-Laviña, J.J. Marizcurrena, O. Volonterio, R.P. de León and S. Castro-Sowinski. 2017. Hydrolytic enzyme-producing microbes in the Antarctic oligochaete *Grania* sp. (Annelida). Polar Biol. 40: 947–953. doi: 10.1007/s00300-016-2012-0.

Hollants, J., F. Leliaert, O. De Clerck and A. Willems. 2013. What we can learn from sushi: a review on seaweed–bacterial associations. FEMS Microbiol. Ecol. 83: 1–16. doi: 10.1111/j.1574-6941.2012.01446.x.

Huovinen, P. and I. Gómez. 2013. Photosynthetic characteristics and UV stress tolerance of Antarctic seaweeds along the depth gradient. Polar Biol. 36: 1319–1332. doi: 10.1007/s00300-013-1351-3.

Iken, K., M. Quartino, E. Barrera Oro, J. Palermo, C. Wiencke and T. Brey. 1998. Trophic relations between macroalgae and herbivores in Potter Cove (King George Island, Antarctica). pp. 258–262. *In*: C. Wiencke, G. Ferreyra, W. Arntz and C. Rinaldi [eds.]. The Potter Cove coastal ecosystem, Antarctica. Berichte zur Polarforschung No. 299.

Ivanova, E.P., I.Y. Bakunina, T. Sawabe, K. Hayashi, Y.V. Alexeeva, N.V. Zhukova, et al. 2002. Two species of culturable bacteria associated with degradation of brown algae *Fucus evanescens*. Microb. Ecol. 43: 242–249. doi: 10.1007/s00248-001-1011-y.

Kientz, B., H. Agogué, C. Lavergne, P. Marié and E. Rosenfeld. 2013. Isolation and distribution of iridescent *Cellulophaga* and other iridescent marine bacteria from the Charente-Maritime coast, French Atlantic. Syst. Appl. Microbiol. 36: 244–251. doi: 10.1016/j.syapm.2013.02.004.

Kizhakkekalam, V.K. and K. Chakraborty. 2019. Pharmacological properties of marine macroalgae-associated heterotrophic bacteria. Arch. Microbiol. 201: 505–518. doi: 10.1007/s00203-018-1592-1.

Krieg, N.R., W. Ludwig, J. Euzéby and W.B. Whitman. 2010. Phylum XIV. Bacteroidetes phyl. nov. pp. 25–469. *In*: N.R. Krieg, J.T. Staley, D.R. Brown, B.P. Hedlund, B.J. Paster, N.L. Ward, W. Ludwig and W.B. Whitman [eds.]. Bergey's Manual® of Systematic Bacteriology: Volume Four The Bacteroidetes, Spirochaetes, Tenericutes (Mollicutes), Acidobacteria, Fibrobacteres, Fusobacteria, Dictyoglomi, Gemmatimonadetes, Lentisphaerae, Verrucomicrobia, Chlamydiae, and Planctomycetes. Springer New York, New York, NY.

Krumhansl, K.A. and R.E. Scheibling. 2012. Detrital subsidy from subtidal kelp beds is altered by the invasive green alga *Codium fragile* ssp. *fragile*. Mar. Ecol. Prog. Ser. 456: 73–85. doi: 10.3354/meps09671.

Lachnit, T., D. Meske, M. Wahl, T. Harder and R. Schmitz. 2011. Epibacterial community patterns on marine macroalgae are host-specific but temporally variable. Environ. Microbiol. 13: 655–665. doi: 10.1111/j.1462-2920.2010.02371.x.

Lafleur, J.E., S.K. Costa, A.S. Bitzer and M.W. Silby. 2015. Draft genome sequence of *Cellulophaga* sp. E6, a marine algal epibiont that produces a quorum-sensing inhibitory compound active against *Pseudomonas aeruginosa*. Genome Announc. 3: e01565–14. doi: 10.1128/genomeA.01565-14.

Lamilla, C., M. Pavez, A. Santos, A. Hermosilla, V. Llanquinao and L. Barrientos. 2017. Bioprospecting for extracellular enzymes from culturable Actinobacteria from the South Shetland Islands, Antarctica. Polar Biol. 40: 719–726. doi: 10.1007/s00300-016-1977-z.

Lavin, P., J. Gallardo-Cerda, C. Torres-Diaz, G. Asencio and M. Gonzalez. 2013. Antarctic strain of *Bacillus* sp. with extracellular agarolitic and alginate-lyase activities. Gayana 77: 75–82. doi: 10.4067/S0717-65382013000200001.

Lavín, P., C. Atala, J. Gallardo-Cerda, M. Gonzalez-Aravena, R. De La Iglesia, R. Oses, et al. 2016. Isolation and characterization of an Antarctic *Flavobacterium* strain with agarase and alginate lyase activities. Pol. Polar Res. 37: 403–419. doi: 10.1515/popore-2016-0021.

Leiva, S., P. Alvarado, Y. Huang, J. Wang and I. Garrido. 2015. Diversity of pigmented Gram-positive bacteria associated with marine macroalgae from Antarctica. FEMS Microbiol. Lett. 362: fnv206. doi: 10.1093/femsle/fnv206.

Li, Z., M. Xing, W. Wang, D. Wang, J. Zhu and M. Sun. 2016. Phylogenetic diversity of culturable bacteria in surface seawater from the Drake Passage, Antarctica. Chin. J. Oceanol. Limn. 34: 952–963. doi: 10.1007/s00343-016-5132-z.

Lo Giudice, A. and C. Rizzo. 2018. Bacteria associated with marine benthic invertebrates from polar environments: unexplored frontiers for biodiscovery? Diversity 10: 80. doi: 10.3390/d10030080.

Loque, C.P., A.O. Medeiros, F.M. Pellizzari, E.C. Oliveira, C.A. Rosa and L.H. Rosa. 2010. Fungal community associated with marine macroalgae from Antarctica. Polar Biol. 33: 641–648. doi: 10.1007/s00300-009-0740-0.

Martin, M., D. Portetelle, G. Michel and M. Vandenbol. 2014. Microorganisms living on macroalgae: diversity, interactions, and biotechnological applications. Appl. Microbiol. Biotechnol. 98: 2917–2935. doi: 10.1007/s00253-014-5557-2.

Marzinelli, E.M., Z. Qiu, K.A. Dafforn, E.L. Johnston, P.D. Steinberg and M. Mayer-Pinto. 2018. Coastal urbanisation affects microbial communities on a dominant marine holobiont. NPJ Biofilms Microbiomes 4: 1. doi: 10.1038/s41522-017-0044-z.

Mayali, X. and F. Azam. 2004. Algicidal bacteria in the sea and their impact on algal blooms. J. Eukaryot. Microbiol. 51: 139–144. doi: 10.1111/j.1550-7408.2004.tb00538.x.

Melo, I., E. Vilela, L. Fleuri, A. Araujo and V. Pellizari. 2009. Lipolytic activity of bacteria associated with the antarctic macroalga *Palmaria decipiens* (Reinsch) RW Ricker. XVII Simpósio Brasileiro sobre Pesquisa Antártica (XVII SBPA), São Paulo. p. 85–87.

Mojib, N., R. Philpott, J.P. Huang, M. Niederweis and A.K. Bej. 2010. Antimycobacterial activity *in vitro* of pigments isolated from Antarctic bacteria. Antonie van Leeuwenhoek 98: 531–540. doi: 10.1007/s10482-010-9470-0.

Nedzarek, A. and S. Rakusa-Suszczewski. 2004. Decomposition of macroalgae and the release of nutrient Admiralty Bay, King George, Antarctica. Polar Biosci. 17: 26–35.

Ogaki, M.B., M.T. de Paula, D. Ruas, F.M. Pellizzari, C.X. García-Laviña and L.H. Rosa. 2019. Marine fungi associated with Antarctic macroalgae. pp. 239–255. *In*: S. Castro-Sowinski [ed.]. The Ecological Role of Micro-organisms in the Antarctic Environment. Springer International Publishing, Cham, Switzerland.

Papaleo, M.C., M. Fondi, I. Maida, E. Perrin, A. Lo Giudice, L. Michaud, et al. 2012. Sponge-associated microbial Antarctic communities exhibiting antimicrobial activity against *Burkholderia cepacia* complex bacteria. Biotechnol. Adv. 30: 272–293. doi: 10.1016/j.biotechadv.2011.06.011.

Parrilli, E., P. Tedesco, M. Fondi, M.L. Tutino, A. Lo Giudice, D. de Pascale, et al. 2019. The art of adapting to extreme environments: the model system *Pseudoalteromonas*. Phys. Life Rev.: doi: 10.1016/j.plrev.2019.04.003.

Pellizzari, F., M.C. Silva, E.M. Silva, A. Medeiros, M.C. Oliveira, N.S. Yokoya, et al. 2017. Diversity and spatial distribution of seaweeds in the South Shetland Islands, Antarctica: an updated database for environmental monitoring under climate change scenarios. Polar Biol. 40: 1671–1685. doi: 10.1007/s00300-017-2092-5.

Quartino, M., H. Klöser, I. Schloss and C. Wiencke. 2001. Biomass and associations of benthic marine macroalgae from the inner Potter Cove (King George Island, Antarctica) related to depth and substrate. Polar Biol. 24: 349–355. doi: 10.1007/s003000000218.

Quartino, M.L. and A.L. Boraso de Zaixso. 2008. Summer macroalgal biomass in Potter Cove, South Shetland Islands, Antarctica: its production and flux to the ecosystem. Polar Biol. 31: 281–294. doi: 10.1007/s00300-007-0356-1.

Quartino, M.L., S.C. Vazquez, G.E.J. Latorre and W.P. Mac Cormack. 2015. Possible role of bacteria in the degradation of macro algae *Desmarestia anceps* Montagne (Phaeophyceae) in Antarctic marine waters. Rev. Argent. Microbiol. 47: 274–276. doi: 10.1016/j.ram.2015.04.003.

Raghukumar, S. 2017. The Macroalgal Ecosystem. pp. 115–141. Fungi in Coastal and Oceanic Marine Ecosystems: Marine Fungi. Springer International Publishing, Cham.

Rahimi, S. 2019. Urgent action on antimicrobial resistance. Lancet. Respir. Med. 7: 208–209. doi: 10.1016/S2213-2600(19)30031-1.

Ramesh, C., N. Vinithkumar and R. Kirubagaran. 2019. Marine pigmented bacteria: a prospective source of antibacterial compounds. J. Nat. Sc. Biol. Med. 10: 104–113. doi: 10.4103/jnsbm.JNSBM_201_18.

Rao, D., J.S. Webb and S. Kjelleberg. 2006. Microbial colonization and competition on the marine alga *Ulva australis*. Appl. Environ. Microbiol. 72: 5547–5555. doi: 10.1128/aem.00449-06.

Reichardt, W. and G. Dieckmann. 1985. Kinetics and trophic role of bacterial degradation of macro-algae in Antarctic coastal waters. pp. 115–122. *In*: W.R. Siegfried, P.R. Condy and R.M. Laws [eds.]. Antarctic Nutrient Cycles and Food Webs. Springer Berlin Heidelberg.

Rieper-Kirchner, M. 1990. Macroalgal decomposition: Laboratory studies with particular regard to microorganisms and meiofauna. Helgoländer Meeresun 44: 397–410. doi: 10.1007/bf02365476.

Roleda, M.Y., K. Zacher, A. Wulff, D. Hanelt and C. Wiencke. 2007. Photosynthetic performance, DNA damage and repair in gametes of the endemic Antarctic brown alga *Ascoseira mirabilis* exposed to ultraviolet radiation. Austral Ecol. 32: 917–926. doi: 10.1111/j.1442-9993.2007.01796.x.

Salaün, S., N. Kervarec, P. Potin, D. Haras, M. Piotto and S. La Barre. 2010. Whole-cell spectroscopy is a convenient tool to assist molecular identification of cultivable marine bacteria and to investigate their adaptive metabolism. Talanta. 80: 1758–1770. doi: 10.1016/j.talanta.2009.10.020.

Sánchez Hinojosa, V., J. Asenjo and S. Leiva. 2018. Agarolytic culturable bacteria associated with three antarctic subtidal macroalgae. World J. Microbiol. Biotechnol. 34: 73. doi: 10.1007/s11274-018-2456-1.

Serebryakova, A., T. Aires, F. Viard, E.A. Serrão and A.H. Engelen. 2018. Summer shifts of bacterial communities associated with the invasive brown seaweed *Sargassum muticum* are location and tissue dependent. PLoS One 13: e0206734. doi: 10.1371/journal.pone.0206734.

Silva, T.R., A.W.F. Duarte, M.R.Z. Passarini, A.L.T.G. Ruiz, C.H. Franco, C.B. Moraes, et al. 2018. Bacteria from Antarctic environments: diversity and detection of antimicrobial, antiproliferative, and antiparasitic activities. Polar Biol. 41: 1505–1519. doi: 10.1007/s00300-018-2300-y.

Singh, J., R.P. Singh and R. Khare. 2018. Influence of climate change on Antarctic flora. Polar Sci. 18: 94–101. doi: 10.1016/j.polar.2018.05.006.

Singh, R.P. and C.R.K. Reddy. 2014. Seaweed–microbial interactions: key functions of seaweed-associated bacteria. FEMS Microbiol. Ecol. 88: 213–230. doi: 10.1111/1574-6941.12297.

Singh, R.P., P. Kumari and C.R.K. Reddy. 2015. Antimicrobial compounds from seaweeds-associated bacteria and fungi. Appl. Microbiol. Biotechnol. 99: 1571–1586. doi: 10.1007/s00253-014-6334-y.

Singh, R.P. and C.R.K. Reddy. 2016. Unraveling the functions of the macroalgal microbiome. Front. Microbiol. 6: 1488. doi: 10.3389/fmicb.2015.01488.

Soliev, A.B., K. Hosokawa and K. Enomoto. 2011. Bioactive pigments from marine bacteria: applications and physiological roles. Evid. Based Complement. Alternat. Med. 2011: 670349. doi: 10.1155/2011/670349.

Stingl, U., J.-C. Cho, W. Foo, K.L. Vergin, B. Lanoil and S.J. Giovannoni. 2008. Dilution-to-extinction culturing of psychrotolerant planktonic bacteria from permanently ice-covered lakes in the McMurdo Dry Valleys, Antarctica. Microb. Ecol. 55: 395–405. doi: 10.1007/s00248-007-9284-4.

Stratil, S.B., S.C. Neulinger, H. Knecht, A.K. Friedrichs and M. Wahl. 2013. Temperature-driven shifts in the epibiotic bacterial community composition of the brown macroalga *Fucus vesiculosus*. Microbiologyopen 2: 338–349. doi: 10.1002/mbo3.79.

Stratil, S.B., S.C. Neulinger, H. Knecht, A.K. Friedrichs and M. Wahl. 2014. Salinity affects compositional traits of epibacterial communities on the brown macroalga *Fucus vesiculosus*. FEMS Microbiol. Ecol. 88: 272–279. doi: 10.1111/1574-6941.12292.

Tian, S., H. Jin, S. Gao, Y. Zhuang, Y. Zhang, B. Wang, et al. 2015. Sources and distribution of particulate organic carbon in Great Wall Cove and Ardley Cove, King George Island, West Antarctica. Adv. Polar Sci. 26: 55–62. doi: 10.13679/j.advps.2015.1.00055.

Torregrosa-Crespo, J., Z. Montero, J.L. Fuentes, M. Reig García-Galbis, I. Garbayo, C. Vílchez, et al. 2018. Exploring the valuable carotenoids for the large-scale production by marine microorganisms. Mar. Drugs 16: 203. doi: 10.3390/md16060203.

Tropeano, M., S. Coria, A. Turjanski, D. Cicero, A. Bercovich, W. Mac Cormack, et al. 2012. Culturable heterotrophic bacteria from Potter Cove, Antarctica, and their hydrolytic enzymes production. Polar Res. 31: 18507. doi: 10.3402/polar.v31i0.18507.

Tropeano, M., S. Vazquez, S. Coria, A. Turjanski, D. Cicero, A. Bercovich, et al. 2013. Extracellular hydrolytic enzyme production by proteolytic bacteria from the Antarctic. Pol. Polar Res. 34: 253–267. doi: 10.2478/popore-2013-0014.

van der Loos, L.M., B.K. Eriksson and J. Falcão Salles. 2019. The macroalgal holobiont in a changing sea. Trends Microbiol. 27: 635–650. doi: 10.1016/j.tim.2019.03.002.

Vazquez, S.C. and W.P. Mac Cormack. 2002. Effect of isolation temperature on the characteristics of extracellular proteases produced by Antarctic bacteria. Polar Res. 21: 63–71. doi: 10.3402/polar.v21i1.6474.

Vila, E., D. Hornero-Méndez, G. Azziz, C. Lareo and V. Saravia. 2019. Carotenoids from heterotrophic bacteria isolated from Fildes Peninsula, King George Island, Antarctica. Biotechnol. Rep. 21: e00306. doi: 10.1016/j.btre.2019.e00306.

Wahl, M., L. Shahnaz, S. Dobretsov, M. Saha, F. Symanowski, K. David, et al. 2010. Ecology of antifouling resistance in the bladder wrack *Fucus vesiculosus*: patterns of microfouling and antimicrobial protection. Mar. Ecol. Prog. Ser. 411: 33–48. doi: 10.3354/meps08644.

Wahl, M., F. Goecke, A. Labes, S. Dobretsov and F. Weinberger. 2012. The second skin: ecological role of epibiotic biofilms on marine organisms. Front Microbiol. 3: 292. doi: 10.3389/fmicb.2012.00292.

Wang, J., S. Leiva, J. Huang and Y. Huang. 2018. *Amycolatopsis antarctica* sp. nov., isolated from the surface of an Antarctic brown macroalga. Int. J. Syst. Evol. Microbiol. 68: 2348–2356. doi: 10.1099/ijsem.0.002844.

Wang, L., X. Liu, S. Yu, X. Shi, X. Wang and X.-H. Zhang. 2017. Bacterial community structure in intertidal sediments of Fildes Peninsula, maritime Antarctica. Polar Biol. 40: 339–349. doi: 10.1007/s00300-016-1958-2.

Wang, X.-J., L. Xu, N. Wang, H.-M. Sun, X.-L. Chen, Y.-Z. Zhang, et al. 2019. *Putridiphycobacter roseus* gen. nov., sp. nov., isolated from Antarctic rotten seaweed. Int. J. Syst. Evol. Microbiol.: doi: 10.1099/ijsem.0.003809.

Webster, N.S. and D. Bourne. 2007. Bacterial community structure associated with the Antarctic soft coral, *Alcyonium antarcticum*. FEMS Microbiol. Ecol. 59: 81–94. doi: 10.1111/j.1574-6941.2006.00195.x.

Wichard, T. 2015. Exploring bacteria-induced growth and morphogenesis in the green macroalga order Ulvales (Chlorophyta). Front. Plant Sci. 6: 86. doi: 10.3389/fpls.2015.00086.

Wiencke, C., J. Rahmel, U. Karsten, G. Weykam and G.O. Kirst. 1993. Photosynthesis of marine macroalgae from Antarctica: light and temperature requirements. Bot. Acta 106: 78–87. doi: 10.1111/j.1438-8677.1993.tb00341.x.

Wiencke, C. and C.D. Amsler. 2012. Seaweeds and their communities in polar regions. pp. 265–291. *In*: C. Wiencke and K. Bischof [eds.]. Seaweed Biology: Novel Insights into Ecophysiology, Ecology and Utilization. Springer Berlin Heidelberg, Berlin, Heidelberg.

Wiencke, C., C. Amsler and M. Clayton. 2014. Macroalgae. pp. 66–73. *In*: C. De Broyer and P. Koubbi [eds.]. Biogeographic Atlas of the Southern Ocean. Scientific Committee on Antarctic Research.

Yarzábal, L.A. 2016. Antarctic psychrophilic microorganisms and biotechnology: history, current trends, applications, and challenges. pp. 83–118. *In*: S. Castro-Sowinski [ed.]. Microbial Models: From Environmental to Industrial Sustainability. Springer Singapore, Singapore.

Zdanowski, M.K. 1998. Factors regulating bacterial abundance in Antarctic coastal and shelf waters. Pol. Polar Res. 19: 169–186.

Zhao, C. and L. Ruan. 2011. Biodegradation of *Enteromorpha prolifera* by mangrove degrading micro-community with physical–chemical pretreatment. Appl. Microbiol. Biotechnol. 92: 709–716. doi: 10.1007/s00253-011-3384-2.

Zhu, B., L. Ning, Y. Sun and Z. Yao. 2019. Chapter 19 – Marine algae–degrading enzymes and their applications in marine oligosaccharide preparation. pp. 417–447. *In*: A. Trincone [ed.]. Enzymatic technologies for marine polysaccharides. CRC Press, Boca Raton.

Zielinski, K. 1990. Bottom marcroalgae of the Admiralty Bay (King George Island, South Shetlands, Antarctica). Pol. Polar Res. 11: 95–131.

10

Antarctic Bacteria in Microbial Mats From King George Island, Maritime Antarctica

Rocío J. Alcántara-Hernández[1,*], Luisa I. Falcón[2], Neslihan Tas[3], Patricia M. Valdespino-Castillo[4,5], Silvia Batista[5], Martin Merino-Ibarra[6] and Julio E. Campo[7]

Introduction

Microbial mats are highly diverse and physiologically active consortia that are often found in glaciers and snow melting water streams of polar environments, representing one of the most active microbial associations in these inland waters (de los Ríos et al. 2004, Fernández-Valiente et al. 2007, Alcántara-Hernández et al. 2014). Their activity and development in polar environments are restricted by the low temperature and solar irradiance, presenting higher growth rates during summer (Sabbe et al. 2004). Microbial mats are formed by diverse microorganisms, mainly Bacteria, embedded in an extracellular polymeric matrix, building complex ecological interactions and metabolic networks along several environmental gradients, including light, oxygen concentration and pH (Vincent et al. 1993, Fernández-Valiente et al. 2007, Komárek and Komárek 2010, Stal 2012), and available nutrients such as nitrogen (N) and phosphorus (P) (Valdespino-Castillo et al. 2018). As a result, their microbial diversity and metabolic coupling favor their successful establishment in extreme environments, such as inland polar ecosystems, where they represent one of the most prosperous bacterial consortia in periodic water biotopes and meltwater ponds in glacier zones (McKnight et al. 2004, Fernández-Valiente et al. 2007, Jungblut et al. 2010, Komárek and Komárek 2010).

The main core composition of microbial mats includes Bacteria, mainly Cyanobacteria and Proteobacteria, Eukarya, comprising green algae, stramenopiles, alveolates, fungi, and nematodes, and Archaea (Fernández-Valiente et al. 2001, Jungblut et al. 2009, Verleyen et al. 2010, Varin et al. 2012, Tytgat et al. 2014, Callejas et al. 2018). Non-surprisingly, there are several studies on Antarctic microbial mats, most of which have focused on lake and shallow pond environments located in Continental Antarctica, including Dry-Valleys, while fewer studies have focused on

[1] Instituto de Geología, Universidad Nacional Autónoma de México, CdMx, Mexico 04510.
[2] Laboratorio de Ecología Bacteriana, Instituto de Ecología, Universidad Nacional Autónoma de México, Parque Científico y Tecnológico de Yucatán, Sierra Papacal, Mexico 97302.
[3] Lawrence Berkeley National Laboratory, EESA Department. Berkeley, CA USA 94720.
[4] Lawrence Berkeley National Laboratory, BSISB, Molecular Biophysics and Integrated Bioimaging Department. Berkeley, CA USA 94720.
[5] Grupo Microbiología Molecular-BIOGEM, Instituto de Investigaciones Biológicas Clemente Estable, Montevideo, Uruguay.
[6] Unidad Académica de Ecología Marina, Instituto de Ciencias del Mar y Limnología, Universidad Nacional Autónoma de México, CdMx., Mexico 04510.
[7] Laboratorio de Biogeoquímica Terrestre y Clima, Instituto de Ecología, Universidad Nacional Autónoma de México, CdMx., Mexico 04510.
Instituto de Geología, Universidad Nacional Autónoma de México, Ciudad Universitaria, Av. Universidad 3000, Del. Coyoacán, 04510, Ciudad de México, Mexico.
* Corresponding author: ralcantarah@geologia.unam.mx, rocio.alcantara.h@gmail.com

microbial mats in Maritime Antarctica (Fernández-Valiente et al. 2007, Callejas et al. 2011, 2018, Valdespino-Castillo et al. 2018). These two biogeographic entities have been long recognized as different ecoregions, showing high degrees of regional endemism (Convey et al. 2008).

The South Shetland Islands are in the northern region of Maritime Antarctica and represent a group of more than 20 islands between the Drake Passage and Bransfield Sea. These landmasses concentrate most of the research activities and scientific stations in Antarctica. Maritime Antarctica is highly affected by global climate change, showing rapid temperature increases during the last fifty years (Turner et al. 2005, Turner et al. 2007, Vaughan et al. 2013), which have caused intense glacier meltdown. Fildes Peninsula, within the King George Island, has a large number of scientific stations, including an aerodrome, and is limited to the north by the receding Bellingshausen Glacier Dome. During the austral summer, this area is essentially ice-free with the presence of temporal water biotopes formed by the melting of snowfields and glacier zones, which increases the inland water's availability.

Despite ecological, climatic, and anthropic alterations within the Fildes Peninsula, there are only a handful of studies describing its microbial diversity. Culture-dependent and culture-independent studies have found psychrotrophic and psychrotolerant Bacteria in soil and sediments, mainly Proteobacteria, Actinobacteria, Acidobacteria, Bacteroidetes, and Firmicutes (Foong et al. 2010, Fan et al. 2013, Wang et al. 2015), and fungi, such as *Geomyces pannorum* and *Mrakia frigida* (Krishnan et al. 2011). Studies have also assessed the cyanobacterial component of microbial mats in the Peninsula. Komárek and Komárek (2010) propose that in the early stage of establishment of the community, filamentous cyanobacteria, including *Phormidium commune* and *Leptolyng biaantarctica*, trichomes start to grow (Komárek and Komárek 2010, Callejas et al. 2011). Also, coccoid cyanobacteria, *Gloeocapsopsis aurea*, have been described in films and crusts on the rocks (Mataloni and Komárek 2004). However, there is not enough information regarding the genetic and functional bacterial diversity associated with these mats (Alcántara-Hernández et al. 2014).

Relevant metabolic features of microbial mats are their carbon dioxide (C-CO_2) and nitrogen (N-N_2) fixation capacities. Both processes represent important sources of C and N to maintain the microbial activity in Antarctic oligotrophic limnetic ecosystems and are mainly carried out by Cyanobacteria and Proteobacteria (van Gemerden 1993, Joye and Lee 2004, Stal 2012). Microorganisms involved in N_2-fixation have traditionally been surveyed by culture-independent methods using *nifH* genes, which encode for the nitrogenase Fe subunit (Zehr and McReynolds 1989, Zani et al. 2000). Cyanobacteria also fuel ATP demands for N_2-fixation by phototrophic activity. In the low ammonium (NH_4^+) concentrations, cyanobacteria assimilate N primarily through the glutamine synthetase (GS)–glutamine:2-oxoglutarate amido-transferase (GOGAT) pathway (GS-GOGAT) (Flores and Herrero 1994). The GS enzyme is encoded by the *glnA* gene, a monophyletic fragment that resembles the 16S rRNA phylogeny, and can be used as a molecular marker to identify the cyanobacterial counterpart in environmental samples (Gibson et al. 2006).

In this chapter, we describe the bacterial diversity of five microbial mats that develop in glacial melt streams and coastal plains in the Fildes Peninsula of King George Island in Maritime Antarctica using an Illumina high-throughput 16S rRNA sequencing approach. We also characterize the diazotrophic diversity (*nifH* genes) and suggest the potential of cyanobacteria in N-assimilation (*glnA* gene). We hypothesize that there will be local changes in the genetic structure of the microbial mats studied in the Fildes Peninsula and overall genetic potential for N_2-fixation and N-assimilation.

Material and Methods

Sampling Sites

Microbial mat samples were collected during an expedition organized by the *Instituto Antártico Uruguayo* to King George Island (January 2012; 62°020 S, 58°210 W) and five microbial mats in different geographical locations in the Fildes Peninsula were studied (Figure 1). Site description and microbial mat biomass sampling were further explained in Alcántara-Hernández et al. (2014). Briefly, Sites 1 and 2 (Mat1, Mat2) were located near the Artigas Station (Uruguay, *ca.* 20 m) and the Great Wall Station (China, *ca.* 30 m), respectively. Site 3 (Mat3) was isolated geographically by abrupt slopes facing the Maxwell Bay. Sites 4 and 5 (Mat4, Mat5) were *ca.* 500 m from the Drake Passage coastline. To characterize the environmental context of the microbial mats, samples of stream water and sediments were also collected.

Microbial mats from each location were collected in three subsamples of *ca.* 10 g, placed in sterile 50 ml Falcon tubes and kept frozen until analysis. Overlying water was collected in acid-washed 60 mL plastic bottles after filtering through 0.45 μm nylon membranes and kept in the dark at 4°C until analysis to determine inorganic dissolved forms. For sediment collection, a clean sterile spatula was used and *ca.* 200 g (wet weight) of upper sediment material (0-5 cm depth), located below the microbial mat, were placed in sterile plastic bags and kept at 4°C until analysis.

Water and Sediment Analyzes

The pH and temperature of the streams where microbial mats were collected were measured *in situ* with a 6231N pH/mV/Temp meter (Jenco Instruments). A Skalar SAN++ segmented flow autoanalyzer (Netherlands) was used to determine NH_4^+, nitrate (NO_3^-), and nitrite (NO_2^-) in field-collected water samples using the standard methods adapted by Grasshoff et al. (1983) and the circuits suggested by Kirkwood (1994).

The sediment samples were sieved in the laboratory (to pass a 2 mm mesh) and a subsample dried at constant weight for water content determination. The remaining sample was used to measure organic C and total N concentrations. Samples were used to determine pH (in pore water) and available N and P concentrations. Mat and sediment C were analyzed in an automated C-analyzer (SHIMADZU 5005A), after grinding a 5 g air-dried subsample screened in a 100 mesh. The concentration of total N was determined from acid digestion in H_2SO_4 concentrated using a NP elemental analyzer (Technicon Autoanalyzer II). Mineral N concentrations (NO_3^-–N + NH_4^+–N) were measured by extracting a 15 g subsample of each sediment sample in 100 ml 2 M KCl (Robertson et al. 1999). The sediment KCl solution was shaken for 1 hour and allowed to settle overnight. A 20 mL aliquot supernatant was transferred into sample vials and frozen for later analysis. The analysis of N from the mineral matrix was done in an autoanalyzer system using procedures to determine NO_3^-–N + NO_2^-–N, which were reported as NO_3^-–N, and the salicylate-hypochlorite procedure for NH_4^+–N. Phosphorus concentrations were determined using a modified Hedley soil P-fractionation method (see Lajtha et al. 1999). The fractionation scheme included the extraction of readily solubilized P-inorganic plus readily mineralized P-organic (available P). Total P (Pi plus Po) in the extract was determined using an acidified (H_2SO_4), ammonium-persulfate digestion (45 minutes) in an autoclave. All extracts were neutralized and diluted prior to P content analysis. The P concentration in all solutions and standards was measured colorimetrically using an autoanalyzer. All chemical determinations were duplicated, rejecting those with a more than 10% difference among them.

DNA Extraction, Amplification and Sequencing

Metagenomic DNA from the microbial mats was extracted following a combined mechanical, chemical, and enzymatic lysis process, explained in detail in Alcántara-Hernández et al. (2014). DNA pellets were further purified using the DNeasy Blood & Tissue kit (Qiagen, Valencia, CA) according to manufacturer's instructions; the eluted DNA was then precipitated and centrifuged to obtain a pellet. After washing with ethanol (80%) and removing residual ethanol by air-drying, pellets were resuspended in 30 μL molecular grade water and kept at –20°C until analysis.

For insights into total bacterial composition, a high-throughput microbial analysis on an Illumina MiSeq platform was employed. The 16S rRNA gene community sequencing was done using primers 515F/806R (region V4) including Bacteria and Archaea by following described protocols

(Caporaso et al. 2010, Caporaso et al. 2012). Subsamples were treated as separate samples, thus each PCR reaction contained a specific Golay reverse primer (Caporaso et al. 2010). Amplicons were purified and sequenced (~20 ng per sample) on an Illumina MiSeq platform (Yale Center for Genome Analysis, CT, USA), resulting in ~250 bp paired-end reads.

For the targeted-primer analysis on the mat nitrogen fixation potential, a cloning-sequencing approach was used employing *nifH* and *glnA* genes as molecular markers. For *nifH*, a nested PCR protocol was used with the primers and PCR conditions reported by Zani et al. (2000). The first amplification round was done with nif4:nif3 primers (Zani et al. 2000) and then followed by a second round with primers nif1:nif2 (Zehr and McReynolds 1989). For *glnA*, a reported nested protocol was also followed (Gibson et al. 2006). Primer pairs GLNcdhpCOF:GLNcdhpGOR and GLNcdhpCIF:GLNcdhpGIR were employed sequentially as suggested. PCR mixture compositions were done depending on the targeted gene following the recommendations of Zani et al. (2000) and Gibson et al. (2006), while DNA polymerization was carried out by Taq DNA polymerase (Invitrogen) in the appropriate PCR buffer provided by the supplier (Invitrogen Life Technologies, Sao Paulo, Brazil).

The obtained PCR products (*nifH* ~360 bp; *glnA* ~700 bp) were inserted into the pCR®2.1 vector using the Original TA Cloning Kit (Invitrogen, Carlsbad, CA) according to the manufacturer's instructions. Chemically competent *E. coli* DHα5 cells were transformed with the constructed vectors and positive clones were selected by α-complementation on LB plates containing ampicillin and 5-bromo-4-chloro-3-indolyl-β-D-galactopyranoside. An ABI 3730xl DNA analyzer (Applied Biosystems, CA, USA) was used for sequencing using M13 primers.

Sequence Analysis

For the 16S rRNA gene community sequencing data, the paired-end sequences were overlapped and merged using FLASH (Magoč and Salzberg 2011). Quality filtering and demultiplexing were performed as previously reported (Caporaso et al. 2012, Bokulich et al. 2013). Operational taxonomic units (OTUs) were defined at 97% sequence identity, and chimeric sequences were detected and removed with USEARCH (Edgar et al. 2011). Taxonomic assignments were done with QIIME (Caporaso et al. 2010) v.1.7.0 using the RDP classifier (Wang et al. 2007) and the Greengenes database (release 13_5). The 16S rRNA sequence data are available from NCBI (SUB1674542: KX703042-KX707435).

For *nifH* and *glnA*, the obtained nucleotide sequences were translated to amino acids using the SeaView program v. 4.2.12 (Gouy et al. 2010), and pseudogenes were removed after detecting unexpected stop codons on the putative reading frame. Chimeras were also detected using USEARCH 6.0 (Edgar et al. 2011) and OTUs assigned with MOTHUR using the furthest-neighbor algorithm to collapse similar nucleotide sequences (Schloss et al. 2009), considering a 5% nucleotide sequence difference cut-off level for *nifH* (Gaby and Buckley 2011) and 2% for *glnA* (Gibson et al. 2006). The assigned OTUs were used for phylogenetic analysis, while reference sequences were selected by comparison with entries in databases using the Standard Nucleotide BLAST (BLASTN 2.2.27, February 2017) (Zhang et al. 2000). Sequence data were deposited in the GenBank under accession numbers KX650296-KX650351.

Statistical Analysis

The microbial diversity of each mat was analyzed after pooling the DNA from each sub replicate. Therefore, each sample represents the local diversity indicated as Mat1-Mat5. To observe major phyla abundance, a heatmap was done using the 'aheatmap' function in the 'NMF' package (Gaujoux and Seoighe 2010). In addition to this, this function uses 'hclust' to cluster the samples based on a dissimilarity structure produced by dist ('stats' package). This clustering analysis was further confirmed with a relative abundance matrix with Bray-Curtis dissimilarity index with the functions 'vegdist' from 'vegan' package (Oksanen 2011) in R 2.15.3 (R Development Core Team 2012). The optimal number of groups (clusters) was determined with the Calinski criterion (Gordon 1999), and two groups were created after clustering. The Shannon-Weaver diversity index was calculated using the vegan R-package with the formula $H = -sum\ p_i\ log(b)\ p_i$.

The datasets from sediments were subjected to a one-way analysis of variance, testing the effect of the sampling site. The honest significant difference (HSD) test of the means was used when statistical differences ($P < 0.05$) were observed among sites.

Results

The Physicochemical Environment

The microbial mats analyzed were sampled in five sites of the Fildes Peninsula in King George Island (Maritime Antarctica), during the summer of 2012 (Figure 1). Mat1, Mat2, and Mat3 were sampled in snow and glacial meltwater streams, while Mat4 and Mat5 were obtained from a coastal plain environment. Mat growth was well established in all cases.

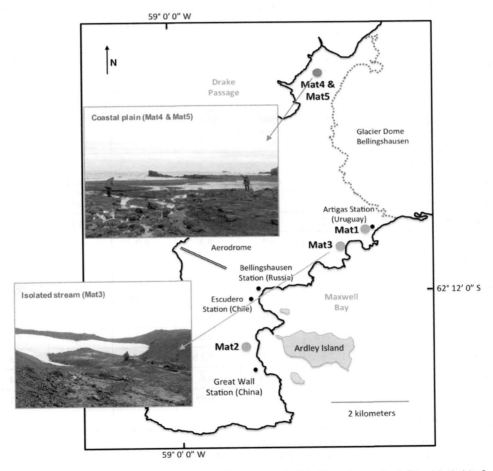

Figure 1. Distribution of sampling points in the Fildes Peninsula (Maritime Antarctica). Sites Mat1, Mat2, and Mat3 are indicated with green dots, while Mat4 and Mat5 located in the coastal plain of the Drake Passage are indicated with pink dots. Insets show the main environmental context of the sampled sites. Black dots indicate reference locations: Artigas Base (Uruguay), Airport Teniente Marsh (Aerodrome), Bellingshausen Station (Russia), Escudero Station (Chile), and the Great Wall Station (China). The blue dotted line shows the limits of the Bellingshausen Glacier Dome.

The temperature of flowing water above microbial mats oscillated between 6.6-8.5°C (Table 1). Water samples above the microbial mats were circumneutral and showed varied concentrations of dissolved inorganic N-forms (DIN = NH_4^+ + NO_3^- + NO_2^-). DIN concentrations ranged from 7.8 to 22.5 µM, where NH_4^+ represented 36 to 79% of total DIN. Organic C and total N concentrations in sediments were low and comparable to those reported for the upper profile of 'Typic Gelaquent' soils in King George Island (Michel et al. 2006). C:N ratios higher than 6.6 in sediments suggest a limited N-availability in the system. In addition to this, low concentrations of available P were also found (4.7 and 6.2 µg g^{-1}) (Table 1). Sediments with faunal excrement deposition, such as in penguin and seal colonies, have shown higher P content than pristine soils in the Fildes Peninsula (Wang et al. 2015). The low mineral N and available P concentrations in the sampled sediments indicate the oligotrophic status of these environments.

TABLE 1: Physicochemical characteristics of the aquatic environment, microbial mats, and sediments in the hyporheic zone at the study sites. Mat4 and Mat5 exhibited lithic contact and no sediment beneath

Aquatic Environment	Mat1	Mat2	Mat3	Mat4[1]	Mat5[1]	Mean
	Stream	Stream	Isolated stream	Coastal plain	Coastal plain	
Temperature (°C)	8.5	7.0	6.6	7.8	7.6	**7.5**
pH	7.0	8.1	8.0	8.3	8.0	**7.9**
NH_4^+ (µM)	3.5	4.9	10.7	9.7	15.7	**8.9**
NO_3^- + NO_2^- (µM)	6.3	2.9	11.8	4.8	4.1	**6.0**
DIN (µM)	9.8	7.8	22.5	14.5	19.8	**14.9**
Microbial mat characteristics						
Total C (mg g^{-1})	153.65	62.30	95.25	82.30	9.77	**80.7**
Total N (mg g^{-1})	20.90	9.05	11.55	13.70	1.47	**11.3**
Total P (mg g^{-1})	3.42	1.70	1.79	1.58	0.59	**1.8**
C:N ratio	7.4	6.9	8.3	6.0	6.7	**7.1**
Sediments in the hyporheic zone						
pH (H$_2$O)	5.6 b[2]	5.7 b	6.3 a	No sediment	No sediment	**5.9**
Organic C (mg g^{-1})	13.2 a	13.6 a	8.1 b			**11.6**
Total N (mg g^{-1})	1.08 b	1.69 a	1.00 b	–	–	**1.3**
C:N ratio	12.2 a	8.0 b	8.1 b	–	–	**9.3**
NO_3^- (µg g^{-1})	4.00 a	0.81 b	n.d.	–	–	**2.4**
NH_4^+ (µg g^{-1})	3.90 a	4.55 a	0.85 b	–	–	**3.1**
Available P (µg g^{-1})	4.7 b	4.7 b	6.2 a	–	–	**5.2**

[1] No sediment, there is lithic contact under microbial mats.
[2] Different letters indicate means are significantly different ($P < 0.05$) among sampling sites.
n.d., non-detectable

Bacterial Diversity in Microbial Mats

The 16S rRNA analysis indicated that the bacterial component of microbial mats was similar at the phylum level with Proteobacteria (44-50%), Planctomycetes (7-17%), Bacteroidetes (8-12%), and Actinobacteria (6-9%) as the main components (Figure 2) with overall composition similar to previously reported polar mats. The sampled mats were classified within two main groups after the cluster analysis. The clustering pattern was found independently of the phylogenetic level used for the analyzes, from phyla to OTU levels (97% cut-off). This classification grouped mats from streams located nearby scientific stations (Mat1 and Mat2) and in a slope (isolated stream Mat3) and in a second cluster the mats located in a coastal plain in the Drake Passage (Mat4 and Mat5). The analysis was supported by the Calinski criteria. The mats from the coastal plain environment (Mat4 and Mat5) were more diverse than the mats sampled in proximity to scientific stations (Mat1, Mat2, and Mat3) according to the Shannon-Weaver (H') index and the number of genera estimated (95% cut-off) (Figure 2).

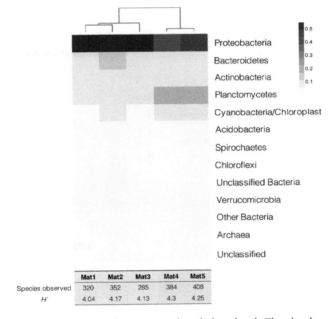

	Mat1	Mat2	Mat3	Mat4	Mat5
Species observed	320	352	285	384	408
H'	4.04	4.17	4.13	4.3	4.25

Figure 2. 16S rRNA genetic composition heatmap at the phylum level. The dendrogram above shows the relationships between microbial mats distributed in two main clusters, the color scale in the right shows the relative proportion of each phylum. The table shows the number of species observed and the Shannon-Weaver (H') diversity index calculated using the vegan R-package [$H' = -sum\ pi\ log(b)\ pi$].

The mats from the coastal plain (Mat4 and Mat5) had a lower proportion of Proteobacteria and a higher number of Planctomycetes (Figure 3c and 3d). Both mats showed a higher proportion of Thaumarchaeota, including nitrifying Archaea such as *Candidatus* Nitrososphaera and Nitrosopumilus (Figure 3e). These Archaea are often found in oceanic waters and are also probably the result of sea-land interactions. Groups including Bacteroidetes showed the lowest representation compared to those near base stations (Mat1 and Mat2) (Figure 3a). Cyanobacteria and chloroplast sequences represented 2.9-9.5% of the relative genetic diversity, Mat2 (Great Wall Station) being the one with the lowest frequency. In contrast, mats from the Drake Passage (Mat4 and Mat5) and the Artigas Station (Mat1) had the highest relative percentage of microbial oxyphototrophs (> 5,8%) (Figure 3a). Microbial mats sampled in the Drake Passage (Mat4 and Mat5) had stramenopiles as a major component of chloroplast diversity (Figure 3a). These two mats also showed a higher genetic diversity of algae, including Rhodophyta, Haptophyceae, and Ulvophyceae. The higher diversity in algae may be related to sea-land interactions in the Drake Passage since Rhodophyta and Ulvophyceae are often found in these oceanic waters.

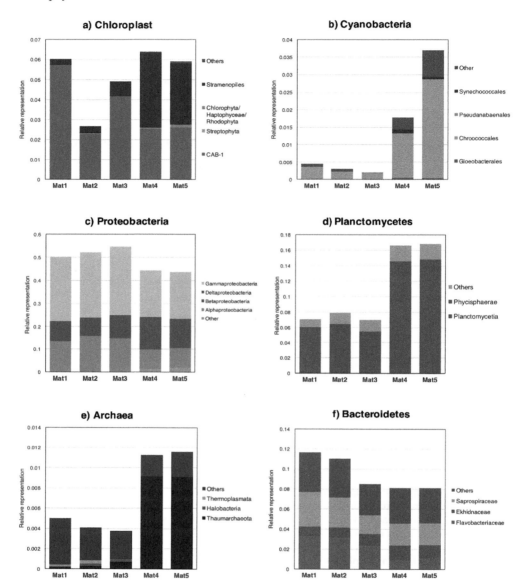

Figure 3. Phylogenetic affiliation of 16S rRNA sequences in microbial mats. (a) Chloroplast, (b) Cyanobacteria, (c) Proteobacteria, (d) Planctomycetes, (e) Archaea, and (f) Bacteroidetes.

Cyanobacteria and Proteobacteria were found as the main diazotrophs in microbial mats after *nifH* analysis (Fig. 4). Cyanobacteria (mainly Nostocales), α- and β-Proteobacteria were the most diverse *nifH* phyla, while δ-Proteobacteria (Geobacter) and Firmicutes (Clostridia and Bacteroidetes) were also present in all localities.

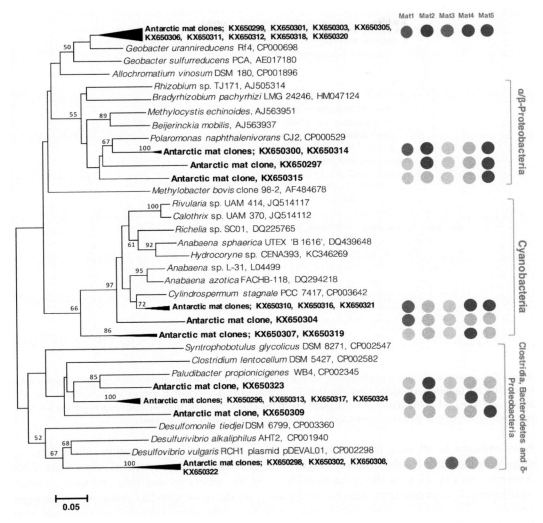

Figure 4. Neighbor-joining (Kimura 2 parameter) phylogenetic tree of N$_2$-fixing (*nifH*) phylotypes found in microbial mats from the Fildes Peninsula. The mid-pointed neighbor-joining tree based on *nifH* gene partial sequences (*ca.* 300 nucleotides); bootstrap values >50% are shown (1,000 replicates). In bold, the phylotypes detected and their accession number. The bright circles represent the microbial mats where the phylotypes were detected. The scale bar represents 5% divergence.

Most *gln* sequences recovered (*n* = 25) are related to Oscillatoriales cyanobacteria, although two sequences (from Drake Passage mats) clustered with *Acaryochloris marina* (Figure 5), the only organism known to contain chlorophyll *d*.

Discussion

Cyanobacteria and microscopic algae play a key role in nutrient cycling in microbial mats from Maritime Antarctica (Komárek and Komárek 2010) as both fix CO$_2$, and cyanobacteria can also fix N$_2$. A recent review by Jungblut and Vincent (2017) report that they are psychrotolerant rather that psychrophilic. This study shows that Cyanobacteria has the potential for N$_2$-fixation and

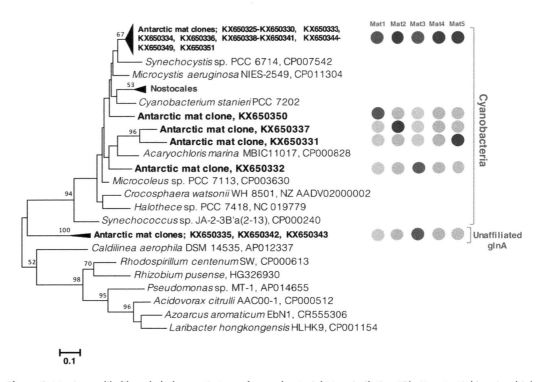

Figure 5. Maximum likelihood phylogenetic tree of cyanobacterial N-assimilation (GlnA) potential in microbial mats. The nucleotide sequences were translated into amino acids; bootstrap values >50% are shown (1,000 replicates). The mid-pointed tree was calculated with the LG model. The bright circles represent the microbial mats where the phylotypes were detected. The scale bar represents 10% divergence.

N-assimilation in all mats reported from the Fildes Peninsula, at temperatures above freezing, suggesting the relevant role of diazotrophs in N-limited Maritime Antarctica microbial mats. These Antarctic mats develop on the upper millimeters of the sediments and soils and have a vertical distribution of microbial taxa, where cyanobacteria are the most abundant group in surface layers associated with photosynthetic metabolisms, whereas many other microbes are most abundant in the deeper layers (Koo et al. 2017). The unique interaction between cyanobacteria and certain heterotrophic Bacteria is also starting to emerge, suggesting a functional coupling between a highly diverse ensemble of microbes (Fernández-Valiente et al. 2007, Koo et al. 2017). The role of oscillatorean and heterocystic cyanobacteria in N_2-fixation, and algal and cyanobacterial mats in N-assimilation has been previously studied in Antarctic glacier meltwater-associated microbial mats (Howard-Williams et al. 1989, Fernández-Valiente et al. 2001).

Understanding the diversity of N_2-fixers in Antarctic aquatic environments is a key feature since N_2-fixation may be a major source of N inputs to the ecosystems (Olson et al. 1998). So far, this is the first study to report the N-N_2 fixer diversity of the Fildes Peninsula microbial mats, while signaling the key role of the cyanobacterial component in N-assimilation. In addition to this, the finding of *Acaryochloris*-related cyanobacteria is interesting since it suggests the relevance of filamentous cyanobacteria for N-assimilation and the ecological importance of *A. marina* in extremely harsh environments, including Antarctic microbial mats. The role of *Acaryochloris* sp. HICR111A as a relevant diazotroph has been already stressed (Pfreundt et al. 2012); here we suggest that *A. marina* (*Acaryochloris* sp. MBIC011017) can also assimilate N.

As clearly stated in the review by Cavicchioli (2015), microorganisms in many Antarctic aquatic environments are poorly understood, although Antarctica might be the most important continent in terms of planetary climate and ocean function. Antarctic microorganisms show several metabolic adaptations oriented toward maximizing light energy utilization, biomass formation, and C- and N-cycling (Priscu et al. 1998, Paerl et al. 2000). In this regard, in cyanobacteria, the coupling of photosynthesis, N-assimilation, and N_2-fixation—the last shared with other microorganisms, mainly Proteobacteria—as shown in this study is fundamental to sustain microbial mat growth in the high biological diverse study region relative to others of Antarctica (Chown and Convey 2007).

Conclusions

Microbial mats in King George Island are highly diverse communities that show a pattern of biogeographic separation between the southern sites of the island and those related to the tidal effect of maritime environments, such as the Drake Passage. All microbial mats analyzed have the genetic potential to fix atmospheric N_2, beckoning research projects that target the measurement of nitrogenase activity. Further, the dual role of cyanobacteria as potential N_2 fixers, and responsible for ammonia assimilation in these communities, plus their role as the main phototrophs is acknowledged. The revealed N_2-fixation potential of α- and β-Proteobacteria in these microbial mats, as well as their large potential for denitrification (Alcántara-Hernández et al. 2014), indicate the relevant role of these groups in the N-cycle. These results also suggest a limited N-availability for these systems as suggested by high C:N ratios in sediments underlying microbial mats. Furthermore, mats collected in the marine-influenced Drake Passage Coast did not have sediment development, another major difference between the microbial mats analyzed. Low mineral N and available P concentrations also indicated the oligotrophic status of these environments, where the biological crust could act as a N-hotspots in the landscape.

Acknowledgments

We gratefully acknowledge *Instituto Antártico Uruguayo* and the staff of *Base Científica Antártica Artigas* for their logistic and technical support. We also thank Osiris Gaona, Fermin S. Castillo-Sandoval, and Enrique Solís (UNAM) for valuable technical support in the experimental and analytical processes. Financial support was provided through the PNUD-URU program, (project No. 13/001). *Secretaría de Relaciones Exteriores*, AMEXCID, Mexico, and AUCI, Uruguay, are acknowledged for logistics and administrative support.

References

Alcántara-Hernández, R.J., C.M. Centeno, A. Ponce-Mendoza, S. Batista, M. Merino-Ibarra, J. Campo, et al. 2014. Characterization and comparison of potential denitrifiers in microbial mats from King George Island, Maritime Antarctica. Polar Biol. 37: 403–416.

Bokulich, N.A., S. Subramanian, J.J. Faith, D. Gevers, J.I. Gordon, R. Knight, et al. 2013. Quality-filtering vastly improves diversity estimates from Illumina amplicon sequencing. Nat. Meth. 10: 57–59.

Callejas, C., P.R. Gill, A.I. Catalán, G. Azziz, S. Castro-Sowinski and S. Batista. 2011. Phylotype diversity in a benthic cyanobacterial mat community on King George Island, maritime Antarctica. World J. Microb. Biot. 27: 1507–1512.

Callejas, C., G. Azziz, E.M. Souza, P.R. Gill and S. Batista. 2018. Prokaryotic diversity in four microbial mats on the Fildes Peninsula, King George Island, maritime Antarctica. Polar Biol. 41: 935–943.

Caporaso, J.G., J. Kuczynski, J. Stombaugh, K. Bittinger, F.D. Bushman, E.K. Costello, et al. 2010. QIIME allows analysis of high-throughput community sequencing data. Nat. Meth. 7: 335–336.

Caporaso, J.G., C.L. Lauber, W.A. Walters, D. Berg-Lyons, J. Huntley, N. Fierer, et al. 2012. Ultra-high-throughput microbial community analysis on the Illumina HiSeq and MiSeq platforms. ISME J. 6: 1621–1624.

Cavicchioli, R. 2015. Microbial ecology of Antarctic aquatic systems. Nat. Rev. Microbiol. 13: 691.

Chown, S.L. and P. Convey. 2007. Spatial and temporal variability across life's hierarchies in the terrestrial Antarctic. Philos. Trans. R. Soc. Lond. B Biol. Sci. 362: 2307–2331.

Convey, P., J.A. Gibson, C.D. Hillenbrand, D.A. Hodgson, P.J. Pugh, J.L. Smellie, et al. 2008. Antarctic terrestrial life – challenging the history of the frozen continent? Biol. Rev. 83: 103–117.

de los Ríos, A., C. Ascaso, J. Wierzchos, E. Fernández-Valiente and A. Quesada. 2004. Microstructural characterization of cyanobacterial mats from the McMurdo Ice Shelf, Antarctica. Appl. Environ. Microb. 70: 569–580.

Edgar R.C., B.J. Haas, J.C. Clemente, C. Quince and R. Knight. 2011. UCHIME improves sensitivity and speed of chimera detection. Bioinformatics 27: 2194–2200.

Fan J., L. Li, J. Han, H. Ming, J. Li, G. Na, et al. 2013. Diversity and structure of bacterial communities in Fildes Peninsula, King George Island. Polar Biol. 36: 1385–1399.

Fernández-Valiente, E., A. Quesada, C. Howard-Williams and I. Hawes. 2001. N$_2$-fixation in cyanobacterial mats from ponds on the McMurdo Ice Shelf, Antarctica. Microb. Ecol. 42: 338–349.

Fernández-Valiente, E., A. Camacho, C. Rochera, E. Rico, W.F. Vincent and A. Quesada. 2007. Community structure and physiological characterization of microbial mats in Byers Peninsula, Livingston Island (South Shetland Islands, Antarctica). FEMS Microbiol. Ecol. 59: 377–385.

Flores, E. and A. Herrero. 1994. Assimilatory nitrogen metabolism and its regulation. pp. 487–517. *In*: D.A. Bryant [ed.]. The Molecular Biology of Cyanobacteria. Springer Netherlands, Dordrecht.

Foong, C.P., C.M. Ling and M. González. 2010. Metagenomic analyses of the dominant bacterial community in the Fildes Peninsula, King George Island (South Shetland Islands). Polar Sci. 4: 263–273.

Gaby, J.C. and D.H. Buckley. 2011. A global census of nitrogenase diversity. Environ. Microbiol. 13: 1790–1799.

Gaujoux, R. and C. Seoighe. 2010. A flexible R package for nonnegative matrix factorization. BMC Bioinformatics 11: 367.

Gibson, A.H., B.D. Jenkins, F.P. Wilkerson, S.M. Short and J.P. Zehr. 2006. Characterization of cyanobacterial *glnA* gene diversity and gene expression in marine environments. FEMS Microbiol. Ecol. 55: 391–402.

Gordon, A. 1999. Classification. 2nd edn Boca Raton. FL: Chapman and Hall/CRC.

Gouy, M., S. Guindon and O. Gascuel. 2010. SeaView version 4: a multiplatform graphical user interface for sequence alignment and phylogenetic tree building. Mol. Biol. Evol. 27: 221–224.

Grasshoff, K., K. Kremlling and M. Ehrhardt. 1983. Methods of Seawater Analysis. VerlagChemie, Weinheim.

Howard-Williams, C., J.C. Priscu and W.F. Vincent. 1989. Nitrogen dynamics in two Antarctic streams. Hydrobiologia 172: 51–61.

Joye, S.B. and R.Y. Lee. 2004. Benthic microbial mats: important sources of fixed nitrogen and carbon to the Twin Cays, Belize ecosystem. National Museum of Natural History, Smithsonian Institution.

Jungblut, A.D., M.A. Allen, B.P. Burns and B.A. Neilan. 2009. Lipid biomarker analysis of cyanobacteria-dominated microbial mats in meltwater ponds on the McMurdo Ice Shelf, Antarctica. Org. Geochem. 40: 258–269.

Jungblut, A.D., C. Lovejoy and W.F. Vincent. 2010. Global distribution of cyanobacterial ecotypes in the cold biosphere. ISME J. 4: 191.

Jungblut, A.D. and W.F. Vincent. 2017. Cyanobacteria in polar and alpine ecosystems. pp. 181–206. *In*: R. Margesin [ed.]. Psychrophiles: From Biodiversity to Biotechnology. Springer International Publishing, Cham.

Kirkwood, D. 1994. Sanplus Segmented Flow Analyzer and its Applications. Seawater Analysis. Skalar, Amsterdam.

Komárek, O. and J. Komárek. 2010. Diversity and ecology of cyanobacterial microflora of Antarctic seepage habitats: comparison of King George Island, Shetland Islands, and James Ross Island, NW Weddell Sea, Antarctica. pp. 515–539. *In*: J. Seckbach and A. Oren [eds.]. Microbial Mats: Modern and Ancient Microorganisms in Stratified Systems. Springer Netherlands, Dordrecht.

Koo, H., N. Mojib, J.A. Hakim, I. Hawes, Y. Tanabe, D.T. Andersen, et al. 2017. Microbial communities and their predicted metabolic functions in growth laminae of a unique large conical mat from Lake Untersee, East Antarctica. Front. Microbiol. 8: 1347.

Krishnan, A., S.A. Alias, C.M. Wong, K-L. Pang and P. Convey. 2011. Extracellular hydrolase enzyme production by soil fungi from King George Island, Antarctica. Polar Biol. 34: 1535–1542.

Lajtha, K., C.T. Driscoll, W.M. Jarrell and E.T. Elliott. 1999. Soil phosphorus: characterization and total element analysis. pp. 115-142. *In*: G.P. Robertson, C. Bledsoe and P. Sollins [eds.]. Standard soil methods for long-term ecological research. Oxford University Press, New York, NY.

Magoč, T. and S.L. Salzberg. 2011. FLASH: fast length adjustment of short reads to improve genome assemblies. Bioinformatics 27: 2957–2963.

Mataloni, G. and J. Komárek. 2004. *Gloeocapsopsisaurea*, a new subaerophyticcyanobacterium from maritime Antarctica. Polar Biol. 27: 623–628.

McKnight, D.M., R.L. Runkel, C.M. Tate, J.H. Duff and D.L. Moorhead. 2004. Inorganic N and P dynamics of Antarctic glacial meltwater streams as controlled by hyporheic exchange and benthic autotrophic communities. J. North Am. Benthological Soc. 23: 171–188.

Michel, R.F., C.E. Schaefer, L.E. Dias, F.N. Simas, V. de Melo Benites and E. de SáMendonça. 2006. Ornithogenic gelisols (cryosols) from maritime Antarctica. Soil Sci. Soc. Am. J. 70: 1370–1376.

Oksanen, J. 2011. Multivariate analysis of ecological communities in R: vegan tutorial. R package version. 1: 11–12.

Olson, J.B., T.F. Steppe, R.W. Litaker and H.W. Paerl. 1998. N_2-fixing microbial consortia associated with the ice cover of Lake Bonney, Antarctica. Microb. Ecol. 36: 231–238.

Paerl, H.W., J.L. Pinckney and T.F. Steppe. 2000. Cyanobacterial-bacterial mat consortia: examining the functional unit of microbial survival and growth in extreme environments. Environ. Microbiol. 2: 11–26.

Pfreundt, U., L.J. Stal, B. Voß and W.R. Hess. 2012. Dinitrogen fixation in a unicellular chlorophyll d-containing cyanobacterium. ISME J. 6: 1367.

Priscu, J.C., C.H. Fritsen, E.E. Adams, S.J. Giovannoni, H.W. Paerl, C.P. McKay, et al. 1998. Perennial Antarctic lake ice: an oasis for life in a polar desert. Science 280: 2095–2098.

R Development Core Team. 2012. R: A language and environment for statistical computing. R Foundation for Statistical Computing, Vienna, Austria. ISBN 3-900051-07-0, URL http://www.R-project.org/.

Robertson, G.P., D. Wedin, P.M. Groffman, J.M. Blair, E.A. Holland, K.J. Nadelhoffer, et al. 1999. Soil carbon and nitrogen availability: nitrogen mineralization, nitrification, and soil respiration potentials. pp. 258–271. In: G.P. Robertson, C.S. Bledsoe, D.C. Coleman and P. Sollins [eds.]. Standard Soil Methods for Long-Term Ecological Research Oxford University Press, New York.

Sabbe, K., D.A. Hodgson, E. Verleyen, A. Taton, A. Wilmotte, K. Vanhoutte, et al. 2004. Salinity, depth and the structure and composition of microbial mats in continental Antarctic lakes. Freshwater Biol. 49: 296–319.

Schloss, P.D., S.L. Westcott, T. Ryabin, J.R. Hall, M. Hartmann, E.B. Hollister, et al. 2009. Introducing mothur: open-source, platform-independent, community-supported software for describing and comparing microbial communities. Appl. Environ. Microb. 75: 7537–7541.

Stal, L.J. 2012. Cyanobacterial mats and stromatolites. pp. 65–125. In: B.A. Whitton [ed.]. Ecology of Cyanobacteria II: Their Diversity in Space and Time, Book XV. Springer Netherlands, Dordrecht.

Turner, J., S.R. Colwell, G.J. Marshall, T.A. Lachlan-Cope, A.M. Carleton, P.D. Jones, et al. 2005. Antarctic climate change during the last 50 years. Int. J. Climatol. 25: 279–294.

Turner, J., J.E. Overland and J.E. Walsh. 2007. An arctic and antarctic perspective on recent climate change. Int. J. Climatol. 27: 277–293.

Tytgat, B., E. Verleyen, D. Obbels, K. Peeters, A. De Wever, S. D'hondt, et al. 2014. Bacterial diversity assessment in Antarctic terrestrial and aquatic microbial mats: a comparison between bidirectional pyrosequencing and cultivation. PloS One 9: e97564.

Valdespino-Castillo, P.M., D. Cerqueda-García, A.C. Espinosa, S. Batista, M. Merino-Ibarra, N. Taş, et al. 2018. Microbial distribution and turnover in Antarctic microbial mats highlight the relevance of heterotrophic bacteria in low-nutrient environments. FEMS Microbiol. Ecol. 94: fiy 129.

van Gemerden, H. 1993. Microbial mats: a joint venture. Mar. Geol. 113: 3–25.

Varin, T., C. Lovejoy, A.D. Jungblut, W.F. Vincent and J. Corbeil. 2012. Metagenomic analysis of stress genes in microbial mat communities from Antarctica and the High Arctic. Appl. Environ. Microb. 78: 549–559.

Vaughan, D.G., J.C. Comiso, I. Allison, J. Carrasco, G. Kaser, R. Kwok, et al. 2013. Observations: cryosphere. pp. 317–382. In: T.F. Stocker, D. Qin, G.-K. Plattner, M. Tignor, S.K. Allen, J. Boschung, et al. [eds.]. Climate Change 2013: The Physical Science Basis. Contribution of Working Group to the Fifth Assessment Report of the Intergovernmental Panel on Climate Change. Cambridge University Press, New York, NY.

Verleyen, E., K. Sabbe, D.A. Hodgson, S. Grubisic, A. Taton, S. Cousin, et al. 2010. Structuring effects of climate-related environmental factors on Antarctic microbial mat communities. Aquat. Microb. Ecol. 59: 11–24.

Vincent, W., M. Downes, R. Castenholz and C. Howard-Williams. 1993. Community structure and pigment organisation of cyanobacteria-dominated microbial mats in Antarctica. Eur. J. Phycol. 28: 213–221.

Wang, N.F., T. Zhang, F. Zhang, E.T. Wang, J.F. He, H. Ding, et al. 2015. Diversity and structure of soil bacterial communities in the Fildes Region (maritime Antarctica) as revealed by 454 pyrosequencing. Front. Microbiol. 6: 1188.

Wang, Q., G.M. Garrity, J.M. Tiedje and J.R Cole. 2007. Naive Bayesian classifier for rapid assignment of rRNA sequences into the new bacterial taxonomy. Appl. Environ. Microb. 73: 5261–5267.

Zani, S., M. Mellon and J. Collier. 2000. Expression of *nifH* genes in natural microbial assemblages in Lake George, New York, detected by reverse transcriptase PCR. Appl. Environ. Microb. 66: 3119–3124.

Zehr, J.P. and L.A. McReynolds. 1989. Use of degenerate oligonucleotides for amplification of the *nifH* gene from the marine cyanobacterium *Trichodesmiumthiebautii*. Appl. Environ. Microb. 55: 2522–2526.

Zhang, Z., S. Schwartz, L. Wagner and W. Miller. 2000. A greedy algorithm for aligning DNA sequences. J. Comput. Biol. 7: 203–214.

11

Physiological and Metabolic Basis of Microbial Adaptations Under Extreme Environments

Saritha M, Praveen Kumar*, Nav Raten Panwar and Uday Burman

Introduction

In an anthropocentric view, cells are reported to function optimally at a pressure of 1 atm, at near-neutral pH, and in an isotonic solution having a temperature of 30-37°C. This delineates the term 'extreme' which is often synonymous with being severe, harsh, and hostile. However, microbial life blurs this distinctness by inhabiting domains over a range of environmental factors, such as temperature, acidity, alkalinity, pressure, and salt concentrations. The organisms which thrive in environments that are usually considered 'unfavourable for growth' are termed extremophiles. They are known to inhabit diverse and unusual habitats, like hot deserts, frozen soils and rocks, hot sulfur springs, solfatara, the deep sea, black smokers, alkaline lakes, extremely acidic environments like the acid mine ores and mine-refuse piles, natural salt lakes, artificial saltpans, or even salt crystals (Prins et al. 1990, Rampelotto 2013). Interestingly, most of these extremophilic organisms require such extreme conditions for their growth and survival.

Extremophiles include members from all the three domains of life – bacteria, archaea, and eukarya. Archaea forms the main group with microorganisms like *Methanopyrus kandleri* strain 116 that grows at 122°C and the genus *Picrophilus torridus* which has the ability to grow at a pH of –0.06 (Schleper et al. 1995, Takai et al. 2008). Among bacteria, the most adapted group is the cyanobacteria, capable of forming microbial mats on ice, hot springs and even under xerophilic conditions, the enterobacteria (*Bacillus, Clostridium*) and the Thio bacteria (*Thiobacillus*) (Vieille and Zeikus 2001, Jha 2014). The most impressive extremophilic organisms are the eukaryotic tardigrades, which can survive temperatures from –272°C to 151°C, the pressure of 6,000 atm, and even high doses of radiation (Seki and Toyoshima 1998, Jönsson et al. 2008, Rampelotto 2013). Hence, extremophiles exhibit great phylogenetic diversity and at the same time, it is intriguing that the organisms which are widely dispersed in the phylogenetic tree display similar ecological specialization.

These organisms are mostly 'polyextremophiles', capable of living in habitats characterized by one or more environmental extremes. They may be broadly grouped in two categories: the extremophilic organisms, which require one or more extreme environmental conditions for growth and survival, and the extremotolerant organisms, which are able to tolerate one or more environmental extremes, but grow optimally under 'normal' conditions (Trotsenko and Khmelenina 2002, Merino et al. 2019). In order to be able to survive under such extreme limits of physicochemical environmental

Division of Integrated Farming System, ICAR-Central Arid Zone Research Institute, Jodhpur, Rajasthan 342 003, India.
* Corresponding author: Praveen.Kumar@icar.gov.in

parameters, these organisms manifest specialized physiological adaptations and possess unique metabolic machinery with specifically modified biomolecules, which confer the organisms with requisite genetic and physiological adaptations.

Physiological Responses of Microorganisms to Extreme Environmental Parameters

Microbial growth is greatly affected by the physicochemical nature of their surroundings that include temperature, pH of the medium, salt concentration, and pressure. Microorganisms are mostly adapted to a range of levels of these factors wherein they grow, rather sub-optimally, and may also remain viable. Microorganisms are categorized based on their responses to the abovementioned environmental factors as given below (Table 1). It is interesting to note that besides the adaptations that allow microorganisms to adjust their physiology and metabolism within their desired ecological range, they have also evolved adaptive strategies that protect them outside of these ranges.

TABLE 1: Classification of Bacteria Based on Their Growth Responses to Physical and Chemical Environmental Factors

Category	Characteristics	Examples	Reference
1. Classification Based on Temperature Requirement			
Psychrophile	Cold-loving microorganisms are able to grow at temperatures ranging from 0°C to 20°C but with optimum growth at 15°C or lower temperatures	*Sporosarcina psychrophila, Colwellia psychrerythraea, Moritella marina*	Kautharapu and Jarboe 2012, Yan et al. 2016, Techtmann et al. 2016
Psychrotroph	Microorganisms are able to grow at temperatures ranging from 0°C to 35°C but with optimum growth at 20°C to 30°C	*Arthrobacter globiformis, Pseudomonas fluorescens, Psychrobacter immobilis*	Berger et al. 1996, Guillou and Guespin-Michel 1996, Pacova et al. 2001
Mesophile	Microorganisms are able to grow optimally at temperatures ranging from 20°C to 45°C	*Escherichia coli, Staphylococcus aureus, Streptococcus pneumoniae*	Tettelin et al. 2001, Willey et al. 2011, Tong et al. 2015
Thermophile	Microorganisms are able to grow at 55°C or higher temperatures with optimum growth at 55°C to 65°C	*Geobacillus stearothermophilus, Chaetomium thermophilum, Thermobispora bispora*	Nazina et al. 2001, Liolios et al. 2010, Willey et al. 2011
Hyperthermophile/ Caldoactive bacteria	Microorganisms surviving under the extreme hot conditions with an optimal growth temperature between 85°C to 113°C	*Methanopyrus kandleri, Thermotoga maritima, Pyrolobus fumarii*	Huber et al. 1986, Blöchl et al. 1997, Takai et al. 2008
2. Classification Based on Response to pH			
Acidophile	Microorganisms, which grow optimally at pH values, ranging from 0 to 5.5	*Sulfolobus acidocaldarius, Acidithiobacillus thiooxidans, Picrophilus oshimae*	Angelov et al. 2011, Jha 2014, Sharma et al. 2012
Neutrophile	Microorganisms are able to grow at neutral pH conditions ranging from 6.5 to 7.5	*Escherichia coli, Pectobacterium carotovorum, Salmonella* sp.	Toth et al. 2003, Cheminay et al. 2004, Willey et al. 2011
Alkaliphile	Microorganisms that grow optimally at a pH between 8.0 and 11.5	*Bacillus firmus, Thermococcus alcaliphilus, Pseudomonas alcaliphila*	Guffanti et al. 1980, Keller et al. 1995, Yumoto et al. 2001

TABLE 1: Contd....

TABLE 1: Contd.

Category	Characteristics	Examples	Reference
3. Classification Based on Salt Requirement			
Halotolerant	Microorganisms are able to grow in the absence as well as the presence of high salt concentrations	*Exiguobacterium* sp., *Staphylococcus aureus*, *Micrococcus* sp.	Larsen 1986, Remonsellez et al. 2018
Halophile	Salt-loving microorganisms, which require high salt concentrations to grow • Slightly halophilic: require 0.2-0.8 M NaCl • Moderately halophilic: require 0.8-3.4 M NaCl • Extremely halophilic: require 3.4-5.1 M NaCl	*Halobacterium* sp., *Ectothiorhodospira variabilis*	Larsen 1986, Jha 2014
4. Classification Based on High Pressure Requirement			
Barophile/ Piezophile	Microorganisms which grow rapidly at high hydrostatic pressures equal or above 10 MPa	*Shewanella benthica*, *Thermococcus piezophilus*	Willey et al. 2011, Dalmasso et al. 2016

Response of Microorganisms to Temperature Extremes

Thermophiles and psychrophiles entail greater structural and functional stability of their bio-macromolecules under extremely high/low temperatures, respectively (Koga 2012, Dhakar and Pandey 2020). The physiological and biochemical modifications which confer them with tolerance to such temperature extremes are detailed below.

Modifications of Cell Membrane and Its Components

Membrane lipids are known to exhibit phase transitions, depending upon their composition, with respect to specific temperatures. For instance, a membrane may be highly restrictive at temperatures below the phase transition temperature of membrane lipids which make them too rigid, while at temperatures above the limits of phase transition, the membrane may be highly permeable due to disorganization of membrane barriers (Kogut 1980). This phenomenon in a way also governs the upper and lower limits of temperature tolerance in microorganisms.

A striking feature in the cell envelope composition of thermally-adapted microorganisms is the presence of unusual ether-linked lipids with long isoprenoid hydrocarbon side chains and a glycerol-1-phosphate backbone in place of the usual ester-linked lipids with fatty acid chain and a glycerol-3-phosphate backbone found in mesophiles (Figure 1) (Kogut 1980, Koga 2012). The ether-linked lipids are known to be the most thermotolerant and are present in thermophilic bacteria (e.g., bacterial lipids like 15,16-dimethyl-30-glyceryloxytriacontanoic acid; 1,2-di-hydroxynonadecane; 15,16-dimethyltriacontandioic acid) and archaea (e.g., diphytanylglycerol; digeranylgeranylglycerophosphate; caldarchaeol) (Koga 2012, Jha 2014). Besides the stability of lipids, extremophiles also display altered fatty acid composition in order to remain viable under harsh temperature ranges. At higher temperatures, more saturated and straight-chain fatty acids prevail which raise the melting point of the membrane and decrease membrane fluidity (Innis and Ingraham 1978, Mansilla et al. 2004). While, at lower temperatures, the proportion of unsaturated and branched-chain fatty acids increases leading to greater membrane fluidity (Mansilla et al. 2004, Koga 2012). These characteristics provide the appropriate level of membrane permeability that is needed for providing an optimum intracellular environment needed for the metabolism and normal functioning of the organism (Russell and Fukunaga 1990).

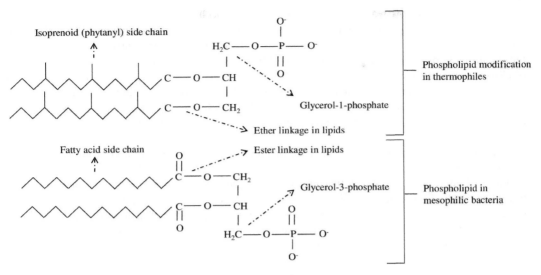

Figure 1. Cell envelop modification in thermopiles.

Nucleic Acid Modifications

In a broad sense, temperature stresses may alter the nucleic acid conformations, which may lead to specific modifications in translated proteins that then confer the required intrinsic thermostability.

Microorganisms with higher growth temperatures usually have greater G + C content (nucleotide guanine-cytosine pairs) which accounts for the enhanced stability of structural RNAs (Hurst and Merchant 2001). Although the high guanine-cytosine content was earlier considered to be an adaptation to high temperature, recent studies prove that a direct correlation between optimal growth temperature of microorganism and the content of G–C pairs exists only in rRNAs and tRNAs and not the whole genomic DNA (Hurst and Merchant 2001, Wang et al. 2006). Stabilization of nucleic acids is mainly attributed to their association with cationic proteins that provide additional spiralization or supercoiling (Jha 2014). Moreover, hyperthermophiles possess a reverse gyrase (ATP-dependent topoisomerase I) enzyme that induces positive supercoils in DNA (the DNA of all other organisms being negatively supercoiled) (Schulze-Makuch and Irwin 2008). Besides, post-transcriptional nucleoside modifications also contribute to nucleic acid stabilization. The tRNA of the hyperthermophiles, *Thermus thermophilus* and *Pyrococcus furiosus*, contain 5-methyl-2-thiouridine or 2-thioribothymidine, wherein the extent of thiolation and the thermostability of the tRNA increase with the increase in temperature (Watanabe et al. 1976, Shigi 2018). Similarly, in *Geobacillus stearothermophilus*, tRNA showed increased methylation of ribose at 70°C (Stenesh 1976). Such post-transcriptional modifications have also been observed in psychrophiles. They contain 40 to 70% more dihydrouridine than that of the mesophile *E. coli* (Dalluge et al. 1997). These modifications are essential for protein synthesis at extreme temperatures.

Protein Modifications

The exposure of microorganisms to temperature extremes induces the expression of a class of proteins known as heat shock proteins (Hsps) or colds shock proteins (Csps), which influence the cellular processes of transcription, translation, protein folding, and regulation of membrane fluidity (Phadtare 2004, D'Amico et al. 2016). These Csps and Hsps, usually seen in mesophiles, have their homologues in extremophiles. Hsps are classified based on their molecular weight in kDa as Hsp60, Hsp70, Hsp90, etc., and they function as molecular chaperons, which keep the proteins in an active form, preventing protein aggregation and providing refolding of denatured proteins (Laksanalamai and Robb 2004). Hsp60 was found to be induced in the hyperthermophilic *Sulfolobus* sp. and *Pyrodictium* sp., wherein it accounted for 80% of the total cellular proteins at 108°C (Rossi and Guagliardi 2011). Similarly, cold-acclimation proteins (Caps) are constitutively expressed at low temperatures in psychrophiles (D'Amico et al. 2016). Some of these are the antifreeze proteins that bind to ice crystals due to their larger surface area, thereby reducing the effects of ice nucleation (Jha 2014). Further, cryoprotectants like trehalose also help in preventing protein denaturation and aggregation (Phadtare 2004). Psychrophiles also synthesize glycoproteins and peptides with antifreeze properties which significantly decrease the freezing point of the fluid in cytoplasm and organelles (Jha 2014).

Langridge (1968) has illustrated the relationship between temperature and protein structure by using a mutant *E. coli* enzyme, β-galactosidase, which had an altered thermostability with single amino acid alterations. Increased thermostability of proteins in extremophiles has been attributed to higher ratios of basic/acidic amino acid residues, greater hydrophobic interactions, and reduced surface area to volume ratio (Kogut 1980, Hough and Danson 1999). Furthermore, extremophiles, especially archaea, also contain a very diverse polyamine repertoire, which plays a protector role *in vivo* and helps in the enzyme activation (Michael 2016).

Response of Microorganisms to pH Extremes

Acidophiles and alkaliphiles have mechanisms to keep their internal pH close to neutral, which helps them to cope with external acidic or alkaline conditions. This is achieved through the following mechanisms of pH homeostasis:

Membrane Characteristics

Acidophiles maintain a large chemical proton gradient across their membranes by reversing their membrane potential by restricting the proton influx into the cytoplasm (Guffanti et al. 1984). This is accomplished by the presence of an impermeable membrane made of ether-linked lipids and a bulky isoprenoid core in common with thermophiles (Figure 2) (Kogut 1980). In thermoacidophiles like *Alicyclobacillus acidocaldarius* and *Thermoplasma acidophilum*, active proton pumping has been observed to remove the excess protons out of the cell (Baker-Austin and Dopson 2007). In another mechanism, acidophiles like *Acidithiobacillus thiooxidans* are seen to develop an internal positive potential ($\Delta\Psi$) (Donnan potential) in contrast to the negative $\Delta\Psi$ of neutrophiles by a greater influx of potassium ions (Fig. 2) (Suzuki et al. 1999, Baker-Austin and Dopson 2007).

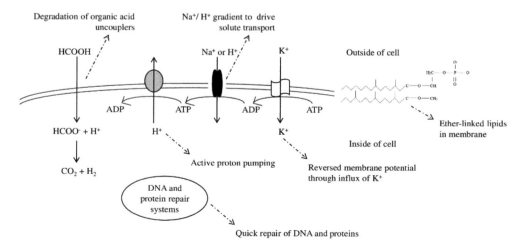

Figure 2. Cell response of acidophiles to low pH conditions.

Although studies on alkaliphiles are not as abundant as acidophiles, a few general trends in adaptation have been observed. Replacement of protons with Na$^+$ ions in transmembrane electrogenic processes is observed as "the protons, upon transport from the cell, cannot effectively generate the membrane proton motive force required for cellular respiration, after their return into the cytoplasm, as they would move against the concentration gradient" (Jha 2014). The presence of Na$^+$/H$^+$ (K$^+$/H$^+$) antiporters, to maintain proton driving force is thus observed in alkaliphiles (Grant et al. 1990). Alkaliphiles also exhibit greater variability of membrane lipids with bis-mono-acylglycerophosphate and cardiolipins in high quantities (Krulwich 2011). Another alkaliphilic characteristic is the presence of squalenes, tetrahydrosqualenes, or other polyisoprenes in the membrane, which reduces lipid motility and proton leakiness (Hauss et al. 2002, Siliakus et al. 2017).

Protection of Nucleic Acid and Protein

Sequencing of acidophilic genomes has revealed a large proportion of DNA to be protein repair genes (Crossman et al. 2004). Genes encoding the enzymes of organic acid degradation pathways have been found in the genomes of extreme acidophiles because organic acids act as uncouplers of the respiratory chain and become deleterious to the organism (Sharma et al. 2016). Also, chaperones that aid in protein refolding have been reported to be highly expressed in an acid mine drainage biofilm community and constituted 11% of the total expressed proteome (Ram et al. 2005). Besides, acid-stability of enzymes and proteins has been attributed to the abundance of glutamic and aspartic surface residues that generate a negative charge at low pH (Huang et al. 2005). Golyshina et al. (2006) noted that the enzyme system in the obligate acidophile *Ferroplasma acidiphilum* was active at a much lower pH of 1.7-4.0, while the intracellular pH was 5.6. This property was later ascribed to a high proportion of iron proteins that acted as 'iron rivet' in maintaining the 3-D conformation and contributed to the pH stability of the enzymes (Ferrer et al. 2007). Protein adaptations to alkaline pH are mostly indistinct as the cell envelope features usually maintain a near-neutral cytoplasm. However, the presence of glycosylated proteins and the presence of a high proportion of acidic residues have been found to help in maintaining a neutral intracellular pH (Reed et al. 2013). The ecological and biotechnological relevance of the extremophilic microorganisms with wide pH tolerance has recently been recognized (Dhakar and Pandey 2016).

Response of Microorganisms to Environmental Salt Concentrations

The survival of microorganisms under saline/hypersaline conditions requires the maintenance of their osmotic balance. This is achieved through the following cellular and protein adaptations:

Cellular Mechanisms

Halophiles have evolved two main strategies to counter the excessive water loss under high salt conditions: (i) accumulation of large amounts of inorganic salt intracellularly so that its concentration matches the external salt concentration; (ii) accumulation of compatible organic solutes or osmoprotectants. The former mechanism, called the high 'salt-in' strategy involves Cl$^-$ pumps that transport Cl$^-$ from the environment into the cytoplasm so that the counter ion K$^+$ gets accumulated inside (Edbeib et al. 2016). This has been observed in many genera, like *Halobacteria*, *Halococcus*, *Haloanaerobium*, and the uptake of K$^+$ via the K$^+$ uniport is driven by the electrical potential generated by the Na$^+$/H$^+$ antiporter (Oren 2006, Edbeib et al. 2016). A more flexible means to attain osmotic balance is through the accumulation of organic solutes (like amino acids), polyols (like glycine betaine, ectoine, glycerol, glutamine), and sugars (like sucrose and trehalose), which do not affect the metabolic processes (Martin et al. 1999, Jha 2014). These solutes may be synthesized *de novo* or may be accumulated from the medium, and they help to maintain the turgor and stabilize microbial membranes.

Protein Adaptations

Negative protein charges and increased hydrophobicity are the two main factors responsible for the salt-adaptation of halophilic enzymes (Bayley et al. 1978). The abundance of negatively charged acidic amino acid residues has been reported in *Halobacterium* sp. (Ng et al. 2000), *H. marismortui*, and *Planococcus* sp. (Huang et al. 2015). The negatively charged protein surface improves its ability to compete with ions for water molecules (Edbeib et al. 2016). The acidic residues on halophilic proteins also bind to hydrated cations, which would maintain a shell of hydration around the protein (Reed et al. 2013). Bioinformatics analysis of halophilic protein sequences has revealed them to contain considerably fewer polar residues, like serines, as they provide little or no protection against dehydration (Zhang and Ge 2013). Recent research has shown that the halophiles utilize the salts for protein function and salt-dependent protein folding has been reported by Müller-Santos et al. (2009). Moreover, the polyextremophilic characteristics render the haloactive enzymes to remain active under even alkaliphilic and thermophilic conditions (de Lourdes Moreno et al. 2009).

Response of Microorganisms to Environmental Pressure

Due to the technical difficulties in obtaining piezophilic microorganisms, most of the studies on the effects of high hydrostatic pressure are based on mesophiles or piezotolerant microorganisms. These studies have revealed that under high pressure the organisms alter the membrane composition, especially the amount of monounsaturated fatty acids in lipids to increase the membrane fluidity (Simonato et al. 2006). Interestingly, a study by Lauro et al. (2007) on the molecular phylogenies of several piezophiles has revealed unique 16S rRNA genes with elongated helices that increase with the extent of adaptation to growth at elevated pressure. These helix changes are believed to improve ribosome function under deep sea conditions (Lauro et al. 2007). In a study on gene expression in the psychrophilic piezophile, *Photobacterium profundum* SS9, it was found that the expression of OmpH protein, which is responsible for the transport of nutrients into the cell, was found to be maximum under 28 MPa (Welch and Bartlett 1996). This regulation helps in survival under conditions of nutrient fluctuation in high pressure environments. With respect to proteins, the general adaptations for archaeal and bacterial piezophiles are a compact, dense hydrophobic core with the prevalence of smaller hydrogen-bonding amino acids and increased multimerization (Di Giulio 2005, Reed et al. 2013). In *Pyrococcus abyss*, a large number of small amino acids leads to a reduction in the number of large hydrophobic residues in the core, which allows for tighter packing, creating a more pressure stable protein (Di Giulio 2005). The formation of dimers or multimers enhances the stability of proteins and provides less chance for water to penetrate the core of the protein at high pressures, which would otherwise disrupt the protein structure (Rosenbaum et al. 2012).

Metabolic Adaptations of Microorganisms to Extreme Environmental Conditions

Apart from the intrinsic properties, stability in extremophiles is also induced by an appropriate milieu provided by metabolic substrates, coenzymes, salts, and ions (Ljungdahl and Sherod 1976). In other words, each biomolecule must be adapted to the conditions in which the microorganism grows. Among the various environmental factors, the temperature is the most important as it crosses physical barriers making it difficult for the organisms to efficiently protect themselves in the way they can protect themselves from extreme external pH or salinity by maintaining steep concentration gradients over biological membranes (Engqvist 2018). Metabolism is known to vary in extremophiles, and metabolic pathways vary even among members of the same class of extremophiles. The three common glycolytic pathways of EMP (Embden-Meyerhoff-Parnas), ED (Entner-Doudoroff), and the pentose phosphate pathway, usually encountered in mesophiles are also seen in thermophilic bacteria. However, archaea exhibit certain modifications. For instance, the classical EMP pathway is seen in the hyperthermophilic bacterium, *Thermotoga maritima*, while the hyperthermophilic heterotrophic archaea, *Pyrococcus furiosus*, utilizes a modified EMP pathway (Figure 3), which makes use of ADP-dependent kinases, a unique glyceraldehyde-3-phosphate: ferredoxin oxidoreductase, and ADP-forming acetyl-CoA synthetase, which are not encountered in bacteria (Kengen et al. 1996). Similarly, studies on *Thermus thermophilus* have revealed alternative pathways of amino acid synthesis, wherein lysine is synthesized by alpha-aminoadipate pathway instead of the diaminopimelate pathway (Kosuge and Hoshino 1998).

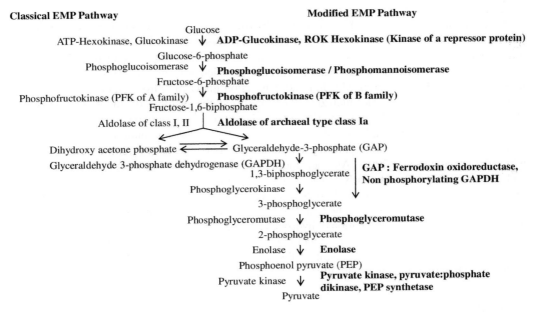

Figure 3. Metabolic differences in terms of the glycolytic (EMP) pathway in mesophilic bacteria and extremophiles (Adapted from Bräsen et al. 2014).

Recently, Engqvist (2018) has identified pathways with enzymes whose occurrence strongly correlates with growth temperatures which state that many of the metabolic pathways are under evolutionary pressure in temperature adaptation. In psychrophiles, the oxidative metabolism involving glycolysis, the tricarboxylic acid cycle (TCA), and the electron transport chain is generally depressed due to increased oxygen solubility at low temperatures, generating reactive oxygen species (ROS) that cause oxidative stress (Tribelli and López 2018). Genome analysis of *Psychrobacter arcticus* 273-4 has shown that it lacks glycolysis genes and a phosphotransferase system but possesses the gluconeogenic enzymes, fructose-1,6-bisphosphatase, and phosphoenolpyruvate synthase, suggesting that although it is not able to utilize sugars, other oxidized carbon sources are preferred (Ayala-del-Río et al. 2010). Also, the transcriptomic analysis of *Pseudomonas extremaustralis* revealed a downregulation of the genes involved in primary metabolism and an upregulation of the genes involved in ethanol oxidation (Tribelli et al. 2015). Though lower in energy yields, these secondary pathways and intermediate pathways which bypass the central metabolism are essential for survival under cold conditions.

Similarly, the metabolism of mesophiles is also altered when challenged with pH changes. The natural habitats of many acidophiles (e.g., solfatara) are devoid of complex carbon sources, necessitating the process of autotrophy (Counts et al. 2017). Their energy demand is often met chemolithotrophically by the oxidation of basic ions and molecules such as Fe^{2+}, H_2, and sulfur compounds (Christel 2018). Sulfur metabolism is a key feature of acidophiles and it is proposed that the organisms either physically associates with sulfur particles or that at elevated temperatures sulfur becomes sufficiently soluble to support the growth of thermoacidophiles (Counts et al. 2017). Acidophiles also possess enzymes like the sulfur oxygenase reductase that is capable of utilizing the elemental sulfur and its derivatives from the environment (Kletzin 1989). Besides the oxidation of elemental sulfur ($S°$) and inorganic sulfur compounds, acidophiles like *Acidithiobacillus ferroxidans* can catalyze dissimilatory redox transformations of sulfur and iron, wherein sulfur is used as the electron donor and ferric iron as the electron acceptor (Osorio et al. 2013). Metabolic strategies of microorganisms when challenged with alkali, like the upregulation of deaminases and fermentation that produce metabolic acids to spare the use of energy-dependent antiport, have been suggested (Counts et al. 2017) but they require further validation.

A remarkable overall switch of cell metabolism as an adaptive strategy is prominent in piezophiles. This adaptation involves major cellular processes, such as the transport of solutes like amino acids, amino acid metabolism, and pathways involved in energy production (Vannier et al. 2015). Recently, Michoud and Jebbar (2016) have revealed several gene clusters and metabolic pathways involved in energy production and conversion under high hydrostatic pressure by using transcriptomic and proteomic analyzes. They analyzed the genome of an obligate piezophile, *Pyrococcus yayanosii* CH1 and found that it lacked important biosynthetic pathways, like the basic and aromatic amino acid synthesis pathways. This may be a strategy to omit the high cost incurring pathways like the biosynthetic pathway for the synthesis of the largest amino acid, tryptophan, which is aromatic (Swire 2007). Another finding was that the genes involved in energy production and conversion were found nearly twice as often (17%) as in other strains (9-11%), and the genes involved in information storage and processing were overrepresented (Michoud and Jebbar 2016). One of the main pathways regulated in *P. yayanosii* under stressful pressures was the energy pathway wherein the downregulation of hydrogenases and formate metabolism and the upregulation of oxidoreductases constituted a large metabolic shift in the piezophile (Michoud and Jebbar 2016). This metabolic shift might be a result of the adaptive mechanism of cell membrane-embedded proteins to changing membrane fluidity.

Conclusions

Extremophiles are distinct from other microorganisms in their ability to survive under extreme and hostile environmental conditions. The apparent diversity of these microorganisms has driven research in this area in order to unravel the molecular machinery central to their survival. It is noteworthy that the experimental procedures during sampling, culturing, choice of materials and preparatory techniques used are often as extreme as the habitats studied (Prins et al. 1990). In spite of the technical difficulties, the field of study on extremophiles continues to excite researchers and this has led to the unveiling of their unique physiology and discovery of unusual biochemical pathways that confer the organisms with the ability to cope with their harsh environments. In this way, extremophiles also serve as unique substrates for the identification of novel biomolecules that can work under a range of physical and chemical ambient conditions, hence have huge biotechnological potential. For instance, the thermostable *Taq* DNA polymerase routinely used in Polymerase Chain Reactions (PCR) has been obtained from the thermophilic bacterium, *Thermus aquaticus*. The majority of the industrial enzymes used today—like the alkaline proteases used in detergent industry, thermophilic cellulases used in paper and pulp industry, the acid-stable amylase, and glucose isomerase used in starch-based industries—are all derived from extremophiles. The biologically active substances derived from extremophiles have found their way even to medicine and cosmetology. Recent advances in the field of extremophiles have even paved the way for developments in astrobiology. The search for extraterrestrial life on Mars is based on the hypothesis that microorganisms that survive in extreme environments on Earth like the Atacama Desert and the Antarctic Dry Valleys could be able to withstand the conditions prevailing on the Martian surface and subsurface.

Deciphering the physiological and metabolic basis of adaptations in extremophiles is a beginning in the research on these fascinating microorganisms. Despite our advances in exploring the vast microbial diversity, a major portion of the Earth's 'inhospitable' surfaces remain unexplored. Furthermore, microbial communities in extreme environments tend to evolve faster than those populating 'moderate' environments. Through the application of advanced 'omics' strategies, exciting discoveries in the microbial world seem possible. This would further broaden our understanding of the molecular changes underlying their adaptive phenomena and give insights into the capabilities of these microorganisms to cope with the complex effects of environmental fluctuations.

Acknowledgments

The authors are thankful to the Director, ICAR-Central Arid Zone Research Institute, Jodhpur, India, for providing the facilities to prepare the manuscript.

References

Angelov, A., J. Voss and W. Liebl. 2011. Characterization of plasmid pPO1 from the hyperacidophile *Picrophilus oshimae*. Archaea Volume 2011, Article ID 723604, 4 pages.

Ayala-del-Río, H.L., P.S. Chain, J.J. Grzymski, M.A. Ponder, N. Ivanova and P.W. Bergholz, et al. 2010. The genome sequence of *Psychrobacter arcticus* 273-4, a psychroactive Siberian permafrost bacterium, reveals mechanisms for adaptation to low-temperature growth. Appl. Environ. Microbiol. 76(7): 2304–2312.

Baker-Austin, C. and M. Dopson. 2007. Life in acid: pH homeostasis in acidophiles. Trends Microbiol. 15(4): 165–171.

Bayley, S.T., R.A. Morton and J.K. Lanyi. 1978. Recent developments in the molecular biology of extremely halophilic bacteria. CRC Crit. Rev. Microbiol. 6(2): 151–206.

Berger, F., N. Morellet, F. Menu and P. Potier. 1996. Cold shock and cold acclimation proteins in the psychrotrophic bacterium *Arthrobacter globiformis* SI55. J. Bacteriol. 178(11): 2999–3007.

Blöchl, E., R. Rachel, S. Burggraf, D. Hafenbradl, H.W. Jannasch and K.O. Stetter. 1997. *Pyrolobus fumarii*, gen. and sp. nov., represents a novel group of archaea, extending the upper temperature limit for life to 113 degrees C. Extremophiles 1(1): 14–21.

Bräsen, C., D. Esser, B. Rauch and B. Siebers. 2014. Carbohydrate metabolism in archaea: current insights into unusual enzymes and pathways and their regulation. Microbiol. Mol. Biol. Rev. 78(1): 89–175.

Cheminay, C., D. Chakravortty and M. Hensel. 2004. Role of neutrophils in murine salmonellosis. Infect. Immun. 72(1): 468–477.

Christel, S. 2018. Function and Adaptation of Acidophiles in Natural and Applied Communities. Ph.D. Thesis, Linnaeus University Press, Växjö.

Counts, J.A., B.M. Zeldes, L.L. Lee, C.T. Straub, M.W.W. Adams and R.M. Kelly. 2017. Physiological, metabolic and biotechnological features of extremely thermophilic microorganisms. Wiley Interdiscip. Rev. Syst. Biol. Med. 9: e1377.

Crossman, L., M. Holden, A. Pain and J. Parkhill. 2004. Genomes beyond compare. Nature Rev. Microbiol. 2: 616–617.

Dalluge, J.J., T. Hamamoto, K. Horikoshi, R.Y. Morita, K.O. Stetter and J.A. McCloskey. 1997. Posttranscriptional modification of tRNA in psychrophilic bacteria. J. Bacteriol. 179(6): 1918–1923.

Dalmasso, C., P. Oger, G. Selva, D. Courtine, S. L'Haridon, A. Garlaschelli, et al. 2016. *Thermococcus piezophilus* sp. nov., a novel hyperthermophilic and piezophilic archaeon with a broad pressure range for growth, isolated from a deepest hydrothermal vent at the Mid-Cayman Rise. Syst. Appl. Microbiol. 39(7): 440–444.

de Lourdes Moreno, M., M.T. García, A. Ventosa and E. Mellado. 2009. Characterization of *Salicola* sp. IC10, a lipase- and protease-producing extreme halophile. FEMS Microbiol. Ecol. 68: 59–71.

Dhakar, K and A. Pandey. 2016. Wide pH range tolerance in extremophiles: towards understanding an important phenomenon for future biotechnology. Appl. Microbiol. Biotechnol. 100: 2499–2510.

Dhakar, K and A. Pandey. 2020. Microbial ecology from the Himalayan cryosphere perspective. Microorganisms 8: 257.

Di Giulio, M. 2005. A comparison of proteins from *Pyrococcus furiosus* and *Pyrococcus abyssi*: barophily in the physicochemical properties of amino acids and in the genetic code. Gene 346: 1–6.

D'Amico, S., T. Collins, J.C. Marx, G. Feller and C. Gerday. 2016. Psychrophilic microorganisms: challenges for life. EMBO Rep. 7(4): 385–389.

Edbeib, M.F., R.A. Wahab and F. Huyop. 2016. Halophiles: biology, adaptation, and their role in decontamination of hypersaline environments. World J. Microbiol. Biotechnol. 32: 135.

Engqvist, M.K.M. 2018. Correlating enzyme annotations with a large set of microbial growth temperatures reveals metabolic adaptations to growth at diverse temperatures. BMC Microbiol. 18: 177.

Ferrer, M., O.V. Golyshina, A. Beloqui, P.N. Golyshin and K.N. Timmis. 2007. The cellular machinery of *Ferroplasma acidiphilum* is iron-protein-dominated. Nature 445(7123): 91–94.

Golyshina, O.V., P.N. Golyshin, K.N. Timmis and M. Ferrer. 2006. The 'pH optimum anomaly' of intracellular enzymes of *Ferroplasma acidiphilum*. Environ. Microbiol. 8(3): 416–425.

Grant, W.D., W.E. Mwatha and B.E. Jones. 1990. Alkaliphiles: ecology, diversity and applications. FEMS Microbiol. Lett. 75(2–3): 255–269.

Guffanti, A.A., R. Blanco, R.A. Benenson and T.A. Krulwich. 1980. Bioenergetic properties of alkaline-tolerant and alkalophilic strains of *Bacillus firmus*. J. Gen. Microbiol. 119: 79–86.

Guffanti, A.A., M. Mann, T.L. Sherman and T.A. Krulwich. 1984. Patterns of electrochemical proton gradient formation by membrane vesicles from an obligately acidophilic bacterium. J. Bacteriol. 159(2): 448–452.

Guillou, C. and J.F. Guespin-Michel. 1996. Evidence for two domains of growth temperature for the psychrotrophic bacterium *Pseudomonas fluorescens* MF0. Appl. Environ. Microbiol. 62(9): 3319–3324.

Hauss, T., S. Dante, N.A. Dencher and T.H. Haines. 2002. Squalane is in the midplane of the lipid bilayer: implications for its function as a proton permeability barrier. Biochim. Biophys. Acta. 1556(2–3): 149–154.

Hough, D.W. and M.J. Danson. 1999. Extremozymes. Curr. Opin. Chem. Biol. 1: 39–46.

Huang, X., J. Lin, X. Ye and G. Wang. 2015. Molecular characterization of a thermophilic and salt- and alkaline-tolerant xylanase from *Planococcus* sp. SL4, a strain isolated from the sediment of a soda lake. J. Microbiol. Biotechnol. 25(5): 662–671.

Huang Y., G. Krauss, S. Cottaz, H. Driguez and G. Lipps. 2005. A highly acid-stable and thermostable endo-glucanase from the thermoacidophilic archaeon *Sulfolobus solfataricus*. Biochem. J. 385: 581–588.

Huber, R., T.A. Langworthy, H. König, M. Thomm, C.R. Woese, U.B. Sleytr, et al. 1986. *Thermotoga maritima* sp. nov. represents a new genus of unique extremely thermophilic eubacteria growing up to 90°C. Arch. Microbiol. 144: 324–333.

Hurst, L.D. and A.R. Merchant. 2001. High guanine-cytosine content is not an adaptation to high temperature: a comparative analysis amongst prokaryotes. Proc. Biol. Sci. 268(1466): 493–497.

Innis, W.E. and J.L. Ingraham. 1978. Microbial life at low temperatures: mechanisms and molecular aspects. pp. 73–99. *In*: D.J. Kushner [ed.]. Microbial Life in Extreme Environments. Academic Press, London.

Jha, P. 2014. Microbes thriving in extreme environments: how do they do it? Int. J. Appl. Sci. Biotechnol. 2(4): 393–401.

Jönsson, K.I., E. Rabbow, R.O. Schill, M. Harms-Ringdahl and P. Rettberg. 2008. Tardigrades survive exposure to space in low Earth orbit. Curr. Biol. 18(17): R729–R731.

Kautharapu, K.B. and L.R. Jarboe. 2012. Genome sequence of the psychrophilic deep-sea bacterium *Moritella marina* MP-1 (ATCC 15381). J. Bacteriol. 194(22): 6296–6297.

Keller, M., F.J. Braun, R. Drmeier, D. Afenbradl, S. Burggraf, R. Rachel, et al. 1995. *Thermococcus alcaliphilus* sp. nov., a new hyperthermophilic archaeum growing on polysulfide at alkaline pH. Arch. Microbiol. 164: 390–395.

Kengen, S.M., A.J.M. Stams and W.M. de Vos. 1996. Sugar metabolism of hyperthermophiles. FEMS Microbiol. Rev. 18: 119–137.

Kletzin A. 1989. Coupled enzymatic production of sulfite, thiosulfate, and hydrogen sulfide from sulfur: purification and properties of a sulfur oxygenase reductase from the facultatively anaerobic archaebacterium *Desulfurolobus ambivalens*. J. Bacteriol. 171: 1638–1643.

Koga, Y. 2012. Thermal adaptation of the archaeal and bacterial lipid membranes. Archaea Volume 2012, Article ID 789652, 6 pages.

Kogut, M. 1980. Are there strategies of microbial adaptation to extreme environments? Trends Biochem. Sci. 5(2): 15–18.

Kosuge, T. and T. Hoshino. 1998. Lysine is synthesized through the α-aminoadipate pathway in *Thermus thermophilus*. FEMS Microbiol. Lett. 169(2): 361–367.

Krulwich, T.A., J. Liu, M. Morino, M. Fujisawa, M. Ito and D.B. Hicks. 2011. Adaptive mechanisms of extreme alkaliphiles. pp. 119–139. *In*: K. Horikoshi [ed.]. Extremophiles Handbook. Springer, Tokyo.

Laksanalamai, P. and F.T. Robb. 2004. Small heat shock proteins from extremophiles: a review. Extremophiles 8: 1–11.

Langridge, J. 1968. Genetic and enzymatic experiments relating to the tertiary structure of β-galactosidase. J. Bacteriol. 96: 1711–1717.

Larsen, H. 1986. Halophilic and halotolerant microorganisms–an overview and historical perspective. FEMS Microbiol. Rev. 39: 3–7.

Lauro, F.M., R.A. Chastain, L.E. Blankenship, A.A. Yayanos and D.H. Bartlett. 2007. The unique 16S rRNA genes of piezophiles reflect both phylogeny and adaptation. Appl. Environ. Microbiol. 73(3): 838–845.

Liolios, K., J. Sikorski, M. Jando, A. Lapidus, A. Copeland, T. Glavina, et al. 2010. Complete genome sequence of *Thermobispora bispora* type strain (R51). Stand. Genomic Sci. 2(3): 318–326.

Ljungdahl, L.G. and D. Sherod. 1976. Extreme Environments: Mechanisms of Microbial Adaptation. Academic Press, London, New York, San Francisco.

Mansilla, M.C., L.E. Cybulski, D. Albanesi and D. De Mendoza. 2004. Control of membrane lipid fluidity by molecular thermosensors. J. Bacteriol. 186: 6681–6688.

Martin, D.D., R.A. Ciulla and M.F. Roberts. 1999. Osmoadaptation in Archaea. Appl. Environ. Microbiol. 65: 1815–1825.

Merino, N., H.S. Aronson, D.P. Bojanova, J. Feyhl-Buska, M.L. Wong, S. Zhang, et al. 2019. Living at the extremes: extremophiles and the limits of life in a planetary context. Front. Microbiol. 10: Article 780, 25 pages.

Michael, A.J. 2016. Polyamines in eukaryotes, bacteria, and archaea. J. Biol. Chem. 291: 14896–14903.

Michoud, G. and M. Jebbar. 2016. High hydrostatic pressure adaptivestrategies in an obligate piezophile *Pyrococcus yayanosii*. Sci. Rep. 6: 27289.

Müller-Santos, M., E.M. de Souza, O. Pedrosa Fde, D.A. Mitchell, S. Longhi, F. Carrière, et al. 2009. First evidence for the salt-dependent folding and activity of an esterase from the halophilic archaea *Haloarcula marismortui*. Biochim. Biophys. Acta. 1791(8): 719–729.

Nazina, T.N., T.P. Tourova, A.B. Poltaraus, E.V. Novikova, A.A. Grigoryan, A.E. Ivanova, et al. 2001. Taxonomic study of aerobic thermophilic bacilli: descriptions of *Geobacillus subterraneus* gen. nov., sp. nov. and *Geobacillus uzenensis* sp. nov. from petroleum reservoirs and transfer of *Bacillus stearothermophilus*, *Bacillus thermocatenulatus*, *Bacillus thermoleovorans*, *Bacillus kaustophilus*, *Bacillus thermodenitrificans* to *Geobacillus* as the new combinations G. *stearothermophilus*, G. *th*. Int. J. Syst. Evol. Microbiol. 51(Pt 2): 433–446.

Ng, W.V., S.P. Kennedy, G.G. Mahairas, B. Berquist, M. Pan, H.D. Shukla, et al. 2000. Genome sequence of *Halobacterium* species NRC-1. Proc. Natl. Acad. Sci. U.S.A. 97(22): 12176–12181.

Oren, A. 2006. Life at high salt concentrations. pp. 263–282. *In*: E. Rosenberg, E.F. DeLong, S. Lory, E. Stackebrandt and F. Thompson [eds.]. The prokaryotes. Springer, Berlin.

Osorio, H., S. Mangold, Y. Denis, I. Nancucheo, M. Esparza, D.B. Johnson, et al. 2013. Anaerobic sulfur metabolism coupled to dissimilatory iron reduction in the extremophile *Acidithiobacillus ferrooxidans*. Appl. Environ. Microbiol. 79: 2172– 2181.

Pacova, Z., E. Urbanova and E. Durnova. 2001. *Psychrobacter immobilis* isolated from foods: characteristics and identification. Vet. Med. – Czech. 4: 95–100.

Phadtare, S. 2004. Recent developments in bacterial cold-shock response. Curr. Issues Mol. Biol. 6(2): 125–136.

Prins, R.A., W. de Vrij, J.D. Gottschal and T.A. Hansen. 1990. Adaptation of microorganisms to extreme environments. FEMS Microbiol. Rev. 75: 103–104.

Ram, R.J., N.C. Ver Berkmoes, M.P. Thelen, G.W. Tyson, B.J. Baker, R.C. Blake II, et al. 2005. Community proteomics of a natural microbial biofilm. Science 308: 1915–1999.

Rampelotto, P.H. 2013. Extremophiles and extreme environments. Life 3: 482–485.

Reed, C.J., H. Lewis, E. Trejo, V. Winston and C. Evilia. 2013. Protein adaptations in archaeal extremophiles. Archaea Volume 2013, Article ID 373275, 14 pages.

Remonsellez, F., J. Castro-Severyn, C. Pardo-Esté, P. Aguilar, J. Fortt, C. Salinas, et al. 2018. Characterization and salt response in recurrent halotolerant *Exiguobacterium* sp. SH31 isolated from sediments of Salar de Huasco, Chilean Altiplano. Front. Microbiol. 9: Article 2228, 17 pages.

Rosenbaum, E., F. Gabel, M.A. Durá, S. Finet, C. Cléry-Barraud, P. Masson, et al. 2012. Effects of hydrostatic pressure on the quaternary structure and enzymatic activity of a large peptidase complex from *Pyrococcus horikoshii*. Arch. Biochem. Biophys. 517(2): 104–110.

Rossi, M. and A. Guagliardi 2011. Heat-shock response in thermophilic microorganisms. Extremophiles – Vol. I. Encyclopedia of Life Support Systems (EOLSS).

Russell, N.J. and N. Fukunaga. 1990. A comparison of thermal adaptation of membrane lipids in psychrophilic and thermophilic bacteria. FEMS Microbiol. Rev. 75: 171–182.

Schleper, C., G. Puehler, I. Holz, A. Gambacorta, D. Janekovic, U. Santarius, et al. 1995. *Picrophilus* gen. nov., fam. nov.: a novel aerobic, heterotrophic, thermoacidophilic genus and family comprising archaea capable of growth around pH 0. J. Bacteriol. 177(24): 7050–7059.

Schulze-Makuch, D. and L.N. Irwin. 2008. Life in the Universe: Expectations and Constraints. Springer, Berlin.

Seki, K. and M. Toyoshima. 1998. Preserving tardigrades under pressure. Nature 395: 853–854.

Sharma, A., D. Parashar and T. Satyanarayana. 2016. Acidophilic Microbes: Biology and Applications. pp. 215–241. *In*: P.H. Rampelotto. [ed.]. Biotechnology of Extremophiles, Grand Challenges in Biology and Biotechnology 1. Springer International Publishing, Switzerland.

Sharma, A., Y. Kawarabayasi and T. Satyanarayana. 2012. Acidophilic bacteria and archaea: acid stable biocatalysts and their potential applications. Extremophiles 16(1): 1–19.

Shigi, N. 2018. Recent advances in our understanding of the biosynthesis of sulfur modifications in tRNAs. Front. Microbiol. 9: Article 2679, 9 pages.

Siliakus, M.F., J. van der Oost and S.W.M. Kengen. 2017. Adaptations of archaeal and bacterial membranes to variations in temperature, pH and pressure. Extremophiles 21(4): 651–670.

Simonato, F., S. Campanaro, F.M. Lauro, A. Vezzi, M. D'Angelo, N. Vitulo, et al. 2006. Piezophilic adaptation: a genomic point of view. J. Biotechnol. 126(1): 11–25.

Stenesh, J. 1976. Extreme Environments: Mechanisms of Microbial Adaptation. Academic Press, London, New York, San Francisco.

Suzuki, I., D. Lee, B. Mackay, L. Harahuc and J.K. Oh. 1999. Effect of various ions, ph, and osmotic pressure on oxidation of elemental sulfur by *Thiobacillus thiooxidans*. Appl. Environ. Microbiol. 65(11): 5163–5168.

Swire J. 2007. Selection on synthesis cost affects interprotein amino acid usage in all three domains of life. J. Mol. Evol. 64: 558–571.

Takai, K., K. Nakamura, T. Toki, U. Tsunogai, M. Miyazaki, J. Miyazaki, et al. 2008. Cell proliferation at 122°C and isotopically heavy CH_4 production by a hyperthermophilic methanogen under high-pressure cultivation. Proc. Natl. Acad. Sci. U.S.A. 105(31): 10949–10954.

Techtmann, S.M., K.S. Fitzgerald, S.C. Stelling, D.C. Joyner, S.M. Uttukar, A.P. Harris, et al. 2016. *Colwellia psychrerythraea* strains from distant deep sea basins show adaptation to local conditions. Front. Environ. Sci. 4: Article 33, 10 pages.

Tettelin, H., K.E. Nelson, I.T. Paulsen, J.A. Eisen, T.D. Read, S. Peterson, et al. 2001. Complete genome sequence of a virulent isolate of *Streptococcus pneumoniae*. Science 293(5529): 498–506.

Tong, S.Y., J.S. Davis, E. Eichenberger, T.L. Holland and V.G. Fowler Jr. 2015. *Staphylococcus aureus* infections: epidemiology, pathophysiology, clinical manifestations, and management. Clin. Microbiol. Rev. 28(3): 603–661.

Toth, I.K., K.S. Bell, M.C. Holeva and P.R. Birch. 2003. Soft rot erwiniae: from genes to genomes". Mol. Plant Pathol. 4(1): 17–30.

Tribelli, P.M., E.C.S. Venero, M.M. Ricardi, M. Gómez-Lozano, L.J.R. Iustman, S. Molin, et al. 2015. Novel essential role of ethanol oxidation genes at low temperature revealed by transcriptome analysis in the antarctic bacterium *Pseudomonas extremaustralis*. PLoS One 10(12): e0145353.

Tribelli, P.M. and N.I. López. 2018. Reporting key features in cold-adapted bacteria. Life 8(1): 8.

Trotsenko, Y.A. and V.N. Khmelenina. 2002. Biology of extremophilic and extremotolerant methanotrophs. Arch. Microbiol. 177(2): 123–131.

Vannier, P., G. Michoud, P. Oger, V. Thór Marteinsson and M. Jebbar. 2015. Genome expression of *Thermococcus barophilus* and *Thermococcus kodakarensis* in response to different hydrostatic pressure conditions. Res. Microbiol. 166: 717–725.

Vieille, C. and G.J. Zeikus 2001. Hyperthermophilic enzymes: sources, uses, and molecular mechanism for thermostability. Microbiol. Mol. Biol. Rev. 65: 1–43.

Wang, H., X. Xia and D. Hickey. 2006. Thermal adaptation of the small subunit ribosomal RNA gene: a comparative study. J. Mol. Evol. 63: 120–126.

Watanabe, K., M. Shinma, T. Oshima and S. Nishimura. 1976. Heat-induced stability of tRNA from an extreme thermophile, *Thermus thermophilus*. Biochem. Biophys. Res. Commun. 72(3): 1137–1144.

Welch, T.J. and D.H. Barlett. 1996. Isolation and characterization of the structural gene for OmpL, a pressure-regulated porin-like protein for the deep sea bacterium *Photobacterium* species strain SS9. J. Bacteriol. 178: 5027–5031.

Willey, J., L. Sherwood and C.J. Woolverton. 2011. Prescott's Microbiology. McGraw-Hill, International Edition.

Yan, W., X. Xiao and Y. Zhang. 2016. Complete genome sequence of the *Sporosarcina psychrophila* DSM 6497, a psychrophilic *Bacillus* strain that mediates the calcium carbonate precipitation. J. Biotechnol. 226: 14–15.

Yumoto, I., K. Yamazaki, M. Hishinuma, Y. Nodasaka, A. Suemori, K. Nakajima, et al. 2001. *Pseudomonas alcaliphila* sp. nov., a novel facultatively psychrophilic alkaliphile isolated from sea water. Int. J. Syst. Evol. Microbiol. 51(Pt 2): 349–355.

Zhang, G. and H. Ge. 2013. Protein hypersaline adaptation: insight from amino acids with machine learning algorithms. Protein J. 32(4): 239–245.

12

Shaping Microbial Communities in Changing Environments: The Paradigm of Solar Salterns

Alyssa Carré-Mlouka

Introduction

Salt is nowadays considered a common good, a practically inexhaustible product found on every table and a universal tasty supplement to dishes worldwide. Salt production for human consumption, however, is only a marginal fraction of the global salt market. Other uses include chemical or pharmaceutical additives, de-icing leather tanning, water treatment, animal feed, etc. Because salt is essential to cell functions, Mankind has sought to extract it from natural sources for alimentary purposes since the Neolithic era. Salt is chemically defined as an assembly of ions combined to form a neutral compound, but the word is commonly used to designate sodium chloride, the most abundant of these substances.

Salt can be obtained from different types of natural sources. Solid salt can be found directly as rocks (halite), which consist of crystallized salt. Halite can be encountered as large deposits due to evaporation over geological times, sometimes in salt deserts or as huge masses pushed up by tectonic pressure, named salt domes. Salt is exploited from these sources either in salt mines to extract halite or can also be processed by the projection of pressured water over the rocks to produce brines that are then evaporated. Liquid sources of salt are evidently seawater, but other inland areas such as salt marshes, salt lakes, or salted springs are also exploited; total evaporation is reached by heating, drying, or pressuring those brines (Bousiges 2015).

As human populations adopted agriculture as a mode of living and became sedentary, they have been exploiting the surrounding natural brines, and solar salterns have gradually become the most successful salt-making method, still largely used nowadays (Weller 2015). Solar salterns, also called salines or salinas, consist of a series of connected shallow ponds through which brines are pumped and left to reside for a variable time depending on temperature, winds, and rainfalls. The purpose is to evaporate water and concentrate salt in successive basins until reaching the crystallization ponds where NaCl precipitates as halite and can be harvested.

The fields of environmental microbiology and salt production may seem disconnected as salt is known as a compound inhibiting microbial growth. Indeed, one traditional usage of alimentary salt has been long-term food conservation to avoid microbial spoilage. At lower salinities, similar to

Associate Professor, National Museum of Natural History, Molecules of Communication and Adaptation of Microorganisms (MCAM).
Address: CP54, 57 rue Cuvier, 75231 Paris Cedex 05, France.
*Corresponding author: alyssa.carre-mlouka@mnhn.fr

those of seawater entering the ponds of a solar saltern (30-35 g/L dissolved salts) or slightly above, living organisms are abundant with a variety of microorganisms, plants, algae, fishes, mollusks, and crustaceans that serve as the basis of a food web attracting birds. As salinity increases, fewer organisms are able to cope with the high levels of salts, and macroscopic life disappears. These so-called hypersaline environments are thus considered extremophilic, in the sense that they do not allow the development of most living beings. However, modern microbiology has evidenced that hypersaline areas are in fact colonized by numerous microorganisms, which can be defined as halophiles, salt-loving organisms (Javor 1989, Oren 2019).

Solar salterns thus represent very convenient systems for ecological studies examining the effect of a salinity gradient on microbial diversity and the cellular mechanisms of adaptation to osmotic pressure, ranging from ordinary moderate salt concentrations to extreme ones (Javor 2002). Indeed, the environmental conditions are relatively stable over time, and microbial communities including members of the three domains of life are well defined, although biogeographic characteristics may influence their composition. Understanding how microbes thrive in solar salterns also engenders practical applications in the field of salt-making as the presence of microorganisms influences crystallization and purity of halite but also in the land reclamation processes, such as bioremediation or revegetalization (Oren 2015a).

This chapter presents how solar salterns function, what is the diversity and community composition of microorganisms colonizing the ponds, and how microbes adapt to salinity gradients and other environmental changes.

Solar Salterns: Characteristics and Functions

Salt production areas, also called saltworks, are present all over the world in temperate, tropical, and even desert regions. Most salts nowadays are produced in solar salterns, man-made facilities that exploit any liquid source of salt, to extract salt crystals for human use. The principal type of salt exploited is sodium chloride (NaCl), also called halite, which has multiple uses: table salt, but also food, glass, soap, detergent, dyeing, or de-icing industry. At a much smaller scale, other salts can be commercially extracted from solar salterns: calcium salts used in the chemical industry; potassium salts collectively called potash with fertilizer properties; magnesium salts for food or paper industry; lithium salts used in the pharmaceutical industry or as a component of batteries (Pedrós-Alió 2004).

The basic process of solar salterns is to take advantage of environmental (sun, wind, aridity) and geographical (flat shallow depressions) conditions to concentrate salts from a liquid source. The installation of a saltwork is thus imposed to be located close to a source of salted waters and to occupy large areas constituting evaporating basins that shape the landscape. Saltworks are high primary productivity ecosystems that constitute particular habitats, often sheltering complex flora and fauna and providing refuge for migratory birds. Besides these ecological benefits, solar salterns can also furnish ecosystemic services to mankind, such as biodiversity maintenance, carbon sequestration, or eco-tourism (de Melo Soares et al. 2018).

Most solar salterns are localized along coasts, where seawater is pumped as the source of salted water. In those salterns, the waters are called thalassohaline, meaning their proportional ionic composition reflects that of seawater and the most abundant salt being sodium chloride. Saltworks have been constructed along shores in numerous zones all over the world, where they occupy large areas that can be easily spotted with exquisite colors on aerial or satellite photographs (Figure 1).

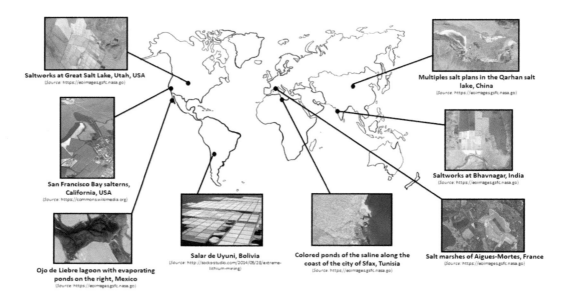

Saltworks at Great Salt Lake, Utah, USA
(Source: https://eoimages.gsfc.nasa.go)

Multiples salt plans in the Qarhan salt
lake, China
(Source: https://eoimages.gsfc.nasa.go)

San Francisco Bay salterns,
California, USA
(Source: https://commons.wikimedia.org)

Saltworks at Bhavnagar, India
(Source: https://eoimages.gsfc.nasa.go)

Ojo de Liebre lagoon with evaporating
ponds on the right, Mexico
(Source: https://eoimages.gsfc.nasa.go)

Salar de Uyuni, Bolivia
(Source: http://socks-studio.com/2014/05/28/extreme-
lithium-mining)

Colored ponds of the saline along the
coast of the city of Sfax, Tunisia
(Source: https://eoimages.gsfc.nasa.go)

Salt marshes of Aigues-Mortes, France
(Source: https://eoimages.gsfc.nasa.go)

Figure 1. World Map Featuring Selected Solar Salterns.

Particular coastal areas exploited for salt extraction include:

✓ Intertidal zones, also called seashores, lie below water level at high tide and above water level
at low tide. They typically constitute ideal zones to implant solar salterns as the salty water
may naturally enter the construction with the tides. Most solar salterns are localized on seashores around the world as is the saline of Sfax, on the south-eastern coast of Tunisia (Figures
1 and 2), or the saltworks at Bhavnagar in India (Figure 1).

✓ Salt marshes are particular intertidal zones, characterized by the nature of the sediments
that are composed of mud and peat. They constitute natural coastal wetlands along low-
energy shores where limited tides allow for growth of salt-resistant plants and accumulation
of decomposing organic matter, thus giving rise to the sediments rich in organic matter.
Salt marshes have been transformed into evaporating ponds for salt extraction in numerous
regions; some have been degraded and are now considered for land restoration (Gedan et al.
2009). The San Francisco Bay salterns in California, USA (Figure 1), or the Aigues-Mortes
marshes in southern France are examples of large scale exploitation of salt marshes (Figure 1).

✓ Coastal lagoons are shallow areas separated from the sea by a low sandbank, a coral reef,
or an atoll. When communication with the open sea is reduced and the dry and hot climate
results in high evaporation, the seawater may naturally concentrate and the lagoon water
becomes hypersaline. This is the case in the Ojo de Liebre lagoon, a hypersaline lagoon
located on the northwestern coast of Mexico, where one of the biggest saltwork plants in the
world is being exploited (Figure 1).

But salt exploitation is not restricted to extraction from seawater. Inland areas can also be
operated into saline ponds. The source of salted water can be thalassohaline, thus including waters
that originate from the sea but have been isolated for a long time. This results in modified ionic
composition due to successive cycles of evaporation and rainfalls; however, the main salt is essentially
sodium chloride as in seawater. Athalassohaline waters can also be used for salt extraction. These
are defined as salted waters for which ionic composition differs greatly from that of seawater that
may have been formed by freshwater solubilization of salts from rocks or volcanic ashes. Inland
solar salterns can be implanted in hypersaline environments such as:

✓ Inland hypersaline lakes are landlocked water bodies that have a salt concentration higher than seawater. They are usually large and deep water plans that are filled with water all year long. Both thalassohaline lakes, like the Great Salt Lake in Utah, USA (Figure 1), and athalassohaline lakes, like the Dead Sea in Israel, are being exploited for salt production. Among athalassohaline lakes, alkaline lakes called soda lakes display a peculiar ionic composition as they lack divalent ions and are rich in carbonate CO_3^{2-} and bicarbonate HCO_3^- ions, which make up for the high pH (10-11). For example, Lake Magadi in Kenya is a soda lake where a solar saltern has been implanted to extract halite and gypsum.

✓ 'Sebkhas', also called 'chotts' in Northern Africa, are the equivalent of the 'salars' of South America or 'playas' in Asia. These environments, sometimes called salt pans or salt flats, are typical of arid desertic regions. They are flat geographical depressions onto which evaporite salts accumulate and that can be temporarily filled with concentrated brines. The edges of the sebkhas are undefined and change depending on the seasons and on the climate annual variations. During the rainy season, they may be filled with water and crossed by boat, while during the dry season they consist of a crust of precipitated salts, sometimes with muddy areas. Depending on the depth of the sebkha, salt evaporites can be collected from large areas, or only at the edges. Among the largest of such environments exploited for salt extraction is Lake Qarhan in China (Figure 1).

✓ Salt deserts are extreme environments that derive from inland hypersaline lakes or sebkhas, and the distinction with the latest is sometimes blurry. They are wide drained surfaces covered with salts, found in arid areas. Some are exploited for mining-specific minerals from the salt crust, which is first resuspended in pumped groundwater, then evaporated as in typical solar saltern to separate the salts that must sometimes be further purified by chemical treatments. Examples of exploited sites for extracting lithium salts are the Atacama desert in Chile or the Salar de Uyuni in Bolivia (Figure 1).

✓ Brine springs, also called saline springs, are springs emerging from saline soils. The salt concentration and composition may vary depending on the nature of the surrounding evaporite rocks. Saline waters spring out at low flow rates, and saltworks exploiting this resource are usually of modest size. The village of La Malaha in Spain shelters a natural brine spring, which is exploited by the construction of successive shallow basins taking advantage of the smooth slope of the hills.

The typical operation in a solar saltern is depicted in Figure 2, illustrated by the functioning of the saline of Sfax in Tunisia. The seawater is brought naturally by tides through communicating doors leading to decantation basins, where floating solid material will eventually be cleared and water stocked before further operation. The process then consists of flowing the brines through a series of shallow ponds, where the high surface-to-volume ratio associated with heating from sunlight and the presence of winds favors evaporation of water. This results in the concentration of ions that will combine to form salts, ultimately reaching a concentration that will allow crystallization. The nature of the salts and the exact concentration at which they precipitate depends on the composition of the source of the salted water, but the three main salts crystallize sequentially in the following order: calcite ($CaCO_3$), at approximately a two-fold water concentration, then gypsum ($CaSO_4$) when salted water has concentrated four-fold, that precipitates to form a crust at the bottom of the ponds, and finally when reaching a ten-fold concentration, halite (NaCl) which is the main collected salt. The resulting brines, depleted in sodium chloride, are called bitterns and contain secondary salts, mainly magnesium salts. In many solar salterns, they are treated as wastes and are returned to the sea, which may cause environmental issues as this temporarily increases salinity. Other salt plants further exploit the bitterns to extract magnesium salts or to be commercialized as brines used as coagulants in the food (tofu production) or dyeing industry.

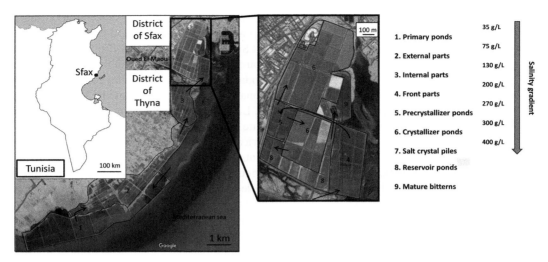

Figure 2. Salt Concentration Along the Ponds in the Saline of Sfax, Tunisia.

The saline of Sfax is a major salt-producing site in Tunisia, exploited by the Cotusal Company. It produces a mean 315,000 tons of salt per year and a few thousand tons of brines. The produced salts include alimentary salts and industrial salts for use in hides, cattle feeding, de-icing, or detergents. Operated since the 19th century, it occupies a surface of 1,700 ha along 13 km of the southern Mediterranean coast of the city of Sfax. The saline belongs to the district of Sfax in the north, and the district of Thyna in the south, separated by a river, the Oued El Maou. Its implantation was facilitated by the natural topology of the area, where a flat and extensive surface runs along the seashore but also by the hot and arid climate, which provokes a large rainfall deficit over the whole period of the year. The saline of Sfax is constituted of series of shallow (20-70 cm deep) ponds with a salinity range of 40-400 g/L. The operating period of the saline is determined by meteorological conditions which must prove optimal for evaporation and concentration of brines (heat, no rain, and constant wind). The salinity of the ponds is carefully monitored and water masses are transferred to subsequent ponds by gravity or pumping, only when they reach the expected density (Ladhar 2014, Trigui et al. 2011a).

In the saline of Sfax, the seawater (salinity 35-40 g/L) flows through the following set of eight series of ponds (the first six series for sodium chloride production and the last two series for magnesium saturated brines production):

1. 17 successive basins, called primary ponds (Figure 2, 1), a total surface of 640 ha, final salinity 75 g/L, that flow into:

2. five parallel circuits, called the external parts (Figure 2, 2), a total surface of 450 ha, final salinity 130 g/L. Brines from the five circuits are then reunited by a canal and distributed into:

3. three parallel circuits, called the internal parts (Figure 2, 3), a total surface of 160 ha, final salinity 200 g/L. Brines from the three circuits are then reunited by a canal and flown through:

4. three parallel circuits, called the front parts (Figure 2, 4), a total surface of 50 ha, final salinity 270 g/L. Brines from the three circuits are then reunited by pumping into the Sidi Salem station and reinjected into:

5. six communicating ponds, called the precrystallizer ponds (Figure 2, 5), a total surface of 20 ha, final salinity 300 g/L. The brines can then feed:

6. the crystallizer ponds (Figure 2, 6), a total surface of 140 ha, where the salinity reaches 400 g/L, thus triggering precipitation of sodium chloride that can then be dried and harvested (Figure 2, 7). The resulting bitterns are then transferred to:

7. the reservoir ponds (Figure 2, 8), where the bitterns are stored during winter, then further evaporated to yield the final brines (Figure 2, 9).

Solar salterns thus represent artificial ecosystems, where one environmental parameter is strictly controlled, can thus be considered as a series of huge semi-closed chemostats with a gradient of salinity allowing to study the impact of salt concentration on microbial communities. This gradient is of special interest to microbial ecologists as it covers habitats ranging from the most common and abundant on Earth (seawater) to conditions considered to be extreme and unsuitable for the development of life (hypersaline waters saturated with salt). Because the setup of solar salterns from different geographical locations is generally similar, they offer an opportunity to investigate the effects of other abiotic factors, such as temperature, light, or the trophic status, on the composition and dynamics of these populations. Most importantly, several studies have demonstrated the stability and reproducibility of solar salterns parameters over time (Khemakhem et al. 2013). While they represent easy to access sampling sites, better knowledge of saltern microbial ecology may result in beneficial applications for mankind, such as understanding the role of microbes in salt-making processes, biodiversity resilience to climate change, and ecosystemic responses to pollution, especially in areas where land reclamation is under consideration (Javor 2002, Oren 2019).

Microbial Residents in Solar Salterns

Solar saltern ponds, therefore, include extremely varied sites, from relatively low-salt areas in which macrofauna (small fishes, crustaceans) and macroflora (algae) can still develop quite easily to areas saturated with salts that are devoid of any form of macroscopic life. Intermediate to hypersaline ponds (150-250 g/L) often shelter important populations of the brine shrimp *Artemia salina*, a small crustacean eaten by migratory birds. Solar salterns are massively colonized by a microbial flora, which confers special colors to the evaporating ponds, from green at lower salt concentrations to red-orange in hypersaline ponds (Figures 1 and 2). Salt-loving microorganisms, called halophiles, can be categorized according to their growth relation to salt:

- Halotolerant microbes tolerate but do not require the presence of salts. They usually develop very well in low-salt water but can be found in ponds with a saline concentration of up to 150 g/L.
- Slightly halophilic strains preferentially grow at a salinity of up to 35 g/L and are typically marine organisms found in seawater entering the salterns.
- Moderately halophilic microorganisms thrive in salterns primary ponds with a salinity up to 150 g/L.
- Extremely halophilic organisms colonize areas with saline concentrations above 150 g/L and can thrive in saturated crystallizer ponds (400-450 g/L depending on temperature and depth). Some are even able to multiply on salt crystals or in saturated brine inclusions within salt crystals.

Halophilic microorganisms can be found in the three domains of life (bacteria, archaea, eukaryotes), all three being accompanied by corresponding haloviruses. Halophily represents a cellular adaptative feature that has emerged at several occurrences during evolution. Halophilic microorganisms usually form polyphyletic groups, but there are a few exceptions like class *Halobacteria* among the archaea for which all members are extreme halophiles.

Microbial habitats in solar salterns are typical of aquatic environments: the water column and the sediments are considered as the principal microbial colonized areas. However, solar salterns habitats come with a few specific features. Indeed, the ponds are generally shallow, which homogenizes the water column that is not stratified as deepest water environments can be. Planktonic microorganisms, besides their different degrees of halophily, are also subjected to various other extreme conditions: direct light and UV irradiation, high temperatures, and oxygen depletion. Indeed, in saltern ponds, the water is left to settle for long periods with little movement which would provide aeration. Besides, oxygen is poorly dissolved at high salt concentration, rendering the final basins micro-aerobic to anaerobic. Salterns planktonic microbes can often be qualified as polyextremophiles. Interestingly, most new halophilic prokaryotic taxa described in the literature, have been isolated from solar salterns, which prove to be a rich source of microbial biodiversity (Oren 2019).

Microorganisms developing on the floors of the saltern ponds can encounter very different situations depending on the nature of the sediment, which can be rich in organic matter as in salt marshes or very poor in many coastal salines. The sediment nature can also be modified due to the process of evaporating. The precipitation of the calcium salt gypsum forms a salt crust at the bottom of the intermediate salinity ponds, which may be several centimeters deep. The salt crusts harbor a complex benthic stratified microbiota organized in mats, synthesizing dense polysaccharidic matrices that help seal the bottom of the pond, avoiding leakage of the brines (Pedrós-Alió 2004, McGenity and Oren 2012).

Microbial inhabitants of solar salterns can be described according to their energy metabolism properties, forming ecological guilds as can be found in other aquatic environments. The variety of colors observed in saltern ponds is mostly due to the development of phototrophic plankton, which produces a diversity of pigments absorbing light spanning the whole spectrum of wavelengths. The green photosynthetic pigments chlorophylls allow photon uptake and electron transfer permitting conversion of light energy into chemical energy. Other pigments involved in photon transfer to chlorophylls and photoprotection may be carotenoids, conferring red, pink, orange, or yellow colorations, and phycobilins providing red (phycoerythrin) or blue (photocyanin) colors (Mulders et al. 2014, Ambati et al. 2019). Heterotrophic microbes may also contribute to the colors of the brines by the accumulation of pigments (Table 1).

PRIMARY PRODUCERS

Both aerobic and anaerobic phototrophs thrive in solar salterns ponds, where light is abundant and easily reaches the bottom of the shallow ponds.

Aerobic phototrophs include eukaryotic microalgae and cyanobacteria, which all contain the green membrane pigment chlorophyll-a. Aerobic photosynthesis uses water as the terminal electron acceptor, oxidizing it to oxygen, thus contributing to maintaining aerobic conditions in stagnant waters in solar saltern ponds. In solar salterns, the most abundant aerobic phototrophs are the green microalgae *Chlorophyceae* belonging to the species *Dunaliella salina* (Figure 3), which is ubiquitous in salty environments. *Dunaliella salina* green single cells are bi-flagellated. They can turn orange due to the accumulation of β-carotene in cytoplasmic granules upon nutritional, osmotic, or light-induced stresses (Oren 2014; Table 1). The pink coloration of certain table salts, like the salt extracted from the Aigues-Mortes salterns, can be attributed to the high cellular concentrations of *Dunaliella salina* in the crystallizer ponds. The salines of Aigues-Mortes have also developed a cosmetic brand named Eclae, *Dunaliella salina* being marketed as the main antioxidant component of their products, which is referred to as "the prodigy of pink waters" (http://eclae.com/fr). Other groups of eukaryotic phytoplankton have been reported in solar saltern ponds, such as diatoms (*Nitzschia, Amphora*), dinoflagellates (*Gymnodinium, Gonyaulax*), and *Euglenophycaea* (Ayadi et al. 2004, Madkour and Gaballah 2012).

Cyanobacteria, bacterial oxygenic phototrophs, are also major occupants of solar salterns. Besides chlorophyll-a, cyanobacteria can also produce significant quantities of specific carotenoids (myxoxanthophylls, echinenone, for example, Table 1) but also typically contain phycobilins. Cyanobacteria do require fairly strong light radiations to grow, and they are very abundant in seawater entering the primary ponds, while some species are halotolerant or halophiles and thrive in upper-salinity basins. Planktonic cyanobacteria, such as species of *Spirulina, Oscillatoria, Leptolyngbia, Aphanotece*, are encountered in solar saltern ponds (Ayadi et al. 2004, Oren 2015b), but most of the primary production attributable to cyanobacteria occurs in benthic environments. The gypsum crust found at the bottom of intermediate salinity ponds harbors benthic cyanobacteria, which constitute the upper layers of the microbial mats. Unicellular species such as *Aphanothece halophytica* usually develop at the top, forming a red-orange layer and filamentous cyanobacteria like *Phormidium* sp. form a green sheet underneath (McGenity and Oren 2012).

TABLE 1: Pigments and Osmoadaptation Strategies of Microorganisms Typical of Solar Salterns

	Salinity growth range (g/L)	Pigments	Osmoadaptation		References
			Strategy	Accumulated solutes	
Primary producers					
Aerobic					
Dunaliella salina	3-300	Chlorophyll-a, β-carotene, β-carotene, lycopene, lutein, zeaxanthin	Salt-out	Glycerol	Wasanasathian and Peng 2007, Chen and Jiang 2009, Gunde-Cimerman et al. 2018
Phormidium sp.	3-300	Chlorophyll-a, β-carotene, lutein, zeaxanthin, echinemone, myxoxanthophyll, canthaxanthin, C-phycocyanin, phycoerythrin	Salt-out	Sucrose, trehalose	Rodrigues et al. 2015 Gunde-Cimerman et al. 2018
Aphanothece sp.	3-300	Chlorophyll-a, phycoerythrin, allophycocyanin	Salt-out	Glycine betaine, proline, ectoine, trehalose	Mares et al. 2013, Oren 2015b, Gunde-Cimerman et al. 2018
Anaerobic					
Chromatium salexigens	30-200	Bacteriochlorophyll a, lycopene, spirilloxanthin, rhodovibrin	Salt-out	Ectoine	Caumette et al. 1988, Gunde-Cimerman et al. 2018
Halorhodospira sp.	90-120	Bacteriochlorophyll a, spirilloxanthin	Salt-out	Glycine betaine, ectoine, trehalose	Hirschler-Réa et al. 2003, Gunde-Cimerman et al. 2018
Heterotrophs					
Aerobic					
Halobacterium salinarum	130-290	Bacterioruberin, bacteriorhodopsin	Salt in	KCl	Zeng et al. 2006, Molaeirad et al. 2015
Hortaea werneckii	35-70	Melanin	Salt-out	Glycerol, erythritol, arabitol, mannitol	Kogej et al. 2007, Gunde-Cimerman et al. 2018
Salinibacter sp.	150-300	Salinixanthin, xanthorhodopsin	Salt in	KCl	Antón et al. 2002, Oren 2013
Halocafeteriaseosinensis	75-300	–	Salt-out	Hydroxyectoine, myoinositol	Park et al. 2006, Gunde-Cimerman et al. 2018
Anaerobic					
Halothermothrix orenii	40-200	–	Salt-in	NaCl, KCl	Cayol et al. 1994, Gunde-Cimerman et al. 2018

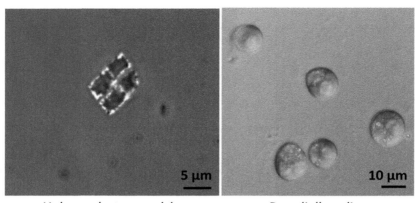

Haloquadratum walsby
(*Source*: http://www.dgburns.net/)

Dunaliella salina
(*Source*: Culture Collection of Autotrophic
Organisms, CCALA)

Figure 3. Electron Micrographs of Microbes Thriving in Solar Salterns.

Anoxygenic phototrophs constitute a fraction of primary producers in solar salterns. Anaerobic photosynthesis, a type of photosynthesis not using oxygen but other mineral elements (sulfates, hydrogen sulfide, elemental sulfur, nitrate) as the final acceptor of electrons, allows growth at lower light intensities in anaerobic environments as can be encountered in sediments of upper-salinity ponds. Some cyanobacterial genera (*Oscillatoria, Aphanotece*) undergo anoxygenic photosynthesis when encountering anaerobic conditions at the bottom of micro-aerobic ponds by shutting down photosystem II and using sulfide as the terminal acceptor of electrons (Oren 2015b). In most microbial mats, developing on gypsum crusts, other anoxygenic phototrophs develop below the aerobic phototrophic cyanobacterial layers. They form a purple sheet constituted of bacteria that accumulate sulfur granules intracellularly (*Halochromatium*) or secrete sulfur (*Halorhodospira*). These bacteria synthesize their own pigments (Table 1) necessary for photosynthesis (bacteriochlorophylls) but also accessory pigments allowing protection against UV rays (carotenoids and xanthophylls). They are also frequently encountered in the eutrophic sediments of anaerobic ponds in salterns operated in salt marshes.

HETEROTROPHS

Heterotrophic microorganisms living in solar salterns are extremely variable in terms of morphology, lifestyle, ecology, metabolic pathways. Unlike primary producers, they require organic matter for growth which therefore relies on the trophic content of the waters and the carbon-based compounds produced by autotrophs. As mentioned for autotrophs, they can develop in aerobic or anaerobic conditions. The diversity of heterotrophs is wide and cannot be covered extensively in this chapter, but the characteristics of some important ecological groups are indicated below.

Aerobic heterotrophs develop mainly in the water column of saltern ponds. They can be bacteria, archaea, or eukaryotes.

The main members of archaea found in solar salterns belong to the class *Halobacteria*. These archaea are considered as the model extreme halophiles with special adaptation to life at high salt concentration. The cells lyze when immersed in a salt-free medium, and they are unable to grow below 150 g/L salinity. They are particularly abundant in crystallizer ponds and can be entrapped in salt crystals. The members of class *Halobacteria* display various cellular morphologies: rods, cocci, triangle, trapezoid, and even square in the case of *Haloquadratum walsbyi* (Burns et al. 2007; Figure 3). Being subjected to intense light, these archaea synthesize carotenoid pigments (lycopene, bacterioruberine, β-carotene, phytoene, Table 1), which confer red-orange tones to the crystallizer ponds together with *Dunaliella*. At low-oxygen concentrations as can be found in some crystallizer ponds, archaea like *Halobacterium salinarum* adapt using a phototrophic pathway: the light-driven proton pump bacteriorhodopsin (Table 1) is a membrane pigment conferring purple color and allowing energy production (MacGenity and Oren 2012).

In the domain bacteria, different taxa harbor halophilic representatives that thrive in solar saltern ponds. Gamma-proteobacteria of the family *Halomonadaceae* are very abundant in the water column of all marine and hypersaline environments in all latitudes and may grow over a wide variety of saline concentrations. The genera *Halomonas* and *Chromohalobacter* are the most widespread (Arahal and Ventosa 2006). *Salinibacter ruber* is a member of the *Bacteroidetes* that has been isolated from Spanish crystallizer ponds of two salterns (Antón et al. 2002). Members of genus *Salinibacter* displays unusual features that resemble those of archaea from the class *Halobacteria* and may have been acquired by horizontal transfer, such as their extreme halophily or the presence of a light-driven pump named xanthorhodopsin. These bacteria also produce special carotenoids (salinixanthin, Table 1) contributing to the colors in crystallizer ponds (Oren 2013).

Aerobic heterotrophic eukaryotes also inhabit saltern basins but are usually abundant only in primary and intermediate salinity ponds. Flagellate, ciliate, and amoeboid protozoa have been documented in saltern ponds. Bacterivorous ciliates (*Strombidium* sp., *Urotricha* sp.) may contribute to regulating bacterial populations in solar salterns waters (Elloumi et al. 2009). A study of a Spanish hypersaline lagoon showed that ciliary cysts usually found in freshwater were present and could be reactivated upon dilution of the water (Esteban and Finlay 2003). The large ciliate *Fabrea salina* is often encountered but saltern management tends to disfavor its growth as it cannot be consumed by *Artemia* ((Harding and Simpson 2018). A few moderate or extremely halophilic nanoflagellates have been isolated from Korean crystallizer ponds (*Halocafeteria seosinensis* and *Pleurostonum flabellatum*, optimal growth at 150 g/Land 300 g/L, respectively) (Park et al. 2006, Park et al. 2007). More recently, halotolerant and halophilic fungi have also been recognized as regular inhabitants of solar salterns. The dominant group is black yeasts (*Hortaea werneckii*, *Aureobasidium pullulans*), but filamentous species (*Aspergillus* sp., *Penicillium* sp.) have also been described (Gunde-Cimerman and Zalar 2014, Chung et al. 2019).

Anaerobic heterotrophs usually develop in anoxic muddy sediments of salt marshes or the gypsum crust of salterns ponds, below the sheets of phototrophic microorganisms. They are always prokaryotes of the domain bacteria or archaea. They may grow using fermentative metabolic pathways or anaerobic respiration, which implies electron acceptors other than oxygen. Firmicutes of the order of *Halanaerobiales* (*Halothermothrix orenii*, Table 1) are halophiles that grow under strictly anaerobic conditions by fermentative routes, producing acetate, ethanol, hydrogen, and carbon dioxide. Some bacteria, such as *Sporohalobacter lortetii*, may form endospores, metabolically inactive but extremely resistant structures. *Haloanaerobacter chitinovorans* is capable of fermenting chitin from the shell of the brine shrimps. Archaea belonging to class *Halobacteria* may also grow using fermentation, like *Halobacterium* species, that can ferment arginine into ornithine, ammonia, and carbon dioxide (McGenity and Oren 2012).

Anaerobic respiration may occur using sulfate as an electron acceptor. Sulfate is particularly abundant in seawater and derived thalassohaline waters. Deltaproteobacteria affiliated with *Desulfohalobiaceae* and *Desulfocella* are abundant sulfate-reducing bacteria in hypersaline sediments from salterns (Lopez Lopez et al. 2010). Other types of anaerobic respiration are also occurring in solar salterns ponds. Denitrification, the dissimilatory reduction of nitrate, can be performed by moderately halotolerant proteobacteria (*Nitrosococcus*, *Nitrosomonas*) and firmicutes (*Bacillus halodenitrificans*), and halophilic archaea belonging to class *Halobacteria* (*Haloferax mediterranei*) (Torregrosa-Crespo et al. 2018). Methanogenesis and acetogenesis are alternative routes for anaerobic respiration, carried out by strictly anaerobic archaea from the families *Methanospirillaceae* and *Methanosarcinaceae* or by members of the *Halanaerobiales* like *Acetohalobium arabaticum*, respectively (McGenity and Oren 2012).

Viruses

In hypersaline environments, the number of viral particles is estimated to be around 10^9/mL, 10 to 100 fold the number of cellular microorganisms. Over one hundred haloviruses, specific to salty environments, are currently known, the vast majority (90%) being archaeal viruses infecting members of the *Halobacteria*. Very little data is available concerning halophilic eukaryotic viruses. One giant virus was described from *Acanthomoeba polyphaga*, a halotolerant amoeba isolated from a Tunisian salt marsh (Boughalmi et al. 2013). All halovirus genomes known so far are composed of double-stranded DNA, which can be linear or circular.

The known haloviruses display the following morphotypes (Atanasova et al. 2015, Atanasova et al. 2016):

- Icosahedral haloviruses with a contractile (myovirus) or non-contractile tail (siphovirus and podovirus); icosahedral viruses with an internal membrane (sphaerolipovirus). Most infect the host cell using a lytic mechanism (for example, the HHTV-1 virus from the archaea *Haloarcula hispanica*), and their host spectrum can be quite broad. The SNJ1 virus is the only icosahedral virus known to follow a lysogenic cycle: it replicates in the cytoplasm of its host in the form of circular double-stranded DNA, like a plasmid. Icosahedral viruses may infect archaea or halophilic bacteria.
- Pleomorphic viruses (pleolipovirus), of variable form, can undergo temperate cycle (lysogenic cycle like SNJ2 that can integrate into the genome of the host in the form of pro-phage) or be persistent. These are produced continuously by the host cell, and usually, this does not slow down its growth. Pleomorphic haloviruses have so far only been described in archaea.
- Lemon-shaped viruses are very abundant in hypersaline environments. They are the most common viruses in halophilic archaea. The His1 virus infecting *Haloarcula hispanica* can change its morphology to form thin tubes, which probably allows for injecting its DNA into the host cell.

Adaptation to Changing Environments and Population Dynamics

Microorganisms are in their vast majority composed of single cells; some species of algae or fungi display filamentous morphologies but the number of cells is limited. Microbes are therefore subjected to highly stressful situations dues to changes in the physico-chemical parameters of their living environment. Hence, they had to develop means of protection and adaptation. In solar salterns, microbes have to cope with multiple changing conditions. Salinity evidently comes first during the process of brine evaporation and also relates to rainfalls and temperature variations. Several studies have demonstrated that it is the main factor driving community composition and dynamics (Eloumi et al. 2009, Ventosa et al. 2014, Oren 2019). However, microbes also have to cope with daily and seasonal modifications of light, temperature, availability of nutrients, and at an increasing frequency, to pollution by toxic compounds.

Among these abiotic factors influencing microbial community composition, several would seem to be correlated with geographical localization, which evidently influences temperature, light, and rainfall. The concept of biogeographical regionalization implies that certain taxa are restricted to specific areas depending on environmental parameters and dispersal barriers. It has been widely investigated by ecologists for animals and plants, but few studies have addressed this issue for microorganisms. With a certain overlap in community composition, microbial assemblages from similar salterns in different regional areas have been shown to be distinct (Zhaxybayeva et al. 2013). However, this is not the case when high rank taxa are considered (families or higher, Clarke et al. 2017). Archaeal communities present in halite from salterns over the world have been investigated by Illumina sequencing and have shown no evidence of biogeographical regionalization (Clarke et al. 2017). Besides, antimicrobial interactions between halophilic bacterial and archaeal strains from several salterns have been shown to extend over geographical distances (Atanasova et al. 2013). It would thus seem that geographical localization in itself does not influence microbial communities in solar salterns. This could be attributed to passive dispersal via seawater, commercial salts, and also carriage by migratory birds feeding in saltern ponds (Clarke et al. 2017).

High salt concentration implies osmotic pressure, generally due to the main salt, sodium chloride. Cells are separated from their environment by a biological membrane, consisting of permeable to water phospholipid bilayer. Osmotic equilibrium will, therefore, tend to be achieved by passive diffusion of water on either side of the membrane. A hypo-osmotic surrounding environment induces a massive income of water, which can lead to lysis by excessive swelling of the cell. When the medium is hyper-osmotic, water leaks out of the cell which will decrease cytoplasmic volume. Halophilic microorganisms have, therefore, developed osmoadaptation strategies to cope with their surrounding osmotic pressure. Osmoregulation allows short-term adaptation during periodic changes in the osmotic pressure, which are frequent in solar salterns due to variations in rainfall, temperature, or sunshine. In order to maintain their cytoplasmic volume in hyper-osmotic fluids, halotolerant and halophilic microbes accumulate compounds in the cytoplasm. These osmolytes allow osmotic balance. The osmolytes can be organic or inorganic compounds and may be retrieved from their environments or be newly synthesized by microorganisms (McGenity and Oren 2012).

Two types of osmoadaptation strategies have been described:

Salt-Out Strategy: Accumulation of Organic Solutes

This strategy consists of accumulating organic molecules, also called compatible solutes or osmoprotectants, allowing them to maintain the osmotic balance. Compatible solutes are small polar molecular mass compounds. High cytoplasmic concentrations of compatible solutes do not interfere with intracellular enzymatic reactions. In addition, they stabilize the three-dimensional structure of proteins. Compatible solutes can be carbohydrates (trehalose, sucrose, sulfotrehalose), amino acids (proline, ectoin, glycine betaine), or polyols (glycerol, arabitol, mannitol). A halophilic microorganism generally uses a cocktail of different compatible solutes (Table 1).

The *salt-out* strategy is costly in energy but extremely widespread among microorganisms. During periodic changes in saline concentration, osmoregulation is carried out by import, biosynthesis, excretion, or degradation of compatible solutes. This strategy allows for great flexibility at a wide range of salt concentrations. Halotolerant microorganisms all use the *salt-out* strategy as do the cyanobacteria and most of the halophilic bacteria, fungi, microalgae as well as methanogenic archaea. The most common osmoprotective solutes are ectoin and glycine betaine (Table 1), mostly found in prokaryotes. Halophilic eukaryotes like *Dunaliella* preferentially use glycerol or other polyols instead (Table 1).

Salt-In Strategy: Accumulation of Inorganic Solutes

The *salt-in* strategy derives from cytoplasmic accumulation of inorganic anions and cations, usually K^+ and Cl^-, although some halophilic microorganisms like members of the *Halanaerobiales* also accumulate Na^+ and Cl^- (Table 1). Multiple active membrane ion transporters, which may be coupled to transport of other molecules such as amino acids, function to maintain high cytoplasmic ion concentration. The cytoplasm of such organisms is therefore extremely salty, which implies that the cellular machinery must be adapted to function at these high ionic concentrations. Indeed, the charges of the ions interfere with the electrostatic interactions maintaining the three-dimensional structure of proteins and nucleic acids. Enzymes from extreme halophilic organisms can function *in vivo* at very high salinity and will on the contrary be denatured if they are dissolved in water. Halophile proteins have a deficit in their basic amino acid composition (lysine and arginine) but are rich in acidic amino acids (glutamic and aspartic acids). This excess of negative charges, distributed over the surface of the protein structures, allows stabilization by solvation with K$^+$ ions. The low content of hydrophobic amino acids also improves the structural flexibility of halophilic enzymes. Some halophilic archaea also sustain adaptation to osmotic pressure by reinforcing their cell wall with the presence of a S-layer.

These osmoadaptation mechanisms are apparently less costly in energy for the microbial cells. However, they imply poor reactivity to short-term variations in salinity. The *salt-in* strategy is, therefore, adapted to relatively stable extremely saline conditions. In fact, the *salt-in* strategy is generally encountered in extreme halophilic prokaryotes, thriving in intermediate salinity or crystallizer ponds of solar salterns. These include archaea of the class *Halobacteria*, anaerobic bacteria of the order *Halanaerobiales*, and *Salinibacter ruber* (Table 1).

A vast number of studies have focused on the impact of salinity on microbial population dynamics in solar salterns, considered as model ecosystems, and have been reviewed recently (Ghai et al. 2011, Ventosa et al. 2014, Oren 2019). Most have focused on prokaryotic communities, while eukaryotic and viral inhabitants of saltern ponds have beneficiated from less attention. Only a summary of the general characteristics inferred from these works is depicted below, with a focus on studies performed in the saline of Sfax (Figure 4), for which bacteria, archaea, phytoplankton, flagellates, ciliates, and viruses communities have been investigated (Ayadi et al. 2004, Elloumi et al. 2009, Baati et al. 2008, Baati et al. 2010, Trigui et al. 2011a, Boujelben et al. 2012, Boujelben et al. 2014).

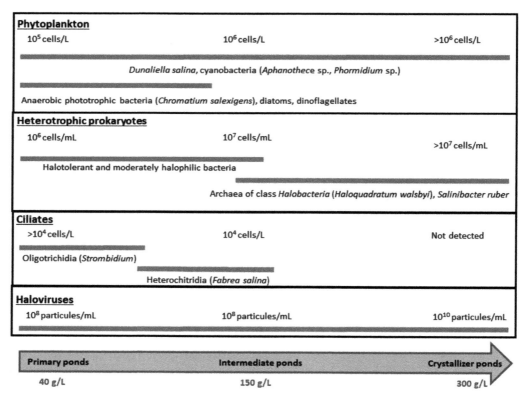

Figure 4: Abundance and Composition of Microbial Groups Growing at Different Salt Concentrations in the Saline of Sfax, Tunisia.

Whatever the microbial groups examined, the trend over the salinity gradients encountered in solar salterns demonstrates a drastic collapse in species diversity. At low salinity, close to that of seawater, heterotrophic prokaryotes are extremely diverse and dominated by bacterial taxa (98%) belonging to the proteobacteria, firmicutes, and actinobacteria. Very few archaea are present. As salinity increases, heterotrophic prokaryotic diversity decreases in favor of the archaea to reach approximately 90% of the prokaryotic community in crystallizer ponds. At high salinity, the genera *Haloquadratum* and *Halorubrum* are the most frequently encountered representatives of the archaea, while *Salinibacter* is the main bacterial genera. Similarly, in primary ponds phytoplankton is widely diverse, including cyanobacteria but mostly dominated by diatoms and flagellates. Usually, these ponds display green colors due to high amounts of chlorophyll-a that has been demonstrated to be at its maximum at salinity 80-120 g/L (Pedrós-Alió 2004). Upon reaching crystallizer ponds, the phytoplankton community gradually shifts to be overwhelmed by *Dunaliella,* which represents in many cases the main or the sole primary producer. The presence of *Dunaliella* supports the development of *Artemia*, which feeds on the algae but also allows the growth of heterotrophic plankton relying on *Dunaliella* compatible solute glycerol as the main consumed carbon compound. High salt ponds including crystallizer ponds are generally colored in red-orange attributed to various amounts of carotenoids (Table 1) from *Dunaliella* (β-carotene), archaea (bacterioruberin), and *Salinibacter* (salinixanthin), bacteriorhodopsin from halophilic archaea (Oren 2014, Oren 2019).

Ciliates are usually present in all ponds except the crystallizer ponds (Eloumi et al. 2009). Whenever their cellular hosts are present, viruses can also be found and are therefore ubiquitous in saltern ponds. Virus-to-cell ratios in hypersaline environments have been reported to reach a value of 100, considerably higher than in most environments. In the saline of Sfax, spherical viruses were dominant but decreased with salinity to the benefit of tailed and lemon-shaped viruses (Boujelben et al. 2012). Both ciliates through algivorous and bacterivorous grazing, and viruses that may produce lysis of their host upon multiplication, would represent important population regulators of microbial communities in solar saltern ponds. This statement must, however, be understated in high salinity basins, where most halophilic viruses are not lytic (Pedrós-Alió 2004, Boujelben et al. 2012).

Solar saltern ponds are high productivity areas. Both primary production values and the activity of heterotrophic plankton vary largely depending on salterns and the seasons at which they are measured (Pedrós-Alió 2004). Abundance and biomass of the microbial communities in ponds of different salinities from the saline of Sfax have been examined (Figure 4). Phytoplankton abundance was 10^5 cells per L in primary ponds and could reach over 10^6 cells per L in the crystallizer ponds due to the dominance of *Dunaliella*. Heterotrophic prokaryotes were more abundant, from 10^6 cells per mL at low salinity to 10^7 cells per mL in the crystallizer ponds. Flagellates and ciliates were present (10^5 cells per mL and 10^4 cells per L, respectively) in low and intermediate salinity ponds but absent from crystallizer ponds. Viruses were ubiquitous, with abundances ranging from 10^8 to 10^{10} particles per mL. Virus-to-cell ratio increased with salinity, reaching approximately 50 in crystallizer ponds (Ayadi et al. 2004, Eloumi et al. 2009, Boujelben et al. 2012).

Sediments and salt crystals from the saline of Sfax have also been examined for prokaryotic composition. Salt crystals were colonized by archaea of class *Halobacteria*, bacteroidetes, and alpha-and gamma-proteobacteria (Baati et al. 2010). Sediments from intermediate and high salinity ponds were compared and showed a higher diversity of prokaryotes, mostly of archaea, at high salinity (Baati et al. 2010). New taxa have been described from the saline of Sfax: *Paludifilum halophilum*, an actinobacterium isolated from sediments of intermediate salinity (Frikha-Dammak et al. 2016) and *Halorubrum sfaxense*, halophilic archaea originating from a crystallizer pond (Trigui et al. 2011b). Several investigations suggest that other yet undescribed taxa are present in the ponds and sediments of the saline of Sfax (Ayadi et al. 2004, Eloumi et al. 2009, Baati et al. 2010, Boujelben et al. 2012). Interesting potential applications of microbes from the solar saltern of Sfax, such as the production of protease, pigments, plant-growth promoting, or antimicrobial compounds have also been reported (Frikha-Dammak et al. 2016, Ghanmi et al. 2016, Boukhris et al. 2017, Ghanmi et al. 2019, Ghanmi et al. 2019, Masmoudi et al. 2019).

Besides salinity demonstrated to be the principal abiotic factor shaping microbial communities in salterns but also in many other physical environments (Lozupone and Knight 2007), only a handful of studies have addressed the specific impact of other environmental parameters. In solar salterns, light irradiation may be considered, besides salinity, as a major element subjected to environmental disturbances. A recent study has examined the effect of light intensities on microbial communities in the ponds of a Spanish saltern by using regular shading decreasing light intensity 37-fold. The high-light adapted communities were dominated by *Haloquadratum* and *Salinibacter*, rich in carotenoids (Table 1), and were very stable and change-resistant. The low-light communities were less resistant to modifications of light intensities, presumably because the previous long-term adaptation of high-light specific taxa had occurred. Overall, these investigations demonstrated that microbial community transitions were influenced by deterministic processes, which explained the major part of population shifts upon light transitions. Microbial salterns communities proved to be highly resilient to environmental changes with salinity and light being major determinants for the settlement of recurrent communities adapted to local conditions (Viver et al. 2019). Similarly, one study addressed the effect of seasons, highly correlated with the effect of light but also of temperature, on the microbial community in Argentinian salterns. Microbial abundances were positively correlated with temperature. Differences in the composition of microbial communities were observed along the seasons, however, at least 40% of taxa proved to be prevalent through the whole year, thus demonstrating important stability and resilience of the microbial assemblages to variations of environmental parameters along the seasons (Di Meglio et al. 2016).

Good knowledge and understanding of the impact of abiotic factors on the microbial communities in salterns have led to the development of a model based on spatiotemporal variations of the physicochemical parameters of the brines in the solar salterns of Sfax (Baati et al. 2012, Khemakhem et al. 2013). In fact, the production of high quantities of good quality salts relies on the presence of a healthy community, although the interactions between saltern microbes and salt crystallization are not totally unveiled (Javor 2002, Oren 2010, Oren 2019). In crystallizer ponds, the presence of dense communities saturated with carotenoid pigments absorbs light, which increases temperature and decreases the albido, thus favoring evaporation. The presence of halophilic archaea has been suggested to improve salt crystallization. Excessive growth of benthic microbial mats, especially the unicellular cyanobacterium *Aphanothece halophytica,* may be deleterious as it produces massive amounts of slime resulting in bad-quality salt. Other problems caused by the overgrowth of mats may be their lifting from the bottom and floating as rafts at the surface of the water due to excessive anaerobic conditions. The rafts decrease evaporation and result in putrefaction odors (Javor 2002).

Models predicting microbial communities depending on environmental parameters could, therefore, constitute good management tools for salt production in solar salterns. An ecological model was established for phytoplankton calibrated with 4-years field data using as variable the nutrient status (nitrogen and phosphorus), light, and temperature. The model was accurate to predict realistic changes in the phytoplankton biomasses (Khemakhem et al. 2013). Another model attempted to predict parameters evolution and microbial density during the process of evaporation. The results showed a significant correlation between factors such as salinity, pH, and dissolved oxygen and the density of *Dunaliella* and prokaryotes (Baati et al. 2012). Altogether, even if these models need to be refined and completed, they provide the basis for predicting the impact of environmental shifts on microbial communities through scenarios of climate change or saltern restoration.

Concluding Remarks: Environmental Issues and Sustainable Development

Human activities often induce pollution of aquatic ecosystems, especially emphasized in closed basins with a poor renewal of waters as is the case in solar salterns. Among those, heavy metal pollution is a severe environmental problem as these compounds are considered toxic to most aquatic organisms. When solar salterns are contaminated with heavy metals, those may be retrieved in the collected table salts, thus resulting in serious alimentary safety issues (Voica et al. 2016). The saline of Sfax is known to have been subjected to heavy metal pollution, by iron, copper, lead, cadmium, nickel, and zinc, inducing threats to human health (Bahloul 2019). Several halophilic bacteria from the saline of Sfax (*Bacillus*) revealed good resistance to heavy metals. Archaeal strains from an Indian saltern pan have shown metal resistance against nickel, aluminum, and mercury and good biosorption of nickel and aluminum (Williams et al. 2012). Heavy metal resistance is documented in a number of halophilic bacterial and archaeal strains. While the presence of heavy metals may impact communities in solar salterns, the bioremediation capacity of halophilic microbes may be exploited in metal-polluted saline soils or saline wastewaters from industry (Voica et al. 2016).

Hydrocarbons are among the most difficult pollutants to degrade. Petroleum is a complex mixture of hydrophobic and chemically very stable compounds (alkanes, cycloalkanes, mono- and poly-aromatic compounds), some of which are carcinogenic, geno- or eco-toxic. Hydrocarbons from petroleum are the most common organic pollutants on Earth. Several studies suggest an excellent capacity of halophilic microorganisms to degrade hydrocarbons with a high saline concentration. For example, archaea of the genus *Haloarcula* are capable of degrading saturated (tetradecane, hexadecane) or aromatic (phenanthrene) hydrocarbons (Barreteau et al. 2019). Many described species isolated from solar salterns have demonstrated the capacity to degrade various types of hydrocarbons, although the exact biochemical mechanisms remain unclear.

Conclusions

Solar salterns are unique ecosystems with special geochemical characteristics and adapted microbial communities. They represent a valuable source of microbial diversity, new taxa being regularly identified, and displaying special features such as the production of a prized compound or environmental remediation capacities. Climate change projections have identified coastal ecosystems as decisive spots to be impacted (Khemakhem et al. 2013). In fact, the singularity of solar salterns ecosystems is threatened by anthropogenic activities like pollution due to sewage effluents, industrial wastewaters and agricultural runoff, urbanization, and eutrophication (Paul and Mormile 2017). The harmful effects might imbalance the equilibrium of these ecosystems, thus irremediably affecting their microbial diversity and consequently other living beings such as birds. Efforts must be undertaken to preserve solar salterns as precious peculiar area, providing a wide diversity of ecosystemic services to mankind.

Acknowledgements

This work was supported by the ATM Microorganisms of the National Museum of Natural History, Paris, France.

References

Ambati, R.R., D. Gogisetty, R.G. Aswathanarayana, S. Ravi, P.N. Bikkina, L. Bo, et al. 2019. Industrial potential of carotenoid pigments from microalgae: current trends and future prospects. Crit. Rev. Food. Sci. Nutr. 59: 1880–1902.

Antón, J., A. Oren, S. Benlloch, F. Rodríguez-Valera, R. Amann and R. Rosselló-Mora. 2002. *Salinibacter ruber* gen. nov., sp. nov., a novel, extremely halophilic member of the *Bacteria* from saltern crystallizer ponds. Int. J. Syst. Evol. Microbiol. 52: 485–491.

Arahal, D.R. and A. Ventosa. 2006. The Family *Halomonadaceae*. pp. 811–835. *In:* M. Dworkin, S. Falkow, E. Rosenberg, K.-H. Schleifer and E. Stackebrandt [eds.]. The Prokaryotes: Volume 6: Proteobacteria: Gamma Subclass. Springer, New York, NY.

Atanasova, N.S., M.K. Pietilä and H.M. Oksanen. 2013. Diverse antimicrobial interactions of halophilic archaea and bacteria extend over geographical distances and cross the domain barrier. Microbiology Open 2: 811–825.

Atanasova, N.S., H.M. Oksanen and D.H. Bamford. 2015. Haloviruses of archaea, bacteria, and eukaryotes. Curr. Opin. Microbiol. 25: 40–48.

Atanasova, N.S., D.H. Bamford and H.M. Oksanen. 2016. Virus-host interplay in high salt environments. Environ. Microbiol. Rep. 8: 431–444.

Ayadi, H., O. Abib, J. Elloumi, A. Bouaïn and T. Sime-Ngando. 2004. Structure of the phytoplankton communities in two lagoons of different salinity in the Sfax saltern (Tunisia). J. Plankton Res. 26: 669–679.

Baati, H., S. Guermazi, R. Amdouni, N. Gharsallah, A. Sghir and E. Ammar. 2008. Prokaryotic diversity of a Tunisian multipond solar saltern. Extremophiles 12: 505–518.

Baati, H., S. Guermazi, N. Gharsallah, A. Sghir and E. Ammar. 2010. Novel prokaryotic diversity in sediments of Tunisian multipond solar saltern. Res. Microbiol. 161: 573–582.

Baati, H., R. Amdouni, C. Azri, N. Gharsallah and E. Ammar. 2012. Brines Modelling Progress: A Management Tool for Tunisian Multipond Solar Salterns, Based on Physical, Chemical and Microbial Parameters. Geomicrobiology J. 29: 139–150.

Bahloul, M. 2019. Pollution characteristics and health risk assessment of heavy metals in dry atmospheric deposits from Sfax solar saltern area in southeast of Tunisia. J. Environ. Health Sci. Eng. 17: 1085–1105.

Barreteau, H., M. Vandervennet, L. Guédon, V. Point, S. Canaan and S. Rebuffat. 2019. *Haloarcula sebkhae* sp. nov., an extremely halophilic archaeon from Algerian hypersaline environment. Int. J. Syst. Evol. Microbiol. 69: 732–738.

Boughalmi, M., H. Saadi, I. Pagnier, P. Colson, G. Fournous, D. Raoult, et al. 2013. High-throughput isolation of giant viruses of the Mimiviridae and Marseilleviridae families in the Tunisian environment. Environ. Microbiol. 15: 2000–2007.

Boujelben, I., M. Gomariz, M. Martínez-García, F. Santos, A. Peña, C. Lopez, et al. 2012. Spatial and seasonal prokaryotic community dynamics in ponds of increasing salinity of Sfax solar saltern in Tunisia. Antonie Van Leeuwenhoek 101: 845–857.

Boujelben, I., M. Martínez-García, J. van Pelt and S. Maalej. 2014. Diversity of cultivable halophilic archaea and bacteria from superficial hypersaline sediments of Tunisian solar salterns. Antonie Van Leeuwenhoek 106: 675–692.

Boukhris, S., K. Athmouni, I. Hamza-Mnif, R. Siala-Elleuch, H. Ayadi, M. Nasri, et al. 2017. The potential of a brown microalga cultivated in high salt medium for the production of high-value compounds. Biomed. Res. Int. 2017: 4018562.

Bousiges, E. 2015. Le sel à travers les âges, élément convoité mais controversé: quels effets sur l'homme? Ph.D. Thesis, University of Montpellier, France.

Burns, D.G., P.H. Janssen, T. Itoh, M. Kamekura, Z. Li, G. Jensen, et al. 2007. *Haloquadratum walsbyi* gen. nov., sp. nov., the square haloarchaeon of Walsby, isolated from saltern crystallizers in Australia and Spain. Int. J. Syst. Evol. Microbiol. 57: 387–392.

Caumette, P., R. Baulaigue and R. Matheron. 1988. Characterization of *Chromatium salexigens* sp. nov., a halophilic chromatiaceae isolated from Mediterranean salinas. Syst. Appl. Microbiol. 10: 284–292.

Cayol, J.L., B. Ollivier, B.K. Patel, G. Prensier, J. Guezennec and J. L. Garcia. 1994. Isolation and characterization of *Halothermothrix orenii* gen. nov., sp. nov., a halophilic, thermophilic, fermentative, strictly anaerobic bacterium. Int. J. Syst. Bacteriol. 44: 534–540.

Chen, H. and J.-G. Jiang. 2009. Osmotic responses of *Dunaliella* to the changes of salinity. J. Cell. Physiol. 219: 251–258.

Chung, D., H. Kim and H.S. Choi. 2019. Fungi in salterns. J. Microbiol. 57: 717–724.

Clarke, D., M. Mathieu, L. Mourot, L. Dufossé, G.J.C. Underwood, A.J. Dumbrell, et al. 2017. Biogeography at the limits of life: Do extremophilic microbial communities show biogeographical regionalization? Glob. Ecol. and Biogeogr. 26: 1435–1446.

de Melo Soares, R.H.R., C.A. de Assunção, F. de Oliveira Fernandes and E. Marinho-Soriano. 2018. Identification and analysis of ecosystem services associated with biodiversity of saltworks. Ocean Coast. Manag. 163: 278–284.

Di Meglio, L., F. Santos, M. Gomariz, C. Almansa, C. López, J. Antón, et al. 2016. Seasonal dynamics of extremely halophilic microbial communities in three Argentinian salterns. FEMS Microbiol. Ecol. 92(12): fiw184.

Elloumi, J., W. Guermazi, H. Ayadi, A. Bouain and L. Aleya. 2009. Abundance and biomass of prokaryotic and eukaryotic microorganisms coupled with environmental factors in an arid multi-pond solar saltern (Sfax, Tunisia). J. Mar. Biol. Assoc. U.K. 89: 243–253.

Esteban, G.F. and B.J. Finlay. 2003. Cryptic freshwater ciliates in a hypersaline lagoon. Protist 154: 411–418.

Frikha-Dammak, D., M.-L. Fardeau, J.-L. Cayol, L. Ben Fguira-Fourati, S. Najeh, B. Ollivier, et al. 2016. *Paludifilum halophilum* gen. nov., sp. nov., a thermoactinomycete isolated from superficial sediment of a solar saltern. Int. J. Syst. Evol. Microbiol. 66: 5371–5378.

Gedan, K.B., B.R. Silliman and M.D. Bertness. 2009. Centuries of Human-Driven Change in Salt Marsh Ecosystems. Ann. Rev. Mar. Sci. 1: 117–141.

Ghai, R., L. Pašić, A.B. Fernández, A.-B. Martin-Cuadrado, C.M. Mizuno, K.D. McMahon, et al. 2011. New abundant microbial groups in aquatic hypersaline environments. Sci. Rep. 1: 135.

Ghanmi, F., A. Carré-Mlouka, M. Vandervennet, I. Boujelben, D. Frikha, H. Ayadi, et al. 2016. Antagonistic interactions and production of halocin antimicrobial peptides among extremely halophilic prokaryotes isolated from the solar saltern of Sfax, Tunisia. Extremophiles 20: 363–374.

Ghanmi, F., A. Carré-Mlouka, Z. Zarai, H. Mejdoub, J. Peduzzi, J. Maalej, et al. 2019. The extremely halophilic archaeon *Halobacterium salinarum* ETD5 from the solar saltern of Sfax (Tunisia) produces multiple halocins. Res. Microbiol. (in press).

Gunde-Cimerman, N. and P. Zalar. 2014. Extremely halotolerant and halophilic fungi inhabit brine in solar salterns around the globe. Food Technol. Biotechnol. 52: 170–179.

Gunde-Cimerman, N., A. Plemenitaš and A. Oren. 2018. Strategies of adaptation of microorganisms of the three domains of life to high salt concentrations. FEMS Microbiol. Rev. 42: 353–375.

Harding, T. and A.G.B. Simpson. 2018. Recent Advances in Halophilic Protozoa Research. J. Eukaryot. Microbiol. 65: 556–570.

Hirschler-Réa, A., R. Matheron, C. Riffaud, S. Mouné, C. Eatock, R.A. Herbert, et al. 2003. Isolation and characterization of spirilloid purple phototrophic bacteria forming red layers in microbial mats of Mediterranean salterns: description of *Halorhodospira neutriphila* sp. nov. and emendation of the genus Halorhodospira. Int. J. Syst. Evol. Microbiol. 53: 153–163.

Javor, B. 1989. Solar Salterns. pp. 189–204. *In*: B. Javor [ed.]. Hypersaline Environments: Microbiology and Biogeochemistry, Brock/Springer Series in Contemporary Bioscience. Springer Berlin Heidelberg, Berlin, Heidelberg.

Javor, B. 2002. Industrial microbiology of solar salt production. J. Ind. Microbiol. Biotechnol. 28: 42–47.

Khemakhem, H., J. Elloumi, H. Ayadi, L. Aleya and M. Moussa. 2013. Modelling the phytoplankton dynamics in a nutrient-rich solar saltern pond: predicting the impact of restoration and climate change. Env. Sci. Pollut. Res. Int. 20: 9057–9065.

Kogej, T., M. Stein, M. Volkmann, A.A. Gorbushina, E.A. Galinski and N. Gunde-Cimerman. 2007. Osmotic adaptation of the halophilic fungus Hortaea werneckii: role of osmolytes and melanization. Microbiology 153: 4261–4273.

Ladhar, C. 2014. Etude de la dynamique, de la composition biochimique et de la variabilité génétique des copépodes et des *Artemia* d'un écosystème extrême: la saline de Sfax (Tunisie). Ph.D Thesis, University of Sfax, Tunisia, and University du Maine, France.

Lopez Lopez, A., P. Yarza, M. Richter, A. Suarez-Suarez, J. Antón, H. Niemann, et al. 2010. Extremely halophilic microbial communities in anaerobic sediments from a solar saltern. Environ. Microbiol. Rep. 2: 258-271.

Lozupone, C.A. and R. Knight. 2007. Global patterns in bacterial diversity. Proc. Natl. Acad. Sci. 104: 11436–11440.

Madkour, F.F. and M.M. Gaballah. 2012. Phytoplankton assemblage of a solar saltern in Port Fouad, Egypt. Oceanologia 54: 687–700.

Mares, J., P. Hrouzek, R. Kaňa, S. Ventura, O. Strunecký and J. Komárek. 2013. The primitive thylakoid-less cyanobacterium *Gloeobacter* is a common rock-dwelling organism. PLoS One 8: e66323.

Masmoudi, F., N. Abdelmalek, S. Tounsi, C.A. Dunlap and M. Trigui. 2019. Abiotic stress resistance, plant growth promotion and antifungal potential of halotolerant bacteria from a Tunisian solar saltern. Microbiol. Res. 229: 126331.

McGenity, T.J. and A. Oren. 2012. Hypersaline environments. pp. 402–437. *In*: E. Bell [ed.]. Life at Extremes: Environments, Organisms and Strategies for Survival. CABI Publishing, Wellington, Australia.

Molaeirad, A., S. Janfaza, A. Karimi-Fard and B. Mahyad. 2015. Photocurrent generation by adsorption of two main pigments of Halobacterium salinarum on TiO_2 nanostructured electrode. Biotechnol. Appl. Biochem. 62: 121–125.

Mulders, K.J.M., P.P. Lamers, D.E. Martens and R.H. Wijffels. 2014. Phototrophic pigment production with microalgae: biological constraints and opportunities. J. Phycol. 50: 229–242.

Oren, A. 2010. Industrial and environmental applications of halophilic microorganisms. Environ. Technol. 31: 825–834.

Oren, A. 2013. *Salinibacter*: an extremely halophilic bacterium with archaeal properties. FEMS Microbiol. Lett. 342: 1–9.

Oren, A. 2014. The ecology of *Dunaliella* in high-salt environments. J. Biol. Res. 21: 23.

Oren, A. 2015a. Halophilic microbial communities and their environments. Curr. Opin. Biotechnol. 33: 119–124.

Oren, A. 2015b. Cyanobacteria in hypersaline environments: biodiversity and physiological properties. Biodivers. Conserv. 24: 781–798.

Oren, A. 2019. Solar salterns as model systems for the study of halophilic microorganisms in their natural environments. pp. 41–56. *In*: J. Seckbach and P. Rampelotto [eds]. Model Ecosystems in Extreme Environments, Astrobiology Exploring Life on Earth and Beyond. Academic Press.

Park, J.S., B.C. Cho and A.G.B. Simpson. 2006. *Halocafeteria seosinensis* gen. et sp. nov. (Bicosoecida), a halophilic bacterivorous nanoflagellate isolated from a solar saltern. Extremophiles 10: 493–504.

Park, J.S., A.G.B. Simpson, W.J. Lee and B.C. Cho. 2007. Ultrastructure and phylogenetic placement within *Heterolobosea* of the previously unclassified, extremely halophilic heterotrophic flagellate *Pleurostomum flabellatum* (Ruinen 1938). Protist 158: 397–413.

Paul, V.G. and M.R. Mormile. 2017. A case for the protection of saline and hypersaline environments: a microbiological perspective. FEMS Microbiol. Ecol. 93(8): fix091.

Pedrós-Alió, C. 2004. Trophic ecology of solar salterns. pp. 33–48. *In*: A. Ventosa [ed]. Halophilic Microorganisms. Springer Berlin Heidelberg, Berlin, Heidelberg.

Rodrigues, D.B., C.R. Menezes, A.Z. Mercadante, E. Jacob-Lopes and L.Q. Zepka. 2015. Bioactive pigments from microalgae *Phormidium autumnale*. Food Res. Int. 77: 273–279.

Torregrosa-Crespo, J., L. Bergaust, C. Pire and R.M. Martínez-Espinosa. 2018. Denitrifying haloarchaea: sources and sinks of nitrogenous gases. FEMS Microbiol. Lett. 365.

Trigui, H., S. Masmoudi, C. Brochier-Armanet, A. Barani, G. Grégori, M. Denis, et al. 2011a. Characterization of heterotrophic prokaryote subgroups in the Sfax coastal solar salterns by combining flow cytometry cell sorting and phylogenetic analysis. Extremophiles 15: 347–358.

Trigui, H., S. Masmoudi, C. Brochier-Armanet, S. Maalej and S. Dukan. 2011b. Characterization of *Halorubrum sfaxense* sp. nov., a new halophilic archaeon isolated from the solar saltern of Sfax in Tunisia. Int. J. Microbiol. 2011: 240191.

Ventosa, A., A.B. Fernández, M.J. León, C. Sánchez-Porro and F. Rodriguez-Valera. 2014. The Santa Pola saltern as a model for studying the microbiota of hypersaline environments. Extremophiles 18: 811–824.

Viver, T., L.H. Orellana, S. Díaz, M. Urdiain, M.D. Ramos-Barbero, J.E. González-Pastor, et al. 2019. Predominance of deterministic microbial community dynamics in salterns exposed to different light intensities. Environ. Microbiol. 21: 4300–4315.

Voica, D.M., L. Bartha, H.L. Banciu and A. Oren. 2016. Heavy metal resistance in halophilic Bacteria and Archaea. FEMS Microbiol. Lett. 363(14): fnw146.

Wasanasathian, A. and C.-A. Peng. 2007. Algal photobioreactor for production of lutein and zeaxanthin. pp. 491–505. *In*: S.-T. Yan [ed]. Bioprocessing for Value-Added Products from Renewable Resources. Elsevier, Amsterdam.

Weller, O. 2015. First salt making in Europe: an overview from Neolithic times. Documenta Praehistorica 42: 185–196.

Williams, G.P., M. Gnanadesigan and S. Ravikumar. 2012. Biosorption and bio-kinetic studies of halobacterial strains against Ni^{2+}, Al^{3+} and Hg^{2+} metal ions. Biores. Technol. 107: 526–529.

Zeng, C., J.-C. Zhu, Y. Liu, Y. Yang, J.-Y. Zhu, Y.-P. Huang, et al. 2006. Investigation of the influence of NaCl concentration on *Halobacteriumsalinarum* growth. J. Therm. Anal. Calorim. 84: 625–630.

Zhaxybayeva, O., R. Stepanauskas, N.R. Mohan and R.T. Papke. 2013. Cell sorting analysis of geographically separated hypersaline environments. Extremophiles 17: 265–275.

13

Metabolic Pathways in Biodegrading Psychrotrophic Bacteria Under Cold Environments

Shruti Sinai Borker, Swami Pragya Prashant, Anu Kumari,
Sareeka Kumari, Rajni Devi and Rakshak Kumar*

Introduction

The Increasing Problem of Solid Waste (Worldwide/India)

The world economy has been growing leaps and bounds since the last five decades. The growth fueled by technological advancements and consumerism has, however, also been accompanied by growth in the generation of waste. The World Bank reports that the world generated 2.01 billion metric tons of urban municipal waste and if no urgent actions are taken to control such generation, it could grow to 3.4 billion metric tones—an estimated 70% growth—in the next thirty years (Silpa et al. 2018). India is no exception to the global trend and generates 1.45 lakh metric tons of solid waste in urban areas and about 62 million tons annually (Swaminathan 2018). Municipal solid waste has its sources in domestic waste, market waste, community waste, etc. The majority of this waste, nearly 51%, is biodegradable and can be converted into farm and agricultural input via composting/ vermicomposting, biomethanation, etc. Composting, which is the microbial breakdown of organic biomass, is the cheapest treatment of the organic fraction of municipal solid waste practiced in the country as it requires no special machinery.

Challenges to Waste Management in Cold/High Altitude Hilly Regions

Composting, as a process, is dependent on some important parameters, such as temperature, availability of moisture, and microbial population. In cold hilly regions, due to predominantly low temperatures and limited availability of moisture and microbial population, composting is much slower in natural conditions as compared to other regions. The Indian Himalayan regions, comprising 11 states [nine fully and two partially (three districts from Assam and two from West Bengal)] and two Union territories, house a population of nearly 50 million. The region generates a total of 4,558 metric tons of municipal solid waste per day (CPCB Annual Report 2016-17). The current scenario of solid waste management facilities in the region is, unfortunately, insufficient and the region faces a huge threat of pollution due to solid waste. With nearly 2,300 metric tons of organic waste generated per day in the area, the composting and vermicomposting facilities are rendered inactive

High Altitude Microbiology Laboratory, Department of Biotechnology, CSIR-IHBT Post Box No. 6, Palampur, Kangra 176061.
*Corresponding author: rakshak@ihbt.res.in, rakshakacharya@gmail.com

during the harsh winters. Previous studies have reported that psychrotrophic bacterial inoculation helps in reducing the mesophilic phase of composting by secreting extracellular enzymes, which help in the degradation of biomass at lower temperatures (Hou et al. 2017). With the reduction in the mesophilic phase, the thermophilic phase is extended and a maximum temperature is attained. This results in better degradation of the organic substrates leading to a better quality of compost with higher quantities of available nitrogen, phosphorus, and potassium. Let us have a detailed look at the role of psychrotrophic bacteria in composting in cold regions and the metabolic pathways involved.

Psychrotrophs: Resident Warriors of the Cold

General Process Flow of Composting, Different Phases, Limitations at Low Temperature and the Role of Psychrotrophs as Decomposers

Composting is a biochemical process involving mineralization of organic substance to CO_2, NH_3, H_2O, and partial humification, subsequently giving a stabilized product with fewer pathogenic microorganisms and reduced toxicity (Onwosi et al. 2017). For efficient composting, the temperature is the main parameter that regulates the process. Being an exothermic process as observed by (Kulikowska 2016), an effective composting process depends on ambient temperature as well as the nature of the substrates. A consequential temperature gradient has also been noted occurring in the composting pile during the process due to the non-uniformity of the microbial mass and energy balances (Zhang et al. 2012, Awasthi et al. 2014). The microbial mass varies within the compost pile during the process according to the physicochemical conditions of the different phases of composting (Hou et al. 2017). Such variation shows the significance of temperature in defining the benefit of one microbial population over the other, thus accrediting the effect of temperature on physicochemical characteristics of compost and ensuring bioavailability of the substrates to the decomposers. Furthermore, based on temperature gradient, the process is categorized into four different phases viz. mesophilic, thermophilic, cooling, and maturation phase (Figure 1) (Onwosi et al. 2017). The thermophilic stage is indicated by the rise in temperature ranging between 50-65°C, highlighting the active phase of the process, while the maturation stage is characterized by temperature drop reaching to an ambient level (Lazcano et al. 2008). The mesophilic phase plays a key role for the onset of the thermophilic phase as during this phase the initial microbial load begins to decompose the organic matter, thus raising the temperature (40-50°C) (Hou et al. 2017).

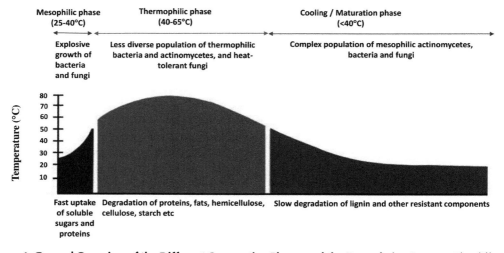

Figure 1. General Overview of the Different Composting Phases and the Degradation Process: The different phases of the composting process are indicated in green (mesophilic phase), orange (thermophilic phase), and brown (cooling or maturation phase). (Adapted from Alsanius et al. 2016).

The thermophilic phase also favors the elimination of pathogens, weeds, and parasites (Zhang et al. 2012, Hou et al. 2017). In cold regions, the low temperature poses a big hurdle in composting due to the low availability of the microbial load. This results in an extended mesophilic phase, thereby shortening the thermophilic phase eventually impacting compost maturity parameters (Hou et al. 2017). The decomposition in such a cold climate (below 20°C), therefore displays the crucial role of psychrotrophs in the composting process. The problem of lower microbial load at the beginning of the composting in the cold region can be resolved with the addition of the psychrotrophic bacteria; also, it accelerates the rise in temperature at the onset of composting and by this it means promoting the maturation of the final product (Singh et al. 2000, Hou et al. 2017). Table 1 represents the physicochemical parameters of a matured compost.

TABLE 1: Physicochemical Parameters of Compost at the End of Maturity Phase

Parameters	Maturity Value	References
C:N ratio	Between 10:1 and 21:1	Harshitha et. al. 2016
Moisture Content	15-30%	Xiao et. al. 2009
pH	Alkaline	Harshitha et. al. 2016
EC	< 4 mS/m	Shilev et. al. 2007
NH_4^+ - N concentration	400 mg/kg compost sample	Zhang et. al. 2015
NH_4^+ - N /NO_3^- - N	< 0.16	Bernal et. al. 2009
Weight reduction	Organic matter must be reduced at least 60% by weight	Karnchanawong and Suriyanon 2011
CEC:TOC	> 1.7	Jeyapandiyan et. al. 2017
C_w: N_{org}	< 0.55	Jeyapandiyan et. al. 2017
Humification index (HI)	> 30	Jeyapandiyan et. al. 2017

The biodegradation process can be further classified into two categories based on the oxygen requirement i.e, aerobic and anaerobic. The major chemical reactions involved in the aerobic process include hydrolysis, oxidation, cell synthesis/biomass generation, and endogenous respiration (Ramana and Singh 2000). Microorganisms degrade or hydrolyze the substrate available using extracellular enzymes secreted and provide building blocks for cellular oxidation to obtain energy for the synthesis of new biomass. The growth rate of microorganisms is therefore defined by their ability to degrade the organic matter/substrate. In endogenous respiration as the process proceeds, further nutrients begin to deplete in the medium and the microorganisms start to consume their protoplasm causing cell death and lysis (Zavala et al. 2002).

The anaerobic degradation process produces CO_2, H_2O, CH_4, NH_4^+, and nitrogen. The types of bacteria involved in this process include hydrolytic bacteria, fermentative bacteria, acetogenic bacteria, CO_2-reducing methanogens (lithotrophic methanogens), and acetoclastic methanogens. The major steps involved in anaerobic degradation are hydrolysis (carried out by extracellular enzymes), acidogenesis, acetogenesis, and methanogenesis (Ramana and Singh 2000) (Figure 2). The lower temperature and low pH values highly affect the obligate anaerobes (methanogens), which are involved in the conversion of organic acids into methane and carbon dioxide, while the facultative anaerobes are least affected by temperature. The rate of methanogenesis is in turn dependent on pH and temperature.

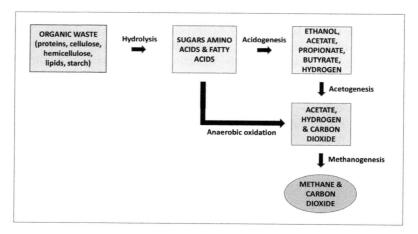

Figure 2. Steps Involved in Anaerobic Biodegradation: Extracellular enzymes produced by bacteria hydrolyze organic substrates into simpler components i.e., amino acids, sugars, and fatty acids. These are then converted into organic acids by fermentative bacteria. CO_2 and CH_3COO^- are oxidation products of anaerobic acid-forming bacteria. Acetogenic bacteria then oxidize higher acids to hydrogen and CH_3COO^- to provide the substrate for methanogens.

During the fermentative process, a minor amount of heat and Gibbs free energy is known to occur. Although the process is exothermic, a larger fraction of energy is entrapped in gaseous products and a little amount of energy is available for microorganisms growth and hence the process is one of the potential sources of renewable energy.

Types of Different Waste Substrate: Starch, Cellulose, Hemicellulose-Xylan, Pectin, and the Enzymes Involved in Their Breakdown

Starch

Starch, also known as amylum, is a polymer of glucose units linked together through the glycosidic bond. Starch is the most significant form of carbon reserve in all higher plants, and it is an important food product and biomaterial used for different purposes worldwide such as a source of energy in the human diet, medicine, paper, chemicals, agriculture, textile, food additives for enhancing the quality of food and preservation, etc.

The starch molecule consists of two types of molecules, a linear molecule of amylose and branched molecule of amylopectin. Amylose has mainly a linear chain structure consisting of D-glucose residues bound together with α-(1, 4)-glycosidic bond and with a few α-(1, 6)-branches. It constitutes up to 30% in the starch molecule and is insoluble in water. On the other hand, amylopectin is a branched structure that collectively makes the major component of the starch. Its structure consists of α-(1, 4)-glucan chain linked by α-(1, 6)-glycosidic bonds. In contrast to amylose, amylopectin makes up 70% of starch molecule and is soluble in water (Bertoft 2017, Eliasson 2004).

Enzymes Involved in Biodegradation of Starch

Starch-degrading enzymes occur universally in plants, eukaryotes, and a variety of microorganisms belonging to the archaea as well as the bacteria (Pandey et al. 2000, Liu et al. 2010). Degradation of starch involves the action of four groups of enzymes i.e., amylase, Glucoamylase/α-glucosidase, cyclodextrin glycosyltransferases (CGTases), and debranching enzymes, pullulanase and isoamylase (Horvathova 2000), which hydrolyzes starch molecules into oligomers of glucose (Figure 3).

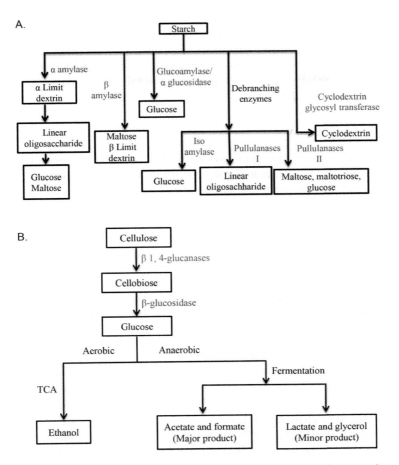

Figure 3. Biodegradation Pathway of Starch and Cellulose: (A) Enzymes and intermediaries involved in the conversion of starch to its degradation products (adapted from Horvathova et al. 2000); (B) systematic representation of synergistic action of enzymes involved in the degradation of cellulose. (Adapted from Chandra et al. 2015).

Starch-degrading enzymes mainly differ in their capability to catalyze starch molecules. These enzymes have diverse applications in different industries, such as in food and bakery (Couto et al. 2006), biofuel and textile (Li et al. 2014b), detergent, and starch processing industries (De et al. 2014, Egharevba 2019). It represents approximately 30% of the world's enzyme production.

There are two types of amylase i.e., alpha (α) amylase and beta (β)-amylase. Both are classified based on the glycosidic bond present between the glucose molecules that are attacked. The α-amylase belongs to a family of endo-amylase and can cleave α-(1, 4)-glycosidic bonds present in the inner part (endo-) of the starch into oligosaccharides, such as maltose and maltotriose from amylose and maltose, glucose, and 'limit dextrin' from amylopectin. Beta (β) amylase belongs to a family of exo-amylase. It hydrolyzes α-(1, 4)-glycosidic linkage present on external glucose residues and produces glucose or maltose and β-limit dextrin. The reducing product is present in β-form, hence it is called beta-amylase (Kossmann and Lloyd 2000). Glucoamylase/α-glucosidase are another group of starch-degrading enzymes that can attack the α-(1, 6)-linkage present in the starch molecule into a glucose unit. Pullulanase and isoamylase belong to debranching enzymes that act specifically on α-(1, 6)-bonds present on branched glucan chains of starch (amylopectin) and branched limit dextrins and produce linear oligosaccharide, maltose, maltotriose, and glucose.

Pullulanase is classified into two types: pullulanase type I and pullulanase type II. Pullulanase type I attack α-(1, 6)-linkage and pullulanase type II can catalyze α-(1, 4)- the linkage of starch. The fourth group of starch-degrading enzymes is cyclodextrin glycosyltransferases (CGTase). This peculiar enzyme performs intra- and intermolecular reactions on starch and produces cyclodextrins, a ring-like structure consisting of 6, 7, or 8 glucose units bound by α-(1, 4)- linkage (Horvathova 2000, Bertoldo and Antranikian 2002).

Cellulose

Cellulose is the most abundant biopolymer synthesized on earth. It is a major structural component of the plant cell wall and plays a crucial role in plant growth and development. In higher plants, cellulose is synthesized by plasma membrane-localized rosette cellulose synthase complexes (Li et al. 2014). Molecular structure of cellulose consists of repeating unit of β-D-glucopyranose molecules, which together forms a linear chain of cellulose structure. Depending on the arrangement of β-D-glucopyranose, molecule cellulose can have a crystallographic or an amorphous form. Cellulose is insoluble in water but can dissolve in ionic liquids (Figueiredo et al. 2010). Various microorganisms including bacteria have cellulose hydrolyzing proteins that initiate depolymerization of glucan unit yielding mono-, di-, or oligosaccharides. In 2017, Hou et al. reported the role of two psychrotrophic bacterium, *Flavobacterium* sp. and *Arthobacter* sp. in low temperature composting having cellulose-degrading capability. Also, in 2020, Zang et al. reported psychrotrophic *Pseudomonas* sp. with cellulose-degrading activity.

Enzymes Involved in Biodegradation of Cellulose

The classical scheme for cellulose degradation involves the synergistic action of three types of enzymes: Endo-(1, 4)-β-glucanases, Exo-(1, 4)-β-glucanases, and β-glucosidases. Since endo-enzymes generate reducing and non-reducing chain these enzymes act synergistically (Figure 3). Exo-enzymes act on these chains and release cellobiose which further converted into glucose by the help of β-glucosidases (Wood et al. 1979, Merino et al. 2007, Kostylev and Wilson 2012). All these enzymes cleave glycosidic bonds by the addition of a water molecule (Davies and Henrissat 1995).

Endo-(1, 4)-β-glucanase randomly cleaves the bonds present inside the cellulose chain. These enzymes may be processive or non-processive. In processive enzymes, the association of enzyme-substrate is followed by several consecutive cleavages at the active site of a single polysaccharide chain (Horn et al. 2006, Horn et al. 2012, Payne et al. 2011). Exo-(1, 4)-β-glucanase attacks the reducing or non-reducing ends of the cellulose polymer. It also acts as processive exo-(1, 4)-β-glucanases and is referred to as cellobiohydrolase. Cellobiohydrolases are one of the most abundant components in natural and commercial cellulase. β-glucosidase or cellobiase converts the cellobiose, a major product of the endo- and exo-glucanase mixture, to glucose. Furthermore, glucose in the presence of aerobic conditions enters into the TCA cycle producing ethanol as an end product and in case of anaerobic conditions enters into the fermentation process yielding acetate and formate as major and lactate and glycerol as minor products.

Hemicellulose

The hemicelluloses constitute the second most abundant naturally existing compound after cellulose and 35% hemicellulose has been observed in hardwood plant species (Timell 1967). Hemicelluloses are in conjunction with cellulose and lignin and constitute both the primary and secondary layers of the plant cell wall. By forming hydrogen bonds, hemicelluloses can join strongly to cellulose microfibrils (Roland et al. 1989). Hemicelluloses consist of non-cellulosic β-D-glucans, polygalacturonans (a pectic substance), and various complex heteropolysaccharides that are rich in xylose (arabinoglucurono and glucuronoxylans), galactose (arabinogalactans), and mannose (glucomannans). In herbaceous crops and hardwoods, glucuronoxylan is the primary hemicellulose whereas glucomannan is most dominant in softwood. Traditionally, hemicelluloses can be categorized into four groups: xylan, xyloglucans, glucomannans, and mixed linked β-glucans.

i. **Xylans:** Almost all plant species consist of xylan in their cell wall. A core framework of β-(1, 4)-D-xylose units supplemented with side chains that consist of diverse sugar and sugar-acid residues are present in xylans. Side chains composed of sugars like arabinose, glucose, galactose. Also, glucuronic acid, rhamnose, and galacturonic acid are present in few amounts.

ii. **Xyloglucans:** Xyloglucansare, composed of a backbone of β-(1, 4)-linked D-glucopyranose units, same as that pf the cellulosic components of the cell wall, and it is continuously branched by α-D-xylose residues.

iii. **Glucomannans:** These are the hemicellulosic component in the cell walls of some plant species. A backbone of β-(1, 4)-D-mannose branched with D-glucose units is present in glucomannans. These are water-soluble polysaccharides.

iv. **Mixed Linkage β-glucans**: These are the rest of the hemicelluloses that are composed of mixed linkage i.e. β-(1, 3)- and β-(1, 4)-glucans. These are present only in the grass and some pteridophytes (Ozyurt and Otles 2016).

Enzymes Involved in Biodegradation of Hemicelluloses

Extracellular enzymes targeting hemicelluloses are hydrolytic and collectively known as hemicellulases and hemicellulolytic enzymes (Dekker 1979, Dekker and Lindner 1979, Dekker and Richards 1976). The glycans that form the backbone chain of the hemicelluloses are degraded by these enzymes (glycan hydrolases). Therefore, the main hemicellulases are β-D-galactanases, β-D-mannanases, and β-D-xylanases. Other than this set of enzymes, there are specific exoglycosidases like α- and β-D-galactosidases, α-D-mannosidases, α-L-arabinosidases, and β-D-xylosidases. These enzymes are effective in degrading low molecular weight glycosides as well as the short-chain or monosaccharide units branching from the backbone chain. The action of glycosidases leads to a total breakdown of hemicelluloses into monosaccharides since these enzymes work in synergy with the hemicellulases. Hemicellulases hydrolyze their substrate in two ways, exo- and endo- hydrolytic attack. An exoenzyme hydrolyzes polysaccharide by continuously removing the terminal oligosaccharide units and moving forward in a sequential manner, normally from the non-reducing end of the polysaccharide chain. Exo-enzymes that attack hemicelluloses are two exo-β-mannanases (McCleary 1982, Araki and Kitamikado 1982). Endo-enzymes randomly hydrolyze polysaccharides. The large molecule is successively degraded into smaller fragments until nonreducible products are formed. Endo-type constitutes the most prevalent class of hemicellulases.

Hemicellulases are secreted by fungi, bacteria from terrestrial and marine ecosystems, rumen microorganisms (protozoa, bacteria, and fungi), yeasts, marine algae, snails crustaceans, insects, and seeds of terrestrial plants. Fungi, yeast, and some bacteria produce hemicellulases extracellularly, although intracellular hemicellulolytic enzymes and cell wall-attached enzymes have also been studied. The hemicellulolytic enzymes are generally secreted as more than one isoenzymes (enzymes with apparently the same specific catalytic function existing in more than one physical form that differ from each other in some properties as optimum pH, temperature, etc.); some have cellulose-binding domains, some are part of the multi-domain protein in which other domains degrade cellulose, while in some organisms, xylanase production is augmented by the presence of cellulose instead of xylan. Hemicellulases of microbial origin have been studied to be secreted both inductively and constitutively (Dekker and Richards 1976). Psychotropic or cold-adapted bacteria i.e., *Burkholderia* isolated from Antarctic islands produces β-glucosidase and polygalactouronase which causes the breakdown of hemicelluloses components of the cell wall (Tomova et al. 2014). The psychrotrophic *Eupenicilliumcrustaceum*, *Paceliomyces* sp., *Bacillus atropheus* and *Bacillus* sp. have been reported to be efficient in the degradation of lignocellulolytic material (Shukla et al. 2016).

The biodegradation pathways employed by microbial enzymes for degradation of hemicelluloses are discussed below. Mode of action of these extracellular enzymes on hemicelluloses is proposed to occur in the following two ways:

(a) The exoglycosidases i.e., xylanase, galactanase, and β-mannanase, attack and eliminate the side branches, thereby uncovering the backbone glycan chain. As a result of this, the hemicellulases can easily target the glycan chain as steric hindrance caused by the side chain residues is decreased.

(b) On the other hand, the endohemicellulases degrade those parts of the glycan chain that are not or less branched by residues. The attack of the endohemicellulases results in a consortium of oligosaccharides of the mixed constitution, which are further reduced by both exoglycosidases and endohemicellulases.

Pathways for degradation of heteroxylan, heterogalactan, and heteromannans are described below.

Biodegradation Pathway for Heteroxylan

The plant cell wall of terrestrial plants consists of heteroxylan and it constitutes around 35%, depending upon plant species. Heteroxylan consists of a group of complex carbohydrates made of a backbone chain of (1, 4)-β-linked D-xylose monomers to which various adjuncts are connected. These might be D-glucuronic acid, L-arabinose, and diverse small oligosaccharide chains constituted by galactose, L-arabinose, D-xylose, and D-glucuronic acid units. While enzymes degrading the heteroxylan are known as xylanases, they also require complementary actions of additional enzymes i.e., β-xylosidases, α-arabinosidases, α-glucuronidases and certain esterases for total hydrolysis (Figure 4) Heteroxylan polysaccharides, like arabinoglucuronoxylans, arabinoxylans and glucuronoxylans are hydrolyzed into xylan, arabinose, glucuronic acid, and xylose oligosaccharides of a varying constitution in the presence of extracellular enzymes i.e., α-L-arabinosidase, α-D-glucuronidase, and endo-β-xylanase. Xylan and xylose oligosaccharides of mixed constitution further undergo degradation through the combined action of enzymes i.e. endo-β-xylanase, β-D-xylosidase, α-L-arabinosidase, and α-D-glucuronidase. The degradation pathway as shown in Figure 5 results in the formation of xylose, arabinose, and glucuronic acid as end products.

Biodegradation Pathway for Heteromannans

The depolymerization strategy of microbes for heteromannans employs the combined action of at least two enzymes – β-D-mannanases and β-mannosidases. There are two types of β-D-mannanases based on their mode of action, the exo- and endo- types. Both these types have characterized and hydrolyzed the β-(1, 4)-D-mannopyranosyl linkages of branched mannans, copolymer mannans, and linear D-mannans (Dekker 1979, Dekker and Richards 1976). Some additional enzymes, such as α-galactosidases, are also needed along with extracellular enzymes (β-mannanases and β-mannosidases) to remove the galactosyl side chains that inhibit the potential of β-mannanases to hydrolyze the β-mannan backbone. Apart from the conventional microbial strategy for hydrolysis of β-mannan that requires the action of endo-β-mannanases to succeed by the depolymerization of manno-oligosaccharides by β-mannosidases, few enzymes also exhibited the potential to aptly hydrolyze both β-mannans and manno-oligosaccharides, indicated as exo-β-mannanases. Exo-β-enzymes from the GH2 family, extracted from *Xanthomonas axonopodis* (XacMan2A), can effectively degrade both manno-oligosaccharides and β-mannan (Domingues et al. 2018). The action of exo-β-mannanaselargely removes D-mannose as the hydrolysis product from mannose oligosaccharides as shown in Figure 4. Only linear β-D-mannans are degraded, while branched and copolymer β-D-mannans are not targeted by exo-β-mannanase. It is observed that exo-type hydrolases terminate the degradation process when they encounter branch points. Galactoglucomannan hydrolyzes L-mannose, galactose, and various mannose oligosaccharides through the synergistic action of endo-β-mannanases and α-D-galactosidase. The action of endo-β-mannanases on glucomannan has also resulted in D-glucose in addition to mannose and mannose oligosaccharides, some of which contained glucose (Ishihara and Shimizu 1980). Various mannose oligosaccharides of mixed constitutions hydrolyzed by the action of extracellular enzymes i.e., endo-β-mannanase, β-D-glucosidase, α-D-mannosidase, and β-D-mannosidase and resulted in galactose, glucose, and mannose as end products (Figure 4).

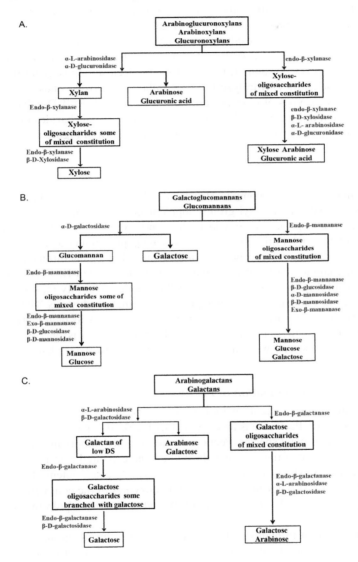

Figure 4. Biodegradation Pathway of Hemicellulose: (A) Diverse potential enzymes are shown on the side of arrows catalyzing main chain and side chain of heteroxylan molecules and resulted into various end products i.e., xylose, arabinose, and glucuronic acid (adapted from Dekker 1985); (B) hydrolysis of heteromannan, a component of hemicellulose is carried out by different enzymes, which are shown on the side of arrows catalyzing main chain and side chain which has resulted into mannose, glucose, and galactose end products (adapted from Dekker 1985); (C) hydrolysis of heterogalactans by different enzymes and their actions on various substituents are indicated by arrows. The pathway resulted in galactose and arabinose end products (adapted from Dekker 1985).

Biodegradation Pathway for Heterogalactans

Extracellular hydrolytic enzymes that cause degradation of D-galactans and L-arabino-D-galactans are known as galactanases. Two types of endo-β-D-galactanases are observed (Dekker and Richards 1976, Dekker 1979). Although, D-galactanases of the exo-type have not been identified, an uncommon β-(1,4)-D-galactanase obtained from *Bacillus subtilis,* removing mainly galactose from β-(1, 4)-D-galactan and also handful other galactose oligosaccharides, was observed to exhibit both exo- and endo- activity (Labavitch et al. 1976). D-galactose and galactose oligosaccharides, some of which are made of L-arabinose residues, are produced when endo-β-galactanases cause random degradation of D-galactans (Figure 4) (Bauer et al. 1977, Dekker 1979). Endo-galactanases release D-galactose and various (1, 3)- and (1, 6)-linked β-D-galactose oligosaccharides, some of which consist of L-arabinose as they hydrolyze β- (1, 3) linked galactosyl bonds of D-arabinogalactans (Hashimoto 1971). Such types of enzymes are known as endo-(1, 3)-β-D-galactanases (Figure 4). β-D-galactosidase and α-L-arabinosidase act on galactose oligosaccharides of mixed constituents and result in end products viz. galactose and arabinose as shown in Figure 4.

Pectin

Pectin is a high-molecular-weight complex structural polysaccharide that contains long galacturonic acid chains with residues of carboxyl groups supplemented with varying degree of methyl esters (Voragen et al. 2009). D-galacturonic acid is the major component of pectin and considered an important source of carbon for microbes living on decaying plant materials. The various structurally well-characterized pectic polysaccharide motifs are homogalacturonan, xylogalacturonan, apiogalacturonan, rhamnogalacturonan-I, and rhamnogalacturonan-II. Pectin due to its gelling property, biocompatibility, non-toxic nature as well as biodegradability makes it suitable to use in food, medical, pharmaceutical, textile, paper, pulp, and cosmetic industries. The chief pool of pectin is found in the primary cell wall of terrestrial plants, which makes its degradation and metabolism by microbes essential for the natural turnover of biomass. Pectin is one of the most important components present in the cell wall, middle lamella of all terrestrial plants, and also present in vegetables and fruits.

Enzymes Involved in Biodegradation of Pectin

Pectinolytic enzymes mostly occur in higher plants and microbes such as molds, yeasts, and bacteria (Whitaker 1990). Pectinase is an enzyme that breaks down pectin, a polysaccharide found in the cell wall of various plant species. Pectic enzymes, mainly comprising pectolyase, pectozyme, and polygalacturonase, one of the most widely studied and exploited commercial pectin-degrading enzymes. Such enzymes show maximum relevance with greater potential to offer to industry (Dayanand and Patil 2006). They are very important enzymes in the commercial sense, especially in the juice, vegetable, food (Kashyap et al. 2000), and in the paper/pulp industry (Viikari et al. 2001).

Pectin molecule present in plants is degraded by a heterogeneous group of enzymes known as pectinases. According to their mode of action, they are distinguished into three categories. The first one comprises polygalacturonase which brings out the hydrolysis process. The second category of the enzyme consists of pectinlyase, which undergo lyase/trans-elimination process and the third category of enzyme involves pectin esterase (Yadav et al. 2009, Chen et al. 2015). Pectin, a polysaccharide largely consists of chains of uronates, the microbial pathways responsible for the degradation of pectin resulting in the production of the chief metabolite i.e., 2-keto-3-deoxygluconate as shown in Figure 5. Metabolism of pectin is commenced by polysaccharide lyases, which cleave the backbone of polysaccharide in endo-action by using a β-elimination mechanism, thus resulting in oligouronates with 4, 5-unsaturated non-reducing ends. In pectinolytic organisms, the conversion of 4, 5-unsaturated galacturonate to 2-keto-3-deoxygluconate occurs through three steps and gives out two intermediates i.e., 5-keto-4-deoxyuronate and 2, 5-diketo-3-deoxygluconate. The last two steps are executed by an isomerase and a reductase. The pathway converges at 2-keto-3-deoxygluconate, which in turn results into pyruvate and 3-phosphoglyceraldehyde via the Entner–Doudoroff pathway (Pattat et al. 2001), ultimately giving energy in a readily usable form for an organism (Figure 5). Among the microorganisms that are able to degrade pectin, the filamentous fungi are the most efficient degraders. Therefore, commercially available pectinolytic enzymes are mostly produced from *Aspergillus* sp., *Trichoderma* sp., and *Penicillium* sp. Antarctic marine sponges are an attractive source of pectinolytic producing fungi. Pectinolytic enzymes produced by *Geomyces* sp. F09-T3-2 could be promising for biotechnological approaches, which involve cold-active pectinases (Poveda et al. 2018).

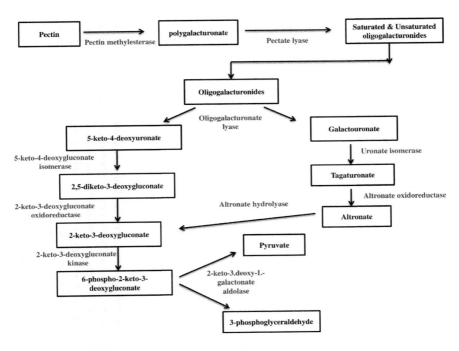

Figure 5. Biodegradation Pathway for Pectin: The different steps are catalyzed by the action of different pectinases. These enzymes located in the periplasm and in the outer membrane of bacteria act on the pectin molecule, which is further catabolized in the cytoplasm and resulted in pyruvate and 3-phosphoglyceraldehyde as end products (adapted from Pattat et al. 2001).

Secretion Systems Involved in Waste Degradation

The biodegradation of complex starch molecules by bacteria requires a cocktail of enzymes to depolymerize it to oligosaccharides and monomer sugars, such as glucose and maltose. The endo-acting enzymes such as alpha-amylase initially hydrolyze the internal linkages of the starch molecule randomly liberating linear and branched oligosaccharides. The psychrophilic extracellular alpha-amylase is produced as a preproenzyme with signal sequence, the mature enzyme, and a C-terminal propeptide. The translocation of the preproenzyme through the inner membrane likely occurs via the Sec-pathway followed by cleavage of the signal peptide exposing the amino-terminal end of the proenzyme. The subsequent proenzyme released extracellularly remains inactive until the extracellular protease cleaves the C-terminal domain during the late exponential stage, liberating the mature enzyme. Tutino et al. (2002) reported a similar secretion system in Antarctic bacteria *Pseudoalteromonas haloplanktis* bacterium.

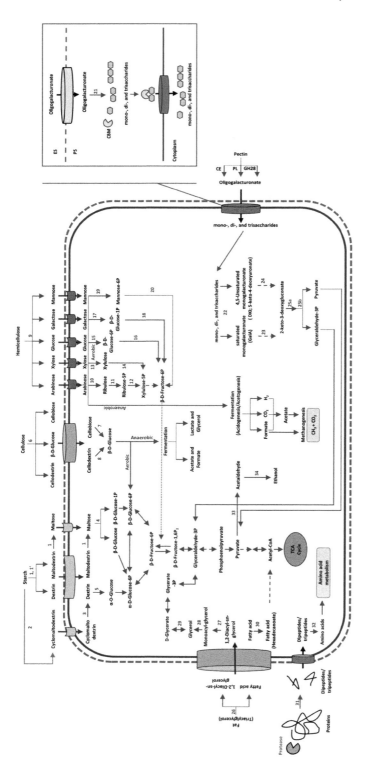

Figure 6. Overview of Degaradation Pathway of Starch, Cellulose, Hemicellulose, Pectin, Lipids, and Proteins: The major enzymes involved in degradation processes are as follows: 1, alpha-amylase; 1', pullulanase; 2, cyclomaltodextringlucanotransferase; 3, beta-endoglucanase; 4, beta-glucosidase; 5, Oligo-1,6-glucosidase; 6, cellulases; 7, beta-1,4-endoglucanase; 8, cellodextrinphosphorylase; 9, endo-α, 4-β-xylanase and xylan 1,4-β-xylosidase; 10, araisomerase; 11, L-ribulokinase; 12, ribo-5-P-epimerase; 13, X-isomerase; 14, xylulose kinase; 15, glucose-6-Phosphatase; 16, phosphoglucoseisomerase; 17, Galactokinase; 18, phosphoglucomutase; 19, hexokinase; 20, mannose-6-P isomerase; 21, polysaccharide lyase and polygalacturonase; 22, pectatelyase and oligogalacturonatelyase; 23, uronate isomerase (UxaC), D-tagaturonatereductase (UxaB), and D-altronatedehydratase (UxaA); 24, 4-deoxy-L-threo-5-hexosulose-uronate ketol-isomerase (KduI) and 2-keto-3-deoxy-D-gluconate dehydrogenase (KduD); 25a, 2-keto-3-deoxygluconate kinase (KdgK); 25b, 2-keto-3-deoxy-6-phosphogluconate aldolase (KdgA); 26, lipase; 27, diacylglycerol(DAG) lipase; 28, monoacylglycerol (MAG) acyl hydrolase; 29, phosphoglycerate kinase and GAP dehydrogenase; 30, long-chain acylCoAsynthetase; 31, Protease; 32, peptidase; 33, Pyruvate decarboxylase; 34, Alcohol dehydrogenase

The pullulanase enzyme specifically cleaves alpha-(1,6)-linkage in pullulan and branched oligosaccharides releasing maltodextrin and linear oligosaccharides. These products are then delivered to the cell cytoplasm through specific transporters for further breakdown. Inside the cells, enzymes such as beta-endoglucanase, beta-glucosidase, oligo-1,6-glucosidase, alpha-amylase attack the substrate producing smaller oligosaccharides and monomeric sugars (Bertoldo and Antranikian 2002). Finally, the monomeric sugars like glucose enter into the glycolysis pathway and ultimately to the TCA cycle (Figure 6) generating energy for cellular growth.

For the utilization of cellulosic substrates, psychrotrophic bacteria requires the ABC transporters specific for the hydrolytic product such as cellobiose, cellodextrin, β-D-glucose. The bacterium produces three different types of cellulase enzymes viz. endo- β-(1,4)-D-glucanase, exo- β-(1, 4)-D-glucanase, and β-glucosidase for the depolymerization of celluloses. These enzymes are either secreted using the type II secretion system or general secretion system by the bacteria. Cellulases cleaves the β-(1,4)-glycosidic linkages within the cellulose polymer releasing cellobiose, glucose, and cellodextrin (Figure 6), which are then transported inside the cell via specific transporters. The β-(1,4)-endoglucanase and cellodextrin phosphorylase then convert cellobiose and cellodextrin into β-D-glucose molecule (Figure 6).

Hemicelluloses are biodegraded by the supportive action of a variety of hydrolytic enzymes. Since xylan is the major polysaccharide found in hemicelluloses, endo-β-(1, 4)-xylanase and xylan β-(1, 4)-xylosidase plays a key role in its degradation. The endo-β-(1, 4)-xylanase acts on xylan polymers generating smaller oligosaccharides and xylan β-(1, 4)-xylosidase further cleaves them to produce xylose monomers. In addition to this, other enzymes such as xylanesterases, ferulic, p-coumaricesterases, α-L-arabinofuranosidases, and α-4-O-methyl glucuronosidases also act synergistically to completely hydrolyze hemicellulose into soluble monosaccharides, such as xylose, arabinose, galactose, and mannose (Chandra 2016). These monosaccharides are then delivered into the cells for the metabolism with the help of sugar ABC transporters (Figure 6).

The extracellular pectin depolymerization carried out by pectinolytic bacteria is associated with the action of pectinases secreted. Initial extracellular pectin depolymerization involves a consortium of enzymes. The major pectinases involved in the process of depolymerization include polysaccharide lyases (PLs), polygalacturonases (GH28s), and carbohydrate esterases (CE8). These enzymes first cleave the pectic network into pectic fragments, which diffuse into the periplasmic space through anionic porins (Abbott et al. 2010). Within the periplasmic space, carbohydrate-binding proteins (CBM) bind and retain the fragments. The PLs and GH28s then further cleaves the pectic substrates into mono-, di-, and trisaccharides (Figure 6, Insight). These are then transported into the cell facilitated by integral membrane systems. In the interior of the cell, mono-, di-, and trisaccharides are converted into saturated monogalacturonate (GalA) and a 4,5-unsaturated GalA-like monosaccharide (DKI; 5-keto-4-deoxyuronate), the reaction catalyzed by the combined action of pectate lyase and Oligogalacturonate lyase. The GalA and DKI are then converted into 2-keto-3-deoxy-D-gluconate (KDG) by a parallel series of chemical reactions involving uronateisomerase (UxaC), D-tagaturonatereductase (UxaB), D-altronatedehydratase (UxaA), 4-deoxy-L-threo-5-hexosulose-uronate ketol-isomerase (KduI), and 2-keto-3-deoxy-D-gluconate dehydrogenase (KduD). The uronateisomerase converts the GalA to D-tagaturonate that is then reduced by D-tagaturonatereductase to D-altronate. The D-altronatedehydratase then forms KDG (Kuivanen et al. 2019). The DKI is converted to 2,5-diketo-3-deoxygluconate (DKII) by KduI which is further converted to KDG by KduD. KDG is subsequently catabolized to form pyruvate and glyceraldehyde-3-phosphate (GAP) by the enzymes 2-keto-3-deoxygluconate kinase (KdgK) which first phosphorylates KDG to generate 2-keto-3-deoxy-6-phosphogluconate (KDPG), which is cleaved by KDPG aldolase (KdgA) to pyruvate and GAP (Figure 6) (Hugouvieux-Cotte-Pattat 2016).

Oil from domestic waste contains long-chain triglycerides arising from vegetable oils and animal fats. Triglyceride degradation involves lipolytic bacterial species, which utilizes available fat as carbon sources for its growth (Janßen and Steinbüchel 2014). In the initial phases of triglyceride degradation, lipases hydrolyze it to fatty acids and 1,2-Diacyl-sn-glycerol (DAG), which are then delivered into the cell. Within the cells, DAG molecules are further cleaved into shorter monoacylglycerol (MAG) and subsequently take part in the oxidation pathway (Figure 6). The secretion of cold-active lipases involves ABC type permease proteins. In 2008, Długołęcka et al. reported the involvement of a similar secretion system in psychrotrophic bacterium *Pseudoalteromonas* sp.

Proteins are the major components of waste biomass along with carbohydrates. The amino acids derived from proteins can be used by the bacterium to either build their biomass or to produce a variety of products, such as hydrogen, which is used as biofuel (Cheng et al. 2015). The proteases secreted by bacteria break down larger proteins into smaller fragments of peptides. Through the specific dipeptide/oligopeptide ABC transporter permease/ATP-binding transporters, these molecules are then delivered into the cell where peptidases further break them into amino acids.

Conclusions

Biodegradable organic waste composting has been thoroughly described to highlight the interlink of the starting raw material components, process flow, and potential benefits for soil enrichment. Low temperature regions face delay in the composting process due to limited microbial population on the onset extending the mesophilic phase and shortening the thermophilic stage. The role of psychrotrophic bacteria and the common biochemical pathways to recycle the nutrients in composting have been described in this chapter. This practice is fundamental to recycle the easily available organic resources at the farm, household, or industrial level in winter prone regions. The various enzymes such as cellulase, amylase, pectinase, xylanase, and protease involved in the efficient degradation of the organic matter are also discussed.

Acknowledgments

Shruti SB is thankful to UGC, Govt. of India for 'Research Fellowship' Grant [UGC-Ref. No.:461/ (CSIR-UGC NET DEC. 2016)]. Authors acknowledge the financial support by the CSIR in-house project MLP0137, NMHS project of MoEF&CC (NMHS sanction no. GBPNI/NMHS-2018-19/ SG/178), Science and Engineering Research Board Start-up research grant no. SRG/2019/001071) and DST-TDT project no. DST/TDT/WM/2019/43. We acknowledge the Director of CSIR-IHBT, Palampur, for providing all the facilities. The authors also acknowledge Aman Thakur and Aman Kumar for their technical assistance in figure generation. This manuscript represents CSIR-IHBT communication no. 4580.

References

Abbott, D.W., H.J. Gilbert and A.B. Boraston. 2010. The active site of oligogalacturonate lyase provides unique insights into cytoplasmic oligogalacturonate β-elimination. J. Biol. Chem. 285(50): 39029–39038.
Araki, T. and M. Kitamikado. 1982. Purification and characterization of a novel exo-β-mannanase from Aeromonas sp. F-25. J. Biochem. 91(4): 1181–1186.
Awasthi, M.K., A.K. Pandey, J. Khan, P.S. Bundela, J.W. Wong and A. Selvam. 2014. Evaluation of thermophilic fungal consortium for organic municipal solid waste composting. Bioresour. Technol. 168: 214–221.
Bauer, W.D., D.F. Bateman and C.H. Whalen. 1977. Purification of an endo-β-1, 4-galactanase produced by Sclerotinia sclerotiorum: effects on isolated plant cell walls and potato tissue. Phytopathol. 67: 862–868.
Bernal, M.P., J.A. Alburquerque and R. Moral. 2009. Composting of animal manures and chemical criteria for compost maturity assessment– a review. Bioresour. Technol.100(22): 5444–5453.
Bertoft, Eric. 2017. Understanding starch structure: recent progress. Agronomy 7(3): 56.
Bertoldo, C. and G. Antranikian. 2002. Starch-hydrolyzing enzymes from thermophilic archaea and bacteria. CurrOpin. Chem. Biol. 6(2): 151–160.
Brett, C.T. 2000. Cellulose microfibrils in plants: biosynthesis, deposition, and integration into the cell wall. Int. Rev. Cytol. 199: 161–199.

Chandra, R., S. Yadav and V. Kumar. 2015. Microbial degradation of lignocellulosic waste and its metabolic products. E.W.M. 2015: 249–298.

Chen, J., W. Liu, C.M. Liu, T. Li, R.H. Liang and S.J. Luo. 2015. Pectin modifications: a review. Crit. Rev. Food Sci. Nutr. 55(12):1684–1698.

Cheng, J., L. Ding, A. Xia, R. Lin, Y. Li, J. Zhou and K. Cen. 2015. Hydrogen production using amino acids obtained by protein degradation in waste biomass by combined dark- and photo-fermentation. Bioresour. Technol. 179: 13–19.

Consolidated Annual Report (For the year 216–17) on Implementation of Solid Waste Management Rules, 2016 (As per provision 24(4) of SWM Rules, 2016). Central Pollution Control Board (Ministry of Environment, Forests & Climate Change).

Couto, S.R. and M.A. Sanromán. 2006. Application of solid-state fermentation to food industry–a review. J. Food Eng. 76(3): 291–302.

Da Silva, J.F. and R.J.P. Williams. 2000. The Biological Chemistry of the Elements: The Inorganic Chemistry of Life. Oxford University Press.

Davies G. and B. Henrissat. 1995. Structures and mechanisms of glycosyl hydrolases. Structure 3: 853–859.

Dayanand, A. and S.R. Patil. 2006. Production of pectinase from deseeded sunflower head by Aspergillusniger in submerged and solid-state conditions. Bioresour. Technol. 97(16): 2054–2058.

De, B.C., D.K. Meena, B.K. Behera, P. Das, P.D. Mohapatra and A.P. Sharma. 2014. Probiotics in fish and shellfish culture: immunomodulatory and ecophysiological responses. Fish Physiol. Biochem. 40(3): 921–971.

Dekker, R.F.H. and G.N. Richards. 1976. Hemicellulases: their occurrence, purification, properties and mode of action. Adv. Carbohyd. Chem. Biochem. 32: 277–352.

Dekker, R.F.H. and W.A. Lindner. 1979. Bioutilization of lignocellulosic waste materials. S. Afr. J. Sci. 75: 65–71.

Depluverez, S., S. Devos and B. Devreese. 2016. The role of bacterial secretion systems in the virulence of Gram-negative airway pathogens associated with cystic fibrosis. Front. Microbiol. 7: 1336.

Dlugolecka, A., H. Cieslinski, M. Turkiewicz, A.M. Bialkowska and J. Kur. 2008. Extracellular secretion of Pseudoalteromonassp cold-adapted esterase in Escherichia coli in the presence of Pseudoalteromonas sp. components of ABC transports system. Protein Expr. Purif. 62(2): 179–184.

Domingues, M.N., F.H.M. Souza, P.S. Vieira, M.A.B.D. Morais, L.M. Zanphorlin, C.R.D. Santos, et al. 2018. Structural basis of exo-β-mannanase activity in the GH2 family. J. Biol. Chem. 1: 1–28.

Egharevba, H.O. 2019. Chemical Properties of Starch and Its Application in the Food Industry. In Chemical Properties of Starch. IntechOpen.

Figueiredo J.A., M.I. Ismael, C.M.S. Anjo and A.P. Duarte. 2010. Cellulose and Derivatives from Wood and Fibers as Renewable Sources of Raw-Materials. In: Rauter A., P. Vogel and Y. Queneau [eds.]. Carbohydrates in Sustainable Development I. Top Curr Chem. 294: 117–128.

Harshitha, J., S. Krupanidhi, S. Kumar and J. Wong. 2016. Design and development of indoor device for recycling of domestic vegetable scrap. Environ. Technol. 37(3): 326–334.

Hashimoto, Y. 1971. Studies on the enzyme treatment of coffee beans part V. Structure of coffee arabinogalactan. Nippon Nogei Kagaku Kaishi 45: 147–150.

Horn, S.J., P. Sikorski, J.B. Cederkvist, G. Vaaje-Kolstad, M. Sørlie, B. Synstad, et al. 2006. Costs and benefits of processivity in enzymatic degradation of recalcitrant polysaccharides. PNAS. 103(48): 18089–18094.

Horn, S.J., M. Sørlie, K.M. Vårum, P. Väljamäe and V.G. Eijsink. 2012. Measuring processivity. Meth. Enzymol. 510: 69–95.

Horváthová, V., S. Janecek and E. Sturdik. 2000. Amylolytic enzymes: their specificities, origins and properties. Biologia-Bratislava 55(6): 605–616.

Hou, N., L. Wen, H. Cao, K. Liu, X. An, D. Li and C. Li. 2017. Role of psychrotrophic bacteria in organic domestic waste composting in cold regions of China. Bioresour. Technol. 236: 20–28.

Hugouvieux-Cotte-Pattat, N. 2016. Metabolism and virulence strategies in Dickeya-host interactions. Prog. Mol. Biol. Transl. Sci. 142: 93–129.

Ishihara, M. and K. Shimizu. 1980. Hemicellulases of brown rotting fungus, Tyromycespalustris. IV. Purification and some properties of an extracellular mannanase. MokuzaiGakkaishi 26: 811–818.

Janßen, H.J. and A. Steinbüchel. 2014. Fatty acid synthesis in Escherichia coli and its applications towards the production of fatty acid based biofuels. Biotechnol. Biofuels 7(1): 1–26.

Jeyapandiyan, N., P. Doraisamy and M. Maheswari. 2017. Effect of Composting on Physico-Chemical Properties of Semi-Finished Tannery Sludge. Adv. Biol. Res. (Rennes). 12(2): 1–7.

Kanimozhi, K. and P.K. Nagalakshmi. 2014. Xylanase production from Aspergillus niger by solid state fermentation using agricultural waste as substrate. Int. J. Curr. Microbiol. Appl. Sci. 3(3): 437–446.

Karnchanawong, S. and N. Suriyanon. 2011. Household organic waste composting using bins with different types of passive aeration. Resour. Conserv. Recy. 55(5): 548–553.

Kashyap, D.R., S. Chandra, A. Kaul and R. Tewari. 2000. Production, purification and characterization of pectinase from a Bacillus sp. DT7. World J. Microbiol. Biotechnol. 16: 277–282.

Kaza, Silpa, Yao, C. Lisa, PerinazBhada-Tata and Frank Van Woerden. 2018. What a Waste 2.0: A Global Snapshot of Solid Waste Management to 2050. Urban Development: Washington, DC: World Bank. © World Bank.https://openknowledge.worldbank.org/handle/10986/30317 License: CC BY 3.0 IGO.

Kim, W.C., D.Y. Lee, C.H. Lee and C.W. Kim. 2004. Optimization of narirutin extraction during washing step of the pectin production from citrus peels. J. Food Eng. 63(2): 191–197.

Kossmann, J. and J. Lloyd. 2000. Understanding and influencing starch biochemistry. CRC Crit. Rev. Plant. Sci. 19(3): 171–226.

Kostylev M. and D. Wilson. 2012. Synergistic interactions in cellulose hydrolysis. Biofuels 3: 61–70.

Kuivanen, J., A. Biz and P. Richard. 2019. Microbial hexuronate catabolism in biotechnology. AMB Express 9(1): 16.

Kulikowska, D. 2016. Kinetics of organic matter removal and humification progress during sewage sludge composting. Waste Manage. 49: 196–203.

Labavitch, J.M., L.E. Freeman and P. Albersheim. 1976. Structure of plant cell walls: purification of β-(1,4)-galactanase which degrades a structural component of the primary cell wall of dicots. J. Biol. Chem. 251: 5904–5910.

Lazcano, C., M. Gómez-Brandón and J. Domínguez. 2008. Comparison of the effectiveness of composting and vermicomposting for the biological stabilization of cattle manure. Chemosphere 72(7): 1013–1019.

Li, Q., J.H. Liang, Y. Y. He, Q.J. Hu and S. Yu. 2014. Effect of land use on soil enzyme activities at karst area in Nanchuan, Chongqing, Southwest China. Plant Soil Environ. 60(1): 15–20.

Li, S., L. Bashline, L. Lei and Y. Gu. 2014. Cellulose synthesis and its regulation. ASPB. 12.

Liu, Y., F. Lu, G. Chen, C.L. Snyder, J. Sun, Y. Li and J. Xiao. 2010. High-level expression, purification and characterization of a recombinant medium-temperature α-amylase from Bacillus subtilis. Biotechnol. Lett. 32(1): 119.

McCleary, B.V. 1982. Purification and properties of a β-D-mannosidemannohydrolase from guar. Carbohydr. Res. 101(1): 75–92.

Merino, S.T., and J. Cherry. 2007. Progress and challenges in enzyme development for biomass utilization. Biofuels 108: 95–120.

Onwosi, C.O. V.C. Igbokwe, J.N. Odimba, I.E. Eke, M.O. Nwankwoala, I.N. Iroh, et al. 2017. Composting technology in waste stabilization: on the methods, challenges and future prospects. J. Environ. Manage. 190: 140–157.

Ozyurt, V.H. and S. Otles. 2016. Effect of food processing on the physicochemical properties of dietary fibre. Acta Sci. Pol. 15(3): 233–245.

Pandey, A., P. Nigam, C.R. Soccol, V.T. Soccol, D. Singh and R. Mohan. 2000. Advances in microbial amylases. Biotechnol. Appl. Biochem. 31(2): 135–152.

Pattat, N.H.C., N. Blot and S. Reverchon. 2001. Identification of Tog MNAB, an ABC transporter which mediates the uptake of pectic oligomers in Erwiniachrysanthemi 3937. Mol. Microbiol. 41(5): 1113–1123.

Payne, C.M., Y.J. Bomble, C.B. Taylor, C. McCabe, M.E. Himmel, M.F. Crowley and G.T. Beckham. 2011. Multiple functions of aromatic-carbohydrate interactions in a processive cellulase examined with molecular simulation. J. Biol. Chem. 286(47): 41028–41035.

Poveda, G., C. Gil-Durán, I. Vaca, G. Levicán and R. Chávez. 2018. Cold-active pectinolytic activity produced by filamentous fungi associated with Antarctic marine sponges. Biol. Res. 51(1): 28.

Ramana, K.V. and L. Singh. 2000. Microbial Degradation of Organic Wastes at Low Temperatures. Def. Sci. J. 50(4): 371.

Rascio, N. and N. La Rocca. 2008. Biological nitrogen fixation. pp. 412–419. In: Sven Erik Jørgensen and Brian D. Fath [eds.]. Encyclopedia of Ecology. Academy Press, USA.

Roland, J.C., D. Reis, B. Vian and S. Roy. 1989. The helicoidal plant cell wall as a performing cellulose-based composite. Biol. Cell. 67: 209–220.

Saxena, I.M. and Jr. R.M. Brown. 2005. Cellulose biosynthesis: current views and evolving concepts. Ann. Bot. 96(1): 9–21.

Shilev S., M. Naydenov, V. Vancheva and A. Aladjadjiyan. 2007. Composting of Food and Agricultural Wastes. pp. 283–301. In: V. Oreopoulou and W. Russ. [eds.]. Utilization of By-Products and Treatment of Waste in the Food Industry.

Shukla, L., A. Suman, A.N. Yadav, P. Verma and A.K. Saxena. 2016. Syntrophic microbial system for *ex-situ* degradation of paddy straw at low temperature under controlled and natural environment. J. App. Biol. Biotech. 4(2): 30–37.

Singh, L., M.S. Ram, M.K. Agarwal and S.I. Alam. 2000. Characterization of Aeromonas hydrophila strains and their evaluation for biodegradation of night soil. World J. Microbiol. Biotechnol.16(7): 625–630.

Swaminathan, Mathangi. 2018. How Can India's Waste Problem See a Systemic Change? Economic & Political Weekly. 53: 16.

Timell, T.E. 1967. Recent progress in the chemistry of wood hemicelluloses. Wood Sci. Technol. 1: 45–70.

Tomova, I., G. Gladka, A. Tashyrev and E. Vasileva-Tonkova. 2014. Isolation, identification and hydrolytic enzymes production of aerobic heterotrophic bacteria from two Antarctic islands. Int. J. Environ. Sci. 4(5): 614–625.

Tutino, M.L., E. Parrilli, L. Giaquinto, A. Duilio, G. Sannia, G. Feller and G. Marino. 2002. Secretion of α-amylase from Pseudoalteromonashaloplanktis TAB23: two different pathways in different hosts. JB. 184(20): 5814–5817.

Van der Wurff, A.W.G., J.G. Fuchs, M. Raviv and A.J. Termorshuizen. 2016. Handbook for composting and compost use in organic horticulture. BioGreenhouse.

Viikari, L., M. Tenkanen and A. Suurnakki. 2001. Biotechnology in the pulp and paper industry. 10: 523–546.

Voragen, A.G.J., G. Coenen, R.P. Verhoef and H.A. Schols. 2009. Pectin, a versatile polysaccharide present in plant cell walls. Struct. Chem. 20: 263–275.

W.M. Fogarty and C.T. Kelly (eds.). 1990. Microbial enzymes and biotechnology (2nd ed.), Elsevier Science Ltd., London 1990: 133–176.

Wood, T.M. and S.I. McCRAE. 1979. Synergism between enzymes involved in the solubilization of native cellulose. Advances in Chemistry 181: 181–209.

Xiao, Y., G.M. Zeng, Z.H. Yang, W.J. Shi, C. Huang, C.Z. Fan and Z.Y. Xu. 2009. Continuous thermophilic composting (CTC) for rapid biodegradation and maturation of organic municipal solid waste. Bioresour. Technol. 100(20): 4807–4813.

Yadav, P.K., V.K. Singh, S. Yadav, K.D.S. Yadav and D. Yadav. 2009. In silico analysis of pectin lyase and pectinase sequences. Biochemistry 74(9): 1049–1055.

Zavala, M.A.L., N. Funamizu and T. Takakuwa. 2002. Characterization of feces for describing the aerobic biodegradation of feces. Doboku Gakkai Ronbunshu 2002 (720): 99–105.

Zhang, H., C. Li, G. Li, B. Zang and Q. Yang. 2012. Effect of Spent Air Reusing (SAR) on maturity and greenhouse gas emissions during municipal solid waste (MSW) composting-with different pile height. Procedia. Environ. Sci. 16: 59–69.

Zhang, L.H., G.M. Zeng, J.C. Zhang, Y.N. Chen and M. Yu. 2015. Response of denitrifying genescoding for nitrite (nirK or nirS) and nitrous oxide (nosZ) reductases to different physico-chemical parameters during agricultural waste composting. Appl. Microbiol. Biotechnol. 99: 4059–4070.

14

Extremophiles Thriving in Extreme Ecosystems of Armenia

Panosyan Hovik* and Margaryan Armine

Introduction

Extremophiles are organisms that are able to thrive in ecological niches with extreme conditions, like high or low temperatures, extreme values of pH, high salt and metals concentrations, high pressure, absence of oxygen, and so on. Extremophiles compose deep lineages of the so-called Tree of Life (Forterre 2015). The study of species diversity, adaptation mechanisms, and phylogeny of extremophiles can shed light upon the origin of life on Earth and the problems of the early evolution of living beings (Merino et al. 2019). The study of extremophiles is informative to understand in which range of conditions life is possible.

Extremophiles produce unique biomolecules and enzymes (extremozymes) with a range of biological activities under extreme conditions (Antranikian and Egorova 2007, Kumar et al. 2011, Rampelotto 2013, Orellana et al. 2018). Nowadays, many extremozymes are widely used in many industries (Dumorné et al. 2017). These microbes hold great promises for genetically based medications and industrial chemicals and processes.

Extreme environments, like geothermal hot springs, saline lakes, saline-alkaline soils, solar salterns and subterranean salt deposits, polymetallic mines, and karst caves are widely distributed in various regions of our planet and offer a new source of fascinating microbes with unique properties. The combined approaches of traditional microbiology with molecular biology techniques have widely increased our understanding of the diversity of microbial communities in various extreme environments (Zhang et al. 2007, López-López et al. 2013, Vera-Gargallo et al. 2019).

Despite its small territory, Armenia is rich by ecosystems characterized by extreme conditions, microbial diversity of which remains unexplored yet. The geology of the region where Armenia is situated is complex, owing to the accretion of terrains through plate-tectonic processes and to the ongoing tectonic activity and volcanism (Badalyan 2000, Henneberger et al. 2000). Thus, numerous geothermal springs of different geotectonic origins, various saline and hypersaline environments, polymetallic mines, and solutional (karst) caves are found on the territory of Armenia (Mkrtchyan 1969, Baghdasaryan 1971, Vehuni 2001, Shahinyan 2005, Panosyan et al. 2018).

Recently, some microbiological studies of Armenian geothermal springs, saline-hypersaline environments, polymetallic mines and solutional (karst) caves based on cultivation and culture-independent methods were carried out and several thermophilic, halophilic, methalophilic, and

Department of Biochemistry, Microbiology and Biotechnology, Yerevan State University, A. Manoogian 1,0025 Yerevan, Armenia.
*Corresponding author: hpanosyan@ysu.am

endolithic microbes (both bacteria and archaea) (Panosyan et al. 2018a, b, Margaryan et al. 2019). In this chapter, an attempt has been made to review on microbial diversity in extreme environments of Armenia with special emphasis to its distribution, ecological significance, and biotechnological potential.

Diversity of Thermophiles in Armenian Geothermal Springs

On the territory of Armenia, many geothermal springs with different geotectonic origins are found (Mkrtchyan 1969; Henneberger et al. 2000). The water temperature of hot springs found in Armenia varies from 25.8 to $\geq$ 53°C, pH values are neutral, moderately alkaline or alkaline. The water of studied hot springs is characterized by higher ratios of Na^+ and K^+/Ca^{2+} and Mg^{2+} and relatively high ratios of chloride to bicarbonate (Cl^-/HCO_3^-) or sulfate to bicarbonate (SO_4^{2-}/HCO_3^-) (Mkrtchyan 1969, Henneberger et al. 2000). Some of these springs contain gases, such as hydrogen sulfide, methane, nitrogen, and carbon dioxide (Mkrtchyan 1969). Almost all the hot springs found in Armenia are tourist spots, known for medicinal value and used by the local people for various domestic purposes (Mkrtchyan 1969, Panosyan et al. 2018a). Some of these geothermal springs distributed on the territory of Armenia are shown in Figure 1.

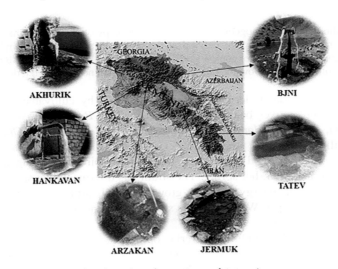

Figure 1. Some geothermal springs distributed on the territory of Armenia.

JERMUK

The spring in Jermuk (53°C) is located in southeast Armenia within the Vayots Dzor region at 39°96′63.90″N, 45°68′52.80″E, 2,080 m above sea level (asl). The spring is circumneutral (pH 7.50), classified as $HCO^{3-}/SO_4^{2-}/Na^+$ type spring, and has a relatively high dissolved mineral content of 4,340 µS/cm (Mkrtchyan 1969). A small pool with a high flow rate exists in the outlet. Biofilms with orange and light green are formed in the spring. Medicinal properties of Jermukare similar to the springs in Karlovy Vary, Czech Republic.

DGGE analysis of the partial bacterial 16S rRNA gene PCR amplicons was used to profile bacterial populations inhabiting Jermuk geothermal spring. The spring in Jermuk was found to be colonized by phylotypes related to sulfur and hydrogen oxidizing chemolithotrophs belonging to Epsilonproteobacteria along with the diversity of Bacteroidetes, Spirochaetes, Ignavibacteriae, and Firmicutes (Panosyan et al. 2017). The 454 GS FLX pyrosequencing of V4-V8 variable regions of the small-subunit rRNA was also applied to analyze the microbial diversity of Jermuk hot spring. Dominant bacterial pyrotags were affiliated with Proteobacteria and Bacteroidetes, and Synergistetes-dominant archaeal pyrotags were affiliated with Euryarchaeota (*Methanosarcinales*, *Methanosaeta*) and the yet-uncultivated Crenarchaeota groups MCG and DHVC1 (Figure 2) (Hedlund et al. 2013). Recently, Illumina HiSeq2500 paired-end sequencing of metagenomic DNA also was used to analyze water/sediment samples of the Jermuk hot spring. Taxonomic analyzes of the metagenomic rRNA sequences revealed the prevalence of Proteobacteria, Firmicutes, and Bacteroidetes. Archaeal community (~1%) was dominated by Euryarchaeota followed by Crenarchaeota, unclassified groups, and a minor fraction of Thaumarchaeota (Poghosyan 2015).

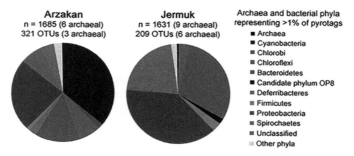

Figure 2. Relative abundance of phylogenetic groups in 'universal' pyrotag datasets. The number (n) of quality-filtered pyrotagsis indicated. OTUs were defined at the 97% identity level and phylum-level taxonomic assignments were given on the basis of comparison to Silva database using the Wang (Bayesian) method with a 50% confidence cut-off; those below the cut-off were binned as 'Unclassified' (Hedlund et al. 2013).

The diversity of bacteria in Jermuk hot spring, studied by cultivable approaches, revealed the presence of representatives belonging to *Anoxybacillus, Bacillus, Geobacillus, Treponema, Desulfomicrobium, Rhodopseudomonas, Desulfovibrio, Rhodobacter*, and *Thiospirillum*. Among thermophilic bacilli, many active producers of extracellular amylases, lipases, and proteases were selected. A strictly anaerobic and moderately thermophilic spirochete, designated *Treponema* sp. J25, was isolated from Jermuk hot spring. The strain has a fermentative sugar-based metabolism using xylan as a carbon and energy source. It grows optimally around 55°C with an upper limit at 60°C and at circumneutral pH. Whole-genome shotgun sequencing of this novel species has been performed (Poghosyan 2015).

Two sulfate-reducing thermophilic bacteria, optimally growing at 55°C and identified as *Desulfomicrobium thermophilum* SRB_21 and *Desulfovibrio psychrotolerans* SRB_141, were also isolated from Jermuk hot spring using lactate as sole carbon source (Poghosyan et al. 2014, Poghosyan 2015).

Paronyan (2003, 2007) isolated moderate thermophilic purple nonsulfur bacterial strains, designated as *Rhodopseudomonas palustris* D-6 (T_{opt} 45°C) and *R. palustris* D-6 (T_{opt} 45°C), using organic compounds as carbon and nitrogen organic source.

The spring in Hankavan (42-44°C; pH 7.0-7.2) is located on the bank of Marmarik river near the village of Hankavanat 40°63'26.50"N, 44°48'46.00"E, 1,900 m asl. In composition, it belongs to the hydrocarbonate-chloride sodium springs containing a high dissolved mineral content of 6,722.9 μS/cm (Mkrtchyan 1969). This geothermal spring is a result of geological drillings carrying water through a pipe. A small pool with a high flow rate exists in the outlet and vigorous degassing is noticeable. Biofilms with yellow, orange, light brown, and light green colors are formed in the spring.

Microbial diversity of Hankavan geothermal spring was studied using cultivation based methods. Several species belonging to *Anoxybacillus, Bacillus, Geobacillus*, and *Brevibacillus* were isolated and screened for extracellular amylase, lipase, and protease production (Panosyan 2017). The thermotolerant and metallotolerant bacilli, *B. circulans* (3A), is able to grow at different concentrations (from 10 to 300 mg/l) of Cd^{2+}, Cu^{2+}, Zn^{2+}, and Ni^{2+} (Margaryan et al. 2010).

Arzakan

Geothermal spring Arzakan (with temperature > 44°C; pH 7.2) is located in central Armenia, near the village Arzakan at 40°27′36.10″N, 44°36′17.76″E, 1,490 masl. The spring has a relatively high dissolved mineral content (4,378.3 µS/cm) and is classified as a hydrocarbonate sodium class of mineral springs and possesses a high concentration of dissolved minerals of which > 20% is HCO_3^- and > 20% is Na^+; HCO_3^-/Na^+ type spring. This geothermal spring is a result of geological drillings with small pool source and high flowrate. The slightly degassing is visible and silicate sands can be seen at the bottom. Biofilms with yellow, light brown, and light green colors are formed in the spring.

Water, sediment, and mat samples from Arzakan hot spring were screened by several metagenomic approaches. DGGE employed to reveal the microbial profile of sediments showed an abundance of bacterial populations related to Proteobacteria (affiliated with the Beta-, Epsilon-, and Gammaproteobacteria), Bacteroidetes, and Cyanobacteria (Panosyan and Birkeland 2014, Panosyan et al. 2017).

Studies based on sequence analysis of 16S rRNA gene clone libraries from the mixed water and sediment sampled from the Arzakan geothermal spring indicated a predominance of Bacteroidetes (48%), Cyanobacteria (35%), Betaproteobacteria (22%), Gammaproteobacteria (13%), Epsilonproteobacteria (9%), Firmicutes (9%), and Alphaproteobacteria (8%) (Panosyan and Birkeland 2014). The archaeal population was presented by Euryarchaeota (methanogenic archaea belonging to *Methanospirillum*, *Methanomethylovorans*, and *Methanoregula*), AOA Thaumarchaeota, *Nitrososphaeragargensis*, and yet-uncultivated Crenarchaeota (MCG and DHVC1 groups) (Hedlund et al. 2013).

Shotgun pyrosequencing of V4 region on 454GS FLX platform revealed highly diverse communities. Dominant bacterial phyla were Cyanobacteria in addition to Proteobacteria, Bacteroidetes, Chloroflexi, and Spirochaeta dominant archaeal pyrotags which were affiliated with Euryarchaeota (*Methanosarcinales* and *Methanosaeta*) and Crenarchaeota (the yet-uncultivated group MCG) (Figure 2) (Hedlund et al. 2013).

Microbial diversity of Arzakan hot spring studied by cultivable approaches revealed the presence of the representatives of genus *Bacillus*, *Anoxybacillus*, *Paenibacillus*, *Sporosarcina*, *Arcobacter*, *Methylocaldum*, *Rhodobacter*, *Rhodopseudomonas*, *Thiocapsa*, and *Methanoculleus*.

All the isolated thermophilic bacilli were tested for enzyme production capacities, such as lipase, protease, amylase, and a number of biotechnologically valuable enzyme producers were selected (Panosyan 2017).

Among thermophilic bacilli, *G. thermodenitrificans* ArzA-6 and *G. toebii* ArzA-8, deposited at the Microbial Depository Center of Armenia (Accession numbers MDC11858 and MDC11859, respectively), could produce extracellular polysaccharides (EPSs) with high molecular mass (≥ 5 × 105 Da) (Panosyan et al. 2018c).

Panosyan and Birkeland (2014) isolated a methanotrophic bacterial strain identified as *Methylocaldum* sp. (96% similarity), and an anaerobic Epsilonproteobacteria identified with 99% similarity as *Arcobacter* sp. (HM584709) based on analysis of the amplified 16S rRNA genes.

Hydrogenotrophic methanogenic archaea designated as *Methanoculleus* sp. Arz-ArchMG-1 (JQ929040) was successfully isolated from Arzakan hot spring.

Akhurik

Geothermal spring Akhurik is located in Shirak region of Armenia, near the village of Akhurik at 40°44′34.04″N, 43°46′53.95″E, 1,490 m asl. The outlet temperature of spring is 30°C, pH about 6.5, and 2,490 µS/cm conductivity and is classified as a hydrocarbonate-sulfate sodium-magnesium type of spring. This geothermal spring is a result of geological drillings where geothermal water comes through a pipe. A shallow pool consists of a small continuous outflow from a 2 m man-made cement-fountain landscape. Slightly degassing from water is visible. Multiple thick biomats (1-5 cm) of various colors (dark brown, red, dark green, and orange) are formed on the fountain, while a dark brown and green filamentous mat can be observed on the bank of the collecting pool.

The microbial diversity of Akhurik hot spring has been studied using culture-dependent traditional methods. Several thermophilic bacilli belonging to *Bacillus* and *Brevibacillus* have been isolated and identified based on phenotypic and phylogenetic characteristics (Panosyan 2010). Some of them were found to be active producers of hydrolytic enzymes (Panosyan 2017).

A lipase producing bacilli strain, identified as *B. licheniformis* Akhurik 107 (KY203975), has been isolated from Akhurikgeothermal spring (Shahinyan et al. 2017). The strain has been deposited at the Microbial Depository Center of Armenia under Accession number MDC11855. The highest lipase activity (0.89 U/ml) of *B. licheniformis* Akhurik 107 strain was observed at pH 6-7 and 55°C temperature. The PCR amplification lipase encoding genes revealed a 600 b.p. sized gene in *B. licheniformis* Akhurik 107 strain. Multiple alignments generated from primary structures of the lipase proteins and annotated lipase protein sequences, conserved regions analysis and amino acid composition, illustrated the similarity (98-99%) of the lipases with GDSL esterase family (family II). It was also shown that lipases contain Zn^{2+} and Ca^{2+} as ligands.

A methanotrophic bacteria identified and designated as *Methylocaldum* sp. AK-K6 (Accession number KP272135) with temperature range for growth of 8-35°C (optimal 25-28°C) and pH range of 5.0-7.5 (optimal 6.4-7.0) was also isolated from Akhurik (Islam et al. 2015).

Tatev

The spring in Tatev (27.5°C; pH 6.0) is located in Syunik region, near Satana's bridge (SataniKamurj) on the bank of Vorotan River at 39°23′76.00″N46°15′48.00″E, 960 m asl in a roughly round-shaped pool (diameter ~3 m, depth ~0.5 m). In composition, it belongs to the carbon-bicarbonate calcium water sources (Mkrtchyan 1969) with the dissolved mineral content of 1,920 μS/cm, many bubbling sources, and no visible outflow. Clays and sands at the bottom and biofilms with the light green color can be seen in this spring. The source is left in its natural form; no trace of human intervention is found. However, there are traces of ancient baths, which testify to the settlements dating BC.

Several bacilli strains as representatives of *Bacillus*, *Anoxybacillus*, and *Geobacillus* have been isolated from Tatev geothermal spring. Some of them were able to produce extracellular lipases, amylases, and proteases. The strains *Geobacillus* sp. Tatev N5 and Tatev N6 showed high lipase activity (70.3 U/ml) at 65°C after 5 hours of incubation (Vardanyan et al. 2015, Shahinyan et al. 2015, Panosyan 2017).

Lipase encoding genes of the strain *Geobacillus* sp. Tatev 4 (Microbial Depository Center of Armenia Accession number: MDC11856) were studied in detail. The PCR amplification of the lipase genes sequenced by using initially designed primer sets revealed the presence of 1,100 b.p. sized genes. Multiple alignments generated from primary structures of the lipase protein and annotated lipase protein sequences, conserved regions analysis, and amino acid composition illustrated the similarity (98-99%) of the lipases with lipase family I. It was also shown that lipases contain Zn^{2+} and Ca^{2+} as ligands (Shahinyan et al. 2017).

The spring in Bjni Tatev (30-37°C; pH 6.2-7.0) is located in Kotayk region, near the village of Bjniat 40°45′94.44″N 44°64′86.11″E, 1,610 m above sea level. In composition, it belongs to the chloride-hydrocarbonate sodium springs (Mkrtchyan 1969) and has a relatively dissolved mineral content of 4,138.3 μS/cm. This geothermal spring is also a result of geological drillings; geothermal water coming through a pipe. Biofilms with yellow, light brown, and light green colors are formed on stones near the spring.

Bacilli from genera *Bacillus*, *Anoxybacillus*, *Geobacillus*, and *Ureibacillus* were isolated from Bjni geothermal spring. The isolates were studied for extracellular amylase, lipase, and protease (Panosyan 2017).

Phototrophic bacteria *Rhodobacter* (*R. sphaeroides*), *Rhodopseudomonas* (*R. palustris*), *Thiocapsa* (*T. roseopersicina*) were also isolated from Bjni geothermal spring and studied for aspartase, aminoacylase, glucoseisomerase, and inulinase activities as a source of proteins, carbohydrates, and vitamins (Paronyan 2002).

Halophiles Diversity in Armenian Saline-Alkaline Soils and Subterranean Salt Deposits

Ararat Plain Soils

Hydromorphic saline-alkaline soils are found in the territory of Armenia (Baghdasaryan 1971; Panosyan et al. 2018b). Soils of more than 29,000 ha territory of Ararat Plain are saline (3% salinity) and alkaline (pH 9-11) (Baghdasaryan 1971). They are characterized by low humus content (<1.0%) and high absorbed sodium content (Figure 3). Hydromorphic saline-alkaline soils are developed in the areas, where subsoil water is mineralized and located close to the surface (1-2 m).

Figure 3. Saline-alkaline soils located in Ararat Plain

Microbial diversity of saline-alkaline soils in Ararat Plain (Armenia), as well as the microbial processes and key microbes involved in soil formation and biogeochemical cycles, has been studied in last few years (Panosyan et al. 2018b) (Figure 2).

The sequence analysis of a 16S rRNA gene clone library and DGGE profiles of the A horizon of natural saline-alkaline soils of Ararat Plain indicated the dominance of Firmicutes (*Halobacillus* (41.2%), *Piscibacillus* (23.5%), *Bacillus* (23.5%), and *Virgibacillus* (11.8%)) populations (Figure 4) (Panosyan et al. 2018b).

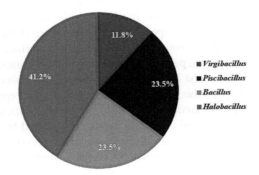

Figure 4. Genus-level distribution of bacilli in saline-alkaline soils of Ararat Plain.

Several halophilic bacilli strains, belonging to *B. alcalophilus* and *B. alcalophilus* subsp. *Halodurns* with optimum growth at 37°C, pH 8.6-10.1, and in presence of 15-17% NaCl were isolated from saline-alkaline soils (Khacaturian et al. 1995). Later, Panosyan (2007) showed that a number of haloalkalophilic bacilli (belonging to *B. subtilis*, *B. licheniformis*, *B. alcalophilus*, and *B. circulans*) were present up to 10^5 CFU/g soil. More recently, eight moderately halophilic bacilli isolates were able to grow up to 25% NaCl (optimum 9-18% NaCl) and were successfully obtained from the enriched cultures of the saline-alkaline soil samples and were characterized by phenotypic and phylogenetic techniques. The 16S rRNA gene sequence analyzes of isolates revealed their affiliation (97.7-99.7% similarity) to the genera *Bacillus* (4 isolates), *Piscibacillus* (2 isolates), and *Halobacillus* (2 isolates) (Panosyan et al. 2018b). Another recent study to identify bacterial diversity of saline-alkaline soils of Ararat Plain revealed the presence of halophilic Actinobacteria. In total five moderately haloalkaliphilic *Streptomyces* strains have been isolated and identified based on phenotypic characteristics and 16S rRNA gene sequence analysis (Hakobyan and Panosyan 2012; Hakobyan et al. 2013). The moderately haloalkaliphilic isolate, designated as *Streptomyces* sp. A3 (KU681239), exhibited optimal growth at 5% NaCl, pH 9, and 37°C, and exhibited antimicrobial activity against Gram-positive bacteria (*B. subtilis, Staphylococcus* sp., and *Enterococcus faecalis*) and yeasts (*Saccharomyces cerevisiae*) (Hakobyan and Panosyan 2012).

S. griseus A5, an actinobacterial isolate, has been identified based on 16S rRNA gene sequence analysis and phenotypic characteristics (Hakobyan et al. 2013). Isolates were able to optimally grow at 5% NaCl concentration with pH 8 and temperature 37°C and were able to utilize glucose, mannitol, xylose, and arabinose as carbon source along with acid production, degrade starch, cellulose, and casein.

Adamyan (2004) reported two halophilic bacilli stains (optimal growth at 12-15% NaCl) designated as *Bacillus* sp. 3849 and *B. sphaericus* 3863 from these saline-alkaline soils with the ability to produce cyclodextrin glucanotransferase (CGTase) with 71 kDa molecular mass.

Several alkalophilic and moderate halophilic phototrophic bacteria were able to grow optimally at 6-8% NaCl and pH 9.0-10.5 and belonging to genera, *Ectothiorhodospira* (*E. mobilis*), *Thiospirillum*, *Chromatium* (*C. vinosum*), *Thiocapsa* (*T. roseopersicina, T. jenense*) and *Lamprocystis*, have been isolated from the saline-alkaline soils of Ararat Plain (Paronyan 2002; 2003). Some of these have been studied for aspartase, aminoacilase, glucose isomerase, and inulinase activities, as well as sources of protein, carbohydrates, and vitamins.

Avan Subterranean Salt Deposits

A hypersaline subterranean salt deposit, found in Armenia, is located on the eastern part of Yerevan and currently referred as Avan subterranean salt deposit (40°13′30.15″N, 44°33′50.32″E) (Figure 5). The major chemical components of the salt stones from deposit are NaCl (95.5%), $CaSO_4$ (1.48%), $CaCl_2$ (0.14%), and $MgCl_2$ (0.08%) with insoluble residues 2.7% (Poghosyan and Khoyetsyan 2008). The Republican Speleological Therapeutic Center established in the subterranean salt deposit is helpful for curing patients suffering from bronchial asthma, respiratory, and some other diseases.

Both culture-based and culture-independent approaches have been used for addressing microbial diversity associated with salt mine. The 16S rRNA gene libraries generated from total DNA extracts and PCR-DGGE fingerprinting method indicated an abundance of representatives belonging to genera *Bacillus, Virgibacillus, Halobacillus, Filobacillus, Anoxybacillus*, and *Streptomyces*. Several sequences of clone libraries were found to be affiliated to uncultured bacteria. DGGE profile was in good agreement with the clone library results indicating the predominance of halophilic bacilli (unpublished data). Sequence analysis of clones in the archaeal library and DGGE profiles indicated that they originated from *Euryarchaeota* and were mainly affiliated with the genera *Haloarcula*, *Halobacterium, Halarchaeum*, and *Natronomonas* (Hakobyan et al. 2015).

Figure 5. Avan subterranean salt deposit.

Several aerobic chemoorganotrophic endospore-forming bacteria were isolated from the salt samples and based on 16S rRNA sequence analyzes identified as representatives of the genus *Halobacillus, Piscibacillus,* and *Bacillus* (Hakobyan et al. 2011).

Archaeal strains related to members of the genera *Haloarcula* (97-99% similarity) and *Halarchaeum* (<97% similarity) were also obtained. These strains required at least 1.5-2.0 M NaCl (optimal growth at 3.5-4 M NaCl) and 10 mM Mg^{2+} for growth and were able to grow at pH 5.5-10 (optimum, pH 7-8) and at 21-45°C (optimum 37°C). Most of the strains grew in the presence of casamino acids, Na-glutamate, D-glucose, L-propionate, L-arginine, and D-cellobiose as carbon sources. Chemotaxonomic studies revealed the presence of the dieter phospholipids phosphatidylglycerol and phosphatidylglycerol phosphate methyl ester derived from C20-C20 archaeol (Hakobyan et al. 2014).

The draft genome *Haloarcula* sp. salt stone-1of the strain was sequenced by using PacBio RS technology and was assembled using the Celera Assembler GATC Biotech, Germany (http://www. gatc-biotech.com) resulting in 4 Mb of unique sequence data distributed into 48 contigs constituting a total of 4,331,370 bp. The G+C content was 61.4%. Gene prediction carried out with the NCBI Prokaryotic Genome Annotation Pipeline, as well as the RAST server (http://rast.nmpdr.org/rast. cgi) identified a total of 2,529 genes, including 4,261 coding DNA sequences and 68 sets of rRNA genes (Azaryan 2018).

Azaryan et al. (2017, 2018b) studied both cell growth and total carotenoid production of the halophilic *Haloarcula* sp. salt stone-1 and *H. japonica* A2 strains. The optimum condition for cell growth and total carotenoid production was observed at 30°C, pH 7.2, and 20% NaCl (w/v) using casamino acids as carbon source. The production of biomass ranged up to 0.7 and the total carotenoids up to 10.2 mg/l. The maximal productivity of carotenoids at the late stationary phase of growth was 14.6 mg/g. Under 24 hours of light conditions, the biomass increased 1.16 fold, while the production of carotenoids increased 1.4 fold, demonstrating the need for light for optimal carotenoids production.

Metallophiles Diversity in Armenian Polymetallic Mines

As part of the Lesser Caucasus mountain chain, Armenia has significant reserves of copper, molybdenum, gold, iron, silver, zinc, antimony, aluminum as well as in scarce and scattered metals (rhenium, selenium, tellurium, cadmium, indium, thallium, bismuth, etc.). With regards to metals, the main metals found included copper and gold, and there are three regions that are especially important (Vehuni 2011), namely:

1. The Alaverdi region in the north with mainly polymetallic (Akhtala mine) and copper (Alaverdi and Shamlug mines) volcanogenic massive sulfide (VMS) type deposits as well as porphyry copper deposits (Teghut mines).
2. The Kapan area in the southeast, which has geology similar to that of the Alaverdi area, thus also hosting copper (Kapan-Central mine) and polymetallic (Shahumyan mine) VMS

deposits. This type of geology can be traced from Georgia through north-eastern Armenia to the Kapan area and into Iran.

3. The area of the Zangezur mountain range in the southwest, stretching from Meghri for some 50 km to the north northwest, where a number of copper and copper-molybdenum deposits are located (Kajaran and Agarak mines).

Gold is sometimes found in the abovementioned type of polymetallic deposits as well as in a number of other locations throughout the country. Furthermore, the Amulsar and other gold projects in VyotsDzor, the Sotk gold mine east of southern Lake Sevan, and a number of gold deposits west of northern Lake Sevan show that the other parts of Armenia are prospective for gold (Vehuni 2011).

Currently, in Armenia, about 670 mines are situated, including 30 base metal and precious metal mines. Among these around 400 mines, including 22 base metal, non-ferrous metal, and precious metal mines are exploited (Suvaryan et al. 2011). In addition to mines, estimated and registered in the state inventory, 115 deposits of various metals have been discovered in the territory of Armenia (Figure 6).

The mining operation in Armenia is leading to the high emission of toxic metals to the environment. There are 15 tailings in the territory of the country, the volumes of which exceed some million m^3 and occupy a total area of around 700 hectares. The production waste, generated as a result of extraction and processing of minerals, accumulated in tailings is not utilized, although they contain a great number of polymetals (Gevorgyan et al. 2016, Suvaryan et al. 2011).

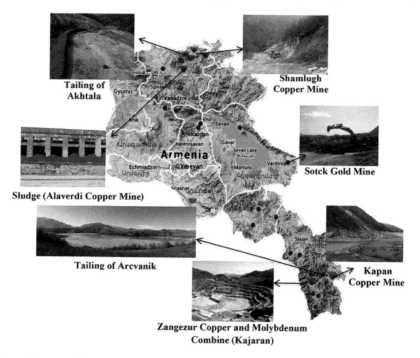

Figure 6. Map of Armenia showing the location of the studied sites (red points on the map are showing the location of mines, yellow semilunar shape figures are showing the tailings).

To characterize the microbiome of Armenian mines and tailing, both molecular and culture-based methods were used, which gave an opportunity to collect information on the composition of microbial communities and to isolate microorganisms capable of colonizing the metal-impacted habits with high resistance potentials (Margaryan et al. 2013, Margaryan 2014, Vardanyan and Vardanyan 2018, Ayvazyan 2018, Margaryan et al. 2019).

TABLE 1: Geographical Location and Culture-Independent Study of Microbial Population in Some Mines and Tailings in the Territory of Armenia

Mine/Tailing	Geographical Location		Culture-Independent Study			References
	GPS location	Altitude (m, Above Sea Level)/pH	Approach	Population Proportion	Accession Number	
Sotk gold mine	45°58′00.76″E 40°13′52.20″N	2,477/6.5-7.0	DGGE	Dominant bacterial phyla were Proteobacteria (50%) and Actinobacteria (50%)	–	Margaryan 2017
Zangezur copper and molybdenum mine	46°08′54.28″E 39°09′13.15″N	350/7.0-7.3	DGGE	Detected bacterial groups were Bacteroidetes (10%) and Actinobacteria (90%)	–	Margaryan 2017
Alaverdi copper smelter	44°39′41.96″E 41°5′58.79″N	1,000/6.8	DGGE	Actinobacteria	–	Margaryan 2017
Shamlugh copper mine	44°44.470′E 41°09.121′N	810/5.6	DGGE	Dominant bacterial groups were Proteobacteria (90%) and Actinobacteria (10%), archaeal* groups were Euryarchaeota (50%) and Crenarchaeota (50%)	–	Margaryan 2017
Akhtala tailing	44°45.948′E 41°09.356′N	740/2.6	Shotgun pyrosequencing of V4 region on 454 GS FLX platform; DGGE	Detected bacterial phyla were Proteobacteria (49%) Bacteroidetes (43%) Saccharibacteria (2%) Verrucomicrobia (1.5%) Gammatimonadetes (1%), archaeal* group was Ferroplasma	SAMN06062036	Margaryan 2017; Margaryan et al. 2019

*The data regarding archaeal 16S rDNA DGGE profiling is not published.

Till date, four mines and one tailing samples have been analyzed using cultivation-independent approaches. Studies based on 16S rDNA denaturing gradient gel electrophoresis (DGGE), profiling of the stone rock samples from Sotk gold mine, Zangezur copper and molybdenum mine, Alaverdi and Shamlugh copper mine as well as sludge sample from Akhtala tailing were performed (Margaryan 2017, Margaryan et al. 2019). It was the first analysis of microbial populations in the mining areas in the territory of Armenia based on cultivation-independent approaches.

DGGE based analysis of the mentioned samples was done to provide a snapshot of the microbial communities' structure in the studied samples. Authors showed that Armenian mines and tailing inhabited mainly with the bacteria belonging to the phyla Proteobacteria, Actinobacteria, and Bacteroidetes (Table 1).

The 16S rDNA DGGE study showed that stone rock samples of Sotk gold mine were rich with the representatives of the phylum Proteobacteria (50%) and Actinobacteria (50%), which involved mainly *Propionibacterium* sp. and uncultured bacterial groups. The bacterial groups belonging to the phyla Bacteroidetes (10%) and Actinobacteria (90%) were found in the samples of Zangezur copper and molybdenum mine. The DGGE gel bands derived from Zangezure copper and molybdenum stone rock samples involved in the clusters with *Nocardiafluminea* and some uncultured bacterial groups. The sample from Alaverdi copper mine was rich with the representatives of the phylum Actinobacteria, which includes bacterial genus of *Micromonospora*, *Geodermatophilus*, *Cryptosporangium*, *Rhodococcus*, *Saccharomonospora* and *Blastococcus*, which are usually found in soil (Margaryan 2017).

The bacterial community in the stone rock sample of Shamlogh copper mine contains bacteria from the Proteobacteria (90%) and Actinobacteria (10%). The authors showed that the phylum Proteobacteriais represented with an anaerobic bacterium of the genus *Georgfuchsia*, aerobic thiosulfate-oxidizing bacteria belonging to aerobic thiosulfate-oxidizing bacteria from the genus *Limnobacter* and *Thiobacillus* and soil bacterium of the genus *Klebsiella* (Figure 7).

Figure 7. Circular phylogram showing the phylogenetic affiliation of 16S rDNA sequences derived from DGGE profiles (●-Sotck gold mine, ○-Zangezur copper and molybdenum mine, ◆-Alaverdi copper mine, ◊-Shamlugh copper mine). The bootstrap values per 100 bootstrap analyzes are presented on the tree. The scale bar represents 0.1 substitutions per site (Margaryan 2017).

The phylum Actinobacteria found in the samples from Shamlugh copper mine is represented only with bacterium from the genus *Rhodococcus* (Margaryan 2017).

Shamlugh copper mine is highlighted also by archaeal community composition in which *Aciduliprofundum* sp., Sulfolobussolfataricus, and uncultured archaea belonging to the phylum Crenarchaeota are found (unpublished data).

Akhtala copper mine tailing is located in Armenia's northern Lori province, near Akhtala town. The tailing characterized by high acidity (pH 2.5) with a relatively high dissolved mineral content (7,700 µS/cm) and high concentration of Fe^{2+}, Zn^{2+}, Cu^{2+}, and Mn^{2+} (6.55, 5.87, 0.498, and 0.98 mg/l, correspondingly). The samples taken from the surface and from 10 cm deep layers of the tailing were analyzed using DGGE fingerprinting. The results showed that the surface layer of Akhtala tailing mainly inhabited by bacteria belonging to the phylum Firmicutes cluster in the genus *Bacillus*. A 10 cm deep layer of the tailing is dominated by bacteria belonging to the uncultured Acidobacteria (Margaryan et al. 2019). The DNA from the mixed samples collected from different layers of Akhtala tailing was also studied by DGGE fingerprinting. Authors indicated that the bacteria found in mixed samples were closely related to the class Gammaproteobacteria, genus *Pseudomonas*.

DGGE fingerprinting was used to detect the archaeal population in the Akhtala copper mine tailing. The results demonstrated that the archaeal community formed in the deep layer of the tailing was represented mainly by archaea belonging to the genus *Ferroplasma* of the phylum Euryarchaeota (unpublished data). The consortium of uncultivated bacterial groups is the main bacterial population of the Akhtala tailing, which indicates the importance of the future exploration of the tailing as an important source of the new and physiologically diverse bacteria.

A more detailed analysis of the bacterial community in the Akhtala tailing was analyzed by tagged 454 pyrosequencing. As a result, a total of 15 bacterial phyla were identified (Figure 8).

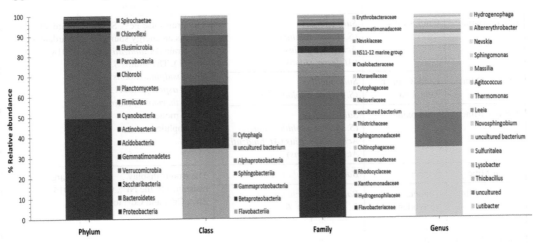

Figure 8. Relative abundances of phylogenetic groups in the sludge sample derived from the Akhtala tailing. The dominant Phylum, Class, Family, and Genus are shown in Figure (Margaryan et al. 2019).

A major part of the sequences, originating from Proteobacteria (49%) and Bacteroidetes (43%), is responsible for biomass degradation and fermentation. Minor bacterial taxa were distributed in phyla Saccharibacteria (2%) (formerly known as candidate division TM7), Verrucomicrobia (1.5%), and Gammatimonadetes (1%). The presence of additional phyla, such as Actinobacteria, Acidobacteria, Cyanobacteria, Firmicutes, and Planctomycetes account for less than 1% based on single-read assignments (Margaryan et al. 2019).

Culture-dependent approaches were applied to the isolates to study heavy metal resistant bacteria in different mining sites in the territory of Armenia. About 17 genera of metal resistant aerobic and anaerobic, mesophilic and moderate thermophilic, heterotrophic, and autotrophic bacterial strains have been isolated from copper, copper-molybdenum, and gold mines in Armenia. The isolates belonged to the genera *Arthrobacter, Algoriphagus, Acidithiobacillus, Bacillus, Brevibacillus, Comamonas, Clostridium, Geobacillus, Leptospirillum, Micrococcus, Methylobacterium, Pseudomonas, Rheinheimera, Stenotrophomonas, Sinomonas, Sulfobacillus,* and *Tepidimicrobium* (Table 2).

TABLE 2: Summary of Thermophilic Bacteria Isolated From Geothermal Springs of the Lesser Caucasus

Mines/ Tailings	Deposit and Metals Mined	Bacterial Genera (Dominant Species)	Comments	References
Sotk Mine	Gold	*Bacillus* (*B. licheniformis, B. subtilis, B. simplex, B. pumilus*), *Geobacillus* (*G. thermodenitrificans*), *Brevibacillus* (*B. thermoruber*)	The draft genome of *B. subtilis* AG4 strain was sequenced (Accession number KY203975). Metal resistance genes encoding copper-translocating P-type ATPase, Lead, cadmium, zinc, and mercury transporting ATPase, copper-translocating P-type ATPase, zinc ABC transporter, cobalt-zinc-cadmium resistance proteins were identified in the genome of the strain.	Margaryan et al. 2013, Margaryan 2014, Margaryan et al. 2018
Drmbon Mine	Gold, Copper	*Sulfobacillus* (*S. thermosulphidoxidans*), *Acidithiobacillus* (*At. ferrooxidans*)	The strains use Fe^{2+}, $S^·$, and FeS_2 as an energy source (the strain *At. tandzuti* 5 use only $S^·$). The strains *S. thermosulphidoxidans* subsp. *asporogenes* 41, 69, and *S. thermosulphidoxidans* 86 are moderate thermophiles.	Vardanyan and Vardanyan 2018
Armanis Mine	Gold, Copper, Zinc, Lead	*Sulfobacillus* (*S. thermosulphidoxidans*)		
Tandzut Mine	Gold oxide	*Sulfobacillus* (*S. thermosulphidoxidans*), *Acidithiobacillus* (*At. tandzuti, At. ferrooxidans*)		
Kapan Mine	Copper	*Bacillus* (*B. subtilis*)	The strains display resistance to the Ni^{2+}, Cu^{2+}, Zn^{2+}, Cr^{6+}, and Cd^{2+} ions.	Margaryan 2014
Zangezur Mine	Copper, Molybdenum	*Bacillus* (*B. licheniformis, B. megaterium, B. subtilis, B. pumilus*)		
Shamlugh Mine	Copper, Zinc, Lead	*Arthrobacter* (*A. oryzae*), *Stenotrophomonas* (*S. maltophilia*), *Bacillus* (*B. thuringiensis, B. cereus, Bacillus* sp.) *Comamonas* (*C. thiooxydans*)		unpublished
Alaverdi Mine	Copper	*Methylobacterium* (*M. phyllostachyos*), *Bacillus* (*Bacillus* sp., *B. firmus, B. cereus*), *Micrococcus* (*Micrococcus* sp.), *Leptospirillum* (*Leptospirillum* sp.)		Margaryan 2014

TABLE 2: Contd....

TABLE 2: Contd.

Mines/ Tailings	Deposit and Metals Mined	Bacterial Genera (Dominant Species)	Comments	References
Teghout Mine	Copper, Molybdenum	*Leptospirillum (L.ferrooxidans)*	The strain uses Fe^{2+}, $S^{\cdot}$, and FeS_2 as energy source.	Vardanyan and Vardanyan 2018
Artsvaniktailing	High contamination with toxic metals is registered	*Rheinheimera (R. soli), Pseudomonas (Pseudomonas sp.), Algoriphagus (A. aquaeductus), Bacillus (B. pumilus, Bacillus sp.)*	*R. soli* AI1, *Pseudomonas* sp. AI2 and *A. aquaeductus* AI3 exhibit hydrolytic activities.	Ayvazyan 2018
Akhtalatailing		*Bacillus (B. licheniformis, B. megaterium), Geobacillus (G. thermodenitrificans, G. pallidus, G. toebii), Brevibacillus (B. thermoruber), Clostridium (C. butyricum), Tepidimicrobium (T. sp.), Sinomonas (Sinamonas sp.), Arthrobacter (Arthrobacter sp.)*	The strain *B. megaterium* AA1 accumulate up to 90% $Cd2^+$ from the growth medium.	Margaryan 2014

The isolated strains display high resistance to Ni^{2+}, Cu^{2+}, Zn^{2+}, Cr^{6+}, Co^{2+}, and Cd^{2+} ions when employed separately as well as mixed in the growth medium.

The draft genome of *B. subtilis* AG4 strain, isolated from Sotk gold mine, was sequenced. Gene prediction identified metal resistance genes encoding copper-translocating P-type ATPase, lead, cadmium, zinc, and mercury transporting ATPase, copper-translocating P-type ATPase, Zinc ABC transporter, cobalt-zinc-cadmium resistance proteins in the genome of *B. subtilis* AG4 (Margaryan et al. 2018). It was shown that *B. subtilis* AG4 accumulated Cu^{2+} and Zn^{2+} equal to 6.8 and 3.0 mg/g wet weight of biomass, respectively. The authors using RT-qPCR showed that the highest expression of the *copA* gene was observed at 1 mM Cu^{2+}, while for the *czcD* expression it was observed at the 0.5 mM Zn^{2+} and 0.05 mM Cd^{2+} (Margaryan et al. 2013, Margaryan et al. 2018).

Acidophilic iron-oxidizing bacteria like *Acidithiobacillus*, *Leptospirillum*, and *Sulfobacillus* are widely found in different mines of Armenian. It is noteworthy that *At. ferrooxidans* and *S. thermosulfidooxidans* are found in copper, copper-molybdenum, and gold polymetallic mines, while *Leptospirillum* spp. dominated in the ores rich in pyrite. These bacteria are extremely acidophilic, using Fe^{2+}, $S^{\cdot}$, and FeS_2 as an energy source and can function in the temperature range from 10 to 55°C (Vardanyan and Vardanyan 2018, Vardanian and Akopyan 2003).

Microbial Diversity in Armenia Karst Caves

Caves are diverse environments differing in both their physical and chemical compositions. There are different classifications of caves like hypogenic/epigenic, anchialine, ice, sea, karst, and lava tube. In general, caves are classified based on (1) the solid rock, (2) groundwater proximity, and (3) the morphology and organization of passages (Engel 2010). The most extensively studied caves for their microbial diversity are karstic or limestone caves. High humidity (90-100%), relatively stable, and low temperature inside with limited temperature fluctuation during the seasons and low to no light conditions are the main characteristics of caves studied so far (Schabereiter-Gurtner et al. 2003, Zhou et al. 2007). From the geomicrobiological aspect, caves are nutrient-limited habitats and have low biotic potential (Barton and Jurado 2007). That is why microorganisms that are able to survive in caves are slow-growing organisms and oligotrophic. Endolithic chemoautotrophs serve as the primary producers in cave environments (Northup et al. 2011). Microbes harboring caves are believed to have adapted and evolved to specific selective pressures (Snider et al. 2009, Engel 2010). Microorganisms inhabiting these ecosystems may produce secondary metabolites valuable for biotechnology.

Despite its limited territory, Armenia is rich in karst caves from a few meters to several kilometers in length, microbial diversity of which has not been thoroughly explored (Shahinyan 2015). The largest karst caves of the Republic of Armenia were found on the left bank of River Arpa from Areni village to Agarakadzor village. They are Archeri, Magel, Mozrovi, Karmir, Vayk, and Simonietc (Figure 9).

Figure 9. Stalagmites and stalactites of Armenian caves: A. Archeri, B. Karmir, and C. Magel.

Recently, cultivation techniques were used to isolate and characterize bacteria from some Armenian karst caves located on Vayots Dzor region. It was the first microbiological investigation of Armenian caves. A total of six bacterial species were isolated from the stalagmites and stalactites sampled from Magel (39°41'12.3"N, 45°18'04.1"E), Mozrovi (39°42'36.3"N, 45°16'06.1"E), and Karmir (39°41'13.4"N, 45°17'26.7"E) caves. In many other caves, Proteobacteria, Firmicutes, and Actinobacteria have been found to be a dominating group of microbes.

All the isolates were identified based on phenotypic and phylogenetic characteristics (16S rRNA gene sequence analysis) as representatives of genera *Bacillus* (*B. zhangzhouensis*), *Microbacterium* (*M. aurantiacum*), *Arthrobacter*, *Pseudomonas* (*P. fulva*), *Leclercia* (*L. adecarboxylata*) (Khachatryan 2018).

Conclusions

Investigation on the diversity of extremophilic microbes colonizing extreme environments is important to understand their role in biogeochemical cycles of biogenic elements in the ecosystem and explore their biotechnological potency. Microbiological studies based on culture-dependent and molecular methods of extreme environments (hot springs, (hyper) saline soils, subterranean salt deposits, polymetallic mines, and karst caves) in Armenia have demonstrated huge diversity of extremophilic bacteria and archaea that thrive in these environments. Many thermophilic, halophilic, and metallophilic microbial strains have been isolated and biotechnological potential of these has been evaluated. Culture-independent studies suggest that a large number of sequences remained taxonomically unresolved, indicating the presence of potential novel microbes. The present work extends available information regarding the extremophiles from various ecological niches of Armenia and their biotechnological applications.

Acknowledgments

This work was supported by research projects of RA MES State Committee of Science №15T-1F399 and № 18T-1F261 to HP, Armenian National Science and Education Fund based in New York, USA, (ANSEF-NS-microbio 2493, 3362, 4676 to HP and 4619 to AM) and partially by grants from the Eurasia program of the Norwegian Center for International Cooperation in Education (CPEA-2011/10081, CPEA-LT-2016/10095 to HP).

References

Adamyan, M.O. 2004. Ecology and biosynthetic peculiarities of halophilic spore-forming bacteria. Ph.D. Thesis, Institute of Microbiology NAS RA, Yerevan, Armenia (in Russian).

Antranikian, G. and K. Egorova. 2007. Extremophiles a unique resource of biocatalysts for industry. pp. 361–406. *In*: C. Gerday and N. Glansdorff [eds.]. Physiology and Biochemistry of Extremophiles. ASM Press, Washington, DC, USA.

Ayvazyan, I. 2018. New heavy metal resistant alkaliphilic bacteria isolated from Artsvanik tailing. Collection of Scientific Articles of YSU SSS. 1.1 (24): 216–223.

Azaryan, A. 2018. Study of wholegenome and some biochemical peculiarities of *Haloarcula* sp. Saltstone-1 archaeal strain isolated from Avansalt deposit. MS. Thesis, Yerevan State University, Yerevan, Armenia (in Armenian).

Azaryan, A., H. Panosyan, A. Trchounian and N.-K. Birkeland. 2018. Optimization of the culture conditions for cell growth and total carotenoid production of *Haloarculajaponica* A2 isolated from the Avan subterranean salt deposit in Armenia. Abstract book of 12th International Congress of Extremophiles. Ischia, Naples, Italy: 63.

Azaryan, A.S., L.S. Gabrielyan, H.H. Panosyan and A.H. Trchounian. 2017. Effect of growth conditions on total carotenoid production by *Haloarcula japonica* A-2 isolated from Avan (Armenia) subterranean deposit. Book of Abstracts and Papers of the Conference. Yerevan, Armenia: 25–33.

Badalyan, M. 2000. Geothermal features of Armenia: a country update. Proceedings World Geothermal Congress. Kyushu-Tohoku, Japan: 71–75.

Baghdasaryan, A.B. 1971. Physical Geography of Armenian SSR. Publishing house of NA of ASSR, Yerevan, Armenia (in Russian).

Barton, H.A. and V. Jurado. 2007. What's up down there? Microbial diversity in caves. Microbe 2: 132–1389.

Deepika. M. and T. Satyanarayana. 2013. Diversity of hot environments and thermophilic microbes. pp. 3–60. *In*: T. Satyanarayana, J. Littlechild and Y. Kawarabayasi [eds.]. Thermophilic Microbes in Environmental and Industrial Biotechnology. Springer, Dordrecht, Heidelberg, New York, London.

Dumorné, K., D.C. Córdova, M. Astorga-Elóand and P. Renganathan. 2017. Extremozymes: a potential source for industrial applications. J. Microbiol. Biotechnol. 27(4): 649–659.

Engel, A.S. 2010. Microbial diversity of cave ecosystem. pp. 219–238. *In*: L.L. Barton, M. Mandl and A. Loy [eds.]. Geomicrobiology: Molecular and Environmental Perspective. Springer Science and Business Media B.V, The Netherlands.

Forterre, P. 2015. The universal tree of life: an update. Front. Microbiol. 6: 1–18.

Gevorgyan, G.A., A.S. Mamyan, L.R. Hambaryan, S.K. Khudaverdyan and A. Vaseashta. 2016. Environmental risk assessment of heavy metal pollution in Armenian river ecosystems: case study of Lake Sevan and Debed river Catchment basins. Pol. J. Environ. Stud. 25(6): 2387–2399.

Hakobyan, A., A. Margaryan and H. Panosyan. 2011. Halophilic aerobic endospore-forming bacteria of alkali-saline soils and salt mine of Armenia. Abstractsbook of XV International Pushchin School-Conference of Youth Scientists "Biolog-Science of XXI century". Pushchino, Russia: 361 (in Russian).

Hakobyan, A. and H. Panosyan. 2012. Antimicrobial activity of moderately haloalkaliphilic *Streptomyces roseosporus* A3 isolated from saline-alkaline soils of Ararat Plain, Armenia. Proceedings of International Young Scientists Conference "Prospectives for Development of Molecular and Cellular Biology-3". Yerevan, Armenia: 89–95.

Hakobyan, A., H. Panosyan and A. Trchounian. 2013. Production of cellulose by the haloalkaliphilic strains of *Streptomyces* isolated from saline-alkaline soils of Ararat Plain, Armenia. El. J. of NAS RA. Biotech. 21(2): 44–46.

Hakobyan, A., H. Panosyan, A. Trchounian and N.-K. Birkeland. 2014. Identification of halophilic archaea from the Avan salt mine in Armenia, using a polyphasic approach including 16S rRNA gene sequence polymorphism. Abstracts Book of International Scientific Workshop on "Trends in Microbiology and Microbial Biotechnology". Yerevan, Armenia: 37.

Hakobyan, A., H. Panosyan and A. Trchounian. 2015. Isolation of halophilic archaea from Avan salt mine in Armenia and their identification using polyphasic approaches. Book of Abstracts 3rd International Scientific Conference "Dialogues on science". Yerevan, Armenia: 27.

Hedlund, B.P., J.A. Dodsworth, J.K. Cole and H.H. Panosyan. 2013. An integrated study reveals diverse methanogens, Thaumarchaeota, and yet-uncultivated archaeal lineages in Armenian hot springs. Antonie Van Leeuwenhoek 104(1): 71–82.

Henneberger, R., D. Cooksley and J. Hallberg. 2000. Geothermal resources of Armenia. Proceedings world geothermal congress, Kyushu-Tohoku, Japan: 1217–1222.

Islam, T., Ø. Larsen, V. Torsvik, L.Ø. vreås, H. Panosyan, C. Murrell, et al. 2015. Novel methanotrophs of the Family *Methylococcaceae* from different geographical regions and habitats. Microorganisms 3: 484–499.

Khacaturian, A.A., N.L. Kazanchian, N.S. Khacaturian, M.O. Adamian and L.O. Khachikian. 1995. About the ecology of extremophilic forms of bacilli in the main types of soils of Armenia. BJA 48(1): 12–18.

Khachatryan, E. 2018. The Study of Microbiome of Some Karst Caves Distributed in Armenia. MS. Thesis, Yerevan State University, Yerevan, Armenia (in Armenian).

Kumar, L., G. Awasthi and B. Singh. 2011. Extremophiles: a novel source of industrially important enzymes. Biotechnology 10: 121–135.

López-López, O., M.E. Cerdán and M.I. González-Siso. 2013. Hot spring metagenomics. Life 3(2): 308–320.

Margaryan, A., H. Panosyan and Y.G. Popov. 2010. Isolation and characterization of new metallotolerant bacilli strains. Biotechnol. Biotechnol. Equip. 24(2): 450–454.

Margaryan, A. 2014. Microbiota of ore rock of Armenian copper and gold mines, isolation, identification and metal resistance mechanisms of heavy metal resistant bacilli. Ph.D. Thesis, Yerevan State University, Yerevan, Armenia (in Russian).

Margaryan, A., H. Panosyan, A. Trchounian and N.-K. Birkeland. 2018. *Bacillus subtilis* AG4 as a prospective tool for bioremediation. 12th International Congress of Extremophiles, Ischia, Italy: 59.

Margaryan, A., H. Panosyan, Ch. Mamimin, A. Trchounian and N.-K. Birkeland. 2019. Insights into the bacterial diversity of the acidic akhtala mine tailing in Armenia using molecular approaches. Curr. Microbio. 76: 462–469.

Margaryan, A.A., H.H. Panosyan, N.-K. Birkeland and A. Trchounian. 2013. Heavy metal accumulation and the expression of the *copA* and *nikA* genes in *Bacillus subtilis* AG4 isolated from the Sotk Gold Mine in Armenia. BJA 3(65): 51–57.

Margaryan, A.A. 2017. Bacterial community structure of Armenian mining territories revealed by PCR-DGGE fingerprinting. BJA 69(4): 56–62.

Merino, N., H.S. Aronson, D.P. Bojanova, J. Feyhl-Buska, M.L. Wong, S. Zhang, et al. 2019. Living at the extremes: extremophiles and the limits of life in a planetary context. Front. Microbiol. 10(780): 1–25.

Mkrtchyan, S. 1969. Geology of Armenian SSR. Publishing house of AS of ASSR, Yerevan (in Russian).

Northup, D.E., L.A. Melim, M.N. Spilde, J.J.M. Hathaway, M.G. Garcia, A. Moya, et al. 2011. Lava cave microbial communities within mats and secondary mineral deposits: implications for life detection on other planets. Astrobiology 11(7): 601–618.

Orellana, R., C. Macaya, G. Bravo, F. Dorochesi, A. Cumsille, R. Valencia, et al. 2018. Living at the frontiers of life: extremophiles in chile and their potential for bioremediation. Front.Microbiol. 9: 2309.

Panosyan, H. 2007. Bacterial population of alkali-saline soils of Ararat plain. Bul. of ASAU 20(4): 20–22.

Panosyan, H. and N.-K. Birkeland. 2014. Microbial diversity in an Armenian geothermal spring assessed by molecular and culture-based methods. J. Basic. Microbiol. 54(11): 1240–1250.

Panosyan, H., A. Margaryan, L. Poghosyan, A. Saghatelyan, E. Gabashvili, E. Jaiani, et al. 2018a. Microbial diversity of terrestrial geothermal springs in Lesser Caucasus. pp. 81–117. *In*: D. Egamberdieva, N.-K. Birkeland, H. Panosyan and W.J. Li [eds.]. Extremophiles in Eurasian Ecosystems: Ecology, Diversity, and Applications. Springer, Singapore.

Panosyan, H., A. Hakobyan, N.-K. Birkeland and A. Trchounian. 2018b. Bacilli community of saline-alkaline soils from the Ararat Plain (Armenia) assessed by molecular and culture-based methods. Syst. Appl. Microbiol. 41: 232–240.

Panosyan, H., P. Di Donato, A. Poliand and B. Nicolaus. 2018c. Production and characterization of exopolysaccharides by *Geobacillus thermodenitrificans* ArzA-6 and *Geobacillustoebii* ArzA-8 strains isolated from an Armenian geothermal spring. Extremophiles 22: 725–737.

Panosyan, H.H. 2010. Phylogenetic diversity based on 16S rRNA gene sequence analysis of aerobic thermophilic endospore-forming bacteria isolated from geothermal springs in Armenia. Biolog. J. Armenia 62(4): 73–80.

Panosyan, H.H. 2017. Thermophilic bacilli isolated from Armenian geothermal springs and their potential for production of hydrolytic enzymes. Int. J. Biotech. Bioeng. 3(8): 239–244.

Panosyan, H.H., A.A. Margaryan and A.H. Trchounian. 2017. Denaturing gradient gel electrophoresis (DGGE) profiles of the partial 16S rRNA genes defined bacterial population inhabiting in Armenian geothermal springs. Biolog. J. Armenia 68(3): 102–109.

Paronyan, A. 2003. Ecology, Biological Peculiarities of Phototrophic Bacteria of Armenia and Prospects its Application. Ph.D. Thesis, Institute of Microbiology NAS of RA, Abovyan, Armenia (in Russian).

Paronyan, A. 2007. Ecophisiological charachteristics of phototrophic bacteria *Rhodopseudomonaspalustris* isolated from mineral geothermal Jermuk. Biol. J. Armenia 59(1–2): 73–77.

Paronyan, A.K. 2002. Ecology and biodiversity of phototrophic bacteria of various ecosystems of Armenia. Biol. J. Armenia 54(1–2): 91–98.

Poghosyan, D.A. and A. Khoyetsyan. 2008. Landscape of the Armenian Highland and Physico-geographical Regions. YSU press, Yerevan (In Armenian).

Poghosyan, L., N.-K. Birkeland and H. Panosyan. 2014. Diversity of thermophilic anaerobes in the geothermal spring Jermuk in Armenia. Abstracts book of international scientific workshop on "Trends in microbiology and microbial biotechnology", Yerevan, Armenia: 83.

Poghosyan, L. 2015. Prokaryotic Diversity in an Armenian Geothermal Spring using Metagenomics, Anaerobic Cultivation and Genome Sequencing. MS. Thesis, University of Bergen, Bergen.

Rampelotto, P.H. 2013. Extremophiles and extreme environtments. Life 3: 482–485.

Schabereiter-Gurtner, C., W. Lubitz and S. Rölleke. 2003. Application of broad-range 16S rRNA PCR amplification and DGGE fingerprinting for detection of tick-infecting bacteria. J. Microbiol. Methods 52(2): 251–260.

Shahinyan, G., A.A. Margaryan, H.H. Panosyan and A.H. Trchounian. 2017. Identification and sequence analysis of novel lipase encoding novel thermophilic bacilli isolated from Armenian geothermal springs. BMC Microbiol. 17(103): 1–11.

Shahinyan, G.S., A.A. Margaryan, H.H. Panosyan and A.H. Trchounian. 2015. Isolation and characterization of lipase-producing thermophilic bacilli from geothermal springs in Armenia and Nagorno-Karabakh. Biol. J. Armenia 67(2): 6–15.

Shahinyan, S. 2005. Caves of Armenia. Zangak Press, Yerevan.

Snider, J.R., C. Goin, R. Miller, P.L. Boston and D.E. Northup. 2009. Ultraviolet radiation sensibility in cave bacteria: evidence of adaptation to the subsurface? Int. J. Speleol. 38(1): 13–22.

Suvaryan, Y., V. Sargsyan and A. Sargsyan. 2011. The problem of heavy metal pollution in the Republic of Armenia: overview and strategies of balancing socioeconomic and ecological development. pp. 309–316. *In*: L. Simeonov, M. Kochubovski and B. Simeonova [eds.]. Environmental Heavy Metal Pollution and Effects on Child Mental Development. NATO Science for Peace and Security Series C: Environmental Security. 1. Springer, Dordrecht.

Vardanyan, G., A. Margaryan and H. Panosyan. 2015. Isolation and characterization of lipase-produvting bacilli from Tatev geothermal spring (Armenia). Collection of scientific articles of YSU SSS: Materials of the scientific session dedicated to the 95th anniversary of YSU, Yerevan: 33–36.

Vardanian, N.S. and V.P. Akopyan. 2003. Diversity of chemolitotrophic bacteria, important of biotechnology, in the sulfide deposits of Armenia. BJA 55: 220–227.

Vardanyan, N.S. and A.K. Vardanyan. 2018. Thermophilic chemolitotrophic bacteria in mining sites. pp. 187–218. *In*: D. Egamberdieva, N.-K. Birkeland, H. Panosyan and W.-J. Li [eds.]. Microorganisms for Sustainability: Extremophiles in Eurasian Ecosystems: Ecology, Diversity, and Applications. Chapter 7, Springer, Singapore.

Vehuni, A. 2001. Geology and Treasures of the Entrails of the Armenian Highland, YSU Press, Yerevan (in Armenian).

Vera-Gargallo, B., T.R. Chowdhury, J. Brown, S.J. Fansler, A. Durán-Viseras, C. Sánchez-Porro, et al. 2019. Spatial distribution of prokaryotic communities in hypersaline soils. Sci. Rep. 9: 1769.

Zhang, H.B., M.X. Yang, W. Shi, Y. Zheng, T. Sha and Z.W. Zhao. 2007. Bacterial diversity in mine tailings compared by cultivation and cultivation-independent methods and their resistance to lead and cadmium. Microb. Ecol. 54(4): 705–712.

Zhou, J., Y. Gu, C. Zou and M. Mo. 2007. Phylogenetic diversity of bacteria in an earth-cave in Guizhou province, southwest of China. J. Microbiol. 45(2): 105–112.

15

Extremophiles Inhabiting Unique Ecosystems in Egypt

Nashwa I. Hagagy[1,5,*], Amna A. Saddiq[2], Hend A. Hamedo[3] and Samy A. Selim[4,5]

Introduction

Egypt occupies the north-eastern side of the African continent with a surface area of approximately one million square kilometers (1,019,600 km^2), equivalent to 0.3% of the total area of Africa. The country lies at the center of the driest and largest desert on the earth. Egypt enjoys a considerable diversity of habitat, especially the extreme one, such as solar saltern in Alexanderia and Sinai, Soda Lakes of Wadi Al-Natrun, hot springs in Sinai, and western and eastern deserts 'Sahara' (Figure 1). The extreme habitats are widely distributed throughout the world and they comprise, rather than present in Egypt, the Arctic and Antarctic ice shields, the permafrost and glaciers, the surface and deep-sea volcanic areas, the thermal, and often acidic springs and lakes. Despite the hostile physico-chemical conditions, these extreme habitats are densely populated by microorganisms, mostly comprise *Archaea* and *Bacteria*, and few *Eukarya*. The biodiversity of extreme ecosystems is a subject of interest due to the significant role of the extremophilic microorganisms in biotechnology, biogeochemical processes, and pharmaceutical fields. This chapter will focus on the biodiversity of extremophiles inhabiting extreme ecosystems in Egypt, such as Soda Lakes, Solar Salterns, and Hot Springs.

Extreme Ecosystems

Soda Lakes of Wadi Al-Natrun

Soda Lakes, which are stable and productive aquatic extreme ecosystems, display pH values around 10 or higher and salt concentrations exceeding 300 g/l (Oren 2002). These alkaline and hypersaline lakes are elsewhere around the world, most in Africa, India, China, and Australia. In Egypt, there is a chain of eight distinct and largest Soda Lakes throughout the world present in a northern province called 'Wadi Al-Natrun' (Figure 2).

[1] Department of Biology, College of Science and Arts at Khulias, University of Jeddah, Jeddah, Saudi Arabia.

[2] Department of Biology, Faculty of Science, University of Jeddah, Jeddah, Saudi Arabia.

[3] Department of Science, Faculty of Science, Al-Arish University, AL-Arish, Egypt.

[4] Department of Clinical Laboratory Sciences, College of Applied Medical Sciences, Jouf University, Sakaka, P.O. 2014, Saudi Arabia.

[5] Department of Botany, Faculty of Science, Suez Canal University, Ismailia, 41522, Egypt.

*Corresponding author: niibrahem@uj.edu.sa

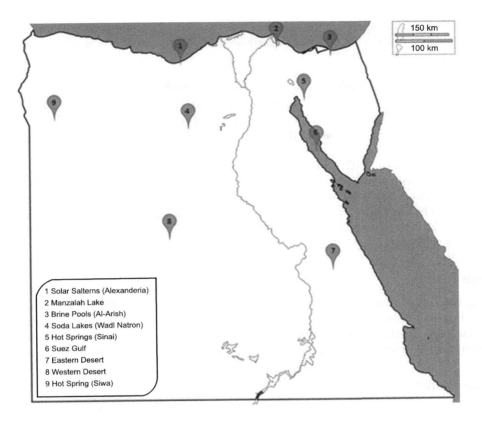

1 Solar Salterns (Alexanderia)
2 Manzalah Lake
3 Brine Pools (Al-Arish)
4 Soda Lakes (Wadl Natron)
5 Hot Springs (Sinai)
6 Suez Gulf
7 Eastern Desert
8 Western Desert
9 Hot Spring (Siwa)

Figure 1. Map of Egypt showing the locations of extreme ecosystems throughout the country.

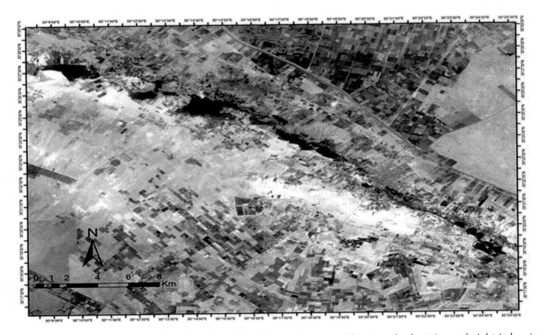

Figure 2. Satellite image of Soda Lakes of Wadi Al-Natrun in Egypt, Showing the locations of eight Lakes in different fascinating shades of red and green. (http://googleearthonline.blogspot.com/).

The Wadi is a depression in the Western Desert located 90 km northwest of Cairo. The Wadi is 50 km long by 10 km wide and runs in north-west and south-east direction, almost parallel to the road and between 10 and 20 km from it. Overall, the altitude over the depression drops to 24 m below sea level (Shortland 2004), but rises gently to the south-west to a ridge some 80-90 m above sea level and eventually up to the heights of Gebel Hadid (185 m) and Gebel Qantara (198 m), which lie some 20-30 km away. Along the valley spreads a chain of eight large alkalines, hypersaline lakes in addition to a number of ephemeral pools. The depth of the lakes ranges between 0.5 and 2 m and is controlled by seasonal changes in flood leakage and evaporation. High evaporation rates and dry climatic conditions during the summer cause the salinity to be above 5 M (Taher 1999).

In mid-September 2002, a joint team from the University of Oxford and Egyptian Geological Survey and Mining Authority (EGSMA) visited the Wadi and recorded their description of the lakes (Shortland 2004). However, we noticed some differences when we visited four lakes in October 2009, i.e., Lakes Hamra, Zugm, Beida, and Fazda (Selim et al. 2012, 2014), and the visit was renewed in September 2014 for the other four lakes, i.e., Ruzania, Um-Risha, Khadra, and Ga'ar (data not published).

Lake Fazda

Lake Fazda is the southern lake, located at (30°19.748′ N, 30°24.245′ E). The lake was nearly dry with little salt depositions; the brine was green-yellowish with a sulphide smell from which the Arabic name 'Fazda', "spoilage or deterioration", is derived since people there suggested that it had a rotting smell. It also had a thin salt crust that covered the sediments (Figure 3A).

Lake Hamra

Lake Hamra is the first lake north of the small town of Bir Hooker, located at 30°23.455′ N, 30°19.640′ E). It has a red colour from which its Arabic name 'birkethamra' (the red lake) is derived. When we visited the lakes in October 2009, the lake was full of brine and shallow, water was colourless, little salts were around them, and a very thin salt crust covered the sediment on the border of the lake (Figure 3B).

Lake Zugm

Lake Zugmis the next lake to the north located at 30°23.917′ N, 30°18.383′ E; it is connected to Hamra Lake by a narrow channel of water. During our visit in October 2009, the lake was almost dry, salt deposited around them, and on the sediment surface, brine was distinctly pink (Figure 3C).

Lake Beida

Lake Beida is the next lake in the chain. 'Beida' is a translation of the Arabic word for 'white', which is the colour reflected in the salt deposition of the lake. When we had visited the lake in October 2009, it was full of brine, was shallow, thick salt had deposited around the brine, and the brine was colourless as well (Figure 3D).

Lake Um-Risha

Lake Um-Risha is the second lake in the chain, located at 30°20.176′ N, 30°22.995′ E. The lake was fully dry, had huge salt deposition, no brine, and a very thick salt crust covered the sediments (Figure 3E).

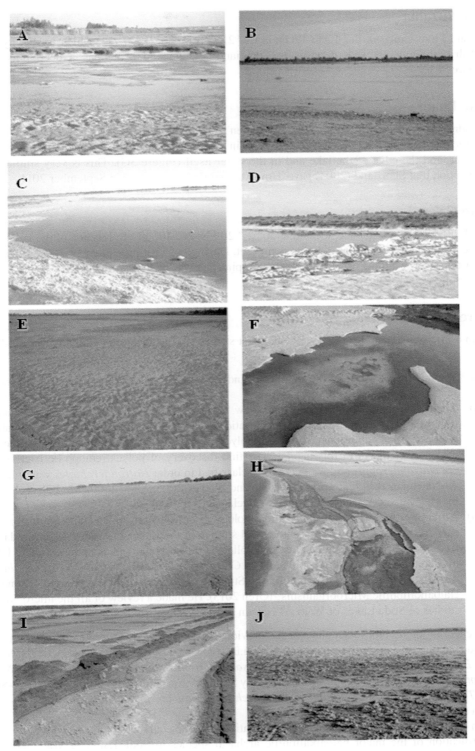

Figure 3. Soda Lakes of Wadi Al-Natrun, A. Lake Fazda, B. Lake Hamra, C. Lake Zugm, D. Lake Bieda,E. Lake Um-Risha, F. Lake Ruzania, G. Lake Khadra, H. Lake Ga'ar. And Solar salterns in Egypt;I. Anthropogenic salterns, located at El-Tina city, Port-Said; J. Natural salterns, located at Romana city, Al-Arish, Sinai. (Photographs by Nashwa I. Hagagy).

Lake Ruzania

Beyond Lake Umrisha lies Lake Ruzania (30°21.748′ N, 30°21.137′ E), which is slightly small, half-dry, shallow, the salt deposits are brilliantly white, and brine has a deep pink colour (Figure 3F).

Lake Khadra

'Khadra' is an Arabic word, translation for 'green', reflecting the green colour of the water. The lake is large and deep, brine is green, and only a thin crust of white salt covers the soil on edge; reeds grow around the edge of the lake and in the water itself (Figure 3G). This is very similar to the description that Shortland (2004) gave. However, our description was in September 2014.

Lake Ga'ar

The last lake in the sequence, located at 30° 27′ 20″ N, 30° 10′ 20″ E; it is one of the largest lakes of the Wadi, a larger amount of halite is present. During our visit in September 2014, the lake was almost dry, very shallow brine, a thick crust of pink salt was on the sediment surface, and the brine was orange-red (Figure 3H).

Extremophiles of Wadi Al-Natrun

From a microbiological point of view, the most striking points of Soda Lakes in Wadi Al-Natrun are:

(i) Having different shades of red, purple, and green which reflect their dense and diverse content of extremophilic microorganisms.

(ii) Their inhabitants can survive and grow optimally under two or more harsh conditions (alkaline pH values, high salt concentrations, high temperatures, intense UV radiation as well as anoxic conditions); so-called 'polyextremophiles', they play a significant role in the cycles of elements in this habitat. The first discovery of polyextremophilic bacteria, genera *Natranaerobius* and *Natronovirga*, classified in order *Natranaerobiales* was from Wadi Al-Natrun by the great work of Mesbah and Wiegel (2005, 2008, 2009), Mesbah et al. (2007a, b), Mesbah and Wiegel (2011), and Wiegel (2011) in the Wadi.

(iii) Attracting the interest of many microbiology scientists from nineteenth-century or maybe earlier; the early known microbiological studies of the Lakes of Wadi Al-Natrun date was from the 1970s; the scientist Oren had collected the early biological studies of these lakes in his comprehensive chapter entitled *"Two Centuries of Microbiological Research in the Wadi Natrun, Egypt*: A Model System for the Study of the Ecology, Physiology, and Taxonomy of Haloalkaliphilic Microorganisms" (Oren 2013). He made a review of all the microbiological studies of Soda Lakes of Wadi Al-Natrun from 1970 to 2011.

Hence, in an attempt to supplement the previous studies with new microbiological data of this unique habitat, this part will be based on two Ph.D. studies: (i) Hagagy (2014), entitled *"Biodiversity and Biotechnological Potential of Archaea Isolated from Soda Lakes of the Wadi Al-Natrun, Egypt"*, under the supervision of Professor Samy Selim (Selim et al. 2012, 2014a) and (ii) El-Shafey (2017), entitled *"Molecular Studies on Haloarchaea Producing Industrially Important Enzymes"* (data not published).

In the first study, a combination of cultivation and molecular methods had allowed for the identification of the major haloalkaliphilic archaeal species in Soda Lakes of Wadi Al-Natrun. The archaeal diversity in the water and sediment samples of four Soda lakes of Wadi An-Natrun, Hamra, Zugm, Beidah, and Fazda was assessed by denaturing gradient gel electrophoresis (DGGE) and quantifying copy numbers of archaeal 16S rRNA genes in the environmental DNA extracts by real-time PCR using *Archaea*-specific primers in addition to enrichment and isolation on ten different complex and defined media. Strains belonging to the genera *Natronococcus, Natronolimnobius, Natronomonas, Natrinema, Natrialba, Haloterrigena, Halobiforma, Halorubrum* as well as uncultured archaeon were identified from sediment and brine samples by enrichment, and the 16S rRNA gene data of the isolated archaeal strains through this study have been deposited in the NCBI and GenBank nucleotide sequence databases under the accession numbers from **HQ658984** to **HQ659003**. Using the culture-independent techniques, all sequences fall into the phyla *Euryarchaeota* and *Crenarchaeota*. The library comprised new phylotypes, which consisted of five clones, that exhibited low 16S rRNA similarity (90-95%). Two clones with 94% and 95% similarity to *Halogranum* and *Halalkalicoccus* that could represent new species as well as two unclassified archaeal phylotypes, which exhibited 97-99% similarity to uncultured archaeon clone. The DGGE analysis and archaeal 16S rRNA gene copy number detected by real-time PCR for the lakes were carried out at microbiology laboratory, Department of Agricultural, Forestry and Food Sciences (DISAFA), University of Torino, Italy, under the supervision of Dr. Roberta Gorra, as previously described by Webster et al. (2006).

In the second Ph.D. study, a first pyrosequencing study for Soda Lakes of Wadi Al-Natrun, Egypt, to our knowledge, was performed; the metagenomics data obtained here provide complete profiling of microbial diversity for one of the largest Soda Lake, Ga'ar Lake, for understanding extremophilic community structure in such unique environment. The results revealed a predominance of domain *Archaea* (81.7%), comprised 10,798 OTUs of prokaryotic community, domain Eubacteria (18%) comprised 2,328 OTUs of microbial community, and less than 0.5% of OTUs were assigned to Viruses. The metagenomics data will provide significant clues in understanding the functional potential of Soda Lake with the probability of obtaining novel genes, as well as extremophilic microorganisms, for both commercial and research purposes. Metagenome sequence data are available on EMBL Metagenomics under the accession no. **PRJEEB18746** (https://www.ebi.ac.uk/ena/data/view/ERR1770058).

Moreover, by culture-dependent technique, fifty haloalkaliphilic archaeal strains were identified within genera *Natrinema, Haloterrigena, Natrialba, Natronococcus, Natronomonas, and Halobiforma* with significantly low similarities (≥ 67%). Therefore, the author can suppose that they may represent new genera or species, and their 16S rRNA gene data have been deposited in the NCBI and GenBank nucleotide sequence databases under the accession numbers from **KT459358** to **KT459407**. Thus, further work is needed for exploring more haloalkaliphilic microorganisms inhabiting this unique ecosystem, and the study of these amazing and unique microorganisms will tell us much about the microbial survival and adaptation to extreme conditions in the next years. The recently obtained genome sequence of one of the representatives of haloalkaliphiles (Selim and Hagagy 2016) will definitely help in this task on such a unique habitat.

Hypersaline Environments

Hypersaline environments are extreme habitats where the salinity is much higher than that of seawater and can be divided into two types depending on whether they are originated from seawater 'thalassolohaline' or not 'athalassohaline'. Example of thalassohaline environments is brine pool in Sinai Peninsula, a solar saltern that was created by human activity (used for the production of salt by evaporation of seawater). Chemically, the thalassohaline habitats are characterized by a predominance of Cl^- and Na^+ in addition to other ions, such as Mg^{2+}, SO_4^{2-}, K^+, Br^-, HCO_3^-, and F^-. The group of extremophilic microorganisms inhabiting hypersaline habitats are called halophiles. Different definitions for what constitutes halophiles were proposed; the most common definition of halophiles is "the microorganisms which require NaCl concentrations more than 0.2 M for optimal growth" (Kushner and Kamekura 1988). They are classified into three groups: extreme halophiles grow at 3.4-5.1 M [20-30% NaCl (w/v)]; moderate halophiles grow in the range 0.85-3.4 M [3-25% NaCl (w/v)]; slightly halophiles grow at 0.2-0.85 M [1-5% NaCl (w/v)]. Halotolerant microorganisms do not require salt for growth but can grow well at high salt concentrations (Ollivier 1994).

In Egypt, some researches were carried out for exploring of halophilic *Archaea* from different hypersaline environments, *Haloarculaquadrata* sp. nov., amotile, square archaeon isolated from a solar saltern, Sinai Peninsula, Egypt (Oren et al. 1999); *Haloferax alexandrines* isolated from solar saltern of Alexandria (Asker and Ohta 2002) and *Halobacterium salinarum* (Yusef and El-Besoumy 2003 and Ghozlan et al. 2004). Screening of moderately halophilic bacteria from saline habitats in Alexandria was also studied by Ghozlan et al. (2006). A molecular genetic study was carried out on the diversity of archaeal communities in two sites, Bashtir and Genka; Manzala Lake revealed the diversity of *Methanothrix soehngenii, Methanospirillum hungatei,* and *Methanocorpusculum labreanum, Crenarchaeota*-like phylotypes affiliated to the thermophilic species *Thermofilum pendens* and *Staphylothermus achaiicus* (ElSaied 2008). New phylotypes of archaeal phyla, methanogen *Euryarchaeota,* and *Thaumarchaeota* by using 16S rRNA gene analysis to explore and evaluate uncultured archaea in a polluted site, El-Zeitia, Suez Gulf (ElSaied 2014). In addition, in my M.Sc thesis (Hagagy 2009), the phenotypic and genotypic characteristics of halophiles inhabiting natural and anthropogenic salterns located at northeastern coast of Egypt were assessed (Figures 3I and J). In such a study, archaeal isolates were diverse and clustered with members of the genera *Haloferax, Haloarcula, Halorubrum,* and *Halococcus.* However, isolates belonging to the domain Bacteria belonged to two main lineages: order *Bacillales* within the low G+C Gram-positive *Firmicutes* (relatives to the genera *Bacillus, Halobacillus, Pontibacillus, Oceanobacillus, Virgibacillus,* and *Marinococcus*) and γ-Proteobacteria (relatives of the genera *Halomonas,* and *Chromohalobacter*) as shown in Figures (4 and 5) (data not published).

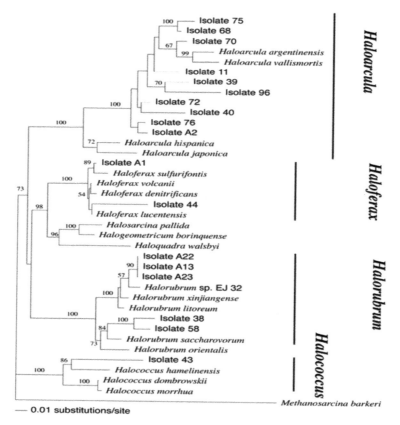

Figure 4. Distance dendogram highlighting the phylogenetic affiliations of archaeal isolates to their most closely validly described relatives in the database. The tree was constructed using Neighbor-joining algorithm with Jukes Cantor corrections. Bootstrap values represent 1000 replicates and are shown for well-supported branches of the tree.

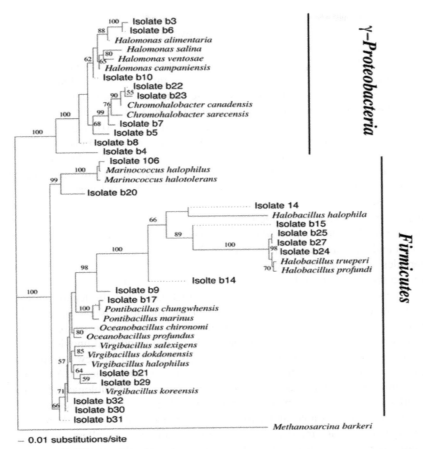

— 0.01 substitutions/site

Figure 5. Distance dendogram highlighting the phylogenetic affiliations of bacterial isolates to their most closely validly described relatives in the database. The tree was constructed using Neighbor-joining algorithm with Jukes Cantor corrections. Bootstrap values represent 1000 replicates and are shown for well-supported branches of the tree. (both dendograms were constructed by Prof. ElShahed M.S, Oklaohoma State University, USA, during his valuable supervision on my master thesis in 2009).

The physiological and biochemical characterization of the isolated strains exhibited interesting results; some of them can grow at pH up to 10.5, and most of them can exhibit growth at temperature up to 65°C. Most strains can produce useful biomolecules, such as alkali and salt-tolerant enzymes, which are uniquely suited for applications requiring such extreme conditions, like detergent compositions, leather tanning, food industries, waste treatment, and textile industries as well as biotransformation.

Hot Springs

Hot springs are widespread all over the world on all continents and even under the oceans and seas. In Egypt, the most of the thermal springs are located along the shores of Gulf of Suez and the Red Sea on the eastern (Oyun Musa 'Musa Spring', Hammam Pharaon 'Pharaon's Bath', and Hammam Musa 'Musa' Bath') and western (Ain Sokhna 'Hot Spring') sides of the Gulf of Suez as shown in Figure 1. In addition to some hot springs present in the Western Desert (in Dakhla, Kharga, Bahariya, and Siwa oases). Hammam Pharaon area considered the hottest spring in Egypt where the water temperature is 70°C; Musa Springs consists of twelve springs (Abdel-Zaher et al. 2010). These places are the most important sites for therapeutic tourism with the highest rate of sulfuric water known in the world (Selim et al. 2014b); they are used in swimming pools and employed in a direct heating greenhouse cultivation (Lund and Boyd 2015, Lashin 2015, Elbarbary et al. 2018).

In Egypt, few researchers have studied the Egyptian hot springs, such as Morgan et al. (1983, 1985), Swanberg et al. (1983), Boulos (1990), Hosney (2000), Abdel Zaher et al. (2010, 2011a, 2011b, 2012a, 2012b, 2018), Lashin (2013, 2015), Chandrasekharam et al. (2015, 2016a, 2016b), Mohamed et al. (2015), and Atef et al. (2016). Unfortunately, these unique ecosystems have been not yet explored in detail from the microbiological point of view; most of the previous investigations have based on geological, geochemical, and geophysical analyses. Salah et al. (2007) have isolated thermophilic cellulases producing bacteria from an Egyptian hot spring at 70°C, *Anoxybacillus flavithermus*, *Geobacillus thermodenitrificans*, and *Geobacillus stearothermophilus*. Selim et al. (2014b) have assessed the microbial community prospering in two hot springs, Hammam Pharaon and Oyun Mossa, by culture-dependent and molecular-based methods; 16S rRNA gene partial sequence results indicated the presence of *Desulfovibrio vulgaris* as well as novel or existing species of *Bacillus*, *Anoxybacillus*, and *Geobacillus*. In addition to this, Abol-Fotouh et al. (2016) have isolated novel strain with high lipolytic activity, *Geobacillus thermoleovorans* DA2, from the geothermal site in southern Sinai and have studied its application in the leather industry.

Thermophiles grow optimally at a temperature of 45 to < 80°C (Kristjansson 1991), and hyperthermophiles grow optimally at ≥80°C or capable of growing at ≥90°C (Blumer-Schuette et al. 2008) and have to survive at high temperatures that cause protein denaturation (Jaenicke and Böhm 1998). There are two mechanisms for stabilizing the proteins and prevent denaturation at high temperatures; the first one is through modification of protein structures to be considerably more compact compared to mesophilic homologs (Berezovsky and Shakhnovich 2005). The second mechanism is sequence-based through modifications that build up few strong interactions for high thermal stability; hence, their structures are not significantly different from their mesophilic homologs (Dominy et al. 2004, Li et al. 2005). The metabolic and physiological strategic adaptations of thermophiles make their great potential in biotechnology by their production of thermozymes; these enzymes are thermostable and can tolerate harsh conditions, like high salinity and the presence of organic solvents and denaturing agents, making it an excellent model for applications of thermophilic organic solid waste degradation. In addition to this, thermonzymes can be widely used in food, detergent, textile, leather, and pharmaceutical industries. Therefore, further studies are needed for exploring novel thermophilic microorganisms and their potential activity in this unique ecosystem in Egypt.

Acknowledgments

Dr. Nashwa I. Hagagy is grateful to all her supervisors of the MSc and Ph.D. theses for their valuable supervisions and offering assistance during her studies, and thank Cultural Affairs and Mission Sector, Ministry of Higher Education, Egypt, for financial support during her Ph.D. study, Suez Canal University, Ismailia, Egypt, in cooperation with the University of Torino, Torino, Italy. Also, the authors are grateful to the University of Jeddah, Jeddah, Saudi Arabia for technical support.

References

Abdel-Zaher, M., S., Ehara and G. El-Qady. 2010. Conceptual model and numerical simulation of the hydrothermal system in Hammam Faraun hot spring, Sinai Peninsula, Egypt. Arab. J. Geosci. 4(1): 161–170.

Abdel Zaher, M., H. Saibi, M. El-Nouby, E. Ghamry and S. Ehara. 2011a. A preliminary regional geothermal assessment of the Gulf of Suez, Egypt. J. Afr. Earth Sci. 60: 117–132.

Abdel Zaher, M., J. Nishijima, G. El-Qady, E. Aboud, O. Masoud, M. Soliman, et al. 2011b. Gravity and magnetotelluric investigations to elicit the origin of Hammam Faraun hot spring, Sinai Peninsula, Egypt. J. Acta Geophys. 59(3): 633–656.

Abdel Zaher, M., H. Saibi and S. Ehara. 2012a. Geochemical and stable isotopic studies of Gulf of Suez's hot springs, Egypt. Chin. J. Geochem. 31(2): 120–127.

Abdel Zaher, M., H. Saibi, J. Nishijima, Y. Fujimitsu, H. Mesbah and S. Ehara. 2012b. Exploration and assessment of the geothermal resources in the Hammam Faraun hot spring, Sinai Peninsula, Egypt. J. Asian Earth Sci. 45: 256–267.

Abdel Zaher, M., H. Saibi, K. Mansour, A. Khalil and M. Soliman. 2018. Geothermal exploration using airborne gravity and magnetic data at Siwa Oasis, Western Desert, Egypt. Renew. Sust. Energ. Rev. 82(3): 3824–3832.

Abol-Fotouh, D.M., R.A. Bayoumi and M.A. Hassan. 2016. Production of thermoalkaliphilic lipase from *geobacillusthermoleovorans* DA2 and application in leather industry. Enzyme Research 2016: http://dx.doi.org/10.1155/2016/9034364.

Asker, D. and Y. Ohta. 2002. *Haloferaxalexandrinus* sp. nov., an extremely halophiliccanthaxanthin-producing archaeon from a solar saltern in Alexandria (Egypt). Int. J. Sys. Evol. Microbiol. 52: 729–738.

Atef, H., A. Abd El-Gawad, M. Abdel Zaher and K. Farag. 2016. The contribution of gravity method in geothermal exploration of southern part of the Gulf of Suez-Sinai region, Egypt. NRIAG J. Astron. Geophys. 5: 173–185.

Berezovsky, I.N. and E.I. Shakhnovich. 2005. Physics and evolution of thermophilic adaptation. Proc. Natl. Acad. Sci. 102(36): 12742–12747.

Blumer-Schuette, S.E., I. Kataeva, J. Westpheling, M.W. Adams and R.M. Kelly. 2008. Extremely thermophilic microorganisms for biomass conversion: status and prospects. Curr. Opin. Biotechnol. 19(3): 210–217.

Boulos, F.K. 1990. Some aspects of the geophysical regime of Egypt in relation to heat flow, groundwater and microearthquakes. *In*: Said, R. [ed.]. The Geology of Egypt. Balkema, Rotterdam.

Chandrasekharam, D., A. Lashin, N. Al-Arifi, A. Al-Bassam and C. Varun. 2015. Evolution of geothermal systems around the Red Sea. Environ. Earth Sci. 73(8): 4215–4236.

Chandrasekharam, D., A. Lashin, N. Al-Arifi, A. Al-Bassam, C. Varun and H.K. Singh. 2016a. Geothermal energy potential of Eastern Desert region, Egypt. Environ. Earth Sci. 75: 697–713.

Chandrasekharam, D., A. Lashin, N. Al-Arifi and A. Al-Bassam. 2016b. Red Sea Geothermal Provinces, Taylor and Francis Group, London, UK, pp. 216.

Dominy, B.N., H. Minoux and C.L. Brooks. 2004. An electrostatic basis for the stability of thermophilic proteins. Protein Struct. Funct. Bioinf. 57(1): 128–141.

El-Shafey, N.F. 2017. Molecular Studies on Haloarchaea Producing Industrially Important Enzymes. Suez Canal University, Ismaillia, Egypt.

Elbarbary, S., M. Abdel Zaher, H. Mesbah, A. El-Shahat and A. Embaby. 2018. Curie point depth, heat flow and geothermal gradient maps of Egypt deduced from aeromagnetic data. Renew. Sust. Energ. Rev. 91: 620–629.

Elsaied, H.E. 2008. Biodiversity of Archaea in Manzala Lake in Egypt based on 16s rRNA gene. Egypt. J. Genet. Cytol. 37: 57–72.

Elsaied, H.E. 2014. Genotyping of uncultured Archaea in a polluted site of Suez Gulf, Egypt, based on 16S rRNA gene analyses. Egy. J. Aqua. Res. 40: 27–33.

Ghozlan, H., H. Yusef, S. Sabry and R. Khalil. 2004. Isolation and characterization of an extremely halophilicarchaeon, *Halobacteriumsalinarum* SHR from a solar salterm in Egypt. FEBS 13(6): 494–500.

Ghozlan, H., H. Deif, R. Abu Kandil and S. Sabry. 2006. Biodiversity of moderately halophilic bacteria in hypersaline habitats in Egypt. J. Gen. Appl. Microbiol. 52(2): 63–72.

Hagagy, N.I. 2009. Biological Aspects of Archaebacteria isolated from Unique Habitats in Egypt. Suez Canal University, Ismaillia, Egypt.

Hagagy, N.I. 2014. Biodiversity and biotechnological potential of alkaliphilicarchaea isolated from Soda Lakes of WadiAnatru, Egypt. Suez Canal University, Ismaillia, Egypt.

Hosney, H.M. 2000. Geophysical Parameters and Crustal Temperatures Characterizing Tectonic and Heat Flow Provinces of Egypt. ICEHM2000 Cairo University, Egypt, pp. 152–166.

Jaenicke, R. and G. Böhm. 1998. The stability of proteins in extreme environments. Curr. Opin. Struct. Biol. 8(6): 738–748.

Kristjansson, J.K. 1991. Thermophilic bacteria. CRC Press, Boca Raton, FL.

Kushner, D.J. and M. Kamekura. 1988. Physiology of halophilic eubacteria. pp. 109–138. *In*: Rodriguez-Valera F. [ed.]. Halophilic Bacteria, Vol 1. CRC, Boca Raton.

Lashin, A. 2013. A preliminary study on the potential of the geothermal resources around the Gulf of Suez, Egypt. Arab. J. Geosci. 6: 2807–2828.

Lashin, A. 2015. Geothermal resources of Egypt, country update. *In*: Proceedings World Geothermal Congress, Melbourne, Australia.

Li W., X., Zhou and P. Lu. 2005. Structural features of thermozymes. Biotechnol. Adv. 23(4): 271–281.

Lucas, A. 1912. Natural Soda Deposits in Egypt, Survey Department Press, Cairo.

Lund, J.W. and T.L. Boyd. 2015. Direct utilization of geothermal energy: worldwide review. Geothermics 60: 66–93.

Mesbah, N.M. and J. Wiegel. 2005. Halophilic thermophiles: a novel group of extremophiles. pp. 91–118. *In*: T. Satyanarayana and B.N. Johri [eds.]. Microbial Diversity: Current Perspectives and Potential Applications. I.K. Publishing House, New Delhi.

Mesbah, N.M., S.H. Abou-El-Ela and J. Wiegel. 2007a. Novel and unexpected prokaryotic diversity in water and sediments of the alkaline, hypersaline lakes of the WadiAnNatrun, Egypt. Microb. Ecol. 54: 598–617.

Mesbah, N.M., D.B. Hedrick, A.D. Peacock, M. Rohde and J. Wiegel. 2007b. *Natranaerobiusthermophilus* gen. nov., sp. nov., a halophilic, alkalithermophilic bacterium from soda lakes of the WadiAnNatrun, Egypt, and proposal of *Natranaerobiaceae* fam. nov. and *Natranaerobiales* ord. nov. Int J. Syst. Evol. Microbiol. 57: 2507–2512.

Mesbah, N.M. and J. Wiegel. 2008. Life at extreme limits. The anaerobic halophilicalkali thermophiles. Ann. N. Y. Acad. Sci. 1125: 44–57.

Mesbah, N.M. and J. Wiegel. 2009. *Natronovirgawadinatrunensis* gen. nov., sp. nov. and *Natranaerobiustrueperi* sp. nov., halophilic, alkalithermophilic micro-organisms from soda lakes of the Wadi An-Natrun, Egypt. Int. J. Syst. Evol. Microbiol. 59: 2042–2048.

Mesbah, N.M. and J. Wiegel. 2011. Halophiles exposed concomitantly to multiple stressors: adaptive mechanisms of halophilicalkalithermophiles. pp. 249–274. *In*: A. Ventosa, A. Oren and Y. Ma [eds.]. Halophiles and Halophilic Environments: Current Research and Future Trends. Springer, Berlin.

Mohamed, H.S., M. Abdel-Zaher, M.M. Senosy, H. Saibi, M. El-Nouby and J.D. Fairhead. 2015. Correlation of aerogravity and BHT data to develop a geothermal gradient map of the northern western desert of Egypt using an artificial neural network. Pure Appl. Geophys. 172(6): 1585–1597.

Morgan, P., K. Boulos and C.A. Swanberg. 1983. Regional geothermal exploration in Egypt. pp. 361–376. *In*: Geophysical Prospecting. 31, EAEG.

Morgan, P., F.K. Boulos, S.F. Hennin, A.A. El-Sherif, A.A. El-Sayed, N.Z. Basta, et al. 1985. Heat flow in eastern Egypt, the thermal signature of a continental breakup. J. Geodyn. 4: 107–131.

Ollivier, B., P. Caumettte, J.L. Garcia and R. Mah, 1994. Anaerobic bacteria from hypersaline environments. Micro. Rev. 58(1): 27–38.

Oren, A., A., Ventosa, C. Gutierrez and M. Kamekura. 1999. *Haloarculaquadrata*sp. nov., a square, motile archaeon isolated from a brine pool in Sinai (Egypt). Int. J. Sys. Bacteriol. 49: 1149–1155.

Oren, A. 2002. Halophilic Microorganisms and their Environments. Kluwer Academic, Dordrecht, The Netherlands.

Oren, A. 2013. Two centuries of microbiological research in the WadiNatrun, Egypt: a model system for the study of the ecology, physiology, and taxonomy of haloalkaliphilic microorganisms. pp. 98–119. *In*: J. Seckbach, H. Stan-Lotter and A. Oren [eds.]. Polyextremophiles. Springer Dordrecht Heidelberg New York London.

Salah, A., S. Ibrahim and A.I. El-Diwany. 2007. Isolation and identification of new cellulases producing thermophilic bacteria from an egyptian hot spring and some properties of the crude enzyme. Aust. J. Basic App. Sci. 1(4): 473–478.

Selim, S., N.I. Hagagy, M. AbdelAziz and E.S. El-Meleigy. 2014a. Thermostable alkaline halophilic-protease production by *Natronolimnobiusinnermongolicus* WN18. Natural Product Research. 28(18): 1476–1479.

Selim, S., M. El-Sherif, S. El-Alfy and N.I. Hagagy. 2014b. Genetic diversity among thermophilic bacteria isolated from geothermal sites by using two PCR typing methods. Geomicrobiology 31(2): 161–170.

Selim, S. and N. Hagagy, 2016. Genome sequence of carboxylesterase, carboxylase and xylose isomerase producing alkaliphilichaloarchaeon *Haloterrigenaturkmenica* WANU15. Genomics Data 7: 70–72.

Selim A.S., M.S. El-Alfy, N.I. Hagagy, A. A.I. Hassanin, R.M. Khattab, E.A. El-Meleigy, et al. 2012. Oil-biodegradation and biosurfactant production by haloalkaliphiliccarchaea isolated from Soda Lakes of the Wadi An-Natrun, Egypt. J. of Pure and Appl. Microbiol. 6(3): 1011–1020.

Shortland, A.J. 2004. Evaporites of the WadiNatrun: seasonal and annual variation and its implication for ancient exploitation. Archaeometry 46(4): 497–516.

Swanberg C.A., E. Morgan and F.K. Boulos. 1983. Geothermal potential of Egypt. Tectonophysics 96: 77–94.

Taher, A.G. 1999. Inland saline lakes of Wadi El Natrun depression, Egypt. Int. J. Salt Lake Res. 8: 149–170.

Webster, G., R.J. Parkes, B.A. Cragg, C.J. Newberry, A.J. Weightman and J.C. Fry. 2006. Prokaryotic community composition and biogeochemical processes in deep subseafloor sediments from the Peru Margin. FEMS Microbiol. Ecol. 58: 65–85.

Wiegel, J. 2011. Anaerobic alkaliphiles and alkaliphilicpolyextremophiles. pp. 81–98. *In*: K. Horikoshi [ed.] Handbook of Extremophiles. Springer, Tokyo.

Yusef, H.H. and A.A. EL-Besoumy. 2003. Purification and characterization of *Halobacteriumsalinarum* SHR protease. Adv. Food Sci. 25(1): 21–27.

16

Microbial Life in Deep Terrestrial Continental Crust

Sufia K Kazy[2], Rajendra P Sahu[1], Sourav Mukhopadhyay[1], Himadri Bose[1], Sunanda Mandal[2] and Pinaki Sar[1,*]

Introduction

The life that resides kilometres below the Earth's crust remains enigmatic and yet to be explored in detail. Ever since the presence and function of microbial life within deep subsurface was established, the exploration of nature and function of deep life has become a rapidly growing research field (Kieft 2016, Magnabosco et al. 2019, Lloyd et al. 2019). Compared to its marine counterpart that has been explored for deep life since the past several decades, the deep terrestrial biosphere has been recently appreciated as a populated, metabolically active, dynamic ecosystem interacting with and perhaps controlling global elemental cycles (Momper et al. 2017). The term 'deep biosphere' is referred to intraterrestrial subsurface habitats which remained relatively inaccessible and devoid of photosynthesis or independent from photosynthetically derived organic matter (Kieft 2016, Momper et al. 2017). Qualitatively, the terrestrial subsurface environment represents a crystalline, progressively hot, high-pressure region devoid of organic nutrients and favourable factors for life and represents an extraordinary environment as many of the standard forces of the ecosystem (e.g., dispersal, metabolic activity and flexibility, thermodynamic feasibility, etc.) are either much reduced or not acceptable (Biddle et al. 2012). Studies done in past decades showed that microbial life, which resides in deep terrestrial biosphere (up to several kilometres below surface), represents one of the largest and most diverse ecosystems with enormous impact on both past and present planetary conditions and processes, including the evolution of our planet and the life within (Pedersen 2000, McMahon and Parnell 2018). It is considered that, if other geological (fluid and gas composition and concentration, geological properties of the bedrock, etc.) and environmental conditions (e.g., availability of electron acceptors and donors) remain favourable, life should then be possible to depths of several kilometres in the continental crust (Hallbeck and Pedersen, 2008, Itävaara et al. 2011). Quantitatively, the habitable volume of the continental subsurface environment corresponds to an estimated value of $7 \times 10^8 - 1 \times 10^9$ km^3 and that of the total subsurface of ~2.3×10^9 km^3 (including the groundwater, subseafloor sediments, marine sediment pore water, and marine

---

[1] Environmental Microbiology and Genomics Laboratory, Department of Biotechnology, Indian Institute of Technology Kharagpur, Kharagpur, 721302, West Bengal.

[2] Department of Biotechnology, National Institute of Technology Durgapur, Durgapur, 713209, West Bengal, India.

*Corresponding author: sarpinaki@yahoo.com, psar@bt.iitkgp.ac.in

oceanic crust) and could host ~70% of all bacterial and archaeal cells (amounting to approximately $2\text{-}6 \times 10^{29}$ cells) and potentially over 80% of all the bacterial and archaeal species (Magnabosco et al. 2018, 2019). In terms of total carbon, the global deep subsurface may contain 23-31 petagram carbon or PgC, while its terrestrial part alone could accommodate 11-12 PgC, indicating its impact on the global C cycle.

Deep subsurface environments are devoid of photosynthetically derived organic carbon and with reduced fluid mobility from surface-bound resources. These habitats are identified as oligotrophic with a minimal supply of energy and carbon sources, liquid water, and other favourable factors for life. Under these oligotrophic conditions, presence of chemosynthetic subsurface ecosystems, fuelled by geochemically generated inorganic energy sources, have been considered to be the major metabolic drivers (Kieft 2016, Lopez-Fernandez et al. 2018). Abiotically produced CO_2, CO, small chain hydrocarbons, and H_2 could potentially fuel the deep biosphere. These geogenic inorganic nutrients support subsistence of the methanogens, sulphate-reducers, and anaerobic methane oxidizers, etc., the key players of Subsurface Lithoautotrophic Microbial Ecosystems (SLiMEs) in the deep continental crust (Kieft 2016). Although the research field has matured over the decades, there are still many outstanding questions to resolve. As organized by US NSF sponsored group and delineated by Kieft (2016), some of these broad questions include: (i) how deeply does the life extend into the earth?; (ii) how organisms are distributed in the subsurface in terms of spatial and temporal scales?; (iii) what fuels the deep biosphere?; (iv) how does the interplay between biology and geology shape the subsurface?; (v) what are the physical states of the subsurface microbes, and how do their capacities for carbon, nitrogen, and sulphur cycling vary with different geophysical/ geochemical settings within the subsurface environment?; and (vi) what is the role of lateral gene transfer among the subsurface populations? Pedersen (2008) has reviewed the interests in investigations of intraterrestrial life that are represented by a very diverse array of general social, professional, and industrial motives. In the present chapter, we have reviewed the deep terrestrial biosphere highlighting its extremities and nutrient sources for microbial life, microbial diversity and function as investigated through metagenomics, the outcome of various enrichment based cultivation studies, and implication of deep life in astrobiological research.

Global Perspective on Microbial Diversity

Deep subsurface research has grown extensively in the last few decades and a wide range of habitats, including groundwater, continental subsurface, subseafloor sediments, marine sediment pore water, and marine crust, have been investigated. Early studies mainly relied on cell numbers followed by cultivation based measurements of microbial activities in the deep subsurface, including oil wells followed by specially drilled boreholes for subsurface investigations (Pedersen and Ekendahl 1990, Onstott et al. 1999, 2009a, Onstott 2016). With the development of cultivation-independent molecular microbiology methods, a paradigm shift has occurred on microbiologist's view and approaches to decipher the community composition. From cell counts to measurement of cellular activities, sequencing and analysis of a large collection of clones of phylogenetic marker genes (e.g., 16S rRNA gene) have been a major choice of methods for investigating the deep biosphere till last decade (Figure 1). With the onset of metagenomics and advent of next-generation sequencing-based deep sequencing of environmental samples, researchers were allowed to gain further insights into the microbial community composition, species diversity, and even started identifying the members of the 'rare microbiome' or organisms with ultra-small cell sizes ($<0.02 \mu m$) (Wu et al. 2016, Jousset et al. 2017). In order to get a wider perspective of microbial community diversity and relationship with geochemical conditions within the terrestrial subsurface biomes, 16S rRNA gene datasets obtained from five countries across the two continents and 15 rock types sequenced in last decade have been analysed recently (Soares et al. 2019). This meta-analysis has shown that terrestrial deep biosphere is dominated by a few bacterial taxa likely to be endowed with diverse metabolic strategies. These taxa include Betaproteobacteria, Gammaproteobacteria, and Firmicutes. Evidence for aquifer-specific microbial communities and a core terrestrial subsurface community has been found.

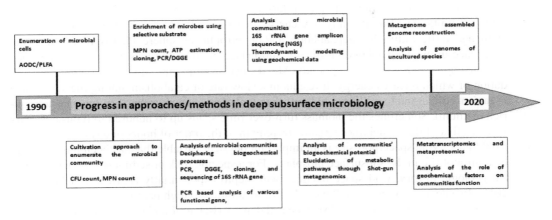

Figure 1. Progress in deep subsurface microbiology research

Terrestrial Deep Subsurface Extreme Habitats and Nutrient (Carbon, Hydrogen, Nitrogen, Sulphur, Phosphorus, and Metals) Sources for Microorganisms

Terrestrial deep subsurface represents one of the most hostile environments for life to thrive. The extremities that interfere with the habitability of the deep biosphere include lack of sunlight, oligotrophy (i.e., scarcity of energy substrates and limited nutrient availability), high temperature, and elevated hydrostatic and lithostatic pressure that are encountered in the subsurface environment (Lau et al. 2016, Dutta et al. 2019a, Govil et al. 2019). Desiccation, extreme pH conditions, radiations, and often the presence of degenerative substances add to the extreme conditions that impinge on the subsurface habitability further (Parkes and Wellsbury 2004, LaRowe and Amend 2015, Rempfert et al. 2017, Takai 2019, Sar et al. 2019). In recent times, less discussed but cardinal aspects of life in the deep subsurface, like ionic and energy limitations, have been brought to light (Hoehler and Jørgensen 2013, Payler et al. 2019). Microbes, in the subsurface, adapt to these limiting extreme conditions through the incorporation of genomic changes and manifest new or altered functional traits, leading to improved fitness (Dutta et al. 2019a).

Extreme Conditions that Delimit Life in Deep Biosphere

The set of extreme conditions that administer the occurrence of life in the deep continental crust are identified to be as follows:

(a) Temperature: Elevated temperature is the most important limitation that challenges life in the subsurface through interfering with its physiological and biochemical functions (Teske et al. 2014). At higher temperature, increased need for metabolic energy has been observed that compensates for degraded biomolecules. Thus, temperature extremity, accompanied by minimal energy flux in the subsurface, makes it unsustainable to live (Head et al. 2003, Hinrichs et al. 2016). Beneath the surface, in oceanic crust and continental crust, the temperature rises at a gradient of 15°C/km and 25°C/km subsequently with depth (Fredrickson and Onstott 1996). Recent studies on deep subsurface of the Deccan traps indicate a temperature gradient of 15°C/km in the granitic crust and 25°C/km in the basaltic crust (Dutta et al. 2018). The putative temperature limit of life is 113°C (Blöchl et al. 1997). IODP T Limit project has been taken up to understand the vertical limit of life on Earth in relation to temperature limit to life and subsequently investigate how thermal decomposition of organic matters in subsurface impact microbial population dynamics (Head et al. 2003, Hinrichs et al. 2016).

(b) Energy Limitations: Significantly low energy flux (i.e., 10^3 fold lower than ideal) is another challenge to life within the deep, aphotic, and oligotrophic biosphere. Photosynthesis does not take place in the deep aphotic biosphere. Subsurface microbiota relies entirely on the buried organic matter and reduced compounds (Mn(II), Fe(II), ammonia, sulphur/sulphide, etc.), oxidation of which releases energy. It results in a slow rate of catabolism among the subsurface organisms (thousand to million-fold lower than that of a model microbe, thriving on a nutrient-rich medium). Sometimes the biomass turn-over time spans up to hundred to thousand years whereas that of the model microorganisms remain in the order of hours or days.

(c) Void Space: Void space/porosity is an important physical variable that has been implicated to be a limit of life in the subsurface biosphere. For a specific rock or sediment type with identical geologic history, void space declines with depth exponentially (Parnell and McMahon 2016). In a deeper horizon, small particle size in addition to high pressure restricts void space and fluid movement, affecting subsurface cell density (Rebata-Landa and Santamarina 2006).

(d) Oxygen Limitation: The subsurface environment can be described as strictly anaerobic, precluding the possibilities of radiolysis or surface recharge of water. Even after radiolysis or surface recharge, the deeper flow is rendered anoxic within a short travel distance due to the presence of reduced organic matter or aerobic metabolism (Lovely and Chapelle 1995, Govil et al. 2019).

(e) Pressure: Hydrostatic and lithostatic pressure are other limiting variables that interfere with subsurface life and determines the indigenous microbial community composition (Lauro and Bartlett 2008, Dutta et al. 2019a). Biogeochemical processes including the formation of macromolecules, required for vitality, are inhibited in elevated pressure. In the deep continental rocky subsurface, the pressure increases at a gradient of 15-28 MPa/km (Rainey and Oren 2006). A recent study has demonstrated a lithostatic pressure gradient of 26.7 MPa/km in Deccan traps (Dutta et al. 2018).

(f) Ionic: One of the extremities, understudied but crucial to subsurface vitality, is the effect of dissolved ions. Changes in it may interfere with microbial replication, leading to inhibition of microbial growth. Dissolved ions can delimit life in subsurface biosphere through changes in water activity, ionic strength, and Chaotropicity (Hallsworth et al. 2007, Tosca et al. 2008, Fox-Powell et al. 2016, Payler et al. 2019).

Nutritional Resources in Deep Biosphere

In the deep continental crust, bacteria and archaea utilise all available energy sources, such as carbon, nitrogen, and sulphur, and contribute to the biogeochemical cycling of elements (Lovley and Chapelle 1995, Hoehler and Jørgensen 2013). Both aerobic and anaerobic (predominant) metabolisms are known to be powering the deep terrestrial biosphere through chemical redox (reduction-oxidation) reactions involving electron donors and acceptors. Table 1 summarises the potential electron donor and acceptors used by deep biosphere microorganisms. The amount of energy released in a metabolic reaction depends on the redox potential of the chemicals involved. Burial of organic matter makes it recalcitrant and almost inaccessible as a substrate to the microorganisms and thereby limiting metabolic activity. Thermal breakdown of the organic matter into volatile makes it difficult for microorganisms to directly utilise these maturation products as a substrate.

TABLE 1: Potential Electron Donors and Acceptors Used by Deep Biosphere Microorganism

Microbial Processes	Electron Donor	Probable Electron Acceptor
Aerobic metabolism	Small organics, hydrocarbons, inorganic substances (NH_3, NO_2, H_2S, H_2, etc.)	O_2
Fermentation	Small organics, hydrocarbons, etc.	Pyruvic acid, acetaldehyde, etc.
Anaerobic sulphide oxidation	S^{2-}	NO^{3-}
Anaerobic iron oxidation	Fe^{2+}	NO^{3-}
Anaerobic methane oxidation	CH_4	SO_4^{2-}/ NO^{3-}
Denitrification	Organic substrate and inorganic e-donors	NO^{3-}
Sulphur/Sulphate reduction	H_2	S/SO_4^{2-}
Hydrogen oxidation	H_2	O_2, NO^{3-}, SO_4^{2-}
Methanogenesis	H_2	CO_2

Microorganisms require carbon, nitrogen, phosphorus, and other macro and micronutrients for their growth and survival. In deep aphotic and oligotrophic biosphere, a number of abiotic processes like hydrothermal activity, magmatic degassing, and thermal metamorphism of carbonates may supply carbon (as CO_2 and CH_4), for biotic use (Sephton and Hazen 2013). Mafic and ultramafic rocks can act as the source for different mineral nutrients and ions, like NO^{3-}, PO_4^{3-}, SO_4^{2-}, Mg, Ca, Fe, K, SiO_2, Al_2O_3, MgO, carbonate, etc. Changes in the composition of ultramafic rocks, due to high temperature, water-rock interactions, etc., are associated with hydrogen release, leading to high reducing conditions in these environments. In turn, these reducing conditions favour the abiogenic production of methane and other organic molecules (Konn et al. 2015). Primary production in a subsurface ecosystem is mainly based on chemolithotrophy rather than photosynthesis, harnessing energy from components released from rock dissolution to carry out metabolic processes required for growth (Parnell and McMahon 2016).

(a) Carbon dioxide and bicarbonate: Abiotic sources of CO_2, carbonate, and bicarbonate molecules in the deep subsurface include dissolution of rocks and rock-forming minerals, volcanic eruptions, seismic activities, hydrothermal fluids, breakdown of basaltic rock, etc. At elevated temperature, rock-forming minerals, such as olivine, orthopyroxene, and clinopyroxene, may undergo dissolution and produce CO_2. Solubility of CO_2 in hydrothermal fluids and partial melts make it available in continental crusts. Studies suggest that the archaean continental lithospheric mantle might have elemental carbon in the form of diamond or graphite, CO_2-bearing components like carbonatite and kimberlite, carbonate-silicate melts, sulphide melts, and CO_2 fluid. Geological carbon sequestration strategies try to store CO_2 underground. This CO_2 gets solubilised via diffusion in water and gets absorbed in rocks (Emerson et al. 2016). CO_2 dissolved in water acts as a weak acid and can help in further dissolution of carbonates in sediments forming bicarbonate ions. Accumulation of bicarbonate in the groundwater can also result from abiological decarboxylation reactions (Lovely et al. 1989). Microbial metabolism helps in dissolution and precipitation of inorganic compounds and also helps with the oxidation of organic matter or small organic molecules to carbon dioxide in the subsurface environments. However, the organic matter that is buried in sediments is a complex mixture of polymeric organic molecules, and microorganisms must use hydrolytic enzymes (often extracellular) to break down the complex organic matter to monomeric components, such as sugars, amino acids, fatty acids, and monoaromatics, prior to metabolism, which can be oxidised to carbon dioxide using different terminal electron acceptors (Lovely and Chapelle 1995).

(b) Methane: Methane is an important compound in the global carbon cycle. One of the sources of abiotic production of CH_4 in the deep biosphere at high pressure and temperature is via catalytic reduction of certain carbon oxides in the presence of H_2. CH_4 is also found to be a major emission by-product of serpentinisation reaction together with H_2 in ultramafic rocks. Under pressure and low temperature, methane forms a thermodynamically stable association with water forming solid compounds known as methane clathrates and hydrates, which are mainly found in permafrost or associated with rocks and mud in deep seafloor environment. Plutonic rocks represent a potentially immense reservoir for abiogenic methane and methane-rich fluids (Kelley 1996 and references therein, Evans 1996). Alternatively, reasonable processes of abiogenic CH_4 formation at high temperature and pressure include the reduction of bicarbonate to graphite and CH_4, thermal decomposition of siderite, and clay mineral-catalysed reactions (McCollom, 2003, Konn et al. 2015). In deep Earth's crust, CH_4 is found in fluid inclusions of metamorphic rocks. Methane is found to emanate from geothermal waters on continents and hot water vents at oceans. In habitats below the Earth's surface, methane is mainly produced by microbial breakdown of organic compounds anaerobically or in found in large quantities within sedimentary formations. In principle, hydrocarbons can be formed either by a break-up of organic matter (e.g., microbially mediated aceticlastic methanogenesis) or by a fusion of carbon and hydrogen bearing molecules (microbial hydrogenotrophic methanogenesis). For biogenic production of methane, mostly archaeal members carry out microbial methanogenesis processes (Kietäväinen and Purkamo 2015).

(c) Other Organic Carbon Compounds: Production of organic carbon in the bedrock can be geogenic or biogenic. Serpentinisation, i.e., hydration of olivine in ultramafic rocks, leads to hydrogen generation and abiotic synthesis of organic matter. Rocks, minerals, and seawater are potential C, H, O, and N sources. Microorganisms, which inhabit hydrothermal vents and the subsurface, can also help in the production of organic compounds by providing C, H, O, and N by two mechanisms: (i) direct production of simple molecules (e.g., CH_4, H_2, acetate, CO_2) and (ii) thermal degradation of the microorganisms themselves if exposed to high-temperature fluids. Organic compounds can either be transported from the surface to deeper subsurface via fluid circulation or after being buried and metamorphosed while remaining trapped in the rock. Acetate, which plays a key role as electron donor and acceptor and serves as a carbon source in oligotrophic deep subsurface environments, can be produced from inorganic carbon by acetogenic microbes. Microorganisms can decompose more complex and sometimes recalcitrant organic matter to smaller, easily utilisable compounds. Microbial groups can contribute significantly to the total organic carbon content and carbon cycling of the deep biosphere (Lever 2012). Organic matter buried in sediments contains a complex mixture of polymeric organic molecules, and microorganisms use hydrolytic enzymes to break down these complex organic matters to monomeric components, such as sugars, amino acids, fatty acids, and monoaromatics prior to metabolism and utilisation. Rock weathering can also contribute to soil development and provide some organic matter in subsurface environments (Burkins et al. 2000, Tang and Lian 2012).

(d) Hydrogen: Hydrogen is one of the prime energy sources for the subsurface microbial community, particularly in habitats that are not well connected to the surface and where alternative electron donors are limited. Hydrogen, as a strongly reduced component, is an integral part of many metabolic pathways and a basal component of the deep biosphere which in turn suggests that the deep biosphere is to some extent dependent on hydrogen (Brazelton et al. 2012). Hydrogen production by abiogenic processes includes radiolysis, graphitisation, cataclasis of silicate minerals, deep crustal outgassing, crystallisation of the basaltic magma, and low-temperature reactions that occur in the shallow crust. Graphitisation is the decomposition of methane to graphite at temperatures above 600°C. Radiolysis of water is the dissociation of water molecules to produce H_2 and H_2O_2 by ionising radiation from radioactive decay of naturally occurring radioactive elements, like uranium, thorium, and potassium.

Cataclasis is the process of hydrogen generation in fault zones by several different processes. Hydromechanical reactions break chemical bonds and form radicals that react with groundwater to produce hydrogen. Reactions between dissolved magmatic gases can also lead to hydrogen formation (Gregory et al. 2019).

- Graphitisation: CH_4 (g) $\rightarrow$ C + $2H_2$ (g)
- Radiolysis of water: $2H_2O \rightarrow H_2O_2 + H_2$ (in the presence of ionising radiation)
- Cataclasis: $Si + H_2O \rightarrow Si + OH + H$
- $2H \rightarrow H_2$

Hydrogen may also be derived from small amounts of water included in minerals in the form of hydroxyls or peroxides.

$$Serpentinisation: 6[(Mg_{1.5}Fe_{0.5})SiO_4] + 7H_2O \rightarrow 3[Mg_3Si_2O_5(OH)_4] + Fe_3O_4 + H_2$$

Hydrothermal vents can be regarded as point sources of elevated hydrogen production. Hydrogen is observed at up to millimolar concentrations in fluids originating from hydrothermal venting sites at mid-ocean ridges (Charlou et al. 2002, Sleep et al. 2004, Proskurowski et al. 2006). Hydrogen production is linked to the capacity of the rock to reduce the aqueous solution, which is given by the reactivity of the minerals (Hellevang 2008, Gregory et al. 2019). Microbial processes such as fermentation and nitrogen fixation can help in the generation of hydrogen. Hydrogen production and consumption are tightly coupled and widely distributed among prokaryotes, which helps in the synthesis of energy-rich ATP and autotrophic carbon fixation (Greening et al. 2016).

(e) Nitrate, Ammonia, Sulphate, Iron, Manganese, and Phosphorus: Most organisms also require nitrogen, phosphorus, iron, sulphur, potassium, and magnesium for cell growth and metabolism (Bennett et al. 2001). Heterotrophic organisms gain energy generated from the transfer of electrons from a reduced carbon substrate to an electron acceptor. In oxic systems, aerobes utilise oxygen as the electron acceptor, while NO_3^-, Fe(III), and SO_4^{2-} can be utilised under anaerobic conditions (Chapelle and Lovely 1990, Chapelle 1993, Ehrlich 1996, Madigan et al. 1997). Endolithic habitats are composed of rocks that act as a source of nutrients for microorganisms. Studies show that these rocks have adequate supplies of inorganic nitrogen as nitrates and ammonium (Greenfield 1988 and references therein). In some cases, it is possible that nutrients required for microbial growth are obtained directly by *in situ* weathering of the host rock. Endolithic microorganisms are involved in physical weathering, which not only lead to the dissolution of minerals but also allow for microbial sequestration of nutrients and trace elements. Bacteria can also react directly with metals owing to their reactive cell surfaces or by changing the redox state of elements to use them as energy sources (Ehrlich 1998). Iron is one of the most abundant chemical elements in the Earth but most of it is in the core. Basalt, gabbro, etc., are mostly found crustal rocks that contain lots of iron. The presence of different iron-bearing minerals, such as iron oxide magnetite, iron sulphide pyrite, iron carbonates (siderite, ankerite), clay minerals (glauconite, chlorite), iron silicates (pyroxenes, olivine, black mica biotite, garnet), etc., make it available as a nutrient source for the inhabiting microorganisms of the terrestrial continental crust. Inorganic phosphate (PO_4^{3-}) is sparse within most subsurface systems, hence acting as a limiting growth factor for the inhabiting microorganisms (Madigan et al. 1997). Phosphorus is required by all organisms for ATP and nucleotide production, and they have evolved different biochemical strategies for bound phosphorus from minerals by acid or ligand production or reduction coupled with different elements (Goldstein 1986). Iron reducers, which utilise ferric ion as an electron acceptor may release phosphate during reductive dissolution (Lovely et al. 1989). Other studies suggest that microorganisms produce a reactive environment at the surface of the mineral that speeds up the dissolution of minerals, releasing limiting phosphate and iron (Bennett et al. 2001).

Microbial Communities Residing in Deep Terrestrial Continental Subsurface Across the Continents

The subsurface biosphere of Earth appears to be far more extensive, phylogenetically diverse, and metabolically complex than previously thought (Amend and Teske 2005, Lin et al. 2006, Pedersen et al. 2008, Kieft 2016, Magnabosco et al. 2018, McMahon and Parnell 2018). Past research on deep intraterrestrial microbial life has provided information that helped in understanding the diversity, function, evolution, and potential origin of life in the deep underground habitats. In the following section, a broad overview of microbial diversity is presented considering the course of investigations and highlighting the major findings from most thoroughly studied study areas. As summarised in Table 2, the deep terrestrial subsurface microbiome is explored across the continental crust (including Precambrain Fennoscandian Shield, Mesoarchean Witwatersrand basin to continental flood basalts of Late Cretaceous Deccan traps, Miocene age Columbia River basalt, and other areas). The primary focus of these exploratory studies has been on the characterisation of deep subsurface microbial communities including the effect of depth, available carbon and other nutrients, and elucidation of metabolic pathways and adaptation mechanisms (Table 2). The presence of several microbial taxa with distinct metabolic functions has been noted, and few of the taxa remained near ubiquitous. Table 3 summarises some of these microbial taxa likely to be involved in distinct metabolic functions.

TABLE 2: Microbial Diversity Studies at Various Terrestrial Deep Biosphere Habitats

S. No.	Site Name	Age	Depth (m)	Material	Aim/Objective	Reference
Fennoscandian Shield						
1	Aspo Hard Rock Laboratory (Aspo HRL)	1,760-1,840 Ma	45-446	Water	Culture-based analysis to decipher the presence of viable, diverse microbial community within the Aspo HRL.	Pedersen et al. 1996, Kotelnikova and Pedersen 1997, Hallbeck and Pedersen 2008, Pedersen et al. 2013, Ionescu et al. 2015
2	Aspo HRL	1,760-1,840 Ma	300-450	Biofilm	Characterisation of the microbial community producing biofilm present within the rock fracture within the terrestrial deep subsurface present.	Wu et al. 2017
3	Aspo HRL	1,760-1,840 Ma	141-448	Water	To check the hypothesis, rapid degradation of dead microbial cells to utilise as a nutrient by other microbes in the deep subsurface.	Lopez-Fernandez et al. 2018

TABLE 2: Contd....

TABLE 2: Contd.

S. No.	Site Name	Age	Depth (m)	Material	Aim/Objective	Reference
4	Aspo HRL	1,760-1,840 Ma	139-380	Water	To check the effect of depth, dissolved organic carbon, and water type on the microbial community of deep aquifer.	Lopez-Fernandez et al. 2018
5	Olkiluoto, Finland	Precambrian	4-545	Water	Relationship of numbers, biomass, and cultivable diversity of microbial populations to depth and borehole-specific conditions.	Haveman et al. 1999, Pedersen et al. 2008, Nyyssönen et al. 2012
6	Olkiluotu	Precambrian	296-978	Water	Characterisation of methanogenic and sulphate reducing microbial communities in the groundwater in Olkiluoto.	Bomberg et al. 2015, 2016
7	Outokumpu	1.19-1.95 Ga	0-2,300	Drill hole (continental crystalline crust) water	Improve the knowledge regarding the compositions and origin of saline fluid and gases in the crystalline rock. To investigate the structural and functional change in the microbial community with respect to various environmental parameters that exhibit within the crystalline rock. Deciphering the possible model regarding metabolism and functional adaption of microbes within this biosphere.	Itavaara et al. 2011, Nyyssönen et al. 2014
8	Outokumpu	1.19-1.95 Ga	180-2,300	Water	To identify the major carbon metabolism pathways beneath the crystalline bedrock environment in Outokumpu.	Purkamo et al. 2015, 2016

TABLE 2: Contd....

TABLE 2: Contd.

S. No.	Site Name	Age	Depth (m)	Material	Aim/Objective	Reference
9	Romuvaara	2.68-2.83 Ga	600	Water	Comparison of microbial community structure and function of deep Archaean bedrock with other sites of Fennoscandian shield.	Purkamo et al. 2018
North America						
1	Columbia river basalt	Miocene	316-1,270	Water	To investigate the microbial community present within this biosphere.	Steven et al. 1995, Fry et al. 1997
2	Henderson Mine	Precambrian	1,004 and 1,044	Water	Characterisation of microbial community present within the deep borehole of Henderson Mine.	Sahl et al. 2007
3	Lupin Au Mine	25,000 year	890 and 1,130	Water	Characterise the microbial community structure and abundance within the deep borehole.	Onstott et al. 2009b
4	Homstake Gold mine	Paleoproterozoic	1,343	Soil	Investigate the microbial community structure within this biosphere.	Rastogi et al. 2009a, 2010
5	Sanford Underground Research Facility (SURF)	Paleoproterozoic	800-4,870 ft	Water	To characterise the microbiome and geochemical parameters to find the possibility of chemolithotrophy within that biosphere.	Osburn et al. 2014
6	SURF	Paleoproterozoic	1,450	Water and Rock	Understanding the ecology of deep terrestrial habitat.	Momper et al. 2017
Witwatersrand Basin						
1	Various deep and ultra-deep mine	2.9 Ga	1,300-3,200	Water, Carbon leader	To detect the structure and diversity of the microbial community in the deep subsurface and their interaction with the surrounding environment.	Takai et al. 2001, Onstott et al. 2003, Kieft et al. 2006, Gihring et al. 2006, Lin et al. 2006, Borgonie et al. 2015

TABLE 2: Contd....

TABLE 2: Contd.

S. No.	Site Name	Age	Depth (m)	Material	Aim/Objective	Reference
Japan						
1	Toyoha Cu-Pb-Zn mine	3 million years	500-550	Water	To decipher the phylogenetic diversity of thermophilic sulphate reducers within effluent of hydrothermal groundwater and microbial mats developed at the Toyoha underground mine, Japan.	Nakagawa et al. 2002
2	Horonobe area	Miocene	296-625	Water	To investigate the microbial community present in the groundwater from Honorobe area.	Shimizu et al. 2006
3	Horonobe Underground Research Laboratory (URL)	Miocene	250-270	Rock	Exploration of indigenous eukaryotic fungi community present in the sedimentary rock of Horonobe URL.	Saitoh et al. 2019
4	Tono Area	10-15 Ma	158-200	Water and Biofilm	To investigate the effect of biofilm on radionuclide migration in the subsurface ecosystem.	Amano et al. 2017
5	Mizunami URL	15-20 Ma	1,148-1,169	Water	To study the geochemical parameter and microbial community within that biosphere.	Fukuda et al. 2010
6	Mizunami URL	15-20 Ma	300	Water	To investigate microbial mediated anaerobic methane oxidation.	Ino et al. 2018
China						
1	Dabie-Sulu ultra-high pressure metamorphic rock	220 Ma	529-3,350	Rock and Drilling fluid	Exploration of microbial life present beneath the ultra-high pressure rock and drilling fluid of China.	Zhang et al. 2005, Zhang et al. 2006

TABLE 2: Contd....

TABLE 2: Contd.

S. No.	Site Name	Age	Depth (m)	Material	Aim/Objective	Reference
India						
1	Deccan trap	65-2,500 Ma	60-1,500	Rock	Exploration of deep microbial life underneath the Deccan Traps.	Dutta et al. 2018
2	Deccan trap	65-2,500 Ma	60-1,500	Rock	Exploration of Achaeal community present beneath deccan trap.	Dutta et al. 2019b
3	Deccan Trap	65-2,500 Ma	1,681-2,990	Drilling fluid	Exploring the deep life related organism present in drilling fluid collected during subsurface drilling.	Bose et al. 2020

TABLE 3: Major Functional Groups of Microbes Detected From Deep Subsurface Environment

Sl. No.	Major Metabolic Function Relevant to Local Biogeochemical Process/ Cycle	Taxa Involved and Detected	References
1.	H_2-Oxidising	Desulfocarbo	Lopez-Fernandez et al. 2018
		Desulfurivibrio	Lopez-Fernandez et al. 2018
		Hydrogenophaga	Momper et al. 2017, Pedersen et al. 2014, Itavaara et al. 2011
		Methanobacterium	Purkamo et al. 2018
		Hydrogenophilaceae	Osburn et al. 2014
		Aquaspirillum delicatum	Zhang et al. 2005
		Thermoanaerobacter ethanolicus	Zhang et al. 2006
		Desulfovibrio	Lopez-Fernandez et al. 2018, Wang et al. 2008
2.	Fe^{2+}-Oxidising	Ferritrophicum	Lopez-Fernandez et al. 2018
		Gallionella	Purkamo et al. 2018, Lopez-Fernandez et al. 2018, Sahl et al. 2007
		Alicyclobacillus	Borgonie et al. 2015
		Spirochitae	Dutta et al. 2018
		Nitrospira	Dutta et al. 2018
		Acidimicrobia	Dutta et al. 2018
		Ferroplasma	Dutta et al. 2019b

TABLE 3: Contd.…

TABLE 3: Contd.

Sl. No.	Major Metabolic Function Relevant to Local Biogeochemical Process/ Cycle	Taxa Involved and Detected	References
3.	S-Oxidising	*Sulfurimonas*	Lopez-Fernandez et al. 2018, Bomberg et al. 2015
		Pseudomonas	Bomberg et al. 2015
		Thiobacteraceae	Bomberg et al. 2016
		Alicyclobacillus	Borgonie et al. 2015
		Thiobacilli	Lopez-Fernandez et al. 2018. Rastogi et al. 2009a, Rastogi et al. 2010
		Thiotrichaceae	Osburn et al. 2014
		Ectothiorhodospiraceae	Osburn et al. 2014
		T. nitratiredecuns	Zhang et al. 2005
		Spirochitae	Dutta et al. 2018
		Nitrospira	Dutta et al. 2018
		Acidimicrobia	Dutta et al. 2018
		Sulfolobaceae	Dutta et al. 2019b
4.	CH_4-Oxidising	*Methanoperedens*	Lopez-Fernandez et al. 2018, Ino et al. 2018
		Methylocystis	Rastogi et al. 2009a
		Methylosinus	Rastogi et al. 2010
		Methylococcaceae	Osburn et al. 2014
		ANME-3	Dutta et al. 2019a
5.	NH_3/NO_2-Oxidising	*Candidatus Nitrosoarchaeum*	Purkamo et al. 2018
		Crenarchaeota	Swanner et al. 2011
		Nitrobacter	Swanner et al. 2011
		Nitrococcus	Swanner et al. 2011
		Nitrospina	Swanner et al. 2011
		Nitrosomonas	Rastogi et al. 2010
		Thaumarchaeota	Osburn et al. 2014, Purkamo et al. 2018
		Nitrospira	Osburn et al. 2014, Purkamo et al. 2018, Swanner et al. 2011, Rastogi et al. 2010, Dutta et al. 2018
		Spirochitae	Dutta et al. 2018
		Acidimicrobia	Dutta et al. 2018
6.	C1 metabolising	*Hydrogenophaga*	Onstott et al. 2009b

TABLE 3: Contd....

TABLE 3: Contd.

Sl. No.	Major Metabolic Function Relevant to Local Biogeochemical Process/ Cycle	Taxa Involved and Detected	References
7.	Fermentative	*Bacteroidetes*	Mangabosco et al. 2014
		Geobacillus	Rastogi et al. 2009b
		Clostridium bifermentans	Zhang et al. 2005
		Crysiogenes arsentia	Zhang et al. 2005
		Cyanobacteria	Momper et al. 2017
		Soehngenia saccharolytica	Zhang et al. 2006
		Caldicellulosiruptor lactoaceticus	Zhang et al. 2006
		Firmicutes	Dutta et al. 2018
		Deltaproteobacteria	Dutta et al. 2018
		Actinobacteria	Dutta et al. 2018
8.	SO_4/Fe/Mn/metal-reducing	*Desulfomicrobium baculatum*	Pedersen et al. 1996
		Deltaproteobacteria	Lopez-Fernandez et al. 2018, Dutta et al. 2018
		Desulfobacteraceae	Bomberg et al. 2016, Pedersen et al. 2014, Fry et al. 1997, Osburn et al. 2014
		Desulfobacula	Pedersen et al. 2014
		Desulfotignum	Pedersen et al. 2014
		Dethiosulfatibacter	Nyyssönen et al. 2014
		Desulfitibacter	Nyyssönen et al. 2014
		Desulfobulbus	Purkamo et al. 2018, Osburn et al. 2014
		Desulfovibrio	Purkamo et al. 2018, Pedreson et al. 1996, Nakagawa et al. 2002
		Desulfococcus multivorans	Fry et al. 1997
		Desulfosarcina variabilis	Fry et al. 1997
		Desulfobotulus sapovorans	Fry et al. 1997
		Shewanella baltica	Onstott et al. 2009b
		Desulfosporoinus	Rastogi et al. 2010, Onstott et al. 2009b
		Peptococcaceae/Desulforudis	Osburn et al. 2014
		Thermodesulforhabdus	Nakagawa et al. 2002
		Desulfotomaculum	Shimizu et al. 2006, Nakagawa et al. 2002, Itavaara et al. 2011, Purkamo et al. 2015, Colwell et al. 1997
		Firmicutes	Dutta et al. 2018
		Actinobacteria	Dutta et al. 2018
		Thermoplasma	Dutta et al. 2019b
		Sulfolobaceae	Dutta et al. 2019b

TABLE 3: Contd....

TABLE 3: Contd.

Sl. No.	Major Metabolic Function Relevant to Local Biogeochemical Process/ Cycle	Taxa Involved and Detected	References
	Fe reducers	*Shewanella*	Pedresen et al. 1996, Onstott et al. 2009b
		Desulfuromonas	Onstott et al. 2009b
		Pseudomonas	Onstott et al. 2009b
		Geobacteriaceae	Osburn et al. 2014
		Thermodesulfobacterium	Nakagawa et al. 2002
	Fe/Mn/Cr	*Pantoea agglomerans*	Zhang et al. 2005
	NO_3^- reducer	*Nitrospirae*	Lopez-Fernandez et al. 2018
		Pseudomonas	Lopez-Fernandez et al. 2018
		Hadesarchaeaeota	Lopez-Fernandez et al. 2018
		Methanoperedens	Lopez-Fernandez et al. 2018
		Comamonadaceae	Purkamo et al. 2018
		Opitutus	Purkamo et al. 2018
		Shewanella baltica	Onstott et al. 2009b
		Alcanivorax	Rastogi et al. 2009a
		Janthinobacterium lividum	Zhang et al. 2005
		Anaeromyxobacter dehalogenas	Zhang et al. 2005
		Pseudomonas pseudoalcaligens	Zhang et al. 2005
		Azoarcus	Zhang et al. 2005
	Metal reducers	*Thermoanaerobacter ethanolicus*	Zhang et al. 2006
		Caldicellulosiruptor lactoaceticus	Zhang et al. 2006
		Anaerobranca gottschalkii	Zhang et al. 2006
9.	CH_4 synthesising	*Methermicoccaceae*	Lopez-Fernandez et al. 2018
		Methanomassiliicoccales	Lopez-Fernandez et al. 2018
		Methanobacteria	Lopez-Fernandez et al. 2018, Purkamo et al. 2016, Gihring et al. 2006, Onstott et al. 2003, Ino et al. 2018, Dutta et al. 2018, Dutta et al. 2019b
		Methanolobus	Gihring et al. 2006
		Methanosarcina	Rastogi et al. 2010, Gihring et al. 2006, Shimizu et al. 2006
		Methanothermobacter thermautotrophicus	Zhang et al. 2006
		Methanospirillum	Ino et al. 2018, Gihring et al. 2006
		Methanosaeta	Ino et al. 2018, Gihring et al. 2006
		Methanoculleus chikugoensis	Shimizu et al. 2006
		ANME-2d	Ino et al. 2018
		Methanomassiliicoccaceae	Ino et al. 2018
		Methanomicrobia	Dutta et al. 2018
		Methanocalculus	Dutta et al. 2019b
10.	CO_2-fixing	*Hydrogenophaga*	Itavaara et al. 2011
		Anaerolinea	Dutta et al. 2018
		Cyanobacteria	Dutta et al. 2018
		Epsilonproteobacteria	Dutta et al. 2018

Fennoscandian Shield

Microbial ecology of deep geobiosphere in Fennoscandian Shield has gained its special importance for the safe, long-term disposal of high-level nuclear waste (Amend and Teske 2005, Pedersen 2012, Pedersen 2013). Groundwater samples recovered from the Precambrian crystalline bedrock in Äspö hard rock laboratory (HRL) (Sweden) and Olkiluoto ONKALO tunnel (Finland) have been studied for decades to understand subsurface microbial communities, their taxonomic composition, and metabolic properties relevant for subsurface biogeochemical processes. During the pre-investigation phase of AspoHRL, diverse populations of active bacteria (1.5×10^4 to 17.5×10^5 cells ml^{-1}) are reported (Pedersen and Ekendahl 1990, Pedersen and Ekendahl 1992a, Pedersen and Ekendahl 1992b, Pedersen 1993). The presence of both autotrophic and heterotrophic microbial populations are observed in these deep crystalline aquifers (Hallbeck and Pedersen 2008, Hallbeck and Pedersen 2012, Nyyssönen et al. 2012, Bomberg et al. 2017, and references cited within). Diverse species of culturable populations [*Acinetobacter*, *Bacillus*, *Desulfovibrio* or *Thiomirospira* and taxa closely related to *Desulfomicrobium baculatum* or *Desulfovibrio* sp. (as sulphate reducer); *Shewanella putrefaciens* (iron reducer), *Pseudomonas* (nitrate reducer), *Hydrogenophega* (H$_2$ oxidizing)] have been reported. Microbial capability of oxygen-, nitrate, iron-, manganese-, sulphate-reduction, and H$_2$ oxidations, as well as acetogenic and methanogenic activities, is noted. To characterise the effect of extraneous H$_2$, acetate or SO$_4$ on *in situ* microbial activities of rock hosted biofilm community flow cells (FCs) have been used (Pedersen et al. 2014). It is demonstrated that sulphate may be the only component needed to trigger a very large community transition in deep sulphate-poor, methane-rich groundwater. In a recent study, groundwater samples from 21 boreholes, representing a range of deep continental habitats from the Äspö HRL, were used to characterise how the different water types influence the microbial communities (Lopez-Fernandez et al. 2018). Many of these taxa are suggested to mediate ferric iron and nitrate reduction. The study indicated that nitrate reduction may be neglected but is an important process in the deep continental biosphere.

Outokumpu Deep Borehole (ODB) drilled within the Palaeoproterozoic (ca. 1.95-1.91 Ga) Outokumpu formation (eastern Finland) of Fennoscandian Shield has provided another unique environment for studying deep subsurface microbiome (Itavaara et al. 2011, Nyyssönen et al. 2014). The first report on microbial diversity of ODB groundwater, sampled up to 1,500 mbsl, has shown the presence of microbial life in the saline oligotrophic deep aquifer, which suggest a relationship between the diversity and geochemical composition of formation water. Sequences of 16S rRNA and *dsr*B genes have indicated the dominance of *Commamonas* and Fimicutes members and potential sulphate reduction by *Desulfotomaculum*. In a subsequent study, drill hole sampling has been extended up to 2,516 mbs, and high throughput sequencing of bacterial and archaeal 16S rRNA genes have shown that microbial communities varied at different depths in response to prevailing lithology and hydrogeochemistry (Nyyssönen et al. 2014). In order to investigate the microbial communities residing in different fracture zones within the crystalline bedrock, while avoiding the chances of mixing of water from each other, inflatable packers have been used in ODB to obtain fracture water from specific depths (500 mbs, 967 mbs, and 2,260 mbs) (Purkamo et al. 2013). Each of the three fracture zones is found to be populated by distinct communities. In subsequent studies, further details of carbon and energy metabolism genes and microbial co-occurrence patterns were specifically studied (Purkamo et al. 2015, Purkamo et al. 2016). The analysis of microbial community structure and function of groundwater from deep Archaean bedrock fracture aquifer in Romuvaara, northern Finland, indicated the dominance of iron-oxidising *Gallionella* and methanogenic, ammonia-oxidising archaea (Purkamo et al. 2015). Some of the other studies in Europe include investigation on the role of plasmids in bacterial adaptation in the subsurface environment (Lubin copper mine, Poland), microbial diversity of high-pressure deep subsurface environment (Pyhäsalmimine, Finland), and assessment of the archaeal diversity of subsurface carbonate/siliciclastic-rock environment (Hainich CZE, Germany), and geomicrobiology of 2,000 m Deep Triassic Rock (Meuse/Haute Marne, France) (Dziewit et al. 2015, Miettinen et al. 2015, Lazar et al. 2017, Leblanc et al. 2019). Microbial ecosystem and metabolic potential in the ultradeep (up to 4.4 km depth) continental bedrock from two Finnish sites (Pyhäsalmi mine environment and geothermal drilling wells in Otaniemi) are recently reported (Purkamo et al. 2020). Diverse communities predominated with *Alkanindiges* and *Pseudomonas* and chemoheterotrophy, fermentation, and nitrogen cycling as potentially significant metabolisms have been identified in these multi-extreme ultradeep environments (Purkamo et al. 2020).

Witwatersrand Basin

The gold mines located on Witwatersrand basin (2.5-2.9 billion years) are the deepest accessible excavations in the world that have provided enormous opportunities for direct exploration of the deep subsurface (Takai et al. 2001a, Onstott et al 2006, Lau et al. 2014). Early studies have estimated 10^4-10^6 cells g^{-1} in a 3.23 km deep freshly mined face (Onstott et al. 1997). Microbial community structure and geochemical properties relating to a brine-bearing, dyke-associated fracture in a gold mine were investigated using a number of cultivation-independent approaches (Takai et al. 2001a, 2001b, Moser et al. 2003). *Desulfotomaculum* spp. have been reported as the major bacterial taxa. Novel clusters of SAGMEG groups of the Euryarchaeota, *Pyroccocus abyssi*, and SAGMCG groups of the Crenarchaeotahave been reported. Moser et al. (2005) investigated the energetic foundation of deep subsurface communities in crystalline aquifers and reported that chemolithoautotrophic mode of nutrition founded upon biological methanogenesis or acetogenesis, supported by geological H_2 production, is a major metabolic resource. In order to investigate the issues related to the age of the microbes in these deep ecosystems and their sources of energy for metabolism, deep (3.1 km below the surface) saline fracture water has been characterised in terms of chemical constituents (including stable isotopes), groundwater age, and indigenous microorganisms (Kieft et al. 2005). These investigators evidenced biological sulphate reduction supporting a sparse microbial community (10^3 cells ml^{-1}), despite the presence of relatively high concentrations of energy-rich compounds (H_2, CH_4, CO, ethane, propane, butane, and acetate). Magnabosco et al. (2014) evaluated the microbial diversity of fracture water from several subsurface sites and thermal springs in the Witwatersrand basin using high throughput sequencing of 16S hypervariable regions. In another study by Borgonie et al. (2015) has reported the presence of Nematodes within the deep subsurface of the Witwatersrand basin. They have suggested that bacteria, archaea, and nematodes present within the fracture waters originated from the ancient seas.

North American Sites

There are several deep subsurface study sites in North America, including basaltic aquifers of Columbia River Basalt (CRB), Piceance Basin, Western Colorado, Snake River Plains, Antrim Shale, Henderson mine, Homstake mine and Lupin Au mine (Canada), etc. Early work on Columbia River Basalt (CRB) has suggested the presence of hydrogen driven subsurface lithoautotrophic microbial ecosystem (SLiME) (Stevens and McKinely 1995). Culture dependent study of late cretaceous and early tertiary rock cores from Piceance Basin has suggested the presence of anaerobic nitrate, sulphate-, ferric iron-, Mn(IV) reducers, fermenters, and methanogens (Colwell et al. 1997). Investigation of fracture water from the deep sub-permafrost microbial ecosystem of Lupin Au mine (Canada) has revealed the dominance of *Desulfosporosinus*, *Halothiobacillus*, and *Pseudomonas* (Onstott et al. 2009b). Interestingly, thermodynamic calculations have shown that sulphate reduction and sulphide oxidation occur via denitrification and not methanogenesis. These investigators have further suggested that this ecosystem might be a Martian analogue and have an implication on the biological origin of methane in Mars. Microbial community analysis up to 1.34 km depth of Homstake Gold mine has been reported (Rastogi et al. 2009a, 2009b, Rastogi et al. 2010). Enrichment of mesophilic, thermophilic cellulose-degrading microbes belonging to genera *Brevibacillus*, *Paenibacillus*, *Bacillus*, and *Geobacillus* and the presence of diverse metabolic functions such as sulphur-oxidation, ammonia-oxidation, iron-oxidation, methane-oxidation, and sulphate-reduction in that ecosystem has been reported. Thermodynamic modelling of geochemical parameters coupled with DNA sequencing of a water sample from Stanford Underground Research Facility (SURF) revealed the presence of diverse groups of chemolithoautotrophs, thermophiles, aerobic and anaerobic heterotrophs, and numerous uncultivated clades, which are fuelled by oxidation of sulphur, iron, nitrogen, methane, and manganese (Osburn et al. 2014). In another interesting study from the same site, *in situ* electrochemical colonisation (ISEC) method has been designed to investigate microbial electron transfer activity (Jangir et al. 2019). With the help of this technique, electrochemically active microorganisms (*Bacillus*, *Anaerospora*, *Comamonas*, *Cupriavidus*, and *Azonexus*) are isolated. The first evidence of microbial life from 2.4 km below the surface of Kidd Creek Mine, Timmins, Ontario, Canada was recently reported (Lollar et al. 2019). Signatures of autotrophic sulphate-reducers and alkane-oxidising sulphate reducers have been detected from this site. Yamamoto et al. (2019) have investigated microbial diversity (including the rare populations) in an aquifer ecosystem of the Mahomet aquifer, USA, by 16S rRNA amplicon deep sequencing. They have concluded that microbial diversity of both abundant (> 0.1% of total abundance) and the rare biosphere (<0.1% of total abundance) correlate with the groundwater geochemistry. Recently, viral diversity from hydraulically fractured wells from five different locations of the USA was reported (Daly et al. 2019). They have experimentally examined the virus-host interaction dynamics and assessed the metabolites from cell lysis in order to understand viral roles in this ecosystem. Their results suggested that diverse and active viral populations play important roles in strain-level microbial community development and resource turnover within the deep terrestrial subsurface ecosystem (Daly et al. 2019).

China, Taiwan, and Japan

Exploration of microbial diversity up to the depth of (529-2,026 mbs) at the Chinese continental drilling site indicated the presence of 5.2-24×10^3 cells per gram rock (Zhang et al. 2005). Presence of Proteobacteria, Bacteriodetes, Actinobacteria, Planctomycetia, and isolation of mesophilic, fermentative, aromatic carbon degrading bacteria showing similarity *Clostridium bifermentans*, *Crysiogenesarsentia*, and thermophilic microorganisms similar to *Clostridium felsineum* was reported. As part of the Taiwan Chelungpu drilling project (TCDP), Wang et al. (2007) reported microbial community present within the Pliocene-Pleistocene sedimentary rocks up to a depth of 1,451 mbs at Chelungupu, Taiwan.

One of the earliest investigations on deep subsurface in Japan has been done by Nakagawa et al. 2002. They have isolated thermophilic sulphate reducing bacteria *Desulfotomaculum*, *Thermodesulforhabdus* from Toyoha Cu-Pb-Zn mine. A thorough characterisation of the microbial community within Miocene hard shale was subsequently done by Shimizu et al. 2006. Coexistence of sulphate reducing bacteria and methanogens has been suggested. A later study on biofilm growing in the aquifer at the depth of 158-200 mbs of Tono area suggested the presence of diverse microorganisms (Amano et al. 2017). Microbial diversity of groundwater from sub-crystalline rock aquifer in Miazunami Underground Research Laboratory was also investigated (Fukuda et al. 2010, Ino et al. 2016).

The Deccan Traps (India)

Investigation on deep life present within the 65 million old Deccan Traps and underneath archean granitic crusts was done using rock cores obtained through deep boreholes drilled at Koyna Intra-plate Seismic zone (Deccan Traps), India (Dutta et al. 2018, 2019a, b, Sar et al. 2019, Bose et al. 2020). With respect to varying temperature, pressure regimes (15°C increase in temperature and

26.7 MPa increase in lithostatic pressure per 1,000 metres below surface in granite) and geochemical conditions, this study provided an excellent prospect for investigating life within the deep igneous terrestrial subsurface. A depth independent distribution of microorganisms predominated by bacteria was reported through qPCR. It was also noted that the abundance of two diagnostics marker genes (*dsr*B and *mcr*A) involved in dissimilatory S-metabolism and methane metabolism are relatively higher in upper, younger basaltic horizon compared to the older and deeper granitic crust. High throughput 16S rRNA gene amplicon sequencing from rock (60-1,500 mbs) metagenomes revealed that bacterial communities are dominated by Alpha-, Beta-, Gamma-Proteobacteria, Actinobacteria, and Firmicutes. Correlation analysis and derived metabolic properties of these interrelated taxa allowed the investigators to hypothesise a possible metabolic roadmap followed by this deep subsurface microbiome. A close association of several taxa previously reported from diverse deep, oligotrophic igneous subsurface and capable of autotrophic carbon fixation (Anaerolinea, Cyanobacteria, Epsilonproteobacteria), lithotrophic nutrition with the oxidation of inorganic nitrogen (NO_2^-), sulphur (S^0) and iron (Fe^{2+}) (Spirochitae, Nitrospira, Acidimicrobia), and organotrophic nutrition with fermentative, denitrifying and sulphate reducing activities (Firmicutes, Actinobacteria, Deltaproteobacteria) is noted. The presence of methanogenic archaea (Methanomicrobia, Methanobacteria) along with Deltaproteobacteria is presented as a clear indication of interspecies metabolite transfer through syntrophic activities. In a subsequent study, diversity and composition of archaeal populations are assessed from the similar samples (Dutta et al. 2019b). It was noted that extreme acid-loving, thermotolerant, sulphur-respiring Thermoplasmataceae, heterotrophic, ferrous-/H-sulphide oxidising Ferroplasmaceae and Halobacteriaceae are more abundant and closely interrelated within basalts. The deeper granitic crust is populated with higher proportions of Thaumarchaeota, Euryarchaeota and Bathyarchaeota affiliated to Methanomicrobia, SAGMCG-1, FHMa11 terrestrial group, AK59. Acetoclastic methanogenic Methanomicrobia, autotrophic SAGMCG-1, and MCG of Thaumarcheaota are present as the signature groups within the organic carbon lean granitic crust. The patterns of archaeal communities across the basaltic and granitic horizons highlighted the significance of local rock geochemistry, particularly the availability of organic carbon, Fe_2O_3, and other nutrients as well as physical constraints (temperature and pressure) in niche-specific colonisation of extremophilic archaeal communities. Recently, this group has further accessed 3,000 m deep granitic crust through Koyna Pilot borehole and analysed the deep life communities residing within this extreme realm through rocks as well as the drilling fluids (DF) (Sar et al. 2019, Bose et al 2020). Rock samples are characterised as low biomass content with estimated cell densities 10^2-10^3 cells/g. The dominance of Gamma-Proteobacteria, Actinobacteria, Firmicutes, Alphaproteobacteria, Verrucomicrobia, Cyanobacteria, Deinococcus-Thermus, and Chloroflexi is observed. The presence of strong correlations among the thermophilic, extreme stress-tolerant, anaerobic, fermentative, sulphate-reducing, N2-fixing microorganisms is noted. Geomicrobiology of DF samples indicated signatures of extremophilic and deep subsurface relevant bacterial genera (*Mongolitalea*, *Hydrogenophaga*, *Marinilactibacillus*, *Anoxybacillus*, *Symbiobacterium*, *Geosporobacter*, and *Thermoanaerobacter*) possibly infused into this fluid from deep crust was reported (Bose et al. 2020).

Metagenomics Reveals Secrets of Deep Biosphere

Over the past decades, applications of metagenomics and next-generation sequencing have allowed us to examine the identities and lifestyle of organisms residing in the deep biosphere (Magnabosco et al. 2018). Culture-independent metagenomics has enabled more straight forward and insightful investigation of the deep crustal microbiome in various geological settings (Chivian et al. 2008; Brazelton et al. 2012, Lau et al. 2014, 2016, Nyyssönen et al. 2014, Hubalek et al. 2016, Magnabosco et al. 2016, Wu et al. 2016, Probost et al. 2018, Dutta et al. 2018, Smith et al. 2019, Yang et al. 2019). Metagenomics, the sequencing of total DNA extracted from the entire microbial community of environmental sample, has been applied in several deep biosphere research. This technique has been successfully applied to explore the deep biosphere of continental subsurface in Fennoscandian shield, Witwatersrand basin, Columbia river basin, the Deccan Traps, and other geologically important provinces. The adoption of metagenomic approach has enabled scientists to gain a better knowledge of (a) deep subsurface microbial communities in terms of their taxonomic composition, (b) community metabolic properties at the genomic level, (c) phylogenetic lineages of functional genes, and (d) individual species/population through genome reconstruction. Such genome resolved information is considered to be the key to gain deeper insights into the role of inhabitant microbes and their metabolic reactions possible in specific habitats (Long et al. 2016). Broad objectives of most of these metagenomics work remain focused either on the generalised themes of understanding the taxonomic composition and metabolic function of deep subsurface communities, energy fluxes, metabolism of C, N, S, H_2, and CH_4, or how environmental parameters are responsible for shaping the structure and function of microbial communities. More specific aspects like whether hydrogen and CO_2 driven anaerobic C-fixing Wood-Lungdahl pathway could be the predominant C fixing process in deep aquifers are also considered. The following section describes the highlights from major investigations carried out with samples from Witwatersrand basin, Fennoscandian Shield, SURF, Columbia river basalt, the Deccan Traps, and marine crust from Juan de Fuca.

Metagenomes collected from the fracture water of several deep terrestrial subsurface habitats underneath the Witwatersrand basin (South Africa) have been used for studying the phylogeny and phylogeography of common functional genes (*NarV, NPD, PAPs, NifH, NifD, NifK, NifE,* and *NifN* genes) shared among the samples (Lau et al. 2014). This study has been carried out to investigate ubiquitous functions of pathways of CH_4, N, and S biogeochemical cycling as well as to characterise the biodiversity and subsurface biogeography as represented by the common functional genes. These investigators have observed various functional guilds in the metagenomes and distinctive taxonomic and phylogenetic distribution of such genes in the fracture water samples, which are not correlated with geographical or environmental parameters or residence time of the water. It has been suggested that the ancestral genetic signatures have been preserved in subsurface habitats, which may be useful in tracking the origin and evolution of prokaryotes. In a subsequent study, these researchers have shown that in some of the deep fluid-filled fractures in the Witwatersrand basin taxonomically and metabolically diverse subsurface microorganisms are supported by subsurface lithoautotrophic microbial ecosystem (Lau et al. 2016). This lithoautotrophic ecosystem is dependent on S and H_2 oxidation. Such organisms develop a syntrophic association with methanogens, methane-oxidisers, and sulphate reducers to overcome the thermodynamic constraints in the deep subsurface environment. Metagenomic and thermodynamic analyses of 3 km deep fracture fluid system within the Tau Tona gold mine (Witwatersrand basin) have revealed the occurrence of most dominant, energy-conserving reductive acetyl CoA (Wood-Lungdahl) C-fixation pathway in this Precambrian continental crust (Magnabosco et al. 2016). They have identified the carbon monoxide (CO) dehydrogenase genes that could participate in both autotrophic and heterotrophic metabolisms via reversible oxidation of CO followed by sulphate reduction, in a direct H_2 utilisation as well as in methanogenesis. They have also demonstrated that such an environment is dominated by Firmicutes and Euryarchaeota (22%) members.

Taxonomically and functionally diverse microbial communities in deep crystalline bedrocks of the Fennoscandian shield in Outokumpu, Finland have been elucidated through metagenomics (Nyyssönen et al. 2014). The lack of hydraulic connections, low/negligible exchange of fluids among fracture systems, long residence times, unique geochemical characteristics, and energy limitation have resulted in highly diverse bacterial, archaeal, and viral communities with diverse metabolic strategies for H_2 driven carbon cycling, reduced carbon assimilation, and nutrient cycling. It has also been suggested that phylogenetic and functional diversity of microbial communities and their adaptation in different depths offered them selective advantages and resilience over geologic time scales. Microbial metagenomes from three aquifers (at 171-448 m depth) in the Fennoscandian

Shield (Aspo Hard Rock Laboratory, Sweden) were also analysed by another group of investigators to investigate metabolic partitioning among the microbial populations and phylogenetically distinct microbial community subsets (Wu et al. 2016). Three metagenomes were analysed through the reconstruction of 69 distinct microbial genomes. It was observed that the most dominant microbial populations were the members of Proteobacteria, Candidate divisions, unclassified archaea, and bacteria having smaller dimensions along with reduced genome size, which may be considered as an adaptive mechanism for oligotrophy in such extreme habitat. The microbial community from shallower water may follow organic carbon dependent heterotrophic lifestyle, whereas the deeper, anciently formed saline water is dominated by lithoautotrophic processes associated with H_2 driven deep biosphere. Hubalek et al. (2016) have shown that connectivity to the surface determined the diversity patterns in subsurface aquifers. The most dynamic community of the shallow site is dominated by the S-oxidising genera (*Sulfurovum* or *Sulfurimonas*), whereas the intermediate aquifer remains less dynamic and dominated by Candidate phylum OD 1. Metagenomic analyses of the water samples have shown the presence of important genes for N and C fixation, SO_4^{2-} reduction, S^{2-} oxidation, and fermentation. The deepest water has been found to be least diverse, but the abundance of Cyanobacteria was noticeable. Whether these phototrophs have survived in such a deep environment because of their ability to switch to a fermentative mode of metabolism or they were less prone to degradation was not determined by the investigators. They have suggested that the formation of akinetes may have allowed such organisms for long term survival.

Energy and carbon metabolisms in a microbial community of deep terrestrial subsurface fluid collected from 1,500 m depth within a gold mine in Lead (Sanford Underground Research Facility, SURF, South Dakota, USA) have been investigated through metagenomics by Momper et al. (2017). Reconstruction of 74 genomes from metagenomes has enabled them to identify the common metabolic pathways. It has been reported that Deltaproteobacteria and Candidate phyla bacterial lineages were most abundant in these samples as well as putative SO_4^{2-} and NO_3/NO_2 reduction are the most common energy metabolism processes. Complete autotrophic C-fixation pathways were found in most of the metagenome assembled genomes (MAGs) with the most common reductive acetyl CoA pathway. A major portion (~40%) of the recovered MAGs has been assigned to bacterial phyla without any cultivated members. A small fraction of the MAGs has constituted two novel putative Candidate phyla of bacterial lineages *Abyssubacteria* (related to deep subsurface) and *Aureabacteria* (related to Homestake gold mine). Genome resolved metagenomics, single-cell genomics and geochemical analyses of groundwater from stratified, sandstone-hosted aquifers collected from a CO_2-driven geyser in Colorado Plateau (Utah, USA) have been studied to understand the differential depth distribution of community structure and function (Probst et al. 2018). These samples, collected over three phases of its eruption cycle, have shown a strong difference in microbial community composition and function with depths of a groundwater source. Autotrophic microbes were predominated across the depths, while the Wood-Lungdahl (WL) C-fixation pathway having the lowest energy cost was dominant in the deeper subsurface. The engagement of the WL pathway is linked to the energy limitation within the deep subsurface. The use of this pathway for carbon metabolism in methanogens, archaeal autotrophs, and other deep biosphere community members has long been reported (Bagnoud et al. 2016, Lau et al. 2016, Magnabosco et al. 2016). Probst et al. (2018) have also observed that autotrophic Candidatus 'Alticarchaeum' and *Nanoarchaea* dominated the deepest groundwater. A *Nanoarchaeon* with limited metabolic capacity is identified as a potential symbiont of the Ca. 'Alticarchaeum'. Candidate phyla Radiation bacteria were also found to be present in the deepest as well as in the intermediate depth groundwater. The investigators have found that autotrophic S-oxidising and N_2-fixing *Sulfurimonas* members were predominant in the shallow aquifer.

The study of Smith et al. (2019) has highlighted the potential pathways for carbon and energy metabolisms in the deep anoxic thermal aquifer ecosystem. They have reconstructed the eleven bacterial and archaeal genomes from a subseafloor olivine biofilm obtained from a Juan de Fuca

Ridge basaltic aquifer. It has been found that the dominant carbon fixation pathway was the WL pathway and the anaerobic respiration appeared to be driven by sulphate reduction. It was suggested that the ancient H_2 based chemolithotrophy might remain as one of the dominant metabolisms in the sub-oceanic aquifer. Metagenomic analysis of subglacial sediment from East Antarctica also has demonstrated chemoautotrophic microbial populations supported by H_2 oxidation (Yang et al. 2019). Several [Ni-Fe]-hydrogenases coding genes have been identified in the sediment. In addition, all genes involved in dissimilatory nitrate reduction and denitrification pathways were also present in the subglacial sediment community. In a previous study, Hernsdorf et al. (2017) have explored microbial community structure and metabolic functions of a sediment-hosted ecosystem at the Horonobe Underground Research Laboratory, Hokkaido, Japan. They have demonstrated that the potential of microbial H_2 and metal transformations are associated with novel bacteria and archaea in deep terrestrial subsurface sediments. Hydrogen-based metabolism, dependent on [FeFe] bifurcating hydrogenases and Ruf complexes, is not widely distributed in this environment.

Conclusions

Our understanding of microbial life within deep continental subsurface has been continually augmented by the exploration of new and deeper horizons, adoption of improved methods guided by specific questions pertaining to origin, evolution, and functioning of life in such extreme niches. Life, earlier presumed to be a surface phenomenon on Earth, is now conceived to occur beneath the surface, in deep aphotic and extreme subsurface. With increased understanding of subsistence of life in the deep subsurface, it calls for further research to probe into the nexus of biosphere-geosphere interaction. Due to the challenges of continental subsurface sampling, there is a dearth of data on the nature of the subsurface ecosystems. Consequently, there remain a plethora of unanswered questions regarding the extent of the deep subsurface microbial communities, their genomic variation, and metabolic potentials and propensity. The response of these microscopic, yet enormous life forms to planetary and anthropogenic processes is emerging as one of the most exciting research areas of the present time to maintain sustainability and industrial activities. The volume of the subsurface that has been sampled is infinitesimal in comparison to the total volume of the deep continental biosphere, and there exist myriads of diverse geologic settings and habitat type yet to be investigated. Novel strategies should be devised to explore the pattern of diversity and succession of subsurface microbiome and appreciation of their potential in a better environment, energy, and healthcare. Better insight into the nature of deep life is required as deep subsurface would provide a better spectacle of the early Earth, the origin of life, and evolution of life on Earth or other possible exoplanets. Available sequence data on subsurface microbiomes might be instrumental for the discovery of novel biocatalysts with biotechnological applications. A prime mechanistic level of knowledge is needed to model subsurface green-house gas fluxes that entrail microbial species interaction, competition, and function beneath the surface, which is pertinent to the present scenario of climate change. To bridge this knowledge gap, future research through genomics of representative microbiomes aided by observations of subsurface genomic, transcriptomic, and proteomic responses under short-term and long-term perturbations remains of prime interest.

Acknowledgements

Authors gratefully acknowledge Ministry of Earth Sciences, Government of India for generous support in the past and ongoing deep life research at authors' laboratories. Support and encouragements from Deep Carbon Observatory are also duly acknowledged.

References

Amano, Y., T. Iwatsuki and T. Naganuma. 2017. Characteristics of naturally grown biofilms in deep groundwaters and their heavy metal sorption property in a deep subsurface environment. Geomicrobiology J. 34: 769–783.

Amend, J.P. and A. Teske. 2005. Expanding frontiers in deep subsurface microbiology. pp. 131–155. *In*: N. Noffke [ed.]. Geobiology: Objectives, Concepts, Perspectives. Elsevier, USA.

Bagnoud, A., K. Chourey, R.L. Hettich, I. De Bruijn, A.F. Andersson, O.X. Leupin, et al. 2016. Reconstructing a hydrogen-driven microbial metabolic network in Opalinus Clay rock. Nat. Comm. 7: 12770.

Bennett, P.C., J.R. Rogers and W.J. Choi. 2001. Silicates, silicate weathering, and microbial ecology. Geomicrobiol. J. 18: 3–19.

Biddle, J.F., J.B. Sylvan, W.J. Brazelton, B.J. Tully, K. Edwards, C.L. Moyer, et al. 2012. Prospects for the study of evolution in the deep biosphere. Front. Microbiol. 2: 285.

Blöchl, E., R. Rachel, S. Burggraf, D. Hafenbradl, H.W. Jannasch and K.O. Stetter. 1997. *Pyrolobus fumarii*, *gen.* and sp. *nov.*, represents a novel group of archaea, extending the upper temperature limit for life to 113 C. Extremophiles 1: 14–21.

Bomberg, M., M. Nyyssönen, P. Pitkänen, A. Lehtinen and M. Itävaara. 2015. Active microbial communities inhabit sulphate-methane interphase in deep bedrock fracture fluids in Olkiluoto, Finland. BioMed Res. Int. 979530.

Bomberg, M., T. Lamminmäki and M. Itävaara. 2016. Microbial communities and their predicted metabolic characteristics in deep fracture groundwaters of the crystalline bedrock at Olkiluoto, Finland. Biogeosciences 13: 6031–6047.

Bomberg, M., M. Raulio, S. Jylhä, C.W. Mueller, C. Höschen, P. Rajala, et al. 2017. CO_2 and carbonate as substrate for the activation of the microbial community in 180 m deep bedrock fracture fluid of Outokumpu Deep Drill Hole, Finland. AIMS Microbiol. 3: 846–871.

Borgonie, G., B. Linage-Alvarez, A. Ojo, S. Shivambu, O. Kuloyo, E.D. Cason, et al. 2015. Deep subsurface mine stalactites trap endemic fissure fluid archaea, bacteria, and nematoda possibly originating from ancient seas. Front. Microbiol. 6: 833.

Bose, H., A. Dutta, A. Roy, A. Gupta, S. Mukhopadhyay, B. Mohapatra, et al. 2020. Microbial diversity of drilling fluids from 3000 meter deep Koyna pilot borehole provides insights in to the deep biosphere of continental earth crust. Scientific Drilling. [Accepted].

Brazelton, W.J., B. Nelson and M.O. Schrenk. 2012. Metagenomic evidence for H_2 oxidation and H_2 production by serpentinite-hosted subsurface microbial communities. Front. Microbiol. 2: 268.

Burkins, M.B., R.A. Virginia, C.P. Chamberlain and D.H. Wall. 2000. Origin and distribution of soil organic matter in Taylor Valley, Antarctica. Ecology 81: 2377–2391.

Chapelle, F.H. and D.R. Lovely. 1990. Rates of microbial activity in deep coastal plain aquifers. Appl. Environ. Microbiol. 56: 1865–1874.

Chapelle, F.H. 1993. Ground-water microbiology and geochemistry. Wiley, New York. p. 424.

Charlou, J.L., J.P. Donval, Y. Fouquet, P. Jean-Baptiste and N. Holm. 2002. Geochemistry of high H_2 and CH_4 vent fluids issuing from ultramafic rocks at the Rainbow hydrothermal field (36 14′ N, MAR). Chem. Geol. 191: 345–359.

Chivian, D., E.L. Brodie, E.J. Alm, D.E. Culley, P.S. Dehal, T.Z. DeSantis, et al. 2008. Environmental genomics reveals a single-species ecosystem deep within earth. Science 322: 275–278.

Colwell, F.S., T.C. Onstott, M.E. Delwiche, D. Chandler, J.K. Fredrickson, Q.J. Yao, et al. 1997. Microorganisms from deep, high temperature sandstones: constraints on microbial colonization. FEMS Microbiol. Rev. 20: 425–435.

Daly, R.A., S. Roux, M.A. Borton, D.M. Morgan, M.D. Johnston, A.E. Booker, et al. 2019. Viruses control dominant bacteria colonizing the terrestrial deep biosphere after hydraulic fracturing. Nat. Microbial. 4: 352.

Dutta, A., S.D. Gupta, A. Gupta, J. Sarkar, S. Roy, A. Mukherjee, et al. 2018. Exploration of deep terrestrial subsurface microbiome in Late Cretaceous Deccan traps and underlying Archean basement, India. Sci. Rep. 8: 17459.

Dutta, A., L.M. Peoples, A. Gupta, D.H. Bartlett and P. Sar. 2019a. Exploring the piezotolerant/piezophilic microbial community and genomic basis of piezotolerance within the deep subsurface Deccan traps. Extremophiles 23: 1–13.

Dutta, A., P. Sar, J. Sarkar, S. Dutta Gupta, A. Gupta, H. Bose, et al. 2019b. Archaeal communities in deep terrestrial subsurface underneath the Deccan traps, India. Front. Microbiol. 10: 1362.

Dziewit, L., A. Pyzik, M. Szuplewska, R. Matlakowska, S. Mielnicki, D. Wibberg, et al. 2015. Diversity and role of plasmids in adaptation of bacteria inhabiting the Lubin copper mine in Poland, an environment rich in heavy metals. Front. Microbiol. 6: 152.

Ehrlich, H.L. 1996. Geomicrobiology. 3rd edition, New York: Marcel Dekker. 719.

Ehrlich, H.L. 1998. Geomicrobiology: its significance for geology. Earth-Sci. Rev. 45: 45–60.

Emerson, J.B., B.C. Thomas, W. Alvarez and J.F. Banfield. 2016. Metagenomic analysis of a high carbon dioxide subsurface microbial community populated by chemolithoautotrophs and bacteria and archaea from candidate phyla. Environ. Microbiol. 18: 1686–1703.

Evans, W.C. 1996. A gold mine of methane. Nature 381: 114.

Fox-Powell, M.G., J.E. Hallsworth, C.R. Cousins and C.S. Cockell. 2016. Ionic strength is a barrier to the habitability of Mars. Astrobiology 16: 427–442.

Fredrickson, J.K. and T.C. Onstott. 1996. Microbes deep inside the earth. Sci. Am. 275: 68–73.

Fry, N.K., J.K. Fredrickson, S. Fishbain, M. Wagner and D.A. Stahl. 1997. Population structure of microbial communities associated with two deep, anaerobic, alkaline aquifers. Appl. Environ. Microbiol. 63: 1498–1504.

Fukuda, A., H. Hagiwara, T. Ishimura, M. Kouduka, S. Ioka, Y. Amano, et al. 2010. Geomicrobiological properties of ultra-deep granitic groundwater from the Mizunami Underground Research Laboratory (MIU), central Japan. Microb. Ecol. 60: 214–225.

Gihring, T.M., D.P. Moser, L.H. Lin, M. Davidson, T.C. Onstott, L. Morgan, et al. 2006. The distribution of microbial taxa in the subsurface water of the Kalahari Shield, South Africa. Geomicrobiology J. 23: 415–430.

Goldstein, A.H. 1986. Bacterial solubilization of mineral phosphates: Historical perspective and future prospects. Am. J. Alternative. Agr. 1: 51–57.

Govil, T., N.K. Rathinam, D.R. Salem and R.K. Sani. 2019. Taxonomical diversity of extremophiles in the deep biosphere. pp. 631–656. *In*: Microbial Diversity in the Genomic Era. Academic Press.

Greenfield, L.G. 1988. 3.2 froms of nitrogen in beacon sandstone rocks containing endolithic microbial communities in Southern Victoria Land, Antarctica. Polarforschung 58: 211–218.

Greening, C., A. Biswas, C.R. Carere, C.J. Jackson, M.C. Taylor, M.B. Stott, et al. 2016. Genomic and metagenomic surveys of hydrogenase distribution indicate H_2 is a widely utilised energy source for microbial growth and survival. ISME J. 10: 761.

Gregory, S.P., M.J. Barnett, L.P. Field and A.E. Milodowski. 2019. Subsurface microbial hydrogen cycling: natural occurrence and implications for industry. Microorganisms 7: 53.

Hallbeck, L. and K. Pedersen. 2008. Characterization of microbial processes in deep aquifers of the Fennoscandian Shield. Appl. Geochem. 23: 1796–1819.

Hallbeck, L. and K. Pedersen. 2012. Culture-dependent comparison of microbial diversity in deep granitic groundwater from two sites considered for a Swedish final repository of spent nuclear fuel. FEMS Microbiol. Ecol. 81: 66–77.

Hallsworth, J.E., M.M. Yakimov, P.N. Golyshin, J.L. Gillion, G. D'Auria, F. de Lima Alves, et al. 2007. Limits of life in $MgCl_2$-containing environments: chaotropicity defines the window. Environ. Microbiol. 9: 801–813.

Haveman, S.A., K. Pedersen and P. Ruotsalainen. 1999. Distribution and metabolic diversity of microorganisms in deep igneous rock aquifers of Finland. Geomicrobiology J. 16: 277–294.

Head, I.M., D.M. Jones and S.R. Larter. 2003. Biological activity in the deep subsurface and the origin of heavy oil. Nat. Rev. 426: 344.

Hellevang, H. 2008. On the forcing mechanism for the H_2-driven deep biosphere. Int. J. Astrobiol. 7: 157–167.

Hernsdorf, A.W., Y. Amano, K. Miyakawa, K. Ise, Y. Suzuki, K. Anantharaman, et al. 2017. Potential for microbial H_2 and metal transformations associated with novel bacteria and archaea in deep terrestrial subsurface sediments. ISME J. 11: 1915–1929.

Hinrichs, K.U., F. Inagaki, V.B. Heuer, M. Kinoshita, Y. Morono and Y. Kubo. 2016. Expedition 370 Scientific Prospectus: T-Limit of the Deep Biosphere off Muroto (T-Limit). International Ocean Discovery Program. http://dx. doi. org/10.14379/iodp.sp, 370.

Hoehler, T.M. and B.B. Jørgensen. 2013. Microbial life under extreme energy limitation. Nat. Rev. Microbiol. 11: 83–94.

Hubalek, V., X. Wu, A. Eiler, M. Buck, C. Heim, M. Dopson, et al. 2016. Connectivity to the surface determines diversity patterns in subsurface aquifers of the Fennoscandian shield. ISME J. 10: 2447.

Ino, K., U. Konno, M. Kouduka, A. Hirota, Y.S. Togo, A. Fukuda, et al. 2016. Deep microbial life in high-quality granitic groundwater from geochemically and geographically distinct underground boreholes. Environ. Microbiol. Rep. 8: 285–294.

Itävaara, M., M. Nyyssönen, A. Kapanen, A. Nousiainen, L. Ahonen and I. Kukkonen. 2011. Characterization of bacterial diversity to a depth of 1500 m in the Outokumpu deep borehole, Fennoscandian Shield. FEMS Microbiol. Ecol. 77: 295–309.

Jangir, Y., A.A. Karbelkar, N.M. Beedle, L.A. Zinke, G. Wanger, C.M. Anderson, et al. 2019. *In situ* Electrochemical studies of the terrestrial deep subsurface biosphere at the Sanford Underground Research Facility, South Dakota, USA. bioRxiv: 555474.

Jousset, A., C. Bienhold, A. Chatzinotas, L. Gallien, A. Gobet, V. Kurm, et al. 2017. Where less may be more: how the rare biosphere pulls ecosystems strings. ISME J. 11: 853.

Kelley, D.S. 1996. Methane-bearing fluids in the oceanic crust: gabbro-hosted fluid inclusions from the southwest Indian ridge. J. Geophys. Res. 101: 2943–2962.

Kieft, T.L., S.M. McCuddy, T.C. Onstott, M. Davidson, L.H. Lin, B. Mislowack, et al. 2005. Geochemically generated, energy-rich substrates and indigenous microorganisms in deep, ancient groundwater. Geomicrobiology J. 22: 325–335.

Kieft, T.L. 2016. Microbiology of the deep continental biosphere. *In*: C. Hurst [ed.]. Their World: A Diversity of Microbial Environments. Advances in Environmental Microbiology, vol 1, Springer, Cham.

Kietäväinen, R. and L. Purkamo. 2015. The origin, source, and cycling of methane in deep crystalline rock biosphere. Front. Microbiol. 6: 725.

Konn C., J.L. Charlou, N.G. Holm and O. Mousis. 2015. The production of methane, hydrogen, and organic compounds in ultramafic-hosted hydrothermal vents of the mid-atlantic ridge. Astrobiology 15: 5.

Kotelnikova, S. and K. Pedersen. 1997. Evidence for methanogenic Archaea and homoacetogenic Bacteria in deep granitic rock aquifers. FEMS Microbiol. Rev. 20: 339–349.

LaRowe, D.E. and J.P. Amend. 2015. Power limits for microbial life. Front. Microbiol. 718.

Lau, M.C., C. Cameron, C. Magnabosco, C.T. Brown, F. Schilkey, S. Grim, et al. 2014. Phylogeny and phylogeography of functional genes shared among seven terrestrial subsurface metagenomes reveal N-cycling and microbial evolutionary relationships. Front. Microbiol. 5: 531.

Lau, M.C., T.L. Kieft, O. Kuloyo, B. Linage-Alvarez, E. Van Heerden, M.R. Lindsay, et al. 2016. An oligotrophic deep-subsurface community dependent on syntrophy is dominated by sulphur-driven autotrophic denitrifiers. Proc. Natl. Acad. Sci. U.S.A. 113: E7927–E7936.

Lauro, F.M. and D.H. Bartlett. 2008. Prokaryotic lifestyles in deep sea habitats. Extremophiles 12: 15–25.

Lazar, C.S., W. Stoll, R. Lehmann, M. Herrmann, V.F. Schwab, D.M. Akob, et al. 2017. Archaeal diversity and CO_2 fixers in carbonate-/siliciclastic-rock groundwater ecosystems. Archaea.

Leblanc, V., J. Hellal, M.L. Fardeau, S. Khelaifia, C. Sergeant, F. Garrido, et al. 2019. Microbial and geochemical investigation down to 2000 m Deep Triassic Rock (Meuse/Haute Marne, France). Geosciences 9: 3.

Lever, M.A. 2012. Acetogenesis in the energy-starved deep biosphere–a paradox? Front. Microbiol. 2: 284.

Lin, L.H., J. Hall, T.C. Onstott, T. Gihring, B.S. Lollar, E. Boice, et al. 2006. Planktonic microbial communities associated with fracture-derived groundwater in a deep gold mine of South Africa. Geomicrobiology J. 23: 475–497.

Lloyd, K.G., C.S. Sheik, B. Gracia-Moreno and C.A. Royer. 2019. The Genetics, Biochemistry, and Biophysics of Carbon Cycling by Deep Life pp. 556–584. *In*: Orcutt, B.N., I. Daniel and R. Dasgupta [ed.]. Deep Carbon: Past to Present. Cambridge University Press.

Lollar, G.S., O. Warr, J. Telling, M.R. Osburn and B.S. Lollar. 2019. 'Follow the Water': hydrogeochemical constraints on microbial investigations 2.4 km below surface at the Kidd Creek Deep Fluid and Deep Life Observatory. Geomicrobiology J. 36: 859–872.

Long, P.E., K.H. Williams, S.S. Hubbard and J.F. Banfield. 2016. Microbial metagenomics reveals climate-relevant subsurface biogeochemical processes. Trends. Microbiol. 24: 600–610.

Lopez-Fernandez, M., M. Åström, S. Bertilsson and M. Dopson. 2018. Depth and dissolved organic carbon shape microbial communities in surface influenced but not ancient saline terrestrial aquifers. Front. Microbiol. 9: 2880.

Lovely, D.R., M.J. Baedecker, D.J. Lonergan, I.M. Cozzarelli, E.J.P. Phillips and D.I. Siegel. 1989. Oxidation of aromatic contaminants coupled to microbial iron reduction. Nature 339: 297–300.

Lovely, D.R. and F.H. Chapelle. 1995. Deep subsurface microbial processes. Rev. Geophy. 33: 365–381.

Madigan, M.T., J.M. Martinko and J. Parker. 1997. Brock Biology of Microorganisms. Prentice-Hall. Englewood Cliffs, NJ: 986 p.

Magnabosco, C., M. Tekere, M.C. Lau, B. Linage, O. Kuloyo, M. Erasmus, et al. 2014. Comparisons of the composition and biogeographic distribution of the bacterial communities occupying South African thermal springs with those inhabiting deep subsurface fracture water. Front. Microbiol. 5: 679.

Magnabosco, C., K. Ryan, M.C. Lau, O. Kuloyo, B.S. Lollar, T.L. Kieft, et al. 2016. A metagenomic window into carbon metabolism at 3 km depth in Precambrian continental crust. ISME J. 10: 730.

Magnabosco, C., L.H. Lin, H. Dong, M. Bomberg, W. Ghiorse, H. Stan-Lotter, et al. 2018. The biomass and biodiversity of the continental subsurface. Nat. Geosci. 11: 707–717.

Magnabosco C., J.F. Biddle, C.S. Cockell, S.P. Jungbluth and K.I. Twing. 2019. Biogeography, ecology, and evolution of deep life. pp. 524–555. *In*: Orcutt, B.N., I. Daniel and R. Dasgupta [ed.]. Deep Carbon: Past to Present. Cambridge University Press.

McCollom, T.M. 2003. Formation of meteorite hydrocarbons from thermal decomposition of siderite ($FeCO_3$). Geochim. Cosmochim. Acta 67: 311–317.

McMahon, S. and J. Parnell. 2018. The deep history of Earth's biomass. J. Geol. Soc. London. 175: 716–720.

Miettinen, H., R. Kietäväinen, E. Sohlberg, M. Numminen, L. Ahonen and M. Itävaara. 2015. Microbiome composition and geochemical characteristics of deep subsurface high-pressure environment, Pyhäsalmi mine Finland. Front. Microbiol. 6: 1203.

Momper, L., S.P. Jungbluth, M.D. Lee and J.P. Amend. 2017. Energy and carbon metabolisms in a deep terrestrial subsurface fluid microbial community. ISME J. 11: 2319.

Moser, D.P., T.C. Onstott, J.K. Fredrickson, F.J. Brockman, D.L. Balkwill, G.R. Drake, et al. 2003. Temporal shifts in the geochemistry and microbial community structure of an ultradeep mine borehole following isolation. Geomicrobiology J. 20: 517–548.

Moser, D.P., T.M. Gihring, F.J. Brockman, J.K. Fredrickson, D.L. Balkwill, M.E. Dollhopf, et al. 2005. Desulfotomaculum and methanobacterium spp. dominate a 4-to 5-kilometer-deep fault. Appl. Environ. Microbiol. 71: 8773–8783.

Nakagawa, T., S. Hanada, A. Maruyama, K. Marumo, T. Urabe and M. Fukui. 2002. Distribution and diversity of thermophilic sulphate-reducing bacteria within a Cu-Pb-Zn mine (Toyoha, Japan). FEMS Microbiol. Ecol. 41: 199–209.

Nyyssönen, M., M. Bomberg, A. Kapanen, A. Nousiainen, P. Pitkänen and M. Itävaara. 2012. Methanogenic and sulphate-reducing microbial communities in deep groundwater of crystalline rock fractures in Olkiluoto, Finland. Geomicrobiology J. 29: 863–878.

Nyyssönen, M., J. Hultman, L. Ahonen, I. Kukkonen, P. Laine, P. Laine, et al. 2014. Taxonomically and functionally diverse microbial communities in deep crystalline rocks of the Fennoscandian shield. ISME J. 8: 126.

Onstott, T.C., K. Tobin, H. Dong, M.F. DeFlaun, J.K. Fredrickson, T. Bailey, et al. 1997. Deep gold mines of South Africa: windows into the subsurface biosphere. In Instruments, Methods, and Missions for the Investigation of Extraterrestrial Microorganisms. SPIE 3111: 344–357.

Onstott, T.C., T.J. Phelps, T. Kieft, F.S. Colwell, D.L. Balkwill, J.K. Fredrickson, et al. 1999. A global perspective on the microbial abundance and activity in the deep subsurface. *In*: J. Seckbach [ed.]. Enigmatic Microorganisms and Life in Extreme Environments. Cellular Origin and Life in Extreme Habitats, vol 1. Springer, Dordrecht.

Onstott, T.C., D.P. Moser, S.M. Pfiffner, J.K. Fredrickson, F.J. Brockman, T.J. Phelps, et al. 2003. Indigenous and contaminant microbes in ultradeep mines. Environ. Microbiol. 5: 1168–1191.

Onstott, T.C., L.H. Lin, M. Davidson, B. Mislowack, M. Borcsik, J. Hall, et al. 2006. The origin and age of biogeochemical trends in deep fracture water of the Witwatersrand Basin, South Africa. Geomicrobiology J. 23: 369–414.

Onstott, T.C., F.S. Colwell, T.L. Kieft, L. Murdoch and T.J. Phelps. 2009a. New horizons for deep subsurface microbiology. Microbe. 4: 499–505.

Onstott, T.C., D.J. McGown, C. Bakermans, T. Ruskeeniemi, L. Ahonen, J. Telling, et al. 2009b. Microbial communities in subpermafrost saline fracture water at the Lupin Au Mine, Nunavut, Canada. Microb. Ecol. 58: 786–807.

Onstott, T.C. 2016. Deep Life: The Hunt for the Hidden Biology of Earth, Mars, and beyond. Princeton University Press.

Osburn, M.R., D.E. LaRowe, L.M. Momper and J.P. Amend. 2014. Chemolithotrophy in the continental deep subsurface: Sanford Underground Research Facility (SURF), USA. Front. Microbiol. 5: 610.

Parkes, R.J. and P. Wellsbury 2004. Deep biospheres. *In*: Alan T. Bull [ed]. Microbial Diversity and Bioprospecting. American Society of Microbiology.

Parnell, J. and S. McMahon. 2016. Physical and chemical controls on habitats for life in the deep subsurface beneath continents and ice. Philos. Trans. A. Math. Phys. Eng. Sci. 374: 20140293.

Payler, S.J., J.F. Biddle, B. Lollar Sherwood, M.G. Fox-Powell, T. Edwards, B.T. Ngwenya, et al. 2019. An Ionic Limit to Life in the Deep Subsurface. Front. Microbiol. 10: 426.

Pedersen, K. and S. Ekendahl. 1990. Distribution and activity of bacteria in deep granitic groundwaters of southeastern Sweden. Microb. Ecol. 20: 37–52.

Pedersen, K. and S. Ekendahl. 1992a. Assimilation of CO_2 and introduced organic compounds by bacterial communities in groundwater from southeastern Sweden deep crystalline bedrock. Microb. Ecol. 23: 1–14.

Pedersen, K. and S. Ekendahl. 1992b. Incorporation of CO_2 and introduced organic compounds by bacterial populations in groundwater from the deep crystalline bedrock of the Stripa mine. Microbiology 138: 369–376.

Pedersen, K. 1993. The deep subterranean biosphere. Earth-Sci. Rev. 34: 243–260.

Pedersen, K., J. Arlinger, S. Ekendahl and L. Hallbeck. 1996. 16S rRNA gene diversity of attached and unattached bacteria in boreholes along the access tunnel to the Äspö hard rock laboratory, Sweden. FEMS Microbiol. Ecol. 19: 249–262.

Pedersen, K. 2000. Exploration of deep intraterrestrial microbial life: current perspectives. FEMS Microbiol. Lett. 185: 9–16.

Pedersen, K., J. Arlinger, S. Eriksson, A. Hallbeck, L. Hallbeck and J. Johansson. 2008. Numbers, biomass and cultivable diversity of microbial populations relate to depth and borehole-specific conditions in groundwater from depths of 4–450 m in Olkiluoto, Finland. ISME J. 2: 760.

Pedersen, K. 2012. Subterranean microbial populations metabolize hydrogen and acetate under *in situ* conditions in granitic groundwater at 450 m depth in the Äspö Hard Rock Laboratory, Sweden. FEMS Microbiol. Ecol. 81: 217–229.

Pedersen, K. 2013. Metabolic activity of subterranean microbial communities in deep granitic groundwater supplemented with methane and H_2. ISME J. 7: 839.

Pedersen, K., A.F. Bengtsson, J.S. Edlund and L.C. Eriksson. 2014. Sulphate-controlled diversity of subterranean microbial communities over depth in deep groundwater with opposing gradients of sulphate and methane. Geomicrobiology J. 31: 617–631.

Probst, A.J., B. Ladd, J.K. Jarett, D.E. Geller-McGrath, C. MK Sieber, J.B. Emerson, et al. 2018. Differential depth distribution of microbial function and putative symbionts through sediment-hosted aquifers in the deep terrestrial subsurface. Nat. Microbiol. 3: 328.

Proskurowski, G., M.D. Lilley, D.S. Kelley and E.J. Olson. 2006. Low temperature volatile production at the Lost City Hydrothermal Field, evidence from a hydrogen stable isotope geothermometer. Chem. Geol. 229: 331–343.

Purkamo, L., M. Bomberg, M. Nyyssönen, I. Kukkonen, L. Ahonen, R. Kietäväinen, et al. 2013. Dissecting the deep biosphere: retrieving authentic microbial communities from packer-isolated deep crystalline bedrock fracture zones. FEMS Microbiol. Ecol. 85: 324–337.

Purkamo, L., M. Bomberg, M. Nyyssönen, I. Kukkonen, L. Ahonen and M. Itävaara. 2015. Heterotrophic communities supplied by ancient organic carbon predominate in deep Fennoscandian bedrock fluids. Microb. Ecol. 69: 319–332.

Purkamo, L., M. Bomberg, R. Kietäväinen, H. Salavirta, M. Nyyssönen, M. Nuppunen-Puputti, et al. 2016. Microbial co-occurrence patterns in deep Precambrian bedrock fracture fluids. Biogeosciences 13: 3091–3108.

Purkamo, L., R. Kietäväinen, H. Miettinen, E. Sohlberg, I. Kukkonen, M. Itävaara, et al. 2018. Diversity and functionality of archaeal, bacterial and fungal communities in deep Archaean bedrock groundwater. FEMS Microbiol. Ecol. 94: fiy116.

Purkamo, L., R. Kietäväinen, M. Nuppunen-Puputti, M. Bomberg and C. Cousins. 2020. Ultradeep microbial communities at 4.4 km within crystalline bedrock: Implications for habitability in a planetary context. Life 10: 2.

Rainey, F.A. and A. Oren. 2006. 1 Extremophile Microorganisms and the Methods to Handle them. *In*: Methods in Microbiology. Academic Press.

Rastogi, G., L.D. Stetler, B.M. Peyton and R.K. Sani. 2009a. Molecular analysis of prokaryotic diversity in the deep subsurface of the former Homestake gold mine, South Dakota, USA. J. Microbiol. 47: 371–384.

Rastogi, G., G.L. Muppidi, R.N. Gurram, A. Adhikari, K.M. Bischoff, S.R. Hughes, et al. 2009b. Isolation and characterization of cellulose-degrading bacteria from the deep subsurface of the Homestake gold mine, Lead, South Dakota, USA. J. Ind. Microbiol. Biot. 36: 585.

Rastogi, G., S. Osman, R. Kukkadapu, M. Engelhard, P.A. Vaishampayan, G.L. Andersen, et al. 2010. Microbial and mineralogical characterizations of soils collected from the deep biosphere of the former Homestake gold mine, South Dakota. Microb. Ecol. 60: 539–550.

Rebata-Landa V. and J.C. Santamarina. 2006. Mechanical limits to microbial activity in deep sediments. Geochem Geophys Geosyst. 7: Q11006.

Rempfert, K.R., H.M. Miller, N. Bompard, D. Nothaft, J.M. Matter, P. Kelemen, et al. 2017. Geological and geochemical controls on subsurface microbial life in the Samail Ophiolite, Oman. Front. Microbiol. 8: 56.

Sahl, J.W., R. Schmidt, E.D. Swanner, K.W. Mandernack, A.S. Templeton, T.L. Kieft, et al. 2008. Subsurface microbial diversity in deep-granitic-fracture water in Colorado. Appl. Environ. Microbiol. 74: 143–152.

Saitoh, Y., S.I. Hirano, T. Nagaoka and Y. Amano. 2019. Genetic survey of indigenous microbial eukaryotic communities, mainly fungi, in sedimentary rock matrices of deep terrestrial subsurface. Ecol. Genet. Genom. 12: 100042.

Sar, P., A. Dutta, H. Bose, S. Mandal and S.K. Kazy. 2019. Deep biosphere: microbiome of the deep terrestrial subsurface. *In*: Tulasi Satyanarayana, Bhavdish Narain Johri and Subrata Kumar Das [eds]. Microbial Diversity in Ecosystem Sustainability and Biotechnological Applications Volume 1. Microbial Diversity in Normal & Extreme Environments. Springer, Singapore.

Sephton, M.A. and R.M. Hazen. 2013. On the origins of deep hydrocarbons. Rev. Mineral. Geochem. 75: 449–465.

Shimizu, S., M. Akiyama, Y. Ishijima, K. Hama, T. Kunimaru and T. Naganuma. 2006. Molecular characterization of microbial communities in fault-bordered aquifers in the Miocene formation of northernmost Japan. Geobiology 4: 203–213.

Sleep, N.H., A. Meibom, T. Fridriksson, R.G. Coleman and D.K. Bird. 2004. H_2-rich fluids from serpentinization: geochemical and biotic implications. Proc. Nat. Acad. Sci. 101: 12818–12823.

Smith, A.R., B. Kieft, R. Mueller, M.R. Fisk, O.U. Mason, R. Popa, et al. 2019. Carbon fixation and energy metabolisms of a subseafloor olivine biofilm. ISME J. 13: 1737–1749.

Soares, A., A. Edwards, D. An, A. Bagnoud, M. Bomberg, K. Budwill, et al. 2019. A global perspective on microbial diversity in the terrestrial deep subsurface. bioRxiv. 602672 (preprint).

Stevens, T.O. and J.P. McKinley. 1995. Lithoautotrophic microbial ecosystems in deep basalt aquifers. Science 270: 450–455.

Swanner, E. and A. Templeton. 2011. Potential for nitrogen fixation and nitrification in the granite-hosted subsurface at Henderson Mine, CO. Front. Microbiol. 2: 254.

Takai, K., D.P. Moser, M. DeFlaun, T.C. Onstott and J.K. Fredrickson 2001a. Archaeal diversity in waters from deep South African gold mines. Appl. Environ. Microbiol. 67: 5750–5760.

Takai, K., D.P. Moser, T.C. Onstott, N. Spoelstra, S.M. Pfiffner, A. Dohnalkova et al. 2001b. *Alkaliphilus transvaalensis gen. nov., sp. nov.*, an extremely alkaliphilic bacterium isolated from a deep South African gold mine. Int. J. Syst. Evol. Microbiol. 51: 1245–1256.

Takai, K. 2019. Limits of Terrestrial Life and Biosphere. *In*: Akihiko Yamagishi, Takeshi Kakegawa and Tomohiro Usui [eds.]. Astrobiology from the origins of life to the search for extraterrestrial intelligence, Springer, Singapore.

Tang, Y. and B. Lian. 2012. Diversity of endolithic fungal communities in dolomite and limestone rocks from Nanjiang Canyon in Guizhou karst area, China. Can. J. Microbiol. 58: 685–693.

Teske, A., A.V. Callaghan and D.E. LaRowe. 2014. Biosphere frontiers of subsurface life in the sedimented hydrothermal system of Guaymas Basin. Front. Microbiol. 5: 362.

Tosca, N.J., A.H. Knoll and S.M. McLennan. 2008. Water activity and the challenge for life on early Mars. Science 320: 1204–1207.

Wang, P.L., L.H. Lin, H.T. Yu, T.W. Cheng, S.R. Song, L.W. Kuo, et al. 2007. Cultivation-based characterization of microbial communities associated with deep sedimentary rocks from Taiwan Chelungpu Drilling Project cores. Terr. Atmos. Ocean. Sci. 18: 395–412.

Wu, X., K. Holmfeldt, V. Hubalek, D. Lundin, M. Åström, S. Bertilsson, et al. 2016. Microbial metagenomes from three aquifers in the Fennoscandian shield terrestrial deep biosphere reveal metabolic partitioning among populations. ISME J. 10: 1192.

Yamamoto, K., K.C. Hackley, W.R. Kelly, S.V. Panno, Y. Sekiguchi, R.A. Sanford, et al. 2019. Diversity and geochemical community assembly processes of the living rare biosphere in a sand-and-gravel aquifer ecosystem in the Midwestern United States. Sci. Rep. 9: 1–11.

Yang, Z., Y. Zhang, Y. Lv, W. Yan, X. Xiao, B. Sun, et al. 2019. H_2 Metabolism revealed by metagenomic analysis of subglacial sediment from East Antarctica. J. Microbiol. 57: 1095–1104.

Zhang, G., H. Dong, Z. Xu, D. Zhao and C. Zhang. 2005. Microbial diversity in ultra-high-pressure rocks and fluids from the Chinese Continental Scientific Drilling Project in China. Appl. Environ. Microbiol. 71: 3213–3227.

Zhang, G., H. Dong, H. Jiang, Z. Xu and D.D. Eberl. 2006. Unique microbial community in drilling fluids from Chinese continental scientific drilling. Geomicrobiology J. 23: 499–514.

17

The Insight into Deep Subsurface Microbial Life in Therapeutic Waters of Poland

Agnieszka Kalwasińska* and Edyta Deja-Sikora[2,3]

Introduction

The deep subsurface is the largest microbial ecosystem on Earth (Edwards et al. 2012, Kallmeyer et al. 2012) and the continental biosphere constitutes approximately up to 20% of the Earth's total biomass (McMahon and Parnell 2014). Limited access to this biome due to technical obstacles is the major restriction responsible for the scarce and fragmentary information regarding the occurrence and diversity of life deep inside the lithosphere. Despite this and other difficulties, attempts are made to explore deep subsurface, and the driving force is acquiring the missing knowledge and practical acquisition of microorganisms, their genes, and metabolism's products, which may be useful for humans.

The deep biosphere comprises a variety of subsurface habitats, such as terrestrial deep aquifers, i.e., water-bearing permeable rocks or rock fractures. They can be found at a depth of hundreds to many thousands of meters below the land surface, depending on the geological structure of the specific area. Deep groundwaters comprise intermediate and regional flow systems, irrespectively of deposition depth (Lovley and Chapelle 1995). These waters are hardly connected with the surface (i.e., independent of precipitation events) and are characterized by a low rate of recharge. Some of them, due to total isolation from the surface, are non-renewable. The relatively slow movement of water through the ground and a physicochemical interdependence between water and bedrock in which water is present are the major features of groundwater distinguishing it from the other water reservoirs (Chilton 1996). The slow rate of flow means that residence times in deep groundwaters are generally orders of magnitude longer than in surface waters and dated in hundreds, thousands, or even millions of years (Chapelle 1993).

---

[1] Department of Environmental Microbiology and Biotechnology, Faculty of Biological and Veterinary Science, Nicolaus Copernicus University in Toruń, Lwowska 1, 87-100 Toruń, Poland.

[2] Centre for Modern Interdisciplinary Technologies, Nicolaus Copernicus University in Toruń, Wileńska 4, 87-100 Toruń Poland.

[3] Department of Microbiology, Faculty of Biological and Veterinary Science, Nicolaus Copernicus University in Toruń, Lwowska 1, 87-100 Toruń, Poland.

* Corresponding author: kala@umk.pl

The deep underground habitat can be regarded as poliextreme environment due to several aspects. Firstly, temperature and pressure increase with depth by 1°C per 33.3 m and 1 atm (1,013.25 hPa) per 3.7 m on average (Chowaniec 2013). Depending on the geology, at the depth of 1,000 m they may reach 40°C and 27 MPa. Secondly, due to permanent isolation from the surface, lack of sun limits photosynthesis and in consequence hampers organic biomass production and generation of oxygen. Therefore, subsurface ecosystems are usually anoxic, oligotrophic, and are characterized by chemolithotrophic modes of life under low energy conditions. Thirdly, the water of long residence time in bedrock becomes saturated with various ions e.g., Cl^-, Na^+, and getting saline as well (Oren 2008).

The temperature of around 122°C has been reported to be the upper limit for the growth of microorganisms at great depth (Takai et al. 2008). However, the limits for life due to pressure and salinity have not been established yet. Some microorganisms can still be active at pressure as high as 2 GPa as well as in NaCl solutions up to 36% (Burns et al. 2007, Vanlint et al. 2011). To survive at high salt concentrations, the high temperature adapted thermophiles and high pressure-loving piezophiles have to maintain an osmotic balance with their external environment (Oren 2008). They synthesize molecules (proteins, lipids, and nucleic acids) that are stable at such conditions (Boyd et al. 2011) and increase the internal osmolality of the cells (Santos and Da Costa 2002).

Extremophiles residing in deep aquifers have evolved a variety of pathways for conversing energy to support their growth. They have the metabolic potential to gain energy from sources dissolved in the water and buried in the sediments, i.e., organic matter, hydrogen (H_2), and reduced compounds, like ammonia, sulfide, Mn(II) or Fe(II). Some subsurface ecosystems are fueled mostly by H_2-driven primary production, and microbes are capable of mediating H_2 formation from minerals (Nealson et al. 2005, Parkes et al. 2011). Microbial community composition and diversity are strongly influenced by the availability of specific electron donors and terminal acceptors (nitrate, Mn(IV), Fe(III), sulfate, and CO_2) determining which redox reactions (e.g., nitrate, ferric iron, or sulfate reduction) are thermodynamically favorable in deep subsurface (Flynn et al. 2013, Nyyssönen et al. 2014, Amenabar et al. 2017). Furthermore, metabolic activities of the microbial community strongly affect the chemistry of the environment to make it favorable for the specific microbial lifestyle (Parkes 2011). Research indicates that microorganisms within the communities are interconnected through the exchange of geochemical resources (Anantharaman et al. 2016, Lau et al. 2016) and sulfate-reducing, denitrifying, iron-reducing, and methanogenic organisms coexist in the same subsurface aquifer (Shirokova et al. 2016, Kumar at al. 2017).

Exploitation of Deep Groundwaters in Poland

Mineral water having the total content of dissolved solids greater than 1 g/l has served as the important source of drinking water, water used for therapies, production of salt, and acquiring geothermal energy for several decades. Highly mineralized groundwater of the total mineralization above 35 g/l is called brine. Its main components are chloride, sodium, and calcium ions. Brines are found almost all over Poland. With depth, their mineralization generally increases and they are usually found in zones isolated from the surface by overlaying rocks (Dowgiałło 2007). The content of dissolved solids in groundwater may vary among different locations. It results mainly from geochemical conditions, e.g., the presence of halite, anhydrite, carbonates, etc. Depending on the geology, brines appear at a depth of several dozen to several thousand meters. Their presence at the Earth's surface, or just below the surface, is usually caused by the ascension of water from greater depths. In many spas, mineral water comes from artesian aquifers in which water is trapped surrounded by layers of impermeable rock or clay, which exert positive pressure on the water (Krawiec 2012).

In Poland, brines have been traditionally used to brew medicinal and edible salts. Starting from the end of the 18th century, they were also used in spa treatment mainly in balneology (Bebek 2003). There are currently 44 towns in Poland that have been granted a status of a health resort, and 191 sanatoria and 49 hospitals (GUS 2019) located in their areas are using mainly subsurface water for treatment.

According to Polish legislation, therapeutic water is groundwater which is not contaminated microbiologically and chemically, shows natural physicochemical variability, has mineralization of at least 1 g/l (mineral water), and meets at least one of the conditions relating to temperature or content of the specific compounds (e.g., $Fe^{2+} \geq 10$ mg l/1, F- ≥ 2 mg/L, I- ≥ 1 mg l/1, $S^{2-} \geq 1$ mg/L). Waters having an outlet temperature higher than 20°C are considered thermal (Act of 9 June 2011). Even weakly mineralized water can be considered healing if it contains at least one specific compound listed in the law or has a temperature exceeding 20°C. There is a large variety of different types of therapeutical waters that determines the chemical composition of water.

Chloride waters are among the most frequently occurring waters and contain chloride and sodium as a dominant ion plus magnesium, calcium, potassium, iodine, and bromine. Specific components of pharmacodynamic properties include mainly iodides and occasionally iron (Krawiec 2012). Sulfidic waters contain sulfur (≥ 1 mg/L) in the form of hydrogen sulfide (H_2S), hydrosulfide ions (HS^-), polysulfides (H_2S_x, x = 2-6), and thiosulfate ions ($S_2O_3^{2-}$). Additionally, they are enriched in iodide and fluoride in doses exceeding pharmacodynamic thresholds (Ciężkowski et. al. 2010, Gała 2011). Brine and sulfidic water baths affecting the vascular, vegetative, and endocrine systems have a positive effect on metabolism, blood circulation as well as on bone and joint system (Latour and Drobnik 2016).

Starting from 2007 to 2013, the value of the world spa sector increased by more than one third and was 94 billion dollars (Valeriani et al. 2019). With the increase in interest regarding health tourism, a significant rise in the number of scientific papers related to thermal water (e.g., Wilson et al. 2008, Kochetkova et al. 2011, Chaudhuri et al. 2017), healing waters, and spas (e.g., Chaabna et al. 2013, Fazlzadeh et al. 2016, Krett et al. 2016, Paduano et al. 2017, Pedron et al. 2019) have been observed (Valeriani et al. 2018). Knowledge pertaining to microorganisms in pools and bottled water covering health aspects of therapeutical waters is nowadays more extensive compared to knowledge regarding microbial communities present in deep subsurface aquifers being the source of spa water. Despite many years of investigations using more and more advanced methods, still very little is known on how particular physicochemical characteristics of a subsurface environment contribute to the observed microbial abundance, diversity, and community structure, how this structure is changed due to its transfer to the recipients, and which species contribute to the observed properties of the given medicinal waters (Paduano et al. 2017, Pedron et al. 2019).

Recent studies, based on metataxonomic techniques, have permitted a detailed knowledge of the microbial diversity in mineral water exploited for healing and recreational purposes in many spa towns. Studies from the brine samples taken from boreholes located in Kołobrzeg, Połczyn (Kalwasińska et al. 2019), Ciechocinek (Kalwasińska et al. 2018), Busko, Wełnin, and Dobrowoda (Deja-Sikora et al. 2019) were conducted. In this review, we will focus on the abundance and community composition of bacterial and archaeal assemblages related to specific conditions of mineral waters used in spa towns of North-West, Central, and South-East Poland located within the Paleozoic platform and Carpathians, respectively (Figure 1).

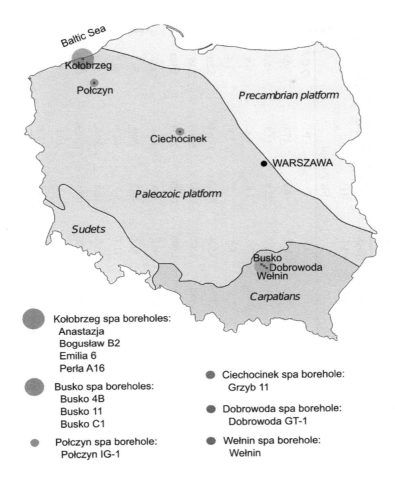

Kołobrzeg spa boreholes:
Anastazja
Bogusław B2
Emilia 6
Perła A16

Ciechocinek spa borehole:
Grzyb 11

Busko spa boreholes:
Busko 4B
Busko 11
Busko C1

Dobrowoda spa borehole:
Dobrowoda GT-1

Połczyn spa borehole:
Połczyn IG-1

Wełnin spa borehole:
Wełnin

Figure 1. Outlook of the Sampling Sites

Stratigraphy and Lithology of Aquifers

Kołobrzg and Połczyn are located in SE Poland in the northern part of the Pomeranian Anticlinorium within the Mid-Polish Trough (MPT). Połczyn IG-1 was the deepest studied intake of 1,248 m extracting brine from the upper Triassic sandstones (Table 1).

Anastazja and Bogusław B2 were the middle deep boreholes having a depth of 354 m and 250 m below ground level and reaching lower Jurassic clays/silts and sandstones, respectively. Emilia of 66 m and Perła A16 of 46 m depth were relatively shallow intakes exploiting mineral water from the middle Jurassic and Quaternary sands or glacial deposits. Grzyb 11, the Central Poland borehole located in Ciechocinek within MPT, was 405 m deep and reached down into cracked series of the middle Jurassic sandstones and limestones. Within the south-east boreholes, located within the northern part of the Carpathian Foredeep, the deepest was Busko C1 having 663 m depth. The second was Busko 17 with 148 m, and the shallowest was Busko 4B reaching 60 m below ground level. The three boreholes drilled in the Busko region were associated with Cretaceous aquifer consisting of marls and sandstones. Relatively deep Dobrowoda GT-1, reaching 300 m inside the lithosphere, was associated mostly with sands of Neogene; however, the additional input of water from Cretaceous and Upper Jurassic deposits was also probable. Borehole Wełnin extracted mineral water from the depth of 170 m within the upper Jurassic marls and limestones.

TABLE 1: Geological and Physicochemical Parameters of Therapeutic Waters from NW, Central, and SE Poland Spas

Parameter	Perła A16	Emilia 6	Bogusław B2	Anastazja	Połczyn IG-1	Grzyb 11	Busko 4B	Busko 17	Busko 1C	Wełnin	Dobrowaoda GT-1
			NW Poland			Cental			SE Poland		
Aquifer stratigraphy	Q^a	J^a	J^a	J^a	Tr^a	J	C^a	C^a	C^a	J^a	$Q\text{-}C\text{-}J^a$
Depth (m b.g.l.)	46	66	250	354	1248	405	60	148	663	170	300
T (°C)	10.0	11.2	10.6	10.0	9.2	19.5	12.4	12.7	22.8	13.5	16.3
pH	7.8	6.7	6.6	6.6	6.9	7.1	7.03	7.0	7.8	6.6	6.9
Mineralization (%)	0.2	5.6	5.9	6.0	7.3	5.1	1.3	1.3	1.3	3.3	1.4
Cl^-	735	33,678	35,804	36,159	41,900	26,500	6,156	5,938	5,743	17,643	6,205
Na^+	475	18,450	19,650	19,950	23,700	39,000	3,950	3,875	4,167	9,303	3,897
SO_4^{2-}	23.5	309	352	340.5	3,170	730	1,918	1,962	1,893	2,303	2,394
HCO_3^-	317	215	233	191	73	490	431	452	408	944	456
HS^-/H_2S	nd	nd	nd	nd	nd	nd	20.6	40.7	23.5	960	59.3
Br^-	5.9	82.4	51.9	87.1	184	89.6	22.2	21.7	9.3	68.2	10.2
NH_4^+	0.6	6.35	6.2	2.84	8.7	23.0	16.7	23.5	18.8	39.6	35.9
NO^{3-}	<0.2	<0.2	<0.2	<0.2	<0.1	7.66	0.57	0.62	0.55	63.23	0.63
I^-	0.05	2.1	2.1	1.7	1.06	3.5	1.6	1.7	1.8	16.1	2.0
TOC	nd	nd	nd	nd	nd	68.3	12.4	4.57	29.4	54.9	1.06

Tr – Triassic, J – Jurassic, C – Cretaceous, Q – Quaternary; ^{a} – artesian water; b.g.l. – below ground level; TOC – total organic carbon; nd – no data

Based on stable isotopes and noble gas analysis most of the waters can be classified as "old" pre-Quaternary paleoinfiltration waters (Zuber and Grabczak 1991, Krawiec 1999, Zuber et al. 2010, Krawiec 2012). This means that they infiltrated in previous geological ages and have a very long residence time in the bedrock due to the lack of connection with the contemporary waters. Accordingly, they can be regarded as poorly renewable or even unrenewable. Water exploited from four boreholes from the south of Poland Busko 4B, 17, C1, and Dobrowoda originates from interglacial infiltration of meteoric waters and compared to the water from Kołobrzeg, Połczyn, and Wełnin, these are relatively younger.

Geochemistry of the Therapeutic Waters

The mineral waters had close-to neutral pH, except Perła and Busko C1, which were slightly alkaline (Table 1). Most waters were cold (T < 20°C), except Busko C1 providing thermal water (22.8°C). Due to the high content of dissolved solids (TDS ≥ 35 g/L) Połczyn IG-1, Anastazja, Bogusław B2, and Grzyb 11 were classified as brines with mineralization between 5.5 and 7.3%. The other waters had a lower concentration of dissolved solids between 0.20 and 3.25%. All waters presented the Cl-Na type of mineralization. The Cl^- and Na^+ concentration varied in the wide ranges between 735 and 41,900 mg/L and from 475 to 39,000 mg/L, respectively. All waters were bromide waters ($Br^- >$ 5 mg/L) with Br^- concentration between 5.9 and 184.0 mg/L and iodine waters ($I^- > 1$ mg/L) with I^- concentration up to 16.1 mg/L, except Perła A16, which had the lowest iodide ion content of 0.05 mg/L. Concentrations of SO_4^{2-} ranged from 24 to 3,170 mg/L and HCO_3^- content varied from 73 to 944.3 mg/L. Dissolution of Badenian gypsum is thought to be the origin of the primary sulfates in the Busko region.

The high concentration of salts in the underground waters is caused by the ascension of highly saline Mesozoic aquifers and mixing with the meteoric water. Interestingly, long-term exploitation of brine from borehole 11 in Ciechocinek (Central Poland) results in lowering the salinity. Pumping out the water from the intake 11 causes depression and local inflow of younger waters from shallower layers that infiltrated in a slightly cooler quarternary climate (Poprawski et al. 1998).

Waters from the Busko area were exceptionally rich in H_2S where content ranged from 20.6 to 960 mg/L, and the high concentration of sulfides is expected to result from microbial redox activities (Zuber et al. 2007, Łebkowska and Karwowska 2010). Ammonium was the major form of N in all the studied waters. Its concentrations between 0.6 to 39.6 mg/L were significantly higher than WHO limits (0.2 mg/L) for drinking waters. Nitrate concentrations varied substantially between sites from values lower than 0.02 mg/L in all NW Poland spas to 63.23 mg/L in the Wełnin sample. The last value was above the authorized concentration of 50 mg/L (WHO, 2008) similar to NH_4^+ content. Values of TOC measured for the Central and SE Poland spas were also relatively high, between 1.06 and 68.3 mg/L, and were higher than the mean concentration of TOC (2 mg/L) in natural European groundwaters (Gooddy and Hinsby 2008).

High ammonium content in shallow groundwater is typically attributed to anthropogenic sources related to agriculture or industry (Wells et al. 2016). However, other causes are also possible. The organic matter, being the source of nitrogen, was accumulated during the sedimentation process and is incorporated during the diagenesis process as ammonium. Under reducing conditions, the organic matter degradation can release high ammonium concentrations as well as HCO_3^- (Holloway and Dahlgren 2002). Therefore, geogenic processes, such as decomposition of organic matter under reducing geochemical conditions, have been attributed to high ammonium concentrations in deep groundwaters in various parts of the world (Mastrocicco et al. 2013, Scheiber at al. 2016, Du et al. 2017).

Abundance of Microorganisms in Therapeutic Waters

The total number of microorganisms in the studied waters was usually between 10^4 and 10^5 cells/mL, except Grzyb 11 where the TNM was as high as 10^7 cells/mL. Cells densities in the range of 10^4-10^5 cells/mL were recorded in many subsurface environments previously (Pedersen and Ekendahl 1990, Beaton et al. 2016, Purkamo et al. 2016, Walczak et al. 2017) and TNM values of 10^4 cells mL were thought to be a limit value in groundwaters with a very low organic carbon concentration (Lovley and Goodwin 1988, Fry et al. 1997). From the data concerning the Central and SE Poland resorts, we can conclude that the abundance of microbial cells was not correlated with the total organic content. While TOC concentrations varied substantially from 1.6 mg/L to 54.9 mg/L in the SE Poland, the TNM was almost constant. This is in agreement with the notion that the main component of the microbial communities in deep subsurface is autotrophic subcommunity, which is mostly independent of organic carbon derived from photosynthesis (Lau et al. 2016, Parnell et al. 2016).

Even an extremely high concentration of H_2S in Wełnin of 960 mg/L did not reduce the number of cells present in mineral waters of SE Poland. Sulfide hydrogen is a known metabolite in prokaryotes and plants (Wang 2011). At high levels, it displays deleterious effects on cells, and the toxicity is mainly related to the ability of H_2S to inhibit oxidative phosphorylation, causing an energy deficit in the cell. Microorganisms present in sulfidic rich environments are, however, well adapted to high H_2S concentration, having a respiratory oxidase cytochrome *bd* that is resistant to H_2S inhibition (Forte et al. 2016). What is more, cytochrome *bd* oxidase in anaerobic bacteria possibly acts as a detoxifier of environmental oxygen to protect the cell from oxidative damage (Giuffrè et al. 2014).

Grzyb 11 was characterized by the highest abundance of microorganisms exceeding the number of cells at other sites by two-three orders of magnitude. It was postulated that the most probable reason for the increased number of cells at this site is the presence of unknown favorable geochemical conditions and/or the inflow of waters from shallower depths containing abundant microbial communities.

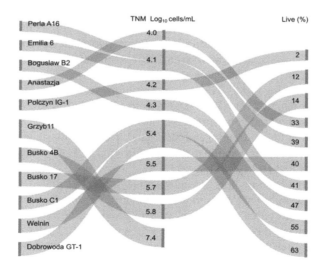

Figure 2. Abundance of Microorganisms Expressed as the Total Number of Microorganisms (TNM) and the Percentage of Live Cells (Live) in Therapeutical Waters in Poland

Quantification of microbial 16S rRNA gene copies revealed that the communities of sulfidic waters were dominated by Bacteria, while Archaea were in a minority. This observation was in line with the results of similar analyzes done for deep subsurface fracture fluids (Beaton et al. 2016, Purkamo et al. 2016) in which Archaea were found to constitute only a small portion of the communities.

Microbial Communities in Deep Subsurface Aquifers

Deep groundwaters investigated so far in Poland belong to two distinct types in terms of their mineralization and chemical composition. The structure of microbial communities found in these waters was shaped by two important drivers of microbial diversity, i.e., high salt concentration and high content of hydrogen sulfide. As salinity is one of the strongest environmental factors in the control of microbial abundance and diversity (Wu et al. 2006, Campbell and Kirchman 2013), the communities found in slightly to moderately saline groundwaters (brines) of NW and Central Poland were described separately from these recognized in slightly saline but sulfide-rich groundwaters of SE Poland. Noticeable similarities, including the common occurrence of bacteria involved in the S-cycle, between these communities were summarized at the end of the chapter.

The structure of subsurface microbiomes presented herein was assessed using a metagenomic approach based on isolation of total environmental DNA, amplification of fragments of bacterial and archaeal 16S rRNA genes using universal primers, high-throughput sequencing of amplicons, and bioinformatics analysis.

Microbial Communities in Deep Subsurface Brines

Cold saline groundwaters of NW Poland with mineralization ranging from 0.2 to 7.3% hosted microbial assemblages of low complexity. The overall biodiversity expressed as the Shannon-Wiener diversity index (H′) and the number of observed species (Sobs) was relatively low, between 1.12 and 2.85 and 19-31, respectively, (Figure 3) and the brines from deep underground strata (Połczyn IG-1 and Anastazja) were characterized by the lower H′ values compared to waters taken from lower depths (Bogusław B2, Emilia, and Perła A16).

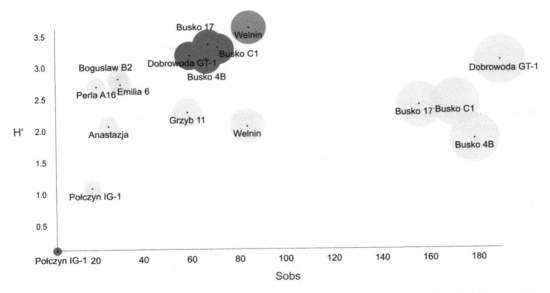

Figure 3. Diversity Indicators (Sobs and H′) of Bacterial (Light Colors) and Archaeal (Dark Colors) Communities in Therapeutic Waters in Poland

The bacterial community consisted mostly of Proteobacteria represented by 48% to nearly 90% of inhabitants (Kalwasińska et al. 2019). The contribution of other bacterial phyla to the community was lower and more variable. The share of Firmicutes ranged from 0.8 to 23%, but Actinobacteria could be found only in some waters while they were absent from the others.

Interesting results were provided by the community structure analysis at the level of the bacterial genus. The communities found in two deepest aquifers at a depth of 354 and 1,248 m b.g.l. and extracted by the boreholes Anastazja and Połczyn IG-1 were strongly dominated by deltaproteobacterial genera involved in sulfur cycling. Sulfate-reducing bacteria (SRB) from the genus *Desufovibrio* constituted from 43% to nearly 67% of total communities in these groundwaters (Figure 4). Along with *Desulfovibrio*, *Desulfotignum* was found to be abundant. In the other saline groundwaters, i.e., Bogusław B2, Emilia 6, and Perła A16, the co-occurrence of additional genera of sulfate reducers were revealed, including *Desulfopila, Desulfotomaculum, Desulfocurvus*, and *Desulfofustis* (Kalwasińska et al. 2019). However, the contribution of SRB was positively correlated with depth and negatively correlated with the dissolved oxygen. This observation suggests that SRB are ubiquitous participants of deep microbial communities and are important for the functioning of subsurface ecosystems (for details see 6.4).

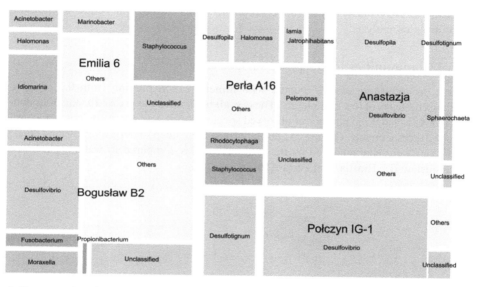

Figure 4. The most abundant bacterial genera (>5% of the total number of 16SrDNA sequences at a given site) in therapeutical waters from NW Poland spas

In the shallowest aquifers (boreholes Emilia and Perła A16 reaching 66 and 46 m b.g.l., respectively) abundant representations of *Idiomarina, Marinobacter, Halomonas, Pelomonas*, and *Acinetobacter* were detected. The presence of these halophilic bacteria, well known for displaying a lot of haloadaptations, was expected as they are common in brackish and saline water habitats, e.g., seas, oceans, salterns, and subsurface brines (Borsodi et al. 2013, Wang et al. 2014, Dong et al. 2015), and pristine waters associated with Permian and Triassic deposits (Kalwasińska et al. 2019). Furthermore, *Halomonas* and *Pelomonas* as denitrifiers may be involved in the N-cycle by coupling organic matter oxidation with NO_3^- reduction. The increased representations of these genera corresponded to the higher content of nitrates in the groundwaters.

The only archaeal phylotype affiliated with the genus *Methanobacterium* was revealed in the deepest brine (borehole Połczyn IG-1). These H_2-consuming (hydrogenotrophic) methanogens grow by reducing one-carbon compounds (e.g., CO_2, CO, methanol) or coal to methane. Methane can be subsequently used as a carbon and energy source by methanotrophs or AOM consortia (i.e., microbes capable of anaerobic oxidation of methane) that frequently occupy deep subsurface environments (Colwell et al. 2008). *Methanobacterium* was formerly detected to be a ubiquitous inhabitant of oxygen-depleted or low-oxygen deep subsurface aquifers, e.g., formation waters in low-porosity continental crust, Outokumpu (Nyyssönen et al. 2014), deep granitic groundwater (Kotelnikova and Pedersen 1998), or subsurface water of the Kalahari Shield, South Africa (Gihring et al. 2006). Furthermore, the presence of *Methanobacterium* in the deep saline aquifer is not surprising as this archaeal genus was also reported to be common in highly saline environments, e.g., commercial salts (Gibtan et al. 2017).

Microbial assemblages revealed in deep aquifers in Kołobrzeg and Połczyn IG-1 consist of slight to moderate halophiles, reaching the growth optima at a salt concentration not exceeding 15%. The content of dissolved solids with sodium and chloride concentration, ranging from 475 to 23,700 mg/L and 735 to 41,900 mg/L respectively, were identified as the major factor accounting for differences in microbial community structures between slightly and moderately saline groundwaters. In our study, the salinity in combination with a low concentration of nitrates (< 0.2 mg/L) was shown to decrease microbial species richness, which is in line with observations made in the other terrestrial (Ventosa et al. 2015) and subsurface habitats (Campbell and Kirchman 2013). The additional environmental factor that significantly and negatively influenced microbial diversity in studied brines (estimated as Shannon-Wiener index, H′) was the content of dissolved oxygen (DO). The lowest species richness and diversity were observed in the deepest and most saline groundwater, i.e., Połczyn in which DO did not exceed 0.9 mg/L (data not shown in Table 1).

Microorganisms living in deep subsurface environments are affected by the very low influx of organic matter, which strongly limits the growth of heterotrophs (Pan et al. 2017). Low input of organic matter in the studied aquifers results from their geological isolation from waters of the local flow (shallow groundwaters, surface waters, and meteoric waters). In Połczyn IG-1, deep subsurface groundwater is physically isolated by the thick layer of low-permeability clay (Ciężkowski et al. 2010) that reduces water recharge and inflow of allochthonous microorganisms. The groundwaters in Kołobrzeg are extracted from artesian aquifers with hindered water exchange with the surface as it was determined by isotopic analysis (Krawiec et al. 2000, Burzyński et al. 2004, Krawiec 2013). Therefore, it is presumed that microbial communities identified in these deep aquifers originate from ancient inhabitants thriving in the former geological periods. Long-lasting isolation, oligotrophic conditions, and factors contributing to species selection like salinity are expected to be the major drivers of the species loss (Kalwasińska et al. 2019).

The brine having mineralization of 5.1% extracted from a deep aquifer in Central Poland (Grzyb 11 reaching 405 m b.g.s.) hosted different microbial communities consisting of Proteobacteria and Bacteroidetes (Kalwasińska et al. 2018). Bacterial genera prevailing in this groundwater included *Idiomarina*, *Sphingobium*, *Chryseobacterium,* and *Sphingomonas*. The share of *Marinobacter* was smaller than in the brines of North Poland but still noticeable. Archaea were not detected in this environment, but it cannot be excluded that the resolution of the analysis was too small to capture the rare members of the biosphere. The high contribution of halophilic *Idiomarina* may indicate the increased availability of dissolved oxygen in this specific brine as these bacteria are described as obligate aerobes. *Idiomarina* is abundantly found in marine waters including deep-sea habitats (Rheaume et al. 2015). However, our studies show that the members of this genus seem to be also frequent inhabitants of deep saline aquifers (Kalwasińska et al. 2018, 2019). In past years, it was resolved that *Idiomarina,* unlike other marine bacteria, presents wide physiological potential, in particular the ability to efficiently grow under a broad range of temperature, salinity, and pH (Martínez-Cánovas et al. 2004).

From Deep Subsurface Into the Surface: How Do Changes in Environmental Conditions Impact the Structure of Microbial Community

The process of subsurface brine concentration in the system of graduation towers is an elegant example showing the impact of changing environmental conditions on the microbial community structure and diversity (Kalwasińska et al. 2018). The graduation tower is a wooden wall-like frame stuffed with bundles of blackthorn twigs (*Prunus spinosa* L.) used to produce the salt. This structure splits the brine running down through the tower into the droplets, which increases the rate of water evaporation and the concentration of mineral salts. In a temperate climate, maximal achievable salt content in the brine is 27% and under favorable weather conditions (sunny day, temperature >20°C) the process of concentration is completed within three days.

The graduation tower park found in Ciechocinek consists of three towers (GT1-GT3) connected by the system of continuous water flow. The GT1 is fed with the deep subsurface brine exploited through the borehole Grzyb 11. After running down through the first graduation tower (i.e., from the top tank into the bottom one), the brine is transferred directly onto the top of the GT2. The last stage of the concentration process takes place in the GT3, and from there, the brine is discharged into the salt works (Kalwasińska et al. 2018).

The microbiome, residing under the stable conditions in the deep subsurface environment, becomes exposed to completely different environmental factors when the brine enters the GT1. Suddenly, the water gets saturated with oxygen, the sunlight energy becomes available, the temperature fluctuates, and salinity gradually increases. It was observed that such dynamic changes in the environmental conditions caused profound alterations in the microbial community composition (Kalwasińska et al. 2018). The flow of the deep brine through the GT1 resulted in the dramatic loss of groundwater-specific bacterial genera, in particular, *Idiomarina* and *Sphingobium* (Figure 5). Increased salt concentration (up to 9%) favored the dominance of *Marinobacter, Roseovarius, Pseudoalteromonas, Alteromonas,* and *Marinomonas*. The ongoing concentration of the brine in the GT2 (up to 16%) caused (i) further increase in the *Roseovarius* share, (ii) an additional prevalence of *Psychroflexus* and *Fabibacter,* and (iii) the resurgence of *Idiomarina*. When the content of salt reached 26-27% in the GT3, the bacterial community became dominated by *Idiomarina* again with an additional co-occurrence of *Fodinibius* and two unclassified genera affiliated to Gammaproteobacteria and Cytophagia. Interestingly, the surface environment in combination with increased salinity caused the growth of archaeal assemblages that were absent from the deep subsurface brine. GT1-GT3 archaeal communities consisted of Halobacteria represented mostly by the genera *Halorubrum, Halohasta, Halonotius,* and *Halolamina* of which *Halorubrum* and *Halolamina* are hyperhalophilic (Cha et al. 2014, Çinar and Mutlu et al. 2016). More concentrated brines from GT2 and GT3 were also abundantly inhabited by genera *Natronomonas, Halobacterium, Haloplanus,* and unclassified genus within Halobacteriaceae (Kalwasińska et al. 2018). Archaeal communities, revealed in the surface brines GT1-GT3, were more similar to each other than bacterial ones. Nevertheless, both assemblages were dominated by typical halophilic microorganisms.

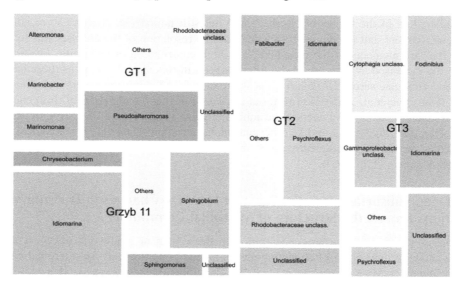

Figure 5. The Most Abundant Bacterial Genera (> 5% of the Total Number of 16S rDNA Sequences at a Given Site) in Therapeutical Waters from Ciechocinek, the Central Poland Spa

Generally, the study by Kalwasińska et al. (2018) indicated that microbial species richness tends to increase with the rising salinity in the surface brine. A similar ecological trend can be observed in solar salterns in crystallizer ponds of moderate salinity (Casamayor et al. 2002). However, more extreme salinity (i.e., 35-37% of salt content) is typically reported to cause dramatic loss of microbial diversity, resulting in the prevalence of only a few highly-tolerant phylotypes (Ventosa et al. 2015, Çinar and Mutlu et al. 2016).

Interesting conclusions come from observing the alteration in microbiome structure, which depended on the brine transfer from the subsurface into the surface. Surely, the microbial community responded to changed environmental conditions in such a way that sub-optimally adapted taxa were replaced by the others, displaying more relevant adaptations to the increased salinity. Based on the study by Kalwasińska et al. (2018), slightly halophilic *Sphingomonas* and *Sphingobium* seem to be more specific to moderately saline groundwater (5% NaCl). The concentration of salt increased by up to 9%, causing the replacement of these taxa by *Alteromonas* and *Pseudoalteromonas* that are known for the higher salt tolerance.

Microbial Communities in Slightly Saline and Sulfide-Rich Groundwaters

Slightly saline, sulfide-rich deep groundwaters, which are found in the vicinity of Busko spa, SE Poland are unique extreme environments due to the surprisingly high concentration of HS$^-$/H$_2$S ranging from 20 to even 960 mg/L. The study by Deja-Sikora et al. (2019) revealed the composition of bacterial and archaeal assemblages thriving in these sulfide-rich subsurface niches.

The analysis of water samples extracted from five independent aquifers located in the close distance to each other indicated the existence of three dissimilar structures of bacterial communities. Most frequently observed structure of the bacterial community, which was found in three groundwaters exploited through the boreholes Busko 17, Dobrowoda IG-1, and Wełnin consisted of a large representation of Proteobacteria (from 77 to nearly 89%) and much smaller share of Bacteroidetes (from 2 to nearly 7%). The similar structure of bacterial communities in these groundwaters was also noticeable at the genus level. The assemblages were dominated by unclassified genus within Comamonadaceae (44 to c.a. 58%), which was accompanied by different deltaproteobacterial sulfate reducers belonging to the genera, e.g., *Desulfopila*, *Desulfomicrobium*, *Desulfovibrio*, SEEP-SRB1, SEEP-SRB2 or MSBL7 (Deja-Sikora et al. 2019).

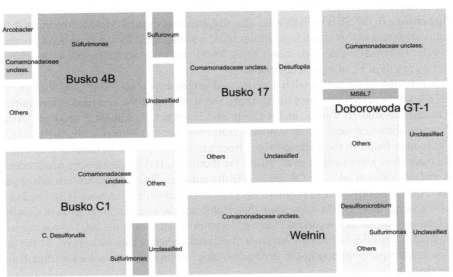

Figure 6. The Most Abundant Bacterial Genera (> 5% of the Total Number of 16S rDNA Sequences at a Given Site) in Therapeutical Waters from SE Poland Spas

As small representations of Comamonadaceae (1.7-5.1%) were also identified in the remaining sulfide-rich groundwaters under the study; the members of this bacterial family may constitute the archetypal microbial species in the deep aquifers of the Carpathian Foredeep. The ubiquity of Comamonadaceae in the other subsurface environments was previously reported. Comamonadaceae were dominant in the aquifers of Outokumpu Deep Drill Hole, Finland (Purkamo et al. 2016), and hard-rock aquifers in Brittany, France (Ben Maamar et al. 2015). It was proposed that Comamonadaceae may constitute the "keystone species" in the deep subsurface communities (Purkamo et al. 2016) as they possess versatile metabolic features important for the functioning of deep oligotrophic environments. Comamonadaceae are presumed to be involved in many geomicrobiological processes, including primary production, denitrification, H_2 oxidation, ferric iron reduction, and possibly sulfur oxidation (Brazelton et al. 2012, Nyyssönen et al. 2014, Rempfert et al. 2017, Deja-Sikora et al. 2019).

The second type of bacterial community was observed in the deepest aquifer (borehole Busko C1, 663 m b.g.l.). This sulfide-rich, anoxic environment was occupied mostly by Firmicutes (73%) with the overwhelming representation of clostridial sulfate reducers from the genus 'Candidatus desulforudis' (71%). Along with this genus, delta- and Epsilonproteobacterial genera noticeably contributed to the total microbiome (c.a. 7 and 6%, respectively), and they were represented among others by *Desulfovibrio*, *Sulfurimonas*, and *Halothiobacillus* (Deja-Sikora et al. 2019). The genus 'C. desulforudis' is frequently encountered in deep aquifers, thus its global distribution in the subsurface environments is concluded. 'C. desulforudis' was revealed in, e.g., Fennoscandian fracture fluids (Purkamo et al. 2016), fracture water from the 2.8-km-deep mine, South Africa (Chivian et al. 2008), 2-km-deep aquifer, Western Siberia (Karnachuk et al. 2019). 'C. desulforudis' displays an extreme metabolic versatility as its genome covers all the functions necessary to form the single-species ecosystem in the deep subsurface aquifer. The study by Chivian et al. (2008) indicated that 'C. desulforudis' is capable of living independently of the presence of other microbial groups.

The structure of bacterial assemblage in the shallowest aquifer (borehole Busko 4B, 148 m b.g.l.) might be shaped by suboxic conditions, resulting from the trace availability of oxygen. The community is composed mainly of sulfur-oxidizing bacteria (SOB), i.e., Epsilonproteobacterial genera, including *Sulfurimonas* (c.a. 65%), *Arcobacter* (5.5%), *Sulfurovum* (5%), and gammaproteobacterial genus *Halothiobacillus* (0.6%). Unlike in other groundwaters, the sulfate reducers were relatively rare in this environment and represented only by *SEEP-SRB1* and unclassified Desulfobacteracae. The surprising change in the SRB to SOB ratio also suggests that oxygen could be the main driver of the community structure in this aquifer. A similar trend was observed for the community occupying the sulfidic water of Frasassi Cave (Macalady et al. 2008), where sulfide to oxygen ratio exceeding 150 favored the growth of Epsilonproteobacteria.

Archea revealed in the sulfide-rich deep groundwaters were assessed to constitute no more than 0.32% of the total microbiome. Not surprisingly, the structure of the archaeal community was very similar across the studied waters, and it comprised Halobacteria almost exclusively. At the genus level, the most abundant were *Natronomonas*, *Halorhabdus*, and *Halorubrum*. Strikingly, the same genera were also found in the surface brines from graduation towers in Ciechocinek but not in the subsurface aquifer (Kalwasińska et al. 2019). Furthermore, Halobacteria were also encountered in low-salt and sulfide-rich lake, SE Oklahoma (Elshahed et al. 2004). It seems that salt concentration not exceeding 1% is sufficient to promote the growth of halophilic Archaea (Elshahed et al. 2004, Deja-Sikora et al. 2019). It is suggested that Halobacteria under anoxic growth conditions can utilize elemental sulfur as a terminal electron acceptor (Elshahed et al. 2004), thus contributing to the formation of sulfides. Although Archaea constitute rather a 'rare biosphere' in the subsurface microbiome, their participation in the geomicrobiological processes may be significant and should not be neglected.

Sulfate-Reducing Bacteria as Common Inhabitants of Deep Aquifers

In the deep subsurface environments, sulfur can be found in all-inorganic forms; however, the proportion of reduced and oxidized sulfur species depends on the specific geomicrobiological conditions. The numerous studies indicated that deep subsurface habitats frequently host the specialized functional groups of microorganisms involved in the biogeochemical cycle of sulfur (Lau et al. 2016, Purkamo et al. 2016, Deja-Sikora et al. 2019, Kalwasińska et al. 2019). These groups include sulfate-reducing bacteria (SRB) and sulfur-oxidizing bacteria (SOB) that under specific environmental conditions catalyze redox transformations of different sulfur forms, thus participating in the complete S metabolism in the subsurface environment. Under the oxygen-depleted conditions, sulfate reducers can degrade organic compounds by utilizing the oxidized inorganic forms of sulfur (i.e., sulfate, sulfite, thiosulfate, or elemental sulfur) as terminal electron acceptors (Lovley et al. 1995). The sulfide (HS^-/H_2S) being a final product of the dissimilatory sulfate reduction pathway along with the intermediates (i.e. S^o and $S_2O_3^{2-}$) can serve as an electron donor/s for sulfur oxidizers in the process of CO_2 fixation (Campbell et al. 2006).

The saline groundwaters in NW Poland, as well as sulfide-rich groundwaters in SE Poland, were enriched in the bacteria of the S-cycle (Deja-Sikora et al. 2019, Kalwasińska et al. 2019). Anoxic conditions found in the subsurface aquifers particularly promoted the growth of multispecies SRB consortia, which consisted of Deltaproteobacteria (e.g., *Desulfovibrio*, *Desulfopila*, and *Desulfomicrobium*) and/or Clostridia ('*Candidatus* desulforudis'). However, each of the studied aquifers was occupied by the site-specific SRB consortium, characterized by the unique combination of different SRB genera. The presence of SRB in slightly to moderately saline subsurface waters is related to their halotolerance. The members of SRB were reported to grow optimally at the salt concentration of 1-4% (Ben Ali Gam et al. 2018). Due to wide adaptation potential, resulting from the plenty of stress response pathways, SRB is capable of inhabiting many extreme habitats, including saline groundwaters (Zhou et al. 2011).

The occurrence of SOB consortia in the sulfide-rich waters in SE Poland indicated that the complete S-cycle may take place in these subsurface niches (Deja-Sikora et al. 2019). SOB consortia were revealed to be less divergent then SRB and consisted of Episoloproteobacteria (mainly *Sulfurimonas* and *Sulfurovum*) and Gammaproteobacteria (*Halothiobacillus*). In the subsurface, the prevalence of SOB is depended on the availability of oxygen or nitrate that is used by these chemoautotrophic microorganisms in the respiration (Burgin and Hamilton 2008). Sulfur oxidation is one of the crucial geomicrobiological processes taking place in the deep habitat as it regenerates the oxidizing power of the environment, where the anaerobic mineralization of organic matter depends on the sulfate pool restoration.

Studying of Deep Subsurface Environments: Perspectives

Undoubtedly, studying of biosphere buried deep down the Earth crust is challenging due to the hindered access to the subsurface environments. Nevertheless, the deep biosphere captures the attention of researches as there is still much to learn about the interrelations between the microbial community composition and underground microbial processes and the geochemistry of the isolated environment. What is more important is that extremophiles isolated from deep subsurface habitats have been proved yet to possess metabolic potential suitable for industrial applications as extremozymes obtained from such microbes are adapted to carry out biochemical transformations under high temperature, pressure and/or high salt concentration. Finally, the knowledge about the deep microbial life gives hints on the evolution of life on Earth and conditions limiting for life occurrence, which can be valuable for astrobiologists looking for extraterrestrial life.

References

Act of 9 Jun 2011 on Geological Work and Mining. Polish.

Amenabar, M.J., E.L. Shock, E.E. Roden, J.W. Peters and E.S. Boyd. 2017. Microbial substrate preference dictated by energy demand rather than supply. Nat. Geosci. 10: 577–581.

Anantharaman, K, C.T. Brown, L.A. Hug, I. Sharon, C.J. Castelle, A.J. Probst, et al. 2016. Thousands of microbial genomes shed light on interconnected biogeochemical processes in an aquifer system. Nature Comm. 7: 13219.

Beaton, E.D., B.S. Stevenson, K.J. King-Sharp, B.W. Stamps, H.S. Nunn and M. Stuart. 2016. Local and regional diversity reveals dispersal limitation and drift as drivers for groundwater bacterial communities from a fractured granite formation. Front. Microbiol. 7: 1933.

Bebek, M. 2003. Physicochemical analysis of salty waters and brines. Prz. Górn. 12: 39–45 [in Polish].

Ben Ali Gam, Z., A. Thioye, J.L. Cayol, M. Joseph, G. Fauque and M. Labat. 2018. Characterization of *Desulfovibrio salinus* sp. nov., a slightly halophilic sulfate-reducing bacterium isolated from a saline lake in Tunisia. Int. J. Syst. Evol. Microbiol. 68(3): 715–720.

Ben Maamar, S., L. Aquilina, A. Quaiser, H. Pauwels, S. Michon-Coudouel, V. Vergnaud-Ayraud, et al. 2015. Groundwater isolation governs chemistry and microbial community structure along hydrologic flowpaths. Front. Microbiol. 6: 1457.

Borsodi, A.K., T. Felföldi, I. Máthé I, V. Bognár V, M. Knáb, G. Krett, et al. 2013. Phylogenetic diversity of bacterial and archaeal communities inhabiting the saline Lake Red located in Sovata, Romania. Extremophiles 17(1): 87–98.

Boyd, E.S., A. Pearson, Y. Pi, W.J. Li, Y.G. Zhang, L. He, et al. 2011. Temperature and pH controls on glycerol dibiphytanyl glycerol tetraether lipid composition in the hyperthermophilic crenarchaeon *Acidilobus sulfurireducens*. Extremophiles 15(1): 59–65.

Brazelton, W., B. Nelson and M. Schrenk. 2012. Metagenomic evidence for H_2 oxidation and H_2 production by serpentinite-hosted subsurface microbial communities. Front. Microbiol. 2: 268.

Burgin, A.J. and S.K. Hamilton. 2008. NO^{3-}-driven SO_4^{2-} production in freshwater ecosystems: implications for N and S cycling. Ecosystems 11(6): 908–922.

Burns, D.G., P.H. Janssen, T. Itoh, M. Kamekura, Z. Li, G. Jensen, et al. 2007. *Haloquadratum walsbyi* gen. nov., sp. nov., the square haloarchaeon of Walsby, isolated from saltern crystallizers in Australia and Spain. J. Syst. Evol. 57(2): 387–392.

Burzyński K, A. Krawiec and A. Sadurski A. 2004. The origin and mobilization of deep brines of deep brines of the aquifer system by considering the circulation systems existing on the polish western coast of the Baltic Sea. pp. 521–530. *In*: L. Aragus, E. Custodio and M. Manzano [eds.]. Groundwater and Saline Intrusion. Selected Papers from the 18th Salt Water Intrusion Meeting. Instituto Geologico y Minero de Espana, Madrid, Spain.

Campbell, B.J., A.S. Engel, M.L. Porter and K. Takai. 2006. The versatile ε-proteobacteria: key players in sulphidic habitats. Nat. Rev. Microbiol. 4: 458–468.

Campbell, B.J. and D.L. Kirchman. 2013. Bacterial diversity, community structure and potential growth rates along an estuarine salinity gradient. ISME J. 7(1): 210–220.

Casamayor, E.O., R. Massana, S. Benlloch, L. Řvreĺs, B. Díez, V.J. Goddard, et al. 2002. Changes in archaeal bacterial and eukaryal assemblages along a salinity gradient by comparison of genetic fingerprinting methods in a multipond solar saltern. Environ. Microbiol. 4: 338–348.

Cha, I.T., K.J. Yim, H.S. Song, H.W. Lee, D.W. Hyun, K.N. Kim, et al. 2014. *Halolamina rubra* sp. nov. a haloarchaeon isolated from non-purified solar salt. Antonie Leeuwenhoek 105: 907–914.

Chaabna, Z., F. Forey, M. Reyrolle, S. Jarraud, D. Atlan, D. Fontvieille, et al. 2013. Molecular diversity and high virulence of *Legionella pneumophila* strains isolated from biofilms developed within a warm spring of a thermal spa. BMC 28(13): 17.

Chapelle, F.H. 1993. Groundwater Microbiology and Geochemistry. J. Wiley and Sons, New York, USA.

Chaudhuri, B., T. Chowdhury and B. Chattopadhyay. 2017. Comparative analysis of microbial diversity in two hot springs of Bakreshwar, West Bengal, India. Genom. Data 12: 122–129.

Chernyh, N.A., A.V. Mardanov, V.M. Gumerov, M.L. Miroshnichenko, A.V. Lebedinsky, A.Y. Merkel, et al. 2015. Microbial life in Bourlyashchy, the hottest thermal pool of Uzon Caldera, Kamchatka. Extremophiles 19(6): 1157–1171.

Chilton J. 1996. Groundwater. pp. 413–510. *In*: D. Chapman [ed.]. Water Quality Assessments – A Guide to Use of Biota, Sediments and Water in Environmental Monitoring - 2nd edition. E & FN Spon, London, UK.

Chivian, D., E.K. Brodie, E.J. Alm, D.E. Culley, P.S. Dehal, T.Z. DeSantis, et al. 2008. Environmental genomics reveals a single-species ecosystem deep within Earth. Science 322: 275–278.

Chowaniec, J. 2013. Obieg wody w skali regionalnej Tatr i Podhala ze szczególnym uwzględnieniem fazy podziemnej. pp. 63–70. *In:* J. Pociask-Karteczka [ed.]. Z badań hydrologicznych w Tatrach. Zakopane: Tatrzański Park Narodowy, Zakopane. Poland.

Ciężkowski, W., J. Chowaniec, W. Górecki, A. Krawiec, L. Rajchel and A. Zuber. 2010. Mineral and thermal waters of Poland. Prz. Geol. 58: 762–773.

Çinar, S. and M.B. Mutlu. 2016. Comparative analysis of prokaryotic diversity in solar salterns in eastern Anatolia (Turkey). Extremophiles 20: 589–601.

Colwell F.S., S.L. Caldwell, J.R. Laidler, E.A. Brewer, J.O. Eberly and S.C. Sandborgh. 2008. Anaerobic oxidation of methane: mechanisms, bioenergetics, and the ecology of associated microorganisms. Environ. Sci. Technol. 42: 6791–6799.

Deja-Sikora, E., M. Gołębiewski, A. Kalwasińska, A. Krawiec, P. Kosobucki and M. Walczak. 2019. Comamonadaceae OTU as a remnant of an ancient microbial community in sulfidic waters. Microb. Ecol. 78(1): 85–101.

Dong, C., X. Bai, H. Sheng, L. Jiao, H. Zhou and Z. Shao. 2015. Distribution of PAHs and the PAH degrading bacteria in the deep-sea sediments of the high-latitude Arctic Ocean. Biogeosciences 12(7): 2163–2177.

Dowgiałło J. 2007. Wody mineralne, lecznicze i termalne oraz kopalniane. pp. 25–33. *In:* B. Płochniewski and A. Sadurski [eds]. Hydrogeologia regionalna Polski, t. II. PIG, Warszawa, Poland.

Du, Y., T. Ma, Y. Deng, S. Shen and Z. Lu. 2017. Sources and fate of high levels of ammonium in surface water and shallow groundwater of the Jianghan Plain, Central China. Environ. Sci. Process. Impacts 19(2): 161–172.

Edwards, K.J., A.T. Fisher and C.G. Wheat. 2012. The deep subsurface biosphere in igneous ocean Crust: Frontier habitats for microbiological exploration. Front Microbiol. 3: 8.

Elshahed, M.S., F.Z. Najar, B.A. Roe, A. Oren, T.A. Dewers and L.R. Krumholz. 2004. Survey of archaeal diversity reveals an abundance of halophilic Archaea in a low-salt, sulfide- and sulfur-rich spring. Appl. Environ. Microbiol. 70: 2230–2239.

Enning, D. and J. Garrelfs. 2014. Corrosion of iron by sulfate-reducing bacteria: new views of an old problem. Appl. Environ. Microbiol. 80(4): 1226–1236.

Fazlzadeh, M., H. Sadeghi, P. Bagheri, Y. Poureshg and R. Rostami. 2016. Microbial quality and physical-chemical characteristics of thermal springs. Environ. Geochem. Health 38(2): 413–422.

Flynn, T.M., R.A. Sanford, H. Ryu, C.M. Bethke, A.D. Levine, N.J. Ashbolt, et al. 2013. Functional microbial diversity explains groundwater chemistry in a pristine aquifer BMC Microbiol. 13: 146.

Forte, E., V.B. Borisov, M. Falabella, H.G. Colaço, M. Tinajero-Trejo, Robert K. Poole, et al. 2016. The terminal oxidase cytochrome *bd* promotes sulfide-resistant bacterial respiration and growth. Sci. Rep. 6: 23788.

Fry, N.K., J.K. Fredrickson, S. Fishbain, M. Wagner and D.A. Stahl. 1997. Population structure of microbial communities associated with two deep anaerobic alkaline aquifers. Appl. Environ. Microbiol. 63: 1498–1504.

Gała, I. 2011. The preliminary diagnosis and characteristic of thermal sulphide waters in the Busko C-1 borehole. Tech. Posz. Geol. 1–2: 339–348.

Gibtan, A., K. Park, M. Woo, J.-K. Shin, D.-W. Lee and H.-S. Lee. 2017. Diversity of extremely halophilic archaeal and bacterial communities from commercial salts. Front. Microbiol. 8: 1–11.

Gihring T.M., D.P. Moser, L.-H. Li, M. Davidso, T.C. Onstott, L. Morgan L, et al. 2006. The distribution of microbial taxa in the subsurface water of the Kalahari Shield, South Africa. Geomicrobiol. J. 23: 415–430.

Giuffrè, A, V.B. Borisov, M. Arese, P. Sarti and E. Forte. 2014. Cytochrome *bd* oxidase and bacterial tolerance to oxidative and nitrosative stress. Biochim. Biophys. Acta. 1837(7): 1178–1187.

Gooddy, D.C. and K. Hinsby. 2008. Organic quality of groundwaters. pp. 59–70. *In:* W.M. Edmunds and P. Shand [eds]. Natural Groundwater Quality. Blackwell Publishing Ltd, Singapore.

GUS, 2019. Consice Statistical Yearbook of Poland. Zakład Wydawnictw Statystycznych, Warszawa, Poland.

Holloway, J.M. and R.A. Dahlgren. 2002. Nitrogen in rock: occurrences and biogeochemical implications. Glob. Biogeochem. Cycles 16(4): 65–71.

Kallmeyer, J., R. Pockalny, R.R. Adhikari, E.C. Smith and S. D'Hondt. 2012. Global distribution of microbial abundance and biomass in sub-seafloor sediment. Proc. Natl. Acad. Sci. USA 109: 16213–16216.

Kalwasińska, A., E. Deja-Sikora, A. Burkowska-But, A. Szabó, T. Felföldi, T. Kosobucki, et al. 2018. Changes in bacterial and archaeal communities during the concentration of brine at the graduation towers in Ciechocinek spa (Poland). Extremophiles 22(2): 233–246.

Karnachuk, O.V., Y.A. Frank, A.P. Lukina, V. Kadnikov., A.V. Beletsky, A.V. Mardanov, et al. 2019. Domestication of previously uncultivated Candidatus Desulforudis audaxviator from a deep aquifer in Siberia sheds light on its physiology and evolution. ISME J. 13(8): 1947–1959.

Kalwasińska, A., E. Deja-Sikora, A. Szabó, A. Krawiec, T. Felföldi, M. Swiontek Brzezinska, et al. 2019. Microbial communities of low temperature, saline groundwater used for therapeutical purposes in North Poland. Geomicrobiol. J. 36(3): 212–223.

Kochetkova, T.V., I.I. Rusanov, N.V. Pimenov, T.V. Kolganova, A.V. Lebedinsky, E.A. Bonch-Osmolovskaya et al. 2011. Anaerobic transformation of carbon monoxide by microbial communities of Kamchatka hot springs. Extremophiles. 15(3): 319–325.

Kotelnikova, S. and K. Pedersen. 1998. Distribution and activity of methanogens and homoacetogens in deep granitic aquifers at Aë spoë Hard Rock Laboratory, Sweden FEMS Microbiol. Ecol. 26: 121–134.

Krawiec, A. 1991. New results of the isotope and hydrochemical investigations of therapeutical waters of Ciechocinek Spa. Przegl. Geolog 47(3): 255–260. Polish.

Krawiec A., A. Rübel, A. Sadurski, S.M. Weise and A. Zuber. 2000. Preliminary hydrochemical isotope, and noble gas investigations of the origin of salinity in coastal aquifers of Western Pomerania, Poland. pp. 87–94. In: A. Sadurski [ed.]. Hydrogeology of the Coastal Aquifers. Proceedings of the 16th Salt Water Intrusion Meeting Miedzyzdroje-Wolin Island. Nicolaus Copernicus University, Toruń, Poland.

Krawiec, A. 2012. Therapeutic waters as geotourism values of the Polish Baltic sea coast Geoturyst. 1–2: 3–12.

Krawiec A. 2013. The origin of chloride anomalies in the groundwaters of the Polish Baltic coast. Wydawnictwo Naukowe UMK, Toruń, Poland.

Krett, G., Z. Nagymáté, K. Márialigeti and A.K. Borsodi. 2016. Seasonal and spatial changes of planktonic bacterial communities inhabiting the natural thermal Lake Hévíz, Hungary. Acta Microbiol. Immunol. Hung. 63(1): 115–130.

Kumar, S., M. Herrmann, B. Thamdrup, V.F. Schwab, P. Geesink, S.E. Trumbore, et. al. 2017. Nitrogen loss from pristine carbonate-rock aquifers of the Hainich Critical Zone Exploratory (Germany) is primarily driven by chemolithoautotrophic anammox processes. Front Microbiol. 8: 1951.

Kutvonen, H., P. Rajala, L. Carpén and M. Bomberg. 2015. Nitrate and ammonia as nitrogen sources for deep subsurface microorganisms. Front. Microbiol. 15, 6: 1079.

Latour, T., and M. Drobnik. 2016. Właściwości biochemiczne wód geotermalnych rozpoznanych w Polsce określające sposób ich wykorzystania do celów leczniczych lub rekreacji. Techn. Posz. Geol. 55(1): 67–74.

Lau, M.C.Y., T.L. Kieft, O. Kuloyo, B. Linage-Alvarez, E. van Heerden, M.R. Lindsay, et. al. 2016. An oligotrophic deep-subsurface community dependent on syntrophy is dominated by sulfur-driven autotrophic denitrifiers. Proc. Natl. Acad. Sci. USA. 113(49): 7927–7936.

Łebkowska, M. and E. Karwowska. 2010. Microorganisms present in sulfide waters. Acta Balneol. 52: 60–63.

Lovley, D.R. and S. Goodwin. 1988. Hydrogen concentrations as an indicator of the predominant terminal electron-accepting reactions in aquatic sediments. Geochim. Cosmochim. Ac. 52: 2993–3003.

Lovley, D.R. and F.H. Chapelle. 1995. Deep subsurface microbial processes. Rev. Geophys. 33: 365–381.

Lovley D.R., J.D. Coates, J.C. Woodward and E.J.P. Phillips. 1995. Benzene oxidation coupled to sulfate reduction. Appl. Environ. Microbiol. 61(3): 953–958.

Macalady, J.L., S. Dattagupta, I. Schaperdoth, D.S. Jones, G.K. Druschel and D. Eastman. 2008. Niche differentiation among sulfuroxidizing bacterial populations in cave waters. ISME J. 2: 590–601.

Martínez-Cánovas, M.J., V. Béjar, F. Martínez-Checa, R. Páez and E. Quesada. 2004. Idiomarina fontislapidosi sp. nov. and Idiomarina ramblicola sp. nov., isolated from inland hypersaline habitats in Spain. Int. J. Syst. Evol. Microbiol. 54: 1793–1797.

Mastrocicco, M., B.M.S. Giambastiani and N. Colombani. 2013. Ammonium occurrence in a salinized lowland coastal aquifer (Ferrara, Italy). Hydrol. Process. 27(24): 3495–3501.

Mays, L.W. 2013. Groundwater resources sustainability: past, present, and future. Water Resour. Manage. 27(13): 4409–4424.

McMahon, S. and J. Parnell. 2014. Weighing the deep continental biosphere. FEMS Microbiol. Ecol. 87(1): 113–120.

Miranda-Herrera, C., I. Sauceda, J. Gonzalez-Sanchez and N. Acuna. 2010. Corrosion degradation of pipeline carbon steels subjected to geothermal plant conditions. Anti. Corros. Method Mater. 57(4): 167–172.

Nealson K.H., F. Inagaki and K. Takai. 2005. Hydrogen-driven subsurface lithoautotrophic microbial ecosystem (SLiMEs): do they exist and why should we care? Trends Microbiol. 13: 405–410.

Nyyssönen, M., J. Hultman, L. Ahonen, I. Kukkonen, L. Paulin, P. Laine, et al. 2014. Taxonomically and functionally diverse microbial communities in deep crystalline rocks of the Fennoscandian shield. ISME J. 8: 126–138.

Oren, A. 2008. Microbial life at high salt concentrations: phylogenetic and metabolic diversity. Saline Syst. 4: 2.

Paduano, S., F. Valeriani, V. Romano-Spica, A. Bargellini, P. Borella and I. Marchesi. 2017. Microbial biodiversity of thermal water and mud in an Italian spa by metagenomics: a pilot study. Water Supply 18(4): 1456–1465.

Pan, D., J. Nolan, K.H. Williams, M.J. Robbins and K.A. Weber. 2017. Abundance and distribution of microbial cells and viruses in an alluvial aquifer. Front. Microbiol. 8: 1199.

Parkes, R.J., C.D. Linnane, G. Webster, H. Sass, A.J. Weightman, E.R.C. Hornibrook, et al. 2011. Prokaryotes stimulate mineral H_2 formation for the deep biosphere and subsequent thermogenic activity. Geology 39: 219–222.

Parkes J., B. Cragg, E. Roussel, G. Webster, A. Weightman and H. Sass. 2014. A review of prokaryotic populations and processes in sub-seafloor sediments, including biosphere: geosphere interactions. Marine Geol. 352: 409–425.

Parnell, J. and S. McMahon. 2016. Physical and chemical controls on habitats for life in the deep subsurface beneath continents and ice. Phil. Trans. R. Soc. A. 374: 20140293.

Pedersen, K., and S. Ekendahl. 1990. Distribution and activity of bacteria in deep granitic groundwaters of southeastern Sweden. Microb. Ecol. 20: 37–52.

Pedron, R., A. Esposito, I. Bianconi, E. Pasolli, A. Tett, F. Asnicar, et al. 2019. Genomic and metagenomic insights into the microbial community of a thermal spring. Microbiome 7: 8.

Poprawski, L., T. Jasiak and M. Wąsik. 1998. Analiza zmian chemizmu wód leczniczych Ciechocinka w trakcie wieloletniej eksploatacji. Prz. Geol. 46: 331–336.

Purkamo, L., M. Bomberg, R. Kietäväinen, H. Salavirta, M. Nyyssönen, M. Nuppunen-Puputti, et. al. 2016. Microbial co-occurrence patterns in deep Precambrian bedrock fracture fluids. Biogeosciences 13: 3091–3108.

Rheaume B.A., S. Mithoefer and K. S. MacLea. 2015. Genome Sequence of the Deep-Sea Bacterium *Idiomarina abyssalis* KMM 227[T]. Genome Announc. 3(5): e01256–15.

Santos, H. and M.S. Da Costa. 2002. Compatible solutes of organisms that live in hot saline environments. Environ. Microbiol. 4(9): 501–509.

Scheiber, L., C. Ayora, E. Vázquez-Suñé, D.I. Cendón, A. Soler and J.C. Baquero. 2015. Origin of high ammonium, arsenic and boron concentrations inthe proximity of a mine: Natural vs. anthropogenic processes. Sci. Total Environ. 15, 541: 655–666.

Shirokova, V.L., A.M.L. Enright, C.B. Kennedy and F.G. Ferris. 2016. Thermal intensification of microbial Fe(II)/Fe(III) redox cycling in a pristine shallow sand aquifer on the Canadian Shield. Water Res. 106: 604–612.

Takai, T., K. Nakamura, T. Toki, U. Tsunogai, M. Miyazaki, J. Miyazaki, et al. 2008. Proliferation at 122 C and isotopically heavy CH_4 production by a hyperthermophilic methanogen under high-pressure cultivation. P. Natl. Acad. Sci. 105(31): 10949–10954.

Valdez, B., M. Schorr, M. Quintero, M. Carrillo, R. Zlatev, M. Stoytcheva, et al. 2009. Corrosion and scaling at Cerro Prieto geothermal field. Anti Corros. Method Mater. 56(1): 28–34.

Valeriani, F., C. Protano, G. Gianfranceschi, E. Leoni, V. Galasso, N. Mucci, et al. 2018. Microflora Thermarum Atlas project: Biodiversity in thermal spring waters and natural SPA pools. Water Sci. Technol. Water Supply 18: 1472–1483.

Valeriani, F., L.M. Margarucci and V.R. Spica. 2019. Recreational use of spa treatment waters: criticisms and perspectives for innovate treatments. pp. 180–191. *In*: E. Leoni [ed.]. Recreational Water Illnesses. MDPI, Basel, Switzerland.

Vanlint, D., R. Mitchell, E. Bailey, F. Meersman, P.F. McMillan, C.W. Michiels, et al. 2011. Rapid acquisition of gigapascal-high-pressure resistance by *Escherichia coli*. mBio. 25; 2(1): e00130–10.

Ventosa, A., R.R. de la Haba, C. Sanchez-Porro and R.T. Papke. 2015. Microbial diversity of hypersaline environments: a metagenomic approach. Curr. Opin. Microbiol. 25: 80–87.

Walczak, M., E. Deja-Sikora, A. Kalwasińska, M. Polatowski and A. Krawiec. 2017. Distribution of bacteria in the mineral waters found in the Polish Lowland. Geol. Q. 61: 177–185.

Wang, R. 2011. Physiological implications of hydrogen sulfide: a whiff exploration that blossomed. Physiol. Rev. 92: 791–896.

Wang, X.X., C. Li, L.B. Zhao, L.Wu, W. An and Y. Chen. 2014. Diversity of culturable hydrocarbons-degrading bacteria in petroleum-contaminated saltern. AMR 1010–1012: 29–32.

Weidler, G.W., M. Dornmayr-Pfaffenhuemer, F.W. Gerbl, W. Heinen and H. Stan-Lotter. 2007. Communities of archaea and bacteria in a subsurface radioactive thermal spring in the Austrian Central Alps, and evidence of ammonia-oxidizing Crenarchaeota. Appl. Environ. Microbiol. 73(1): 259–270.

Wells, N.S., V. Hakoun, S. Brouyere and K. Knoller. 2016. Multispecies measurements of nitrogen isotopic composition reveal the spatial constraints and biological drivers of ammonium attenuation across a highly contaminated groundwater system. Water Res. 98: 363–375.

WHO, 2008. Guidelines for Drinking Water Quality. World Health Organization, Geneva, Switzerland.

Wilson, M.S., P.L. Siering, C.L. White, M.E. Hauser and A.N. Bartles. 2008. Novel archaea and bacteria dominate stable microbial communities in North America's Largest Hot Spring. Microb. Ecol. 56(2): 292–305.

Worden, R.H. 1996. Controls on halogen concentrations in sedimentary formation waters. Mineral. Mag. 60(399): 259–274.

Wu, Q.L., G. Zwart, M. Schauer, M.P. Kamst-van Agterveld and M.W. Hahn. 2006. Bacterioplankton community composition along a salinity gradient of sixteen high-mountain lakes located on the Tibetan Plateau China. Appl. Environ. Microbiol. 72: 5478–5485.

Zhou, J., Q. He, C.L. Hemme, A. Mukhopadhyay, K. Hillesland, A. Zhou et al. 2011. How sulphate-reducing microorganisms cope with stress: lessons from systems biology. Nat. Rev. Microbiol. 9(6): 452–466.

Zuber, A. and J. Grabczak. 1991. On the origin of saline waters in the Mesozoic of central and Northern Poland. pp. 202–208. *In*: Contemporary Problems of Hydrogeology. Warszawa, SGGW-AR, Poland.

Zuber, A., K. Różański and W. Ciężkowski. 2007. Metody Znacznikowe w Badaniach Hydrogeochemicznych. Oficyna Wydawnicza Politechniki Wrocławskiej, Wrocław, Poland.

Zuber, A., B. Porwisz, P. Mochalski, M. Duliński, J. Chowaniec, J. Najman, et al. 2010. Pochodzenie i wiek wód mineralnych rejonu Buska-Zdroju, określone na podstawie znaczników środowiskowych. pp. 125–149. *In*: R. Lisik [ed.]. Wody siarczkowe w Rejonie Buska-Zdroju. Hydrogeotechnika, Kielce, Poland.

18

Multi-Omics Approaches for Extremophilic Microbial, Genetic, and Metabolic Diversity

Tanvi Govil[1,2], Wageesha Sharma[1], David R. Salem[1,2,3,*] and Rajesh K. Sani[1,2,3,*]

Introduction

For much of the last century, microbiologists have known the nature and identity of only a tiny fraction of the inhabitants of the microscopic landscape. The estimate that 99% of environmental microbes are unculturable has made the microbial world and its associated diversity the most massive unexplored niche on the Earth. This challenge appears further heightened in the case of extremophiles, the microbial community with the potential of being metabolically and biochemically operational in extreme and unrelenting environmental conditions, be it temperature, pH, salinity, vacuum pressure, radiation, or microgravity. Most extremophiles are still part of the "microbial dark matter" that has not yet been discovered. Consequently, our understanding of the composition, its functional and physiological dynamics, and the evolution of extremophilic microbial consortia has lagged substantially behind. A research study in 2018 by Lloyd et al. (2018) confirms that highly divergent phylogenetically uncultured microbes, possibly with novel functions, dominate virtually all environments on the Earth (22% to 87% uncultured genera to classes and up to 64% uncultured phyla). Hot springs and hydrothermal vents, for instance, have high frequencies of uncultured phyla identified as both bacteria and archaea (Lloyd et al. 2018). The exceptions are human and human-associated environments, which appear dominated by cultured genera (45% to 97%) (Lloyd et al. 2018). These findings suggest that hunting extremophiles are a challenging pursuit, mainly constrained by the complexity of reaching the microbes' ecological niches and isolating them (Orellana et al. 2018). Sixty to ninety-nine percent of the extremophilic diversity of microbial cells on the Earth belong to phyla with no cultured relatives, which are not easily culturable or are unculturable using standard laboratory techniques (Bernard et al. 2018).

Since diversities in bacterial communities have normally been determined by phenotypic characterization of cultivated strains, conventional characterization of microbial strains has been subjected to debate as it is dependent on the ability of the strains to grow under specific environmental

[1] Department of Chemical and Biological Engineering, South Dakota School of Mines and Technology, Rapid City, USA 57701.

[2] Department of Materials and Metallurgical Engineering, South Dakota School of Mines and Technology, Rapid City, SD 57701, USA.

[3] Composite and Nanocomposite Advanced Manufacturing – Biomaterials Center, Rapid City, SD 57701, USA.

[4] BuG ReMeDEE Consortium, Rapid City, SD 57701, USA.

* Corresponding author: rajesh.sani@sdsmt.edu

conditions. These types of classic microbiological methods are indirect and produce artificial changes in the microbial community structure. Fortunately, pioneering approaches have bypassed the need for culture studies, thereby lifting a blind spot imposed by culture-based investigation to comparative analyzes. These approaches were, first, largely based on the development of 16S rRNA gene sequencing (Schmidt et al. 1991, Barns et al. 1996, Hugenholtz et al. 1998), then on the sequencing of other makers, e.g. gene *rpoB* (Case et al. 2007) or bacterial rhodopsins (Beja et al. 2000), and later on the development of metagenomics (Breitbart et al. 2002, Tyson et al. 2004, Tringe et al. 2005) along with single-cell genomics (Hamm et al. 2019). The commercialization of next-generation sequencing technology has further increased the size of metagenome databases at an impressive speed. As of 2019, more than 390,951 metagenomes are available with more than 1,500 billion sequences in the Metagenomics RAST server (http://metagenomics.anl.gov), whereas this number was around 5,000 at the end of 2014. Largely, metagenomics has emerged as a powerful tool to explore unculturable microbes through the sequencing and analysis of DNA extracted from the environmental samples as well as using experimental methods, such as DNA hybridization, gene expression, proteomics, metabolomics, and enzymatic screening. In addition to this, meta-transcriptomes in which rRNA transcripts are converted to cDNA and sequenced without the use of primers provide a step toward the more conservative goal of identifying populations that are currently active in a mixed community (Lopez et al. 2018).

Altogether, these new tools have also bestowed power in terms of time savings, cost-effectiveness, and data production capability in the field of phylogenetics. The use of modern metagenomics with the "omics" technologies have developed system biology that can provide unprecedented insights to access, characterize, and quantify this untapped diversity of microbes. The omics technologies also empower the discovery of novel genes, metabolic pathways, and essential products with biotechnological, pharmaceutical, and medical relevance. Life has thrived in a swath of harsh environmental conditions and with ever-increasing affordability, advances in high-throughput sequencing are offering new strategies for exploring the microbial universe at a previously unimaginable resolution. Hence, in this chapter, we summarize and discuss the basis of new initiatives, including metagenomics and other omics approaches for the exploration of functions and interactions of the extremophilic communities, and how they link phylogenetic, genetic, metabolic, and functional information about the microbiota of environments dominated by microorganisms that are refractory to cultivation (Figure 1).

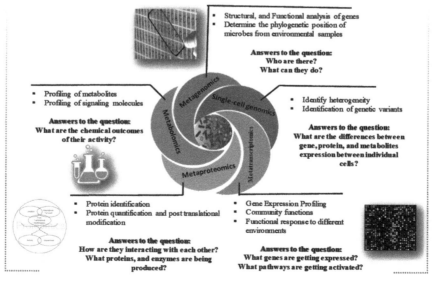

Figure 1. Integrated "omics" for screening microbial, genetic, and metabolic diversity of extremophilic microbiomes

Omics Approaches for Exploring the Unculturable World of Extremophiles

As an alternative to reliance on isolation and cultivation, molecular approaches based on phylogenetic analyzes of rRNA gene marker have long been used to decipher the taxonomic classification and composition of microbial communities in the environmental samples. The sequences of rRNAs (or their genes) from naturally occurring organisms are compared with known rRNA sequences by using techniques of molecular phylogeny. Some properties of an otherwise unknown organism can be inferred based on the properties of its known relatives because representatives of particular phylogenetic groups are expected to share properties common to that group. The same sequence variations that are the basis of the phylogenetic analysis can be used to identify and quantify organisms in the environment by hybridization with organism-specific probes.

Out of the three types of rRNA in prokaryotic ribosomes, classified as 23S, 16S, and 5S, the use of 16S rRNA gene sequences to study bacterial phylogeny and taxonomy has been by far the most common housekeeping genetic marker used for several reasons. These reasons include (i) its presence in all prokaryotes, often existing as a multigene family, or operons; (ii) the function of the 16S rRNA gene over time has not changed, suggesting that random sequence changes are a more accurate measure of time (evolution); (iii) the 16S rRNA gene (1,500 bp) is large enough for informatics purposes (Patel 2001, Janda and Abbott 2007); (iv) the use of 16S rRNA gene sequence informatics provides genus identification for isolates that do not fit any recognized biochemical profiles (Janda and Abbott 2007). Approaches that have been used to obtain rRNA sequences and thereby identify microorganisms in natural samples, without the requirement of laboratory cultivation, include direct sequencing of extracted 16S rRNAs (quantitative PCR), analysis of cDNA libraries of 16S rRNAs (Metatranscriptomics), and analysis of cloned 16S rRNA genes obtained by amplification using the polymerase chain reaction (PCR) (metagenomics) (Schmidt et al. 1991).

PCR Amplification of the 16S rRNA Gene

Almost all of the 16S sequence data in the public repositories to date are the products of PCR amplification of the 16S rRNA gene (Tringe and Hugenholtz 2008). As of November 2019, in EzTaxon (a web-based tool for the identification of prokaryotes based on 16S ribosomal RNA gene sequences), a complete hierarchical taxonomic system is available, containing 62,171 bacteria and 1,879 archaea species/phylotypes, which includes 15,290 valid published names (Yoon et al. 2017). The explosion in the number of recognized taxa is directly attributable to the ease in performance of 16S rRNA gene sequencing studies as opposed to the more cumbersome manipulations involving DNA-DNA hybridization investigations (Janda and Abbott 2007).

The so-called hybridization techniques, including Fluorescent *in situ* hybridization (FISH), allowed tagging of a taxonomic group whose cells could then be counted under a microscope. However, these techniques require developing probes for phylogenetic groups one by one, which is impractical for quantifying highly diverse natural samples that are often comprising thousands of species (Lloyd et al. 2018). Furthermore, hybridization techniques are not always quantitative in all environments due to taxon-specific biases in probe efficacy. However, the analysis of 16S rRNA sequences is also not without limitations, potentially imposing a selection of the sequences that are analyzed.

Methods that copy naturally occurring sequences *in vitro* before cloning potentially select for sequences that interact particularly favorably with 15-25 nucleotide primers broadly targeting bacteria or archaea (Schmidt et al. 1991). However, this approach based on primers is susceptible to biases depending on the level of primer matching in different species (Rosselli et al. 2016); it is known to miss some organisms owing to target mismatches, leading to a considerable amount of diversity that has been overlooked in environmental samples (Tringe and Hugenholtz 2008). Minor constituents of communities may not be detected using the 16S rRNA because only abundant rRNAs can be analyzed. Moreover, the use of the 16S phylogenetic marker is often criticized due to its heterogeneity among operons of the same genome (Acinas et al. 2004) and also because of its lack of resolution at the species level (Pontes et al. 2007, Armougom and Didier 2009). Because of this, it is possible that the identity of taxa's whose rRNA differ from known ones would not be revealed. PCR-based 16S analyzes are criticized for not providing information on the actual physiological relevance of each taxon within an environment (Rosselli et al. 2016). Last but not least, some parts of the conserved region of the bacterial 16S rRNA gene are similar to the conserved regions of plant chloroplasts and eukaryotic mitochondria. Therefore, if DNA contains a large amount of nontarget DNA, this nontarget DNA can be co-amplified and consequently produce useless sequence reads (Nakano 2018).

Nevertheless, the pioneering 16S rRNA gene, still considered as a "gold standard" for bacterial identification, is making a comeback in its own right thanks to a number of methodological advancements, including higher resolution (more sequences), analysis of multiple-related samples (e.g., spatial and temporal series) and improved metadata, and use of metadata (Tringe and Hugenholtz 2008). The deep sequencing of 16S rRNA genes, reaching millions of reads per single run, amplified by universal primers has revolutionized our ability to decode the microbial diversity from complex environmental samples by allowing the characterization of the diversity of the uncultured majority. High-throughput sequencing of the 16S rRNA gene has also increased the size of 16S sequence databases at an impressive speed (Armougom and Didier 2009) and thus improved the ability of 16S sequence identification using frequently updated online databases and tools, such as CAMERA (Sun et al. 2011), the RDP-Ribosomal Database Project (Cole et al. 2009), SILVA (http://www.arb-silva.de), and MG-RAST (Keegan and Meyer 2016), with quality-controlled bacterial and archaeal small subunit rRNA alignments and analysis tools as well as sequence similarity bioinformatics pipelines. Even the National Center for Biotechnology Information (www.ncbi.nlm.nih.gov) houses a nearly complete database of full-length 16S rRNA gene sequences. However, this database is subject to biases because the gene entries have undergone exponential amplification from their initial abundances, and small mismatches between DNA primers and different taxa are magnified during this amplification (Lloyd et al. 2018). Hence, it is claimed that PCR-amplified 16S-rRNA gene sequencing can be useful for straightforward classification of purified isolates. An alternative way to assess accurate taxa quantification in complex environmental samples, like the ones offered by extremophilic environments, is to rely on metagenomic surveys. Whole-genome shotgun sequencing followed by metagenomic analysis adds a more detailed layer of information to the taxonomical characterization of a sample by generating information on the gene composition of the bacteria present. This information can in turn be used to discover new genes and to formulate putative functional pathways and modules, thus providing insight into functional and genetic microbiome variability.

Direct sequencing Without PCR

The main drawback of the existing methodologies, which are all based on PCR, is *a priori* knowledge assumption required for universal primers design. A recent study based on metagenomic data has shown that about 10% of environmental bacterial or archaeal sequences might not be recovered when using a targeted PCR survey with the most common primers for SSU rRNA (Eloe et al. 2016a). As a consequence, the current experimental protocols may determine an altered or incomplete estimation of the true biodiversity of a given environmental sample. Indeed, an uneven primer annealing performance during the first PCR cycles can severely bias the final community member's proportions (Rosselli et al. 2016). Moreover, as primer design is based on existing knowledge, such tools remain self-referenced and do not guarantee the discovery of species whose sequence could diverge from established consensus. To overcome these drawbacks, a research group in 2016 proposed an approach of direct sequencing of 16S ribosomal RNA without any primer- or PCR-dependent step (Rosselli et al. 2016). The method was tested on a microbial community developing in an anammox bioreactor sampled at different time-points with success. The community resulting from direct rRNA sequencing was shown to be highly consistent with the known biochemical processes operative in the reactor. This novel principle is not considered as an alternative but rather as a complementary methodology in microbial community studies. The advantages of the methodology are, first, the approach is independent of PCR and, second, it solves the preferential primer matching issues toward given species. This makes the method also able to detect any rRNA regardless of its evolutionary conservation and/or similarity to any previously known consensus sequence. Nevertheless, dedicated direct approaches are envisaged as necessary to tap on the unknown reservoir of unculturable bacteria with un-amplifiable 16S rRNA genes that could inhabit environments that have hitherto been explored mainly with PCR-based methods (Rosselli et al. 2016).

Metagenome Analysis of Microbial Diversity

In the field of phylogenetics, metagenomics, which involves the random sequencing of all genomic content of a microbiome, represent a winning approach in the field of genomic analysis due to advancements in sequencing technology throughput and capability to profile genes as well as microbiome membership. Since its advent, metagenomics has enabled accessing the reservoir of communities of microbial organisms and the potential novel genes directly in their natural environments, bypassing the need for isolation and laboratory cultivation of individual species. In this approach, to explore this reservoir, DNA from an environmental sample is extracted, cloned into an appropriate vector, and transformed into competent *E. coli* cells (Figure 2).

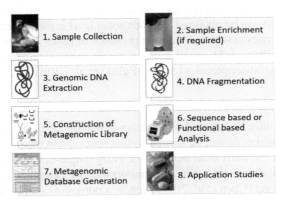

Figure 2: Steps Used in Preparing Metagenomic Library

The resulting transformants in metagenomic libraries are screened for novel physiological, metabolic, genetic features (functional metagenomics), and even particular function or activity (Ngara and Zhang 2018). A functional screen, however, detects genes that produce functional gene products, but not much is revealed about what species the genetic material came from. Alternatively, under sequence-based metagenomics, whole DNA extracted from a sample is sequenced followed by an assembly of sequence reads or by mapping them to a reference database followed by annotation of the genes. Sequence-based metagenomics studies can be used to assemble genomes, identify genes, find complete metabolic pathways, and compare organisms of different communities. The latter approach allows the overall reconstruction of the community structure, potentially revealing metabolic pathways of the whole microbiome and assigning minor or major geoecological roles to community members (Alves et al. 2018). Sequence-based metagenomics can also be used to establish the degree of diversity and the number of different bacterial species existing in a particular sample. Furthermore, it can lead to a better understanding of how specific genes help organisms survive in a particular environment.

Altogether, metagenomics provides a second tier of technical innovation that facilitates the study of the physiology and ecology of environmental microorganisms, including the discovery of novel genes and novel biomolecules through the expression of genes from uncultivated and unknown bacteria in the recipient host cell (Handelsman 2004). And, in contrast to the 16S rRNA gene amplicon approach, metagenomic sequencing can avoid the bias introduced by PCR-based techniques, provides more direct functional prediction information about the system, and ultimately can result in a more precise taxonomy through multigene and whole-genome phylogenetic approaches (Wilkins et al. 2019). Lately, the continuous improvements to DNA sequencing technologies coupled with dramatic reductions in cost have allowed the field of metagenomics to grow at a rapid rate. The introduction of high-throughput sequencing technologies such as i-Seq 100 System, Mini-Seq System, and Next-Seq 550 System for amplicon sequencing has facilitated phylogeny and taxonomy studies of diverse metagenomics samples. Comparatively, NGS platforms can recover up to 5,000 Mb of DNA sequence per day with costs at about 0.50$/Mb, while Sanger sequencing methodology allows about 6 Mb of DNA sequence to be created per day with costs at about 1,000 times higher (Alves et al. 2018).

Future improvements and developments in sequencing technologies, expression vectors, alternative host systems, and novel screening assays will help advance the field further by revealing novel taxonomic and genetic diversity (Culligan and Sleator 2016). The scope for usage has widened in recent years, and the diversity of ecosystems explored using metagenome analysis has increased by multiple folds over the recent past. This is anticipated to hold significant market potential in various life sciences industries. However, while metagenomics is powerful, using it on its own to study a microbiome has limited value. Metagenomics cannot reveal dynamic properties, such as the spatiotemporal activity of the community and the impact of the environment on these activities. The only information that can be obtained at a functional level is the potential of the microbiome to display functional properties associated with the presence of genes with no information about their expression levels or lack thereof. The need to monitor gene expression patterns brings us to the topic of our next section, metatranscriptomics (Aguiar-Pulido et al. 2016).

Single-Cell Metagenomics

Microbiologists traditionally study populations rather than individual cells as it is generally assumed that the status of individual cells will be similar to that observed in the population. However, recent studies have shown that the individual behavior of every single-cell could be quite different from that of the whole population, suggesting the importance of extending traditional microbiology studies to a single-cell level (Chen et al. 2017). Hence, a powerful complementary approach to bulk metagenomics today is single-cell genomics: an approach that addresses and analyzes the genomes of microorganisms one cell at a time and leads to an enhanced understanding of individuality and heterogeneity of microbes in many biological systems. Like bulk metagenomics, single-cell genomics is a useful tool for the study of unknown and uncultivable microorganisms, in particular from extreme environments, such as geothermal areas, where phenotypic plasticity is a common phenomenon. Phenotypic plasticity is a kind of environmental-driven viability where in response to fluctuations in the environment, some individual cells in a microbial subpopulation may express genes that allow them to survive stresses and harsh conditions (Wang et al. 2015). Identification of heterogeneity in the gene, protein, and metabolites expression between individual cells is instrumental in identifying metabolic features, evolutionary histories of the uncultured microbial groups that dominate many environments and biogeochemical cycles as well as in industrial bioprospecting (Stepanauskas 2012). Although the single-cell genome assemblers mentioned above can perform metagenome assembly as well, recent studies have demonstrated that the combined assembly of the metagenome and single-cell genome can greatly improve the assembly continuity and completeness. Hence, several studies today have been combining single-cell genomics and metagenomics to generate much improved bacterial genome assemblies from various bacterial communities. And, with the development of sequencing and computational methods, single-cell metagenomics will undoubtedly broaden its application in various microbiome studies (Xu and Zhao 2018).

Metatranscriptome

Amplicon sequencing of 16 rRNA, as well as metagenomics of environmental samples, has been successful in addressing the question of community membership, profiling taxonomic abundance, and elucidating potential genes and functional pathways (Shakya et al. 2019). However, both of these approaches fail to distinguish the active from inactive members of the microbiome and thus cannot help discriminating those that are contributing to observed ecosystem behavior from those that are merely present, presumably awaiting more favorable conditions. The metagenomic analysis does not distinguish whether this genomic DNA comes from cells that are viable or under what conditions the predicted genes are expressed (Bashiardes et al. 2016).

To obtain more in-depth insights into how a microbial community responds overtime to its changing environmental conditions, microbiome scientists are beginning to employ large-scale metatranscriptomics approaches. While proteomics is also able to provide insight into actively expressed proteins, metatranscriptomics tells us about the gene activity: the genes that are highest expressed in a specific microbial environment. Thus, metatranscriptomics is the profiling of community-wide gene expression (RNA-seq) from environmental samples (Shakya et al. 2019) that can decode the answers for how many different genes are expressed in a microbial community across all species. Which are the highest expressed genes in a specific environmental condition? What is the most important functionality (highly activated pathway) needed in an environment? Which genes show the highest change in expression levels between different conditions (biomarker detection)? By focusing on what genes are expressed by the entire microbial community, metatranscriptomics sheds light on the active functional profile of a microbial community. The metatranscriptome provides a snapshot of the gene expression in a given sample at a given moment and under specific conditions by capturing the total mRNA (Aguiar-Pulido et al. 2016).

This, data from metatranscriptome analyzes complement metagenomics data by elucidating details about populations that are transcriptionally active and not just identify the genetic content of bacterial populations, as shown in the metagenomic analysis (Figure 3). Analyzing these two types of data simultaneously for a sample enables us to conclude the expressed genes versus the potentially existing genes. From such functional data, active metabolic pathways can be identified in the bacterial communities and can be associated with particular environmental conditions (Bashiardes et al. 2016).

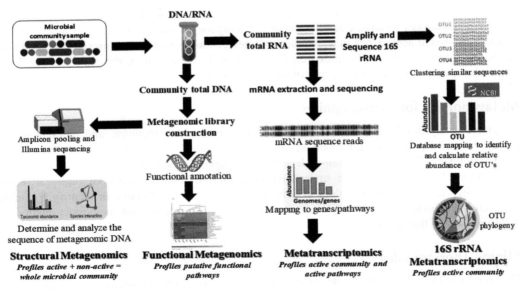

Figure 3. Combined Metagenomics and Metatranscriptomics Workflow

In recent years, the advent of massively parallel sequencing and RNA-seq has provided new and exciting opportunities in the area of metatranscriptome analysis, providing insight and dynamic range previously unimaginable. In a short time, since it was first introduced in the early 2000s, the number of metatranscriptomics projects or the sequencing of RNAs from microbial communities has increased significantly. While in 2010, total metatranscriptomics projects registered in the public repositories were just 10; in 2018, there were a total of 6,990 metatransciptomics datasets (i.e., "runs") deposited in the NCBI Sequence Read Archive (SRA) (Shakya et al. 2019). As such, pipelines integrating metagenomics, metatranscriptomics, metaproteomics, and metabolomics datasets may potentially enable us to gain a holistic view of microbiome composition and function at multiple layers. These merits careful consideration in cases where budget constraints and sample availability are not prohibitive (Bashiardes et al. 2016).

Every methodology has its drawback, and the metatranscriptomics approach that is based entirely on mRNA enrichment, which is then reverse transcribed for sequencing and analysis, may be criticized because this methodology typically requires high quantities of purified mRNA. Furthermore, mRNA is notoriously unstable, compromising the integrity of the sample before sequencing (Aguiar-Pulido et al. 2016). To surmount these limitations, a research group led by Yang et al. (2017) developed a modified RNA-seq method for small subunit ribosomal ribonucleic acid (SSU rRNA)-based microbial community analysis that depends on the direct ligation of a 5′ adaptor to RNA before reverse-transcription. The method requires only a low-input quantity of RNA (10-100 ng) and does not require a DNA removal step (Yan et al. 2017), making it suitable for low biomass environmental samples and providing many opportunities to identify novel bacterial taxa.

Metabolomics and Metaproteomics

A microorganism is not only a composite of its genome but also the multiple expressions of its genotype. Likewise, a microbiome is not just about the microbial communities striving there but also about the influence the microbial community exerts on critical biogeochemical cycles. Metabolomics, which is the comprehensive analysis by which all metabolites (small molecules released by the organism into the immediate environment) are identified and quantified, provides information on the secreted or modulated metabolite composition of the microbiota community, thereby enabling an understanding of the functional dynamics influencing community and host interactions. Similarly, metaproteomics provides an image of the entire protein complement of the microbiota communities studied under different conditions and an insight into the genes expressed and the key metabolic activities characterizing the bacterial communities. Together, these nongenomic-based approaches may complement metagenomics and metatranscriptomics data and add to our understanding of the unexpected diversity in the composition, structure, and regulation of metabolic networks amongst the microbiota communities.

Metagenomics for Extremophiles

Integrated analysis of metagenomes, metatranscriptomes, metaproteomes, and metabolomes is essential to understand the microbial systems biology. Achieving such integration necessitates interdisciplinary efforts and continuous development of appropriate bioinformatics tools to decipher the complex biological networks underlying molecular, functional, and community structure. The in-silico investigation of biological networks could be quite effective in identifying central connected components that could bring, at a later time, more insight on their functionality and dynamics within the system.

Mining Thermophiles

Geothermally heated regions of the Earth, such as terrestrial volcanic areas (e.g., fumaroles, hot springs, and geysers) and deep-sea hydrothermal vents, are of great interest to the general public and scientists alike due to their unique and extreme conditions. Populated by extremophilic archaeal and bacterial microorganisms, it is believed that most of these microbes are often recalcitrant to cultivation (Lloyd et al. 2018). Therefore, ecological, physiological, and phylogenetic studies of these microbial populations have been hampered for a long time. More recently, culture-independent methodologies coupled with the fast development of next-generation sequencing technologies, as well as with the continuous advances in computational biology, have empowered researchers to access the phylogenetic composition and functional potential of microbial consortia thriving within these habitats. Such studies are useful in the sense that they shed light on how extreme physicochemical conditions and biological interactions have shaped such microbial communities with genomic flexibility and metabolic versatility. This makes extreme- and hyper-thermophiles suitable for bioprospecting novel thermostable proteins that can be potentially used in several industrial processes.

The early era of work, based primarily on microscopy and cultivation approaches, successfully led to the discovery and characterization of some beneficial thermophilic bacteria, such as *Thermus aquaticus* "the source of Taq polymerase", *Thermoplasma, Sulfolobus, Thermomicrobium, Thermoanaerobium, Thermodesulfovibrio, Thermodesulfobacterium, Thermoanaerobacter, Anaerobranca, Thermobacteroides, Acidianus,* and *Sulphurococcus* (Brock 2001, Meyer-Dombard et al. 2005). Later, several microbiological surveys of (hyper) thermophilic environments, performed using 16S rRNA gene profiling, led to the advanced microbial characterization of different geothermal areas (Meyer-Dombard et al. 2005, Wang et al. 2014). Using 16S rRNA profiling in 2005, Meyer-Dombard et al. identified members of the *Desulphurococcales, Thermoproteales,* and "uncultured" *Crenarchaeota* in the samples from three hot springs (Obsidian Pool, Sylvan Spring, and 'Bison Pool') in Yellowstone National Park (Wyoming, USA). Also, cloned DNA affiliated with the *Desulphurococcales* (*Aeropyrum, Thermodiscus, Ignicoccus, Staphylothermus, Stetteria,* and *Sulfophobococcus*) and *Thermoproteales* (*Thermoproteaceae* and *Thermofilaceae*) were identified. The bacterial community at all three locations was dominated by members of *Aquificales* and *Thermodesulfobacteriales*, indicating that the "knallgas" reaction (aerobic hydrogen oxidation) may be a central metabolism in these ecosystems (Meyer-Dombard et al. 2005). In yet another study, 16S rRNA sequencing was able to detect signatures of 54 novel bacteria and 36 novel Archaea, including members of the deeply branching Korarchaea, in Obsidian Pool (Mud Volcano area) (Barns et al. 1994, Barns et al. 1996, Hugenholtz et al. 1998). In another remarkable study, phylogenetic analysis of ribosomal RNA sequences obtained from uncultivated organisms of a hot spring in Yellowstone National Park revealed several novel groups of archaea, many of which diverged from the crenarchaeal line of descent prior to previously characterized members of that kingdom (Barns et al. 1994). In fact, the authors claim that the organisms represented by these ribosomal RNA sequences could merit recognition as a new kingdom, provisionally named "Korarchaeota", pending additional data confirmation (Barns et al. 2016).

Furthermore, 16S rRNA sequencing has not just been used for community profiling but also to study the temporal changes of microbial communities in the Tengchong hot springs in response to geochemical changes. Different communities of microbes thrived during different times of the year; *Desulfurella* and *Acidicaldus* in moderate-temperature acidic springs, *Ignisphaera* and *Desulfurococcus* in high-temperature acidic springs, the candidate division OP1 and *Fervidobacterium* in alkaline springs, and *Thermus* and GAL35 in neutral springs. Such studies have geochemical significance in that they enable understanding of the effects that variations in physicochemical parameters, including temperature, pH, and dissolved organic carbon may have, triggering the observed microbial community shifts (Wang et al. 2014).

Over the years, many novel anaerobic thermophilic or hyperthermophilic prokaryotes have been isolated from such hyperthermophilic ecosystems with reliance on just the phylotypic profiling and 16S rRNA sequences. However, it has become apparent that the hyperthermophilic ecosystems do contain many unique 16S rRNA sequences, most of which were not identical to the 16S rRNA sequences of isolates cultivated from the sites (Teske and Reysenbach 2015). Recently developed algorithms to construct whole genomes from environmental samples have further refined, corrected, and revolutionized the understanding of the tree of life. The sequence-based metagenomic (SBM) analysis provides access to the gene composition of a microbial community and its encoded function, giving a much broader and detailed phylogenetic description than the 16S rRNA profiling (Strazzulli et al. 2017). In 2007, using SBM, Bhaya and co-workers identified, for the first time, two different *Synechococcus* populations inhabiting the microbial mats of the Octopus Spring in the Yellowstone National Park (YNP), revealing new facets of microbial diversity. This study revealed extensive genome rearrangements and differences related to the assimilation and storage of several elements, such as nitrogen, phosphorus, and iron, suggesting that the two populations have adapted differentially to the fluxes and gradients of chemical elements (Bhaya et al. 2007). Furthermore, SBM has also been productive in understanding the correlations between function and phylogeny of unculturable microorganisms, allowing the study of evolutionary profiles and their metabolic interactions that may not have been addressed with the basic 16S rRNA gene profiling (Strazzulli et al. 2017). In 2013, identification of novel candidate phyla, i.e., *Geoarchaeota*, within the domain Archaea provided an outstanding opportunity to understand the origin and evolution of Archaea in the early Earth environments that may have essential analogs active in Yellowstone National Park today (Kozubal et al. 2013). Using a similar strategy of stitching replicate *de novo* genome assemblies, Anja Spang and his colleagues reported the discovery in 2015 of the putative archaeal ancestor "Lokiarchaeota", a novel candidate archaeal phylum which forms a monophyletic group with eukaryotes in phylogenomic analyzes, providing strong support for hypotheses regarding the emergence of the eukaryotic host cell from within the archaeal domain of life (Spang et al. 2015). In 2016, a combined approach of these methods was used to identify two novel candidate phyla, *Calescamantes* and *Candidatuskryptonia,* by the analysis of different single amplified genomes and metagenomic databases collected in different high-temperature environments (Eloe-Fadrosh et al. 2016b), proving that, although metagenomics and single-cell genomics are informative of their own, the results of the mixed approach could be greater than the sum of their parts. The discovery of *Kryptonia* within previously studied geothermal springs underscores the importance of globally sampled metagenomic data in the detection of microbial novelty. It highlights the extraordinary diversity of microbial life still awaiting discovery (Eloe-Fadrosh et al. 2016b).

Indeed, very recently, several archaea (*Korarchaeota, Bathyarchaeota,* and *Aciduliprofundum*) and one potentially new species within the bacterial genus *Sulfurihydrogenibium* were taxonomically assigned to groups previously underrepresented in available genome data using assembled draft metagenome-assembled genomes (MAGs) from environmental DNA extracted from two hot springs within an active volcanic ecosystem on the Kamchatka peninsula, Russia (Wilkins et al. 2019). Wilkins and his group recovered 36 MAGs, 29 of medium to high quality, and inferred their placement in a phylogenetic tree consisting of 3,240 publicly available microbial genomes (Wilkins et al. 2019). Such a study adds comprehensive information in the SBM and metagenomic databases, making them an important repository of genes encoding for novel enzymes with potential biotechnological interest. Subsequently, in-silico functional screening of such data banks allows the identification of novel sequences and genes from both cultivable and uncultivable microorganisms that can be exploited without prior sequence information. Alternatively, metagenomic expression libraries can be constructed to perform direct functional screenings of the enzymes of interest with the gene of interest cloned and expressed in mesophilic hosts to produce recombinant enzymes. Ohlhoff et al. (2015) constructed a fosmid metagenomic library from high-temperature compost, which was subsequently screened for novel esterases (Ohlhoff et al. 2015). Smart et al. (2017) exploited one such metagenomic library for screening several classes of accessory lignocellulosic enzymes (Smart et al. 2017). Furthermore, lately, Fortune et al. (2019) characterized novel α-L-arabinofuranosidases (AFases) from the functional screening of the compost metagenome library (Fortune et al. 2019). Hence, recent advances in omics technologies, particularly when applied in a systems biology context, have made significant inroads into the prediction (if not the actual measurement) of *in situ* functionality, making them valid support or even a direct alternative to 16S rRNA profiling. Altogether, meta-omics are highly valuable and embody the best approach for complex communities requiring deeper sequencing.

Mining Psychrophiles

Approximately 80% of our planet's biosphere, which encompass much of the world's oceans, polar regions, mountain regions (the Alps, Himalayas, and the Rocky Mountains), the mesosphere and stratosphere are either periodically or permanently cold, that is at temperatures below 5°C (De Maayer et al. 2014). Taken together, this makes low temperatures the most widespread "extreme" environment, and as such, psychrophiles represent the most abundant, diverse, and widely distributed extremophiles on Earth, many of which form part of the "microbial dark matter" which have not been analyzed using traditional culture-dependent approaches. As discussed earlier, the recent development of metagenomics and related multi-omics strategies have provided a means by which entire microbial communities can be studied directly without the prerequisite of culturing. This progress in modern "omics" methods has provided unprecedented access to the structure and potential function of microbial communities in cold environments, providing increasingly comprehensive insights into the taxonomic richness and functional capacity of the indigenous microorganisms (Casanueva et al. 2010, Siddiqui et al. 2013).

Although there is often less microbial biomass and diversity in permafrost than in overlying active layer soils, which are exposed to seasonal freeze-thaw cycles, several studies utilizing molecular tools show that a variety of microbial phyla reside and are active in permafrost. Taking advantage of the metagenomic binning strategies, several high quality and draft genome sequences of both known and novel psychrophilic methanogens have been reconstructed. This endeavor began in 2011 when a 1.9 Mb genome sequence of a methanogen identified from an Alaskan permafrost site was first drafted (Mackelprang et al. 2011). In 2014, the near-complete genome of a single archaeal phylotype, Candidatus "*Methanoflorensstordalenmirensis*" belonging to the uncultivated lineage "Rice Cluster II" was recovered by the combined use of metagenomics and focused proteomic data, revealing the genes necessary for hydrogenotrophic methanogenesis (Mondav et al. 2014). Similarly, a genomic reconstruction belonging to the genus *Methanosarcina* was assembled from metagenomic data from a methane-producing enrichment of Antarctic permafrost in 2016 (Buongiorna et al. 2016), and deep metagenomic sequencing was then used to determine the functional genes and relate these data to predict the potential role of methanogens under cold conditions. Since then, many such metagenome sequence datasets from various cold habitats have been used for the reconstruction of genomes of some novel psychrophiles (Xue et al. 2019). Novel microbial communities and metabolic functions of the sub-glacial regions were also uncovered by other metagenomic studies, such as in the McMurdo Dry Valleys of Antarctica (Zaikova et al. 2019).

Further, combined metagenomics, metatranscriptomics, metaproteomics, and targeted metabolic profiling for understanding the active physiology of such regions, including the revealing of cold-adapted enzymes and pathways and shedding light on the adaptations of microbes there to sub-freezing conditions in permafrost. Tveit et al. (2015) combined metagenomics, metatranscriptomics, and targeted metabolic profiling of permafrost communities to elucidate the temperature dependence of methane metabolism therein (Tveit et al. 2015). Hultman et al. (2015) were the first to combine all three types of meta-omics information (metagenomics, metatranscriptomics, and metaproteomics) in a single study of thawing permafrost (Hultman et al. 2015). Generally, the attempts at RNA extractions from ancient permafrost studies have not been too successful, suggesting that if gene expression does occur in these samples, it is low (Mackelprang et al. 2016). Nevertheless, substantial strides have been made over the last decade leading to a huge expansion of our understanding of the various environmental, physiological, and molecular adaptations that psychrophilic microorganisms use to thrive under adverse conditions (De Maayer et al. 2014). Altogether, the significant research in the Antarctic continent and deep cold marine habitats through metagenomics and comparative genomics approaches have redefined the previous results and predictions of molecular phylogenetics. Finally, metagenomic bioprospecting of cold environments has also yielded a variety of novel genes and bioactive molecules including cold-shock, and anti-freeze proteins (Aliyu et al. 2017, Liljeqvist et al. 2015, Vester et al. 2014). We expect to see an increasing number of studies that apply multi-omics strategies to periodical frost and permafrost microbial communities with implications for our understanding of psychrophilic microbiology and biogeography and the limits of life in extremely harsh environments.

Mining Halophiles

A massive portion of the Earth's surface is saline. The range of salinity varies to significant extents from the marine and tidal pools with ~4% salinity, hot springs of nearly 10.5% salinity, and soda lakes with salt inclusions of up to 49.7% salinity (Scambelluri et al. 1997). The optimum salinity range for microbial isolation lies between 0 and 35%, and this makes the world of halophilic microorganisms highly diverse. In some ecosystems, salt-loving microorganisms live in such large numbers that their presence can be recognized without the need for a microscope. The brines of saltern crystallizer ponds worldwide are colored pink-red by *Archaea* (*Haloquadratum* and other representatives of the *Halobacteriales*), *bacteria* (*Salinibacter*), and *Eucarya* (*Dunaliella salina*) (Ma et al. 2010). In other saline environments, genera such as *Halobacterium, Haloferax, Haloarcula,* square archaeon *Haloquadratumwaslbyi*, and the bacteroidete *Salinibacterruber* are predominant members that have become popular models for studies in these habitats (Ma et al. 2010). However, to explore to what extent in the halophilic world "everything" is indeed "everywhere" and what degree of variation may be found among different high-salt environments, culture-independent studies based on sequencing of DNA recovered from the hypersaline environments from around the world are being relied upon.

Full random metagenome sequencing was utilized to uncover the novel, abundant, and previously unsuspected microbial groups thriving in the set of hypersaline ponds of the Santa Polasalterns near Alicante (Spain) (Ghai et al. 2011, Fernandez et al. 2014, Pasic et al. 2009), the saltern crystallizer ponds in Australia (salinity 34%) (Oh et al. 2010), Mono Lake and the lower salinity ponds at the Shark Bay saltern in Western Australia (Litchfield and Gillevet 2002), hypersaline Lake Tyrrell in Australia (Heidelberg et al. 2013), Lake Meyghan in Iran (12% salinity) (Naghoni et al. 2017), Spotted Lake (British Columbia, Canada) (Pontefract et al. 2017), Odiel Saltmarshes in Spain (Vera-Gargallo et al. 2018), surface sediments of Siberian soda lakes (Vavourakis et al. 2018), etc. Remarkably, from such studies, many novel dominant lineages previously unknown, including some novel metagenome-assembled genomes (MAGs) belonging to an uncultivated class of halophiles, have been discovered. For instance, a novel uncultivated class, the Nanohaloarchaea, previously detected only once by "classic" 16S rRNA amplicon sequencing was found dominantly in these systems across the world (Ghai et al. 2011, Vavourakis et al. 2018, Hamm et al. 2019). Similarly, the first extremophilic members of the Candidate Phyla Radiation (CPR) were discovered during the sequencing of metagenomic samples from Siberian soda lakes (Vavourakis et al. 2018).

Taking a further dig into the applications of "omics" for understanding extremophilic physiology, detailed functional annotation of a hypersaline metagenome has indicated a generally simplistic C and N biogeochemical cycling capacity with the capacity to use light as an energy source via various bacteriorhodopsins (Fernandez et al. 2014). Further, an extensive and systematic analysis of the genome and proteome composition of halophilic organisms along with a comparative study of non-halophiles have been exploited to characterize the molecular signatures of halophilic adaptation. A study by Paul et al. (2008) has shown that at the protein level, halophilic species are characterized by low hydrophobicity, over-representation of acidic residues, especially Asp, under-representation of Cys, lower propensities for helix formation, and higher propensities for coil structure. At the DNA level, the dinucleotide abundance profiles of halophilic genomes bear some common characteristics, which are quite distinct from those of non-halophiles, and hence may be regarded as specific genomic signatures for salt-adaptation (Paul et al. 2008). And last but not least, using the meta-omics, attempts have been underway to discover or develop novel enzymes with desired characteristics to meet a diversity of applications. Even the halophilic enzymes are now characterized by metagenomics as in the case of extreme halophilic xylanase XylCMS (Ghadikolaeiet al. 2019). Hence, advances in sequencing technology, metagenomic approaches, and analytical tools do provide a more detailed look into the environmental gene pool and functional coding potential of microbes within an ecosystem of biogeochemical relevance.

Mining Acidophiles

As with other extremophilic regions, the microbial communities associated with extreme acidic environments have long been the subject of classical culturing and phylotypic analyzes. Howsoever, the use of the metagenome approach can provide more definitive information on the relative importance of specific genes and organisms in an environment which is unculturable. This is especially important in light of the fact that the vast majority of acidophiles are thought to be uncultivable in the laboratory. Historically, a few bacterial species, such as *Acidithiobacillus* spp. and *Leptospirillum* spp. were believed to be the most ubiquitous and abundant acidophiles. However, current knowledge suggests that acidophilic bacterial diversity stretches over numerous representatives of *Proteobacteria*, *Nitrospirae*, *Firmicutes*, *Antinobacteria*, *Acidobacteria*, *Aquificae*, and *Verrucomicrobia* (Korzhenkovet al. 2019). In these environments, culture-independent molecular techniques suggested the signatures of *Chloroflexi* and various uncultured candidate taxa within *Thermoplasmata* (Mendez-Garcia et al. 2015, Ylton et al. 2013). Groups of uncultured archaea affiliated with the order *Thermoplasmatales* have been predicted in acid mine drainage (AMD), one of which, 'G-plasma', has successfully been isolated, characterized, and described as *Cuniculiplasma divulgatum* (Golyshina et al. 2016). In fact, using "metagenomics and related gene profiling", many more novel archaea have been discovered as reviewed in a research article recently by Korzhenkov et al. (2019). Overall, early studies based on the physiology of the autochthonous microbiota and the growing success of omics-based methodologies have enabled a better understanding of microbial ecology and function in low-pH mine outflows (Korzhenkovet al. 2019).

Research Gaps and Future Potential

By enabling the analysis of populations including many (so-far) unculturable and often unknown microbes, meta-omics is revolutionizing the field of microbiology, and in the case of extremophiles, most omics approaches have been implemented very well for thermophiles, psychrophiles, alkalophiles, and acidophiles (Hodge 2010). As the chance for finding life on other planetary bodies increases, scientists have also been using the results of metagenomics studies to obtain hints about alien biology. However, as the use of metagenomic approaches continues to expand, there are accompanying questions and concerns about different methodologies and philosophies with recent studies revealing various kinds of artifacts present in metagenomics data caused by limitations in the experimental protocols and/or inadequate data analysis procedures, which often lead to incorrect conclusions about a microbial community (Ward 2006, Wooley and Ye 2009). Furthermore, there remains the question of validating predictions of gene function, a bottleneck that will impact and limit the interpretation of metagenomic data in the same way it has for genomic studies of pure cultures (Ward 2006).

The scientific community is increasingly critical of the informatics structure required to run many of the pipelines with software required for metagenomics assays often being extensive and computationally expensive (Forbes et al. 2018). Moreover, the number of individuals employed to analyze the data is not keeping pace with the proliferation of data collected. As a consequence, the costs for bioinformatic data analysis continually goes up (Teeling and Glockner 2012). At the same time, as random shotgun metagenomic projects rapidly multiply and become the dominant source of publicly available sequence data, procedures for the best practices in their execution and analysis become increasingly important. To manage this problem, further development of semi-automated metagenome analysis tools is needed, which will allow scientists to handle the wealth of data with a level of assured quality. There is also a need for a greater number of skilled bioinformatics analysts who can decipher the huge global dataset of metagenomes that has been accumulated. Another important issue is the incorporation of data describing the environment from which metagenomic information has been derived, which has no comprehensive tool for metagenome analysis that incorporates all types of analysis (biodiversity analysis, taxobinning, functional annotation, metabolic reconstruction, and sophisticated statistical comparisons) (Ward 2006). Currently, to fulfill the need to establish commonly accepted data formats for metagenome sequences and their associated contextual (meta) data as well as defined interfaces for data exchange and integration, the Genomic Standards Consortium (GSC, http://gensc.org), supported by the International Nucleotide Sequence Databases Collaboration (INSDC) is addressing this challenge (Teeling and Glockner 2012). However, significant efforts will be required to build the analytical framework as well as the associated bioinformatics pipelines and tools.

The integration of metagenomics with other "omics" has undoubtedly led to new insights in the field of extremophiles. However, large sections of extremophiles such as barophiles and radiophiles have barely been touched by omics approaches. Moreover, it would be of great interest to create the opportunity to apply meta-omics to examine the diversity of airborne microbes. The main challenges of conducting metagenomic studies of airborne microbes remain their low density in the air. Hence, there is considerable research potential in applying omics methodology to these unexplored extremophiles, and conferences or meetings specifically focused on airborne metagenomics could provide the necessary forums for interested parties to discuss potential solutions to current challenges in this area.

Conclusion

Many experts have confirmed that the percentage of documented bacteria is deficient compared to the estimate of bacterial species on our planet. This may be due partially to the impossibility of culturing complex environments or replicating in the laboratory the real conditions in which the microbiome exists. In any case, the reference databases used to classify, and label bacteria are limited to what has been cataloged. The methods based largely on the use of 16 S rRNA gene sequencing typically either discard reads from undocumented microbes or label them based on the closest documented microbe from the database. Thus, inevitably, results have been based on a biased fraction of bacteria present in the samples, representing the first shortcoming of these methods. Secondly, these approaches, have lagged in providing insights into how a microbial community responds over time to their changing environmental conditions. The advent of a sophisticated systems biology approach based on data from multiple omics has filled this gap and has enabled the prediction of *in situ* functionality and at the same time elucidating potential genes and functional pathways. Metagenomics produces a taxonomical profile of the sample, metatranscriptomics helps us to obtain a functional profile, and metabolomics completes the picture by determining which byproducts are being released into the environment.

Overall, each approach provides valuable information separately; when combined, they paint a more comprehensive picture of the diversity and functionality of a microbiome. The 16S studies only directly characterize the taxonomic profile of a microbiome. Yet they remain a cost-effective option to exhaustively capture biodiversity (measuring the maximal dynamic range of relative abundance) of many samples using minimal sequencing, and it should be generally applicable in microbial ecology.

Acknowledgments

This research was supported by the National Science Foundation (Award #1736255, #1849206, and #1920954) and the Department of Chemical and Biological Engineering at the South Dakota School of Mines and Technology. The authors also gratefully acknowledge the support from the CNAM-Bio Center, supported by the South Dakota Governor's Office of Economic Development.

References

Acinas, S.G., L.A. Marcelino, V. Klepac-Ceraj and M.F. Polz. 2004. Divergence and redundancy of 16S rRNA sequences in genomes with multiple rrn operons. J. Bacteriol. 186(9): 2629–2635. doi: 10.1128/jb.186.9.2629-2635.2004.

Aguiar-Pulido, V., W. Huang, V. Suarez-Ulloa, T. Cickovski, K. Mathee and G. Narasimhan. 2016. Metagenomics, Metatranscriptomics, and Metabolomics Approaches for Microbiome Analysis. Evol. Bioinform. 12: 5–16. doi: 10.4137/EBO.S36436.

Aliyu, H., P. De Maayer, S. Sjöling and D.A. Cowan. 2017. Metagenomic analysis of low-temperature environments. pp. 389–421. In: R. Margesin [ed.]. Psychrophiles: From Biodiversity to Biotechnology. Cham: Springer International Publishing.

Alves, L.d. F., C.A. Westmann, G.L. Lovate, G.M.V. de Siqueira, T.C. Borelli and M.-E. Guazzaroni. 2018. Metagenomic approaches for understanding new concepts in microbial science. Int. J. Genomics 2312987–2312987. doi: 10.1155/2018/2312987.

Armougom, F. and R. Didier. 2009. Exploring microbial diversity using 16S rRNA high-throughput methods. J. Comput. Sci. Syst. Biol. 2. doi: 10.4172/jcsb.1000019.

Barns, S.M. R.E. Fundyga, M.W. Jeffries and N.R. Pace. 1994. Remarkable archaeal diversity detected in a Yellowstone National Park hot spring environment. Proc. Natl. Acad. Sci. 91(5): 1609–1613. doi: 10.1073/pnas.91.5.1609.

Barns, S.M., C. F. Delwiche, J.D. Palmer and N.R. Pace. 1996. Perspectives on archaeal diversity, thermophily and monophyly from environmental rRNA sequences. Proc. Natl. Acad. Sci. 93(17): 9188–9193. doi: 10.1073/pnas.93.17.9188.

Bashiardes, S., G. Zilberman-Schapira and E. Elinav. 2016. Use of metatranscriptomics in microbiome research. Bioinform. Biol. Insights 10: 19–25. doi: 10.4137/BBI.S34610.

Beja, O., L. Aravind, E.V. Koonin, M.T. Suzuki, A. Hadd, L.P. Nguyen, et al. 2000. Bacterial rhodopsin: evidence for a new type of phototrophy in the sea. Science 289(5486): 1902–1906. doi: 10.1126/science.289.5486.1902.

Bernard, G., J.S. Pathmanathan, R. Lannes, P. Lopez and E. Bapteste. 2018. Microbial dark matter investigations: how microbial studies transform biological knowledge and empirically sketch a logic of scientific discovery. Genome Biol. Evol. 10(3): 707–715. doi: 10.1093/gbe/evy031.

Bhaya, D., A.R. Grossman, A.-S. Steunou, N. Khuri, F.M. Cohan, N. Hamamura, et al. 2007. Population level functional diversity in a microbial community revealed by comparative genomic and metagenomic analyses. Isme j. 1(8): 703–713. doi: 10.1038/ismej.2007.46.

Breitbart, M., P. Salamon, B. Andresen, J.M. Mahaffy, A.M. Segall, D. Mead, et al. 2002. Genomic analysis of uncultured marine viral communities. Proc. Natl. Acad. Sci. 99(22): 14250–14255. doi: 10.1073/pnas.202488399.

Brock, T.D. 2001. The origins of research on thermophiles. pp. 1–9. In: A.-L. Reysenbach, M. Voytek, and R. Mancinelli [eds.], Thermophiles Biodiversity, Ecology, and Evolution. Boston, MA: Springer US.

Buongiorno, J., J.T. Bird, K. Krivushin, V. Oshurkova, V. Shcherbakova, E.M. Rivkina, et al. 2016. Draft genome sequence of antarctic methanogen enriched from Dry Valley Permafrost. Genome Announc. 4(6): e01362–01316. doi: 10.1128/genomeA.01362-16.

Casanueva, A., M. Tuffin, C. Cary and D.A. Cowan. 2010. Molecular adaptations to psychrophily: the impact of 'omic' technologies. Trends Microbiol. 18(8): 374–381. doi: 10.1016/j.tim.2010.05.002.

Case, R.J., Y. Boucher, I. Dahllöf, C. Holmström, W.F. Doolittle and S. Kjelleberg. 2007. Use of 16S rRNA and rpoB genes as molecular markers for microbial ecology studies. Appl. Environ. Microbiol. 73(1): 278–288. doi: 10.1128/AEM.01177-06.

Chen, Z., L. Chen and W. Zhang. 2017. Tools for genomic and transcriptomic analysis of microbes at single-cell level. Front. Microbiol. 8(1831). doi: 10.3389/fmicb.2017.01831.

Cole, J.R., Q. Wang, E. Cardenas, J. Fish, B. Chai, R.J. Farris, et al. 2009. The ribosomal database project: improved alignments and new tools for rRNA analysis. J. Nucleic Acids D141–D145. doi:10.1093/nar/gkn879.

Culligan, E.P. and R.D. Sleator. 2016. Editorial: from genes to species: novel insights from metagenomics. Front. Microbiol. 7(1181). doi: 10.3389/fmicb.2016.01181.

De Maayer, P., D. Anderson, C. Cary and D.A. Cowan. 2014. Some like it cold: understanding the survival strategies of psychrophiles. EMBO Reports 15(5): 508–517. doi: 10.1002/embr.201338170.

Eloe-Fadrosh, E.A., N.N. Ivanova, T. Woyke and N.C. Kyrpides. 2016a. Metagenomics uncovers gaps in amplicon-based detection of microbial diversity. Nat. Microbiol. 1: 15032. doi: 10.1038/nmicrobiol.2015.32.

Eloe-Fadrosh, E.A., D. Paez-Espino, J. Jarett, P.F. Dunfield, B.P. Hedlund, A.E. Dekas, et al. 2016b. Global metagenomic survey reveals a new bacterial candidate phylum in geothermal springs. Nat. Commun. 7(1): 10476. doi: 10.1038/ncomms10476.

Fernandez, A.B., R. Ghai, A.B. Martin-Cuadrado, C. Sanchez-Porro, F. Rodriguez-Valera and A. Ventosa. 2014. Prokaryotic taxonomic and metabolic diversity of an intermediate salinity hypersaline habitat assessed by metagenomics. FEMS Microbiol. Ecol. 88(3): 623–635. doi: 10.1111/1574-6941.12329.

Forbes, J.D., N.C. Knox, C.-L. Peterson and A.R. Reimer. 2018. Highlighting clinical metagenomics for enhanced diagnostic decision-making: a step towards wider implementation. Comput. Struct. Biotechnol. J. 16: 108–120. doi: 10.1016/j.csbj.2018.02.006.

Fortune, B., S. Mhlongo, L.J. van Zyl, R. Huddy, M. Smart and M. Trindade. 2019. Characterisation of three novel α-L-arabinofuranosidases from a compost metagenome. BMC Biotechnol. 19(1): 22. doi: 10.1186/s12896-019-0510-1.

Ghadikolaei, K.K., E.D. Sangachini, V. Vahdatirad, K.A. Noghabi and H.S. Zahiri. 2019. An extreme halophilic xylanase from camel rumen metagenome with elevated catalytic activity in high salt concentrations. AMB Express 9(1): 86. doi: 10.1186/s13568-019-0809-2.

Ghai, R., L. Pašić, A.B. Fernández, A.-B. Martin-Cuadrado, C.M. Mizuno, K.D. McMahon, et al. 2011. New abundant microbial groups in aquatic hypersaline environments. Sci. Rep. 1: 135–135. doi: 10.1038/srep00135.

Golyshina, O.V., H. Lunsdorf, I.V. Kublanov, N.I. Goldenstein, K.U. Hinrichs and P.N. Golyshin. 2016. The novel extremely acidophilic, cell-wall-deficient archaeon Cuniculiplasma divulgatum gen. nov., sp. nov. represents a new family, Cuniculiplasmataceae fam. nov., of the order Thermoplasmatales. Int. J. Syst. Evol. Microbiol. 66(1): 332–340. doi: 10.1099/ijsem.0.000725.

Hamm, J.N., S. Erdmann, E.A. Eloe-Fadrosh, A. Angeloni, L. Zhong, C. Brownlee, et al. 2019. Unexpected host dependency of Antarctic Nanohaloarchaeota. Proc. Natl. Acad. Sci. 116(29): 14661–14670. doi: 10.1073/pnas.1905179116.

Handelsman, J. 2004. Metagenomics: application of genomics to uncultured microorganisms. Microbiol. Mol. Biol. Rev. 68(4): 669–685. doi: 10.1128/MMBR.68.4.669–685.

Heidelberg, K., W. Nelson, J. Holm, N. Eisenkolb, K. Andrade and J. Emerson. 2013. Characterization of eukaryotic microbial diversity in hypersaline Lake Tyrrell, Australia. Front. Microbiol. 4(115). doi: 10.3389/fmicb.2013.00115.

Hodge, R. 2010. The future of genetics: beyond the human genome project. New York, NY: Facts on File, c2010: Infect. Genet. Evol.

Hugenholtz, P., B.M. Goebel and N.R. Pace. 1998. Impact of culture-independent studies on the emerging phylogenetic view of bacterial diversity. J. Bacteriol. 180(18): 4765–4774.

Hultman, J., M.P. Waldrop, R. Mackelprang, M.M. David, J. McFarland, S.J. Blazewicz, et al. 2015. Multi-omics of permafrost, active layer and thermokarst bog soil microbiomes. Nature 521(7551): 208–212. doi: 10.1038/nature14238.

Janda, J.M. and S.L. Abbott. 2007. 16S rRNA gene sequencing for bacterial identification in the diagnostic laboratory: pluses, perils and pitfalls. J. Clin. Microbiol. 45(9): 2761–2764. doi: 10.1128/jcm.01228-07.

Keegan, K.P., E.M. Glass and F. Meyer. 2016. MG-RAST, a metagenomics service for analysis of microbial community structure and function. Methods Mol. Biol. 1399: 207–233. doi: 10.1007/978-1-4939-3369-3_13.

Korzhenkov, A.A., S.V. Toshchakov, R. Bargiela, H. Gibbard, M. Ferrer, A.V. Teplyuk, et al. 2019. Archaea dominate the microbial community in an ecosystem with low-to-moderate temperature and extreme acidity. Microbiome 7(1): 11–11. doi:10.1186/s40168-019-0623-8.

Kozubal, M.A., M. Romine, R.d. Jennings, Z.J. Jay, S.G. Tringe, D.B. Rusch, et al. 2013. Geoarchaeota: a new candidate phylum in the Archaea from high-temperature acidic iron mats in Yellowstone National Park. Isme j. 7(3): 622–634. doi: 10.1038/ismej.2012.132.

Liljeqvist, M., F.J. Ossandon, C. González, S. Rajan, A. Stell, J. Valdes, et al. 2015. Metagenomic analysis reveals adaptations to a cold-adapted lifestyle in a low-temperature acid mine drainage stream. FEMS Microbiol. Ecol. 91(4): doi: 10.1093/femsec/fiv011.

Litchfield, C.D. and P.M. Gillevet. 2002. Microbial diversity and complexity in hypersaline environments: a preliminary assessment. J. Ind. Microbiol. Biotechnol. 28(1): 48–55. doi: 10.1038/sj/jim/7000175.

Lloyd, K.G., A.D. Steen, J. Ladau, J. Yin and L. Crosby. 2018. Phylogenetically novel uncultured microbial cells dominate earth microbiomes. mSystems 3(5): e00055–00018. doi: 10.1128/mSystems.00055-18.

Lopez-Fernandez, M., D. Simone, X. Wu, L. Soler, E. Nilsson, K. Holmfeldt, et al. 2018. Metatranscriptomes reveal that all three domains of life are active but are dominated by bacteria in the fennoscandian crystalline granitic continental deep biosphere. mBio 9. doi:10.1128/mBio.01792-18.

Ma, Y., E. A. Galinski, W.D. Grant, A. Oren and A. Ventosa. 2010. Halophiles 2010: life in saline environments. Appl. Environ. Microbiol. 76(21): 6971–6981. doi: 10.1128/AEM.01868-10.

Mackelprang, R., M.P. Waldrop, K.M. DeAngelis, M.M. David, K.L. Chavarria, S.J. Blazewicz, et al. 2011. Metagenomic analysis of a permafrost microbial community reveals a rapid response to thaw. Nature 480(7377): 368–371. doi: 10.1038/nature10576.

Mackelprang, R., S.R. Saleska, C.S. Jacobsen, J.K. Jansson and N. Taş. 2016. Permafrost meta-omics and climate change. Annu. Rev. Earth Planet. Sci. 44(1): 439–462. doi: 10.1146/annurev-earth-060614-105126.

Mendez-Garcia, C., A.I. Pelaez, V. Mesa, J. Sanchez, O.V. Golyshina and M. Ferrer. 2015. Microbial diversity and metabolic networks in acid mine drainage habitats. Front Microbiol. 6: 475. doi: 10.3389/fmicb.2015.00475.

Meyer-Dombard, D.R., E.L. Shock and J.P. Amend. 2005. Archaeal and bacterial communities in geochemically diverse hot springs of Yellowstone National Park, USA. Geobiology 3(3): 211–227. doi: 10.1111/j.1472-4669.2005.00052.x.

Mondav, R., B.J. Woodcroft, E.-H. Kim, C.K. McCalley, S.B. Hodgkins, P.M. Crill, et al. 2014. Discovery of a novel methanogen prevalent in thawing permafrost. Nat. Commun. 5(1): 3212. doi: 10.1038/ncomms4212.

Naghoni, A., G. Emtiazi, M.A. Amoozegar, M.S. Cretoiu, L.J. Stal, Z. Etemadifar, et al. 2017. Microbial diversity in the hypersaline Lake Meyghan, Iran. Sci. Rep. 7(1): 11522–11522. doi: 10.1038/s41598-017-11585-3.

Nakano, M. 2018. 16S rRNA gene primer validation for bacterial diversity analysis of vegetable products. J. Food Prot. 81(5): 848–859. doi: 10.4315/0362-028x.jfp-17-346.

Ngara, T.R. and H. Zhang. 2018. Recent advances in function-based metagenomic screening. Genomics Proteomics Bioinformatics 16(6): 405–415. doi: https://doi.org/10.1016/j.gpb.2018.01.002.

Oh, D., Porter, K., Russ, B., Burns, D. and Dyall-Smith M. 2010. Diversity of Haloquadratum and other haloarchaea in three, geographically distant, Australian saltern crystallizer ponds. Extremophiles 14(2): 161–169. doi: 10.1007/s00792-009-0295-6.

Ohlhoff, C.W., B.M. Kirby, L. Van Zyl, D.L.R. Mutepfa, A. Casanueva, R.J. Huddy, et al. 2015. An unusual feruloyl esterase belonging to family VIII esterases and displaying a broad substrate range. J. Mol. Catal. B Enzym. 118: 79–88. doi: https://doi.org/10.1016/j.molcatb.2015.04.010.

Orellana, R., C. Macaya, G. Bravo, F. Dorochesi, A. Cumsille, R. Valencia, et al. 2018. Living at the frontiers of life: extremophiles in chile and their potential for bioremediation. Front. Microbiol. 9(2309). doi: 10.3389/fmicb.2018.02309.

Pasic, L., B. Rodriguez-Mueller, A.B. Martin-Cuadrado, A. Mira, F. Rohwer and F. Rodriguez-Valera. 2009. Metagenomic islands of hyperhalophiles: the case of Salinibacter ruber. BMC Genomics 10: 570. doi: 10.1186/1471-2164-10-570.

Patel, J.B. 2001. 16S rRNA gene sequencing for bacterial pathogen identification in the clinical laboratory. Mol. Diagn. 6(4): 313–321. doi: 10.1054/modi.2001.29158.

Paul, S., S.K. Bag, S. Das, E.T. Harvill and C. Dutta. 2008. Molecular signature of hypersaline adaptation: insights from genome and proteome composition of halophilic prokaryotes. Genome Biol. 9(4): R70-R70. doi: 10.1186/gb-2008-9-4-r70.

Pontefract, A., T.F. Zhu, V.K. Walker, H. Hepburn, C. Lui, M.T. Zuber, et al. 2017. Microbial diversity in a hypersaline sulfate lake: a terrestrial analog of ancient mars. Front. Microbiol. 8(1819). doi: 10.3389/fmicb.2017.01819.

Pontes, D.S., C.I. Lima-Bittencourt, E. Chartone-Souza and A.M. Amaral Nascimento. 2007. Molecular approaches: advantages and artifacts in assessing bacterial diversity. J. Ind. Microbiol. Biotechnol. 34(7): 463–473. doi: 10.1007/s10295-007-0219-3.

Rosselli, R., O. Romoli, N. Vitulo, A. Vezzi, S. Campanaro, F. de Pascale, et al. 2016. Direct 16S rRNA-seq from bacterial communities: a PCR-independent approach to simultaneously assess microbial diversity and functional activity potential of each taxon. Sci. Rep. 6: 32165–32165. doi: 10.1038/srep32165.

Scambelluri, M., G.B. Piccardo, P. Philippot, A. Robbiano and L. Negretti. 1997. High salinity fluid inclusions formed from recycled seawater in deeply subducted alpine serpentinite. Earth Planet. Sci. 148(3): 485–499. doi: https://doi.org/10.1016/S0012-821X(97)00043-5.

Schmidt, T.M., E.F. DeLong and N.R. Pace. 1991. Analysis of a marine picoplankton community by 16S rRNA gene cloning and sequencing. J. Bacteriol. 173(14): 4371–4378. doi: 10.1128/jb.173.14.4371-4378.1991.

Shakya, M., C.-C. Lo and P.S.G. Chain 2019. Advances and challenges in metatranscriptomic analysis. Front. Genet. 10(904). doi: 10.3389/fgene.2019.00904.

Siddiqui, K.S., T.J. Williams, D. Wilkins, S. Yau, M.A. Allen, M.V. Brown, et al. 2013. Psychrophiles. Annu. Rev. Earth Planet. Sci. 41(1): 87–115. doi: 10.1146/annurev-earth-040610-133514.

Smart, M., R.J. Huddy, D.A. Cowan and M. Trindade. 2017. Liquid phase multiplex high-throughput screening of metagenomic libraries using p-nitrophenyl-linked substrates for accessory lignocellulosic enzymes. pp. 219–228. In: W.R. Streit and R. Daniel [eds.]. Metagenomics: Methods and Protocols. New York, NY: Springer New York.

Spang, A., J.H. Saw, S.L. Jorgensen, K. Zaremba-Niedzwiedzka, J. Martijn, A.E. Lind, et al. 2015. Complex archaea that bridge the gap between prokaryotes and eukaryotes. Nature 521(7551): 173–179. doi: 10.1038/nature14447.

Stepanauskas, R. 2012. Single cell genomics: an individual look at microbes. Curr. Opin. Microbiol. 15(5): 613–620. doi: https://doi.org/10.1016/j.mib.2012.09.001.

Strazzulli, A., S. Fusco, B. Cobucci-Ponzano, M. Moracci and P. Contursi. 2017. Metagenomics of microbial and viral life in terrestrial geothermal environments. Rev. Environ. Sci. Biotechnol. 16(3): 425–454. doi: 10.1007/s11157-017-9435-0.

Sun, S., J. Chen, W. Li, I. Altintas, A. Lin, S. Peltier and J. Wooley, 2011. Community cyberinfrastructure for advanced microbial ecology research and analysis: the CAMERA resource. Nucleic Acids Res, 39(Database issue), D546-551. doi: 10.1093/nar/gkq1102.

Teeling, H. and F. O. Glöckner. 2012. Current opportunities and challenges in microbial metagenome analysis–a bioinformatic perspective. Brief. Bioinformatics 13(6): 728–742. doi: 10.1093/bib/bbs039.

Teske, A. and A.-L. Reysenbach. 2015. Editorial: Hydrothermal microbial ecosystems. Front. Microbiol. 6(884). doi: 10.3389/fmicb.2015.00884.

Tringe, S.G., C. von Mering, A. Kobayashi, A.A. Salamov, K. Chen, H.W. Chang, et al. 2005. Comparative metagenomics of microbial communities. Science 308(5721): 554–557. doi: 10.1126/science.1107851.

Tringe, S.G. and P. Hugenholtz. 2008. A renaissance for the pioneering 16S rRNA gene. Curr. Opin. Microbiol. 11(5): 442–446. doi: 10.1016/j.mib.2008.09.011.

Tveit, A.T., T. Urich, P. Frenzel and M.M. Svenning. 2015. Metabolic and trophic interactions modulate methane production by Arctic peat microbiota in response to warming. Proc. Natl. Acad. Sci. USA. 112(19): E2507–2516. doi: 10.1073/pnas.1420797112.

Tyson, G.W., J. Chapman, P. Hugenholtz, E.E. Allen, R.J. Ram, P.M. Richardson, et al. 2004. Community structure and metabolism through reconstruction of microbial genomes from the environment. Nature 428(6978): 37–43. doi: 10.1038/nature02340.

Vavourakis, C.D., A.-S. Andrei, M. Mehrshad, R. Ghai, D.Y. Sorokin and G. Muyzer. 2018. A metagenomics roadmap to the uncultured genome diversity in hypersaline soda lake sediments. Microbiome 6(1): 168–168. doi: 10.1186/s40168-018-0548-7.

Vera-Gargallo, B. and A. Ventosa. 2018. Metagenomic insights into the phylogenetic and metabolic diversity of the prokaryotic community dwelling in hypersaline soils from the odiel saltmarshes (SW Spain). Genes 9(3): 152. doi: 10.3390/genes9030152.

Vester, J.K., M.A. Glaring and P. Stougaard. 2014. Discovery of novel enzymes with industrial potential from a cold and alkaline environment by a combination of functional metagenomics and culturing. Microb. Cell Fact. 13(1): 72. doi: 10.1186/1475-2859-13-72.

Wang, J., L. Chen, Z. Chen and W. Zhang. 2015. RNA-seq based transcriptomic analysis of single bacterial cells. Int. Biol. 7(11): 1466–1476. doi: 10.1039/c5ib00191a.

Wang, S., H. Dong, W. Hou, H. Jiang, Q. Huang, B.R. Briggs, et al. 2014. Greater temporal changes of sediment microbial community than its waterborne counterpart in Tengchong hot springs, Yunnan Province, China. Scientific Reports 4(1): 7479. doi: 10.1038/srep07479.

Ward, N. 2006. New directions and interactions in metagenomics research. FEMS Microbiology Ecology 55(3): 331–338. doi: 10.1111/j.1574-6941.2005.00055.x.

Wilkins, L.G.E., C.L. Ettinger, G. Jospin and J.A. Eisen. 2019. Metagenome-assembled genomes provide new insight into the microbial diversity of two thermal pools in Kamchatka, Russia. Sci. Rep. 9(1): 3059. doi: 10.1038/s41598-019-39576-6.

Wooley, J.C. and Y. Ye. 2009. Metagenomics: facts and artifacts, and computational challenges. J. Comput. Sci. Technol. 25(1): 71–81. doi: 10.1007/s11390-010-9306-4.

Xu, Y. and F. Zhao. 2018. Single-cell metagenomics: challenges and applications. Protein Cell 9(5): 501–510. doi: 10.1007/s13238-018-0544-5.

Xue, Y., I. Jonassen, L. Ovreas and N. Tas. 2019. Bacterial and archaeal metagenome-assembled genome sequences from svalbard permafrost. Microbiol. Resour. Announc. 8(27): e00516-00519. doi: 10.1128/mra.00516-19.

Yan, Y.-W., B. Zou, T. Zhu, W.N. Hozzein and Z.-X. Quan. 2017. Modified RNA-seq method for microbial community and diversity analysis using rRNA in different types of environmental samples. PloS one 12(10): e0186161–e0186161. doi: 10.1371/journal.pone.0186161.

Yelton, A.P., L.R. Comolli, N.B. Justice, C. Castelle, V.J. Denef, B.C. Thomas, et al. 2013. Comparative genomics in acid mine drainage biofilm communities reveals metabolic and structural differentiation of co-occurring archaea. BMC Genom. 14: 485. doi: 10.1186/1471-2164-14-485.

Yoon, S.H., S.M. Ha, S. Kwon, J. Lim, Y. Kim, H. Seo and J. Chun 2017. Introducing EzBioCloud: a taxonomically united database of 16S rRNA gene sequences and whole-genome assemblies. Int. J. Syst. Evol. Microbiol. 67(5): 1613–1617. doi: 10.1099/ijsem.0.001755.

Zaikova, E., D.S. Goerlitz, S.W. Tighe, N.Y. Wagner, Y. Bai, B.L. Hall, et al. 2019. Antarctic relic microbial mat community revealed by metagenomics and metatranscriptomics. Front. Ecol. Evol. 7(1). doi: 10.3389/fevo.2019.00001.

19

Impact of Microbial Genome Sequencing Advancements in Understanding Extremophiles

Garima Suneja[1, 2] and Rajpal Srivastav[3, 4,*]

Introduction

Extremophiles can thrive at extreme environmental conditions, like temperatures near 80°C, pH near zero, and very extreme salinity, where it is difficult for normal life to exist. The extremophilic organisms can be found in hydrothermal vents, hot springs, volcanic fields, and deserts. These organisms can survive in extreme environmental conditions because of their unique genome, which codes for enzymes called extremozymes. These extremozymes can function at extreme physical conditions of temperature, salt, and pH. Extremophiles exhibit different pathways, like metal detoxification pathways, where transcriptional regulation prevents environmental stress. For example, esterase EstATII, an extremozyme produced by thermophiles, helps them in adapting to the harsh environments (Mohamed et al. 2013). The possible reason behind this adaptation could be unique amino acid arrangements and closely packed hydrophobic core in such enzymes (Elleuche et al. 2014).

Extremophilic organisms are important for industrial applications because of unique proteins, metabolites, and thermozymes, like cellulases, polymerases, lipase, etc., which can tolerate harsh conditions of temperature and pH that is generally observed in industrial synthesis procedures. Thermophilic enzyme Taq DNA polymerase, isolated from *Thermus aquaticus*, is widely used in a polymerase chain reaction in laboratories. This polymerase has contributed majorly to the advancement of genetic engineering and recombinant DNA technology. There are other important enzymes isolated from extremophiles including cellulase (Escuder-Rodríguez et al. 2018), xylanase (Liu et al. 2018), lipase (Samoylova et al. 2018), esterase (Mohamed et al. 2013), nitrilase (Dennett and Blamey 2016), transaminase (Ferrandi et al. 2017), and laccase (Brander et al. 2015). However, it is difficult to study extremophiles under normal laboratory setup because of specific requirements of extreme physical parameters for growth. The use of the metagenomic approach provides an alternative to study these extremophiles. By using metagenomics in the last five years, more than 55% of thermophilic endoglucanases and about 18% of thermophilic cellobiosidases and thermophilic β-glucosidases have been discovered (Escuder-Rodríguez et al. 2018). Next-generation sequencing (NGS) and metagenomics have improved the identification and understanding of extremophilic

---

[1] CSIR-Institute of Genomics and Integrative Biology, New Delhi, India.

[2] Academy of Scientific and Innovative Research, Ghaziabad, Uttar Pradesh, India.

[3] Amity Institute of Biotechnology, Amity University Uttar Pardesh, Noida, India.

[4] Department of Science and Technology, Ministry of Science and Technology, Government of India, New Delhi, India.

*Corresponding author: rsrivastav2@amity.edu

organisms at a rapid pace. In this chapter, the impact of microbial genome sequencing advancements on improved understanding of extremophiles has been discussed.

Uniqueness of Extremophiles in Microbial World

Microbes play a significant role in various ecological processes. Extremophilic microbes have evolved to survive in extreme conditions, like low water, high UV radiation, high salinity, and high temperature. These extremophiles can be broadly classified as thermophiles, halophiles, and acidophiles (Table 1). Thermophiles can live at high temperatures ranging 50°C and 64°C, extreme

TABLE 1: Broad Classification of Extremophilic Organisms

Environment	Category	Conditions	Example
Temperature	Hyperthermophiles	Optimum > 80°C	*Aquifex aeolicus*
	Thermophiles	45°C-65°C	*Aeribacillus pallidus TD1*
	Psychrophiles	–20°C to 10°C	*Pseudomonas antarctica PAMC 27494*
	Psychrotolerant	Optimum 20°C-40°C but can grow < 20°C	*Psychrobacter KHI72YL61*
Salinity	Halophiles	Thrives in high salt concentration	
	Slight Halophile	0.3 to 0.8 M	*Dethiosulfovibrio salsuginis*
	Moderate Halophile	0.8 to 3.4 M	*Haloferax volcanii DS2*
	Extreme Halophile	3.4 to 5.1 M	*Halorhodospira halochloris*
pH	Acidophiles	pH 5 or below	*Acidithiobacillus ferrivorans CF27*
	Alkaliphiles	pH 8.5-11	*Alkaliphilus metalliredigens QYMF*
	Obligate	Requires high pH to survive	*Bacillus krulwichiae*
	Facultative	Survives high pH and pH near neutral	*Bacillus pseudofirmus OF4*
	Haloalkaliphiles	Requires high salt concentration	*Halomonas chromatireducens*
Nutrients	Oligotrophs	Low levels nutrients	*Nitrosomonas sp. Is79*
Pressure	Barophile	Thrives high pressure	*Desulfovibrio profundus*
Metal	Metallotolerant	High concentrations of heavy metals like arsenic, cadmium	*Agromyces aureus AR33T*
Radiations	Radioresistance	Resistant to ionizing radiations	*Deinococcus radiodurans*
Temperature and pH	Thermoacidophile	High temperature, low pH	*Acidianus brierleyi DSM-1651*
Water	Xerophiles	Low availability of water	*Trichosporonoides nigrescens*

thermophiles can live between 65°C and 79°C, and hyperthermophiles can survive over 80°C. Some halophiles, like *Halogeometricum, Haloarchaeobius,* and *Halorientalis,* can live in a wide range of salinity and because of these unique features, they are also used in dye production in industries.

Depending on the optimal growth temperature, these extremophiles can be psychrophilic or psychrotolerants. Psychrophilic organisms can grow at low temperatures, between –20°C and 10°C, and are unable to grow at temperatures higher than 15°C. Unlike psychrophiles, psychrotolerant organisms grow at 20-25°C but also can survive at temperatures below 0°C. Various cold-active enzymes have been isolated from these organisms, which have significant industrial applications. Some cold-active enzymes show unique properties, such as superoxide dismutase, isolated from psychrophile *Deschampsia antarctica*, which has optimum temperature 20°C but it can function at 0°C and –20°C and (Rojas-Contreras et al 2015). Cold-adaptive fungi produce various products, like antibiotics, novel enzymes, and pigments, which have various industrial applications. *Penicillium* sp., isolated from the Indian Himalayan region, can produce an orange color pigment at 15°C and pH 3. *Penicillium* sp. produces carotenoid in response to environmental stress at low temperatures (Pandey et al. 2018).

These extremophilic organisms are also important for bioremediation of metals and nuclides due to their unique metabolisms, like *Proteobacteria* from the dry environment of Atacama Desert. Also other isolates from the same place, *Pseudomonas arsenicoxydans*, show tolerance to arsenic and can oxidize As(III) to As(V) (Campos et al. 2010).

Industrial Applications of Extremophiles

Extremophiles have unique enzymes called extremozymes like cellulases, polymerases, amylases, lipases, which can tolerate extreme conditions. There are various industrial applications of extremozymes in industries, cosmetics, and pharmaceuticals, and few applications have been briefly discussed in the following section.

Various commercially available enzymes are of mesophilic origin i.e., they are isolated from the microorganism that grows at a temperature between 20°C and 45°C. In contrast, various industrial processes use harsh conditions, like high temperature and low pH, where extremozymes can efficiently function in comparison to mesophilic origin enzymes. Thermozymes from thermophiles are thermostable and can also tolerate high pressure and denaturing solvents (Zhang et al. 2016). Thermophilic enzymes include lipases from [*Pyrococcus furiosus* (Alqueres et al. 2011) and *Ureibacillus thermosphaericus* (Samoylova et al. 2018)], nitrilases [*Thermotoga maritime MSB8* (Chen et al. 2015) and *Geobacillus pallidus* (Makhongela et al. 2007)], xylanases [*Dictyoglomus thermophilum* (McCarthy et al. 2000) and *Nonomuraea flexuosa* (Hakulinen et al. 2003)], transaminase [*Thermomicrobium roseum* (Mathew et al. 2016) and *Geobacillus thermodenitrificans* (Chen et al. 2016)], and laccase [*Thermus thermophilus* (Miyazaki, K. 2005) and *Aquifex aeolicus* (Fernandes et al. 2007)].

The lipases from extremophiles are used in detergent formulations (Naganthran et al. 2017, Tang et al. 2017), food industry for taste and texture improvement (Raveendran et al. 2018), and paper and pulp industry for the removal of woods hydrophobic components (Ramnath et al. 2017). Lipases can also be used in resolving racemic alcohols or carboxylic acid through transesterification (Braia et al. 2018). The thermozymes, like xylanases, obtained from *Thermotoga neapolitana* (Zverlov et al. 1996), *Nonomuraea flexuosa* (Hakulinen et al. 2003), and *Dictyoglomus thermophilum* (McCarthy et al. 2000) are used in the for the hydrolysis of lignocellulosic biomass into simpler monosaccharides for bioethanol production (Bibra et al. 2018), production of prebiotic oligosaccharides, and preparation of partially digested raw materials of plants for cattle and poultry feeds, which increase the digestibility and nutritional value (Juturu and Wu 2012, Sari et al. 2018).

Cold-adaptive enzymes, like DNase, ligases, alkaline phosphatase, Uracil-DNA glycosylase, are being used in molecular biology. The enzyme is isolated from *Gadusmorhua*, the optimal temperature is between 20-25°C and a half-life of 2 minutes at 40°C (Lee et al. 2009). The enzyme DNA ligase was isolated from a psychrophile *Pseudoalteromonas haloplanktis* can function at a temperature as low as 4°C (Georlette et al. 2000).

The halophiles are important in industries like chemical, biofuel, medical, pharmaceutical because of their ability to produce halophilic enzymes. The genome analysis of *Halomonas smyrnensis* showed that it can produce levan, exopolysaccharides, and polyhydroxyalkanoates

(Diken et al. 2015). Extremophiles can be used in the microbial electrochemical system. Acidophiles, like *Acidiphilium* spp., are used at the anode, and *Acidithiobacillus ferrooxidans* are used as a biocatalyst at the cathode. *Acidiphilum cryptum* was the first acidophile used to generate electricity by oxidizing glucose and reducing ferric iron at pH ≤ 4 (Borole et al. 2008). *Acidiphilium* spp. can transfer the electron generated from glucose metabolism directly to the anode. The polymeric substance produced by the bacterium forms a multilayered biofilm, which contains iron in c-type cytochromes that may be responsible for the transfer of electrons to the anode (Malki et al. 2008). *Shewanella oneidensis* MR-1 is an alkaliphilic bacterium, which transfers the electrons to the anode extracellularly with the help of riboflavin. The amount of riboflavin produced is directly related to electrical output (Yong et al. 2013). *Geoalkalibacter ferrihydriticus* (Badalamenti et al. 2013), *Corynebacterium humireducens MFC-5* (Wu et al. 2011), and *Pseudomonas alcaliphila* are also used in microbial fuel cells at high pH and can produce phenazine1-carboxylic acid, which acts as a mediator to transfer electron during citrate oxidation (Zhang et al. 2011). Biofilm formed by *Thermincolapotens* helps them to transfer electrons to the anode via multiheme c-type cytochromes (Carlson et al. 2012).

S. *solfataricus* is an acidophile, which can degrade phenol and methylphenols (Christen et al. 2011, Zhou et al. 2016). *Chloroflexus aurantiacus* is a thermophilic anoxygenic phototroph that can be used for bioremediation of arsenic-contaminated environments. Acidophiles are used in the bioremediation of polluted soil and water, and they are also essential in the biomining of metals from low-grade sulfur minerals. For the extraction of metals in the bioleaching process, microbial solubilization of metal in acidic conditions is done by using acidophiles (Seeger et al. 1996, Demergasso et al. 2005, Acosta et al. 2017).

Genome Sequencing Advancements Have Improved Identification of Extremophiles

The conventional approaches, for identification and analyzing extremophiles by culturing and phenotypic characterization, are laborious and time-consuming. The new sequencing methods, like Maxam-Gilbert and Sanger sequencing method, had improved microbial genome analysis. The Maxam-Gilbert method was based on the chemical modification and cleavage at specific positions (Maxam and Gilbert 1977), while Sanger sequencing uses a chain termination principle. Lots of improvements have taken place in the last few decades, like fluorescent labeling, capillary electrophoresis, and automation. The limitations of conventional Sanger sequencing were overcome by more advanced sequencing technologies (Buermans and Dunnen 2014). Now, the Multiple Displacement Amplification (MDA) technology where a single microbial genome is being amplified by a billion-fold makes next-generation sequencing powerful tools in microbial genomics (Lasken 2007).

Next-Generation Sequencing (NGS) methods are comparatively advanced than conventional sequencing as NGS provides high throughput genomic data, including whole-genome sequence at high speed and low cost (Zhang et al. 2011, Grada 2013, Gov and Arga 2014). NGS was introduced in the year 2000 with pyrosequencing (Roche 454). Since then, there have been continuous advancements and improvements in NGS technology. Further, sequencing methods have advanced into second and third-generation levels, including single-molecule real-time sequencing (SMRT) and nanopore sequencing (Figure 1). The second-generation sequencing provides about 100-fold high throughput compared to conventional Sanger sequencing. There are various approaches and platforms, like 454, Illumina (Hiseq2000 and MiSeq), Ion Torrent, ABI, etc., associated with next-generation technologies, which are generally user-specific but should be used depending upon experimental and analysis specific requirements as these platforms have different output, read-length, and coverage.

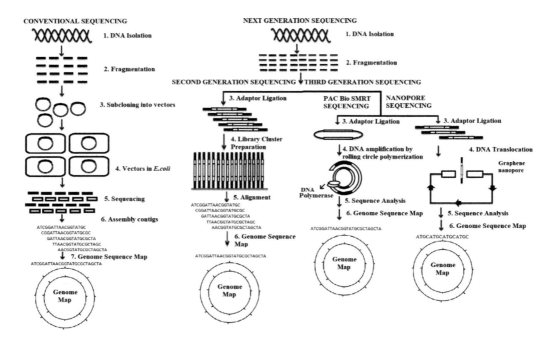

Figure 1. Comparative analysis of workflow of conventional and advanced next-generation microbial genome sequencing

The third-generation sequencing methods like SMRT (DNA molecule is detected per reaction in real-time) and nanopore sequencing (translocation of DNA molecule is detected as a measure of DNA synthesis) gives single-molecule data within a short frame of time. The third-generation platforms like PacBio and MiniION can provide read-length of several thousand base pairs and these are phenomenal in DNA, RNA sequencing, *de novo* assembly of repeated sequence, and complex regions (Ashkenasy et al. 2005, Lundquist et al. 2008, Buermans and Dunnen 2014). Because of these advancements, many extremophilic organisms have been sequenced since the first draft of the genome, and there is a steep increase in the numbers of the draft genomes. The advancement of next-generation sequencing has led to the advancement in the understanding of extremophiles with more insights about their genome and the survival strategies.

The genome analysis of thermophiles *Aquifex aeolicus* and *Thermotoga maritima* has revealed that about one-fourth of their genes were similar to archaea than bacteria. It is speculated that the high rate of evolution of mesophilic bacteria could be the possible reason behind this. It is a known fact that microorganisms cultured under laboratory conditions represent only a minor fraction of the microbial diversity and rest microbes are unculturable (Handelsman 2004). Culture-independent methods and metagenomics have helped in understanding these unculturable microorganisms, including extremophiles. The metagenomics and next-generation sequencing approaches have improved identification and increased the rate of enzyme discovery followed by improved knowledge about adaptations of extremophilic organisms.

The genome of the *Halomonas smyrnensis* AAD6T, a halophilic bacterium was analyzed by next-generation sequencing hybrid approach with platforms Roche 454 GS FLX+ System and Ion Torrent Sequencer. The bacteria have about 2,400 protein-coding genes and the major coding sequences were clustered into 449 subsystems with unique 1,481 coding sequences. The genome of *H. smyrnensis* contains coding sequences responsible for osmotic stress tolerance. These osmotic stress genes are related to choline and glycine betaine uptake and biosynthesis (Diken et al. 2015). The genome analysis showed that the GC content of *H. smyrnensis* AAD6T is relatively higher than *H. elongate* and *C. salexigens* (Copeland et al. 2011, Schwibbert et al. 2011). A phylogenetic study based on whole-genome data shows that *H. smyrnensis*is more closely related to *H. elongate* and *C. salexigens* (Diken et al. 2015). The whole-genome analysis of *H. smyrnensis* suggested that osmotic stress tolerance is due to the presence of genes, which encodes the protein for the influx of betaine and choline. The transport system in halophiles plays an important role in adaptation to the saline environment. There is about 300 ORFs present that allows the intake of important nutrients and excretion of by-products. It also helps to maintain the cytoplasmic proton content and favorable salt concentration for growth and development of bacteria (Kunst et al. 1997, Copeland et al. 2011, Schwibbert et al. 2011, Pastor et al. 2013). Most of the transporter genes in *H. smyrnensis* were categorized into Tripartite ATP-independent periplasmic transporters (Blattner et al. 1997). The growth of the bacterium is related to the presence of Na^+ and Cl^- ions (Tommonaro et al. 2009).

The completed genome sequence analysis of *Methanococcoides burtonii* DSM 6242 revealed that 25 Mb long single circular chromosome harbors genes for signal transduction histidine kinases, RecA-family ATPases and CheY-like response regulators, and numerous transposases (Allen et al. 2009). Genes responsible for cold adaptation were analyzed out by comparing *M. burtonii* genes to archaeal thermophiles. Various clusters of the genes were observed including methyltransferase, restriction endonuclease, specificity genes, hypothetical proteins, divergent AAA domain protein with multiple functions, and variant helicase (Allen et al. 2009). A large amount of RM systems indicate the presence of large amounts of foreign DNA. There are clusters of genes, which are novel ABC transporters. These proteins each possess a signal peptide motif and one transmembrane domain, like ABC-transporter substrate-binding proteins. One of the major differences between the genomes of *M. burtonii* and the *Methanosarcina* is the lack of peptide transporters, which is consistent with the inability of *M. burtonii* to use peptides for growth. *M. burtonii* has a coenzyme F420-dependent sulfite reductase, which allows it to tolerate sulfite in the environment.

The chemolithorophic microorganisms, like *Acidithiobacillus ferrooxidans*, *Sulfobacillus acidophilus*, and *Acidimicrobium ferrooxidans*, which can tolerate low pH, are being used to extract copper from the low-grade ore in a process called as bioleaching that is a cost-effective method in comparison to traditional pyrometallurgy method. These extremophiles can be easily analyzed using the culture-independent approaches. The whole-genome analysis of isolates of *Acidithiobacillus ferrooxidans* from a mining area provided deeper insights about the genetic make-up of chemolithoautotrophs. The genomic analysis showed that genes for electron transport, carbon metabolism, energy control, extracellular polysaccharides, biofilm formation, and detoxification processes make these organisms to survive in extreme environmental conditions (Luo et al. 2009).

There are numerous examples highlighting the impact of microbial genome sequencing on improved understanding and analysis of extremophiles. The recent sequencing advancements led to an increase in the number of complete extremophiles genome sequences (Figure 2). These complete extremophiles genome sequences would provide more insights into many unsolved queries about the molecular adaptation and evolution of extremophilic organisms.

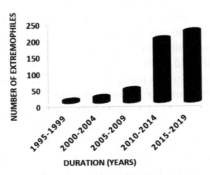

Figure 2. The representation of the trend of extremophiles genome sequenced, using different sequencing technologies over the years (Source: NCBI).

Microbial Genome Sequencing Unraveled Novel Survival Strategies

The advanced next-generation sequencing technologies have led to an increase in the number of draft and complete genome sequences of extremophiles. These genome sequences provide molecular insights about the novel survival, adaptation strategies, and evolution of extremophiles. The complete genome sequence of an alkaliphilic halotolerant *Oceanobacillus iheyensis* HTE831, isolated from deep-sea sediment, was analyzed using whole-genome sequencing by Megabase1000 DNA sequencer (Lu et al. 2001, Takami 2002). The genome of this bacterium is 3.6 Mb and encodes proteins essential for maintaining intracellular pH and osmotic balance (Takami 2002). Genome analysis of *O. iheyensis* showed that about 80% of the genes start with ATG, which is similar to *B. subtilis* and *B. halodurans* (Takami 2002). Sixty proteins are ABC transporters-related protein, which helps in the influx of oligopeptides during growth under alkaline conditions (Takami 2002). *O. iheyensis* have genes to encode for DctA family C4-dicarboxylate carriers and Na$^+$/solute symporter and a Na$^+$-dependent transporter (Janausch et al. 2002). This genome sequence analysis allows us to understand the alkaliphilic adaptation mechanisms and stress responses.

Natranaerobius thermophiles is an anaerobic halophilic alkalithermophiles (Mesbah et al. 2009), and it can tolerate multiple extreme environmental conditions. The genome of *N. thermophiles* was sequenced by using random shotgun sequencing and found to have 3.1 Mb chromosomal genome and two plasmids, including pNTHE02 (Zhao et al. 2011). It has genes for *de novo* synthesis of glycine betaine, L-glutamine synthetase, L-isoaspartate methyltransferase, spermine synthase, 15 genes encoding glycine betaine ABC transporters, four genes encoding for glycine betaine/L-proline ABC transporters, few genes forcarnitine/choline transporters, Na$^+$/proline symporters, Na$^+$/glutamate symporters, and K$^+$ transporters. These symporters and specific K$^+$/H$^+$ antiporter together may help in the regulation of intracellular K$^+$ concentration. For stabilization of DNA and RNA structure at high temperatures, there are genes encoding for rRNAMTases, tRNAMTases, and heat shock proteins in *N. thermophiles* (Zhao et al. 2011). Thus, the genome analysis of *N. thermophiles* showed that multiple genes encode multiple proteins responsible for the adaptation in multiple extreme environments, like high salinity, alkaline pH, and increased temperature.

The biofilm is primarily composed of secreted polysaccharides and biofilm formation is the best adaptive measure to stresses tolerance by extremophiles. The genome analysis of an extremophile *Halomonas smyrnensis* shows that SacB gene encodes for the levansucrase, a 416 aa long protein, which catalyzes the synthesis of levan from sucrose substrate through a process called transfructosylation (Donot et al. 2012). There is a polysaccharide gene cluster, Pel operon, comprising PelA, PelB, PelC, PelD, PelE, PelF, and PelG genes. These genes may be responsible for polysaccharide biosynthesis and PelB shows high similarity with *Pseudomonas mendocina* (99%) and *Pseudomonas aeruginosa* (96%) PelB protein, which is associated with biofilm formation (Diken et al. 2015).

Psychrophilic methanogenic archaeon *Methanococcoides burtonii* can adapt to the cold environment due genomic ability to accommodate altered amino acid content. They harbor various transposases, which play an important role in the evolution of this psychrophile. *M. burtonii* genome has polysaccharide biosynthesis genes, cell wall, membrane, and other signal transduction genes, which have may be responsible for its evolution by nucleotide alteration, gene transfer, and transposition and adaptation to cold environmental changes (Allen et al. 2009).

The extent of pH homeostasis permits the upper limit of growth. Na$^+$ ion cycle plays an important role in pH homeostasis. Na$^+$/H$^+$ antiporter activity maintains low intracellular sodium concentrations in halophilic aerobic bacteria (Nieto et al. 1998). Mrp-encoded antiporter found in *B. halodurans* is an important antiporter for pH homeostasis. A putative protein, OB2392 of *O. iheyensis,* is similar to a sodium channel, which may supply Na$^+$ for the sodium cycle in extreme pH (Chahine et al. 2004). The motility of alkaliphiles is coupled with Na$^+$ and can be found only in cells growing in high pH ranges. *O. iheyensis* flagella proteins facilitate Na$^+$ re-entry at high pH leading to increased antiport and H$^+$ accumulation.

Archaeal extremophiles, like thermophile, psychrophile, and piezophile, can survive the harsh environment due to protein adaptations. Extremophilic proteins and enzymes have various adaptations due to which they maintain their structure and hence can function in extreme environmental conditions. These adaptations include hydrophobic core, disulfide bonds, and ionic interactions, which prevent changes in quaternary structure (Reed et al. 2013). In *Pyrobaculum aerophilum,* citrate synthetase forms a new disulfide bond to make cyclic protein chain and interlinks the monomeric subunits (Boutz et al. 2007). Thermal stability can also be increased by using favorable ionic interactions, which helps in making salt bridges (Reed et al. 2013). The genome size of the *Pyrobaculum* aerophilumis 2,222,430 bp, sequenced using the Applied Biosystem platform (Gibbon et al. 2001). *Thermococcus celer* genome was sequenced by whole-genome sequencing with a combination of Ion Torrent PGM and Illumina Hiseq (Oger 2018). L30e is a ribosomal protein isolated from *T. celer* causing a change in the thermal capacity without any structural change (Chan et al. 2011).

Pyrococcus horikoshii is a hyperthermophile, whose genome is sequenced by the dye terminator method (Kawarabayasi et al. 1998). It was found that *P. abyssi* has a higher number of small amino acids in its proteome, which leads to a decrease in hydrophobic residues like tryptophan and tyrosine in its core. This tighter packing of hydrophobic core provides stability of the protein (Giulio 2005). The presence of multimeric protein like TET3 peptidase (TET3) in *P. horikoshii* may help to tolerate pressure because it forms discreet dodecamer structure rather than barrel shape, which may avoid water penetration in the hydrophobic core of protein (Boonyaratanakornkit et al. 2002, Rosenbaum et al. 2012).

Psychrophiles have the ability to grow at temperatures below 20°C (Tutino et al. 2009). Their proteins can change the conformation because of flexibility at low temperatures (Smalas et al. 2000). Lower protein barriers between the protein conformations may increase flexibility, and this is due to the variations in the amino acid composition related to mesophilic protein. Adaptations seen in these organisms are high glycine residues, which provide conformational mobility to protein, fewer proline residues in the loop region, which provide rigidity, salt bridge, and low arginine residues, which are used in hydrogen bonding, weaker hydrophobic interactions because of the reduced size of non-polar residues in protein core (Feller 2010).

In *Halorubum lacusprofundi,* there is a decrease in the large hydrophobic amino acids like tryptophan and residues for hydrogen bond formation. In β-galactosidase of *H. lacusprofundi*, there is an increase in hydrophobicity of the protein surface, which ultimately replaces the anionic electrostatic interactions, found on the halophilic protein (Dassarma et al. 2013, Karan et al. 2013).

The genome of two psychrophilic methanogens, *Methanogen iumfrigidum* and *Methanococcoides burtonii*, was compared to the other mesophilic or thermophilic methanogens. It was found that cold-adaptive proteins have a low number of charged residues on the surface and with a reduction in the charge of the protein surface, there is an increase in the hydrophobicity of the surface without any aggregation at low temperature (Saunders et al. 2003).

Jeotgalibacillus malaysiensis is a halophilic bacterium, which can tolerate high salt concentration. The reason for these adaptations could be understood by genome analysis using PacBio RSII sequencing system (Goh et al 2015). In saline conditions, *J. malaysiensis* shows a modification in the cell wall, cell membrane, and channel proteins. As cell walls are in direct contact with the environment they act as a first-line defense to osmotic stress. Under stress, channel proteins are highly expressed, which helps the cells to sense and respond to osmotic shock (Yaakop et al. 2016). It was observed that at 20% NaCl concentration, there was an increase in the sporulation, extracellular polysaccharide production, and capsulation which means *J. malaysiensis* generate biofilm in response to the stressful conditions (Yaakop et al. 2016). Proline is an essential osmoprotectant secreted by the bacteria *J. malaysiensis*. Under stressful conditions, these accumulated proteins get broken down into peptides and amino acids (mainly proline and glutamate), which are further used as osmolytes (Yaakop et al. 2016).

Salinibacter ruber maintains the osmotic pressure by accumulating the high concentration of potassium ions (K^+) in the cytoplasm (Sleator and Hill 2001; Oren et al. 2002). *Halobacterium salinarum* tolerates osmotic stress by accumulating more K^+ and exporting Na^+ through Na^+ / K^+ ATPase pump. K^+ ion act as an osmotic solute and activate the intercellular enzymes and helps to accumulate more solutes. Some methanogenic archaea, halophilic algae, halophilic, or halotolerant bacteria (Ma et al. 2010) and some moderate halophiles like *Chromohalobacter salexigens* (Pastor et al. 2013) maintain the osmotic balance by using osmolytes, like sugars, pyrols (glycerol, glucosylglycerol), amino acid (glutamine, proline), quaternary amines (betaines and choline), and ecotines without altering any cellular metabolic pathway (Sleator and Hill 2001, Roberts 2005, Pastor et al. 2013). The secretion of EPS and formations of biofilm to resist desiccation, heavy metal toxicity, and cold stress is another survival and adaptation strategy to tolerate high salt environments.

The codes of the molecular basis of adaptations of extremophiles have been deciphered because of the availability of genome sequences using next-generation sequencing technologies. The applications of next-generation sequencing methods have made the analysis of extremophilic organisms comparatively faster. The novel habitats of extremophiles can be explored and analyzed using metagenomics and next-generation sequencing technologies. The advanced sequencing technologies have illuminated the dark matter of knowledge and widen the horizon of knowledge about extremophilic organisms, which are a very important component of various ecosystems on our earth.

Conclusions

Extremophilic organisms have evolved to survive in extreme environmental conditions, like low water, high UV radiation, high salinity, and high temperature. These organisms can inhabit almost every type of environment, where the existence of normal life is impossible. The next-generation sequencing technologies provide more insights about the genes, genome, adaptations, and the survival strategies of extremophiles. It is quite evident that the advancements of sequencing technologies have led to the advancements in the understanding of extremophiles. Next-generation sequencing provides high throughput genomic data in short duration, which is very useful in genome analysis of extremophiles. The whole-genome sequencing analysis provides the opportunity to do comparative genome analysis of extremophiles.

Acknowledgments

GS acknowledges Council of Scientific and Industrial Research (CSIR), India for fellowship. RS is thankful to Dr. Rakesh Sharma, Principal Scientist, CSIR-IGIB, New Delhi in supporting the concept and work of this article. RS acknowledges Department of Science and Technology (DST), Ministry of Science and Technology, Government of India, New Delhi, India for fellowship and research grant INSPIRE-faculty (IFA16LSBM184).

References

Acosta M., P.A. Galleguillos, M. Guajardo and C. Demergasso. 2017. Microbial survey on industrial bioleaching heap by high-throughput 16S sequencing and metagenomics analysis. Solid State Phenom. 262: 219–223.
Allen M.A., F.M. Lauro, T.J. Williams, D. Burg, K.S. Siddiqui, D. De Francisci, et al. 2009. The genome sequence of the psychrophilic archaeon *Methanococcoides burtonii*: the role of genome evolution in cold adaptation. ISME J. 3(9): 1012–1035.
Alqueres S.M.C., R.V. Branco, D.M.G. Freire, T.L.M. Alves, O.B. Martins, R.V. Almeida. et al. 2011. Characterization of the recombinant thermostable lipase (Pf2001) from Pyrococcus furiosus: Effects of thioredoxin fusion tag and triton X-100.2011. Enzyme Res. 316939.
Ashkenasy, Quesada J.S., H. Bayley and M.R. Ghadiri. 2005. Recognizing a single base in an individual DNA strand: a step toward DNA sequencing in nanopores. Angew Chemie Int Ed. 44(9): 1401–1404.

Badalamenti J.P., R. Krajmalnik-brown and CI. Torres. 2013. Generation of high current densities by pure cultures of anode respiring *Geoalkalibacter spp.* under alkaline and saline conditions in microbial electrochemical cells. mBio 4(3): e00144–13.

Bibra M., V.R. Kunreddy and R.K. Sani. 2018. Thermostable xylanase production by *Geobacillus* sp. strain DUSELR13, and its application in ethanol production with lignocellulosic biomass. Microorganisms 6(3): 93.

Blattner F.R., G. 3rd Plunkett, C.A. Bloch, N.T. Perna, V. Burland, M. Riley, et al. 1997. The complete genome sequence of *Escherichia coli* K-12. Science (80). 277(5331): 1453–1462.

Boonyaratanakornkit B.B., C.B. Park and D.S. Clark. 2002. Pressure effects on intra- and intermolecular interactions within proteins. Biochimica et Biophysica Acta 1595(1–2): 235–249.

Borole A.P., H. O'Neill, C. Tsouris and S. Cesar. 2008. A microbial fuel cell operating at low pH using the acidophile *Acidiphilium cryptum*. Biotechnol. Lett. 30(8): 1367–1372.

Boutz D.R., D. Cascio, J. Whitelegge, L.J. Perry and T.O. Yeates. 2007. Discovery of a thermophilic protein complex stabilized by topologically interlinked chains. J. Mol. Biol. 368(5): 1332–1344.

Braia N., M. Merabet and L. Aribi-Zouioueche. 2018. Efficient access to both enantiomers of 3-(1-hydroxyethyl) phenol by regioselective and enantioselective catalyzed hydrolysis of diacetate in organic media by sodium carbonate. Chirality 30(12): 1312–1320.

Brander S., J.D. Mikkelsen and K.P. Kepp. 2015. TtMCO: A highly thermostable laccase-like multicopper oxidase from the thermophilic *Thermobaculum terrenum*. J. Mol. Catal. B. Enzym. 112: 59–65.

Buermans H.P.J. and J.T. Dunnen. 2014. Next generation sequencing technology: advances and applications. Biochimica. Biophysica. Acta 1842(10): 1932–1941.

Campos V.L., C. Valenzuela, P. Yarza, P. Kämpfer, R. Vidal, C. Zaror, et al. 2010. *Pseudomonas arsenicoxydans* sp, an arsenite-oxidizing strain isolated from the Atacama Desert. Syst. Appl. Microbiol. 33: 193–197.

Carlson H.K., A.T. Iavarone, A. Gorur, B.S. Yeo, R. Tran, R.A. Melnyk, et al. 2012. Surface multiheme c-type cytochromes from *Thermincola potens* and implications for respiratory metal reduction by Gram-positive bacteria. Proc. Natl. Acad. Sci. USA. 109(5): 1702–1707.

Chahine M., S. Pilote, V. Pouliot, H. Takami and C. Sato. 2004. Role of arginine residues on the S_4 segment of the *Bacillus halodurans* Na$^+$ channel in voltage-sensing. J. Membr. Biol. 201: 9–24.

Chan C., T. Yu and K. Wong. 2011. Stabilizing salt-bridge enhances protein thermostability by reducing the heat capacity change of unfolding. PLoS One 6(6): e21624.

Chen Y., D. Yi, S. Jiang and D. Wei. 2016. Identification of novel thermostable taurine-pyruvate transaminase from *Geobacillus thermodenitrificans* for chiral amine synthesis. Appl. Microbiol. Biotechnol. 100: 3101–3111.

Chen Z., H. Chen, Z. Ni, R. Tian, T. Zhang, J. Jia and S. Yang. 2015. Expression and characterization of a novel nitrilase from hyperthermophilic bacterium *thermotoga maritima* MSB8. J. Microbiol. Biotechnol. 25: 1660–1669.

Christen P., S. Davidson, Y. Combet-Blanc and R. Auria. 2011. Phenol biodegradation by the thermoacidophilic archaeon *Sulfolobus solfataricus* 98/2 in a fed-batch bioreactor. Biodegradation 22: 475–484.

Copeland A., K. O'Connor, S. Lucas, A. Lapidus, K.W. Berry, J.C. Detter, et al. 2011. Complete genome sequence of the halophilic and highly halotolerant *Chromohalobacter salexigens* type strain (1H11T). Stand. Genomic. Sci. 5: 379–388.

Dassarma S., M.D. Capes, R. Karan and P. Dassarma. 2013. Amino acid substitutions in cold-adapted proteins from *Halorubrum lacusprofundi*, an extremely halophilic microbe from antarctica. PLoS One 8(3): e58587.

Demergasso C.S., P.A. Galleguillos, G.L.V. Escudero, A.V.J. Zepeda, D. Castillo, E.O. Casamayor, et al. 2005. Molecular characterization of microbial populations in a low-grade copper ore bioleaching test heap. Hydrometallurgy 80: 241–253.

Dennett G.V. and J.M. Blamey. 2016. A new thermophilic nitrilase from an antarctic hyperthermophilic microorganism. Front. Bioeng. Biotechnol. 4: 5.

Diken E., T. Ozer, M. Arikan, Z. Emrence, E.T. Oner, D. Ustek and K.Y. Arga. 2015. Genomic analysis reveals the biotechnological and industrial potential of levan producing halophilic extremophile, *Halomonas smyrnensis* AAD6T. Springerplus 4: 4: 393.

Donot F., A. Fontana, J.C. Baccou and S. Schorr-galindo. 2012. Microbial exopolysaccharides: Main examples of synthesis, excretion, genetics and extraction. Carbohydr. Polym. 87: 951–962.

Díaz-Cardenas C., G. López, J.D. Alzate-Ocampo, L.N. Gonzalez, N. Shapiro, T. Woyke, et al. 2017. Draft genome sequence of *Dethiosulfovibrio salsuginis* DSM 21565T an anaerobic, slightly halophilic bacterium isolated from a Colombian saline spring. Stand. Genomic. Sci. 12: 1–9.

Elleuche S., C. Schröder, K. Sahm and G. Antranikian. 2014. Extremozymes biocatalysts with unique properties from extremophilic microorganisms. Curr. Opin. Biotechnol. 29: 116–123.

Escuder-Rodriguez J.J., M.E. DeCastro, M.E. Cerdan, E. Rodriguez-Belmonte, M. Becerra, M.I. Gonzalez-Siso, et al. 2018. Cellulases from thermophiles found by metagenomics. Microorganisms 6(3): 66.

Feller G. 2010. Protein stability and enzyme activity at extreme biological temperatures. J. Phys. Condens. Matter. 22(32): 323101.

Fernandes A.T., C.M. Soares, M.M. Pereira, R. Huber, G. Grass and L.O. Martins. 2007. A robust metallo-oxidase from the hyperthermophilic bacterium *Aquifex aeolicus*. FEBS J. 274: 2683–2694.

Ferrandi E.E., A. Previdi, I. Bassanini, S. Riva, X. Peng and D. Monti. 2017. Novel thermostable amine transferases from hot spring metagenomes. Appl. Microbiol. Biotechnol. 101: 4963–4979.

Georlette D., Z.O. Jonsson, F. Van Petegem, J.P. Chessa, J. Van Beeumen, U. Hubscher, et al. 2000. A DNA ligase from the psychrophile *Pseudoalteromonas haloplanktis* gives insights into the adaptation of proteins to low temperatures. Eur. J. Biochem. 267: 3502–3512.

Gibbon S.T.F., H. Ladner, U.J. Kim, K.O. Stetter, M.I. Simon, J.H. Miller, et al. 2001. Genome sequence of the hyperthermophilic crenarchaeon *Pyrobaculum aerophilum*. PNAS. 99. 984–989.

Giulio M. Di. 2005. A comparison of proteins from *Pyrococcus furiosus* and *Pyrococcus abyssi*: barophily in the physicochemical properties of amino acids and in the genetic code. Gene. 346: 1–6.

Goh K.M., K.G. Chan, A.S. Yaakop and R. Ee. 2015. Complete genome of *Jeotgalibacillus malaysiensis* D5T consisting of a chromosome and a circular megaplasmid. J. Biotechnol. 204: 13–14.

Gov E., and K.Y. Arga. 2014. Systems biology solutions to challenges in marine biotechnology. Front. Mar. Sci. 1: 1–5.

Grada A. 2013. Next-generation sequencing: methodology and application. J. Invest. Dermatol. 133 (8): e11.

Hakulinen N., O. Turunen, J. Janis, M. Leisola and J. Rouvinen. 2003. Three-dimensional structures of thermophilic β-1,4-xylanases from *Chaetomium thermophilum* and *Nonomuraea flexuosa*: Comparison of twelve xylanases in relation to their thermal stability. Eur. J. Biochem. 270: 1399–1412.

Handelsman J. 2004. Metagenomics: application of genomics to uncultured microorganisms. Microbiol. Mol. Biol. Rev. 68(4): 669–685.

Hua X. and Y. Hua. 2016. Improved complete genome sequence of the extremely radioresistant bacterium *Deinococcus radiodurans* R1 obtained using PacBio single-molecule sequencing. Genome Announc. 4: 2015–2016.

Janausch I.G., E. Zientz, Q.H. Tran, A. Kroger and G. Unden. 2002. C4-dicarboxylate carriers and sensors in bacteria. Biochim. Biophys. Acta 1553(1–2): 39–56.

Juturu V. and J.C. Wu. 2012. Microbial xylanases: engineering, production and industrial applications. Biotechnol. Adv. 30(6): 1219–1227.

Karan R., M.D. Capes, P. Dassarma and S. Dassarma. 2013. Cloning, overexpression, purification, and characterization of a polyextremophilic β-galactosidase from the Antarctic haloarchaeon *Halorubrum lacusprofundi*. BMC Biotechnol. 3: 3.

Kawarabayasi Y., M. Sawada, H. Horikawa, Y. Haikawa, Y. Hino, S. Hamamoto, et al. 1998. Complete sequence and gene organization of the genome of a hyper-thermophilic archaebacterium, *Pyrococcus horikoshii* OT3. Dna Research 5: 55–76.

Kunst F., N. Ogasawara, I. Moszer, A.M. Albertini, G. Alloni, V. Azevedo, et al. 1997. The complete genome sequence of the gram-positive bacterium *Bacillus subtilis*. Nature 390(6657): 249–256.

Lasken, R.S. 2007. Single-cell genomic sequencing using multiple displacement amplification. Curr. Opin. Microbiol. 10(5): 510–516.

Lee M.S., G.A. Kim, M.S. Seo, J.H. Lee and S.T. Kwon. 2009. Characterization of heat-labile uracil-DNA glycosylase from *Psychrobacter* sp. HJ147 and its application to the polymerase chain reaction. Biotechnol. Appl. Biochem. 52(2): 167–175.

Liu, X., T. Liu, Y. Zhang, et al. 2018. Structural insights into the thermophilic adaption mechanism of endo-1,4-β-Xylanase from *Caldicellulosiruptor owensensis*. J. Agric. Food Chem. 66: 187–193.

Lu J., Y. Nogi and H. Takami. 2001. *Oceanobacillus iheyensis* gen. a deep-sea extremely halotolerant and alkaliphilic species isolated from a depth of 1050m on the Iheya Ridge. FEMS. Microbiol. Lett. 205: 291–297.

Lundquist P.M., C.F. Zhong, P. Zhao, A.B. Tomaney, P.S. Peluso, J. Dixon, et al. 2008. Parallel confocal detection of single molecules in real time. Opt. Lett. 33(9): 1026–1028.

Luo H., L. Shen, H. Yin, Q. Li, Q. Chen, Y. Luo, et al. 2009. Comparative genomic analysis of *Acidithiobacillus ferrooxidans* strains using the *A. ferrooxidans* ATCC 23270 whole-genome oligonucleotide microarray. Can. J. Microbiol. 55(5): 587–598.

Makhongela H.S., A.E. Glowacka, V.B. Agarkar, B.T. Sewell, B. Weber, R.A. Cameron, et al. 2007. A novel thermostable nitrilase superfamily amidase from *Geobacillus pallidus* showing acyl transfer activity. Appl. Microbiol. Biotechnol. 75: 801–811.

Malki M., De A.L. Lacey, N. Rodríguez, R. Amils, V.M. Fernandez, M. Malki, et al. 2008. Preferential use of an anode as an electron acceptor by an acidophilic bacterium in the presence of oxygen. Appl. Environ. Microbiol. 74(14): 4472–4476.

Mathew S., K. Deepankumar, G. Shin, E.Y. Hong, B.G. Kim, T. Chung, et al. 2016. Identification of novel thermostable transaminase and its application for enzymatic synthesis of chiral amines at high temperature. RSC Adv. 6: 257–260.

Maxam A.M. and W. Gilbert. 1977. A new method for sequencing DNA. Proc. Acad. Sci. USA. 74(2): 560–564.

Ma Y., E.A. Galinski, W.D. Grant, A. Oren and A. Ventosa. 2010. Halophiles: Life in saline environments. Appl. Environ. Microbiol. 76: 6971–6981.

McCarthy A.A., D.D. Morris, P.L. Bergquist and E.N. Baker. 2000. Structure of XynB, a highly thermostable β-1,4-xylanase from *Dictyoglomus thermophilum* Rt46B.1, at 1.8 Å resolution. Acta Crystallogr. 56: 1367–1375.

Mesbah N.M., G.M. Cook and J. Wiegel. 2009. The halophilic alkalithermophile *Natranaerobius thermophilus* adapts to multiple environmental extremes using a large repertoire of Na^+ $(K^+)/H^+$ antiporters. Mol. Microbiol. 74(2): 270–281.

Miyazaki K. 2005. A hyperthermophilic laccase from *Thermus thermophilus* HB27. Extremophiles 9: 415–425.

Mohamed Y.M., M.A. Ghazy, A. Sayed, A. Ouf and H. El-dorry. 2013. Isolation and characterization of a heavy metal-resistant, thermophilic esterase from a red sea brine pool. Sci Rep. 3: 3358.

Montealegre R. 1993. The Sociedad Minera Pudahuel bacterial thin-layer leaching process at Lo Aguirre. FEMS Microbiol. Rev. 11: 231–235.

Naganthran A., M. Masomian, R.N. Rahman, M. Ali, H. Nooh, A. Naganthran, et al. 2017. Improving the efficiency of new automatic dishwashing detergent formulation by addition of thermostable lipase, protease amylase. Molecules 22(9): 1577.

Nieto N.J., A. Ventosa and A. Oren. 1998. Biology of moderately halophilic aerobic bacteria. Microbiol. Mol. Biol. Rev. 62(2): 504–544.

Oger P.M. 2018. Complete genome sequences of 11 type species from the *Thermococcus* genus of hyperthermophilic and piezophilic archaea. Genome Announc. 6: e00037–18.

Oren A., M. Heldal, S. Norland and E. Galinski. 2002. Intracellular ion and organic solute concentrations of the extremely halophilic bacterium *Salinibacter ruber*. Extremophiles 6: 491–498.

Pandey N., R. Jain, A. Pandey and S. Tamta. 2018. Optimisation and characterisation of the orange pigment produced by a cold adapted strain of *Penicillium* sp. (GBPI_P155) isolated from mountain ecosystem. Mycology 9(2): 81–92.

Pastor J.M., V. Bernal, M. Salvador, M. Argandoña, C. Vargas, L. Csonka, et al. 2013. Role of central metabolism in the osmoadaptation of the halophilic bacterium *Chromohalobacter salexigens*. J. Biol. Chem. 288: 17769–17781.

Ramnath, L., B. Sithole and R. Govinden. 2017. Classification of lipolytic enzymes and their biotechnological applications in the pulping industry. Can. J. Microbiol. 63: 179–192.

Raveendran S., B. Parameswaran, S.B. Ummalyma, A. Abraham, A.K. Mathew, A. Madhavan, et al. 2018. Applications of microbial enzymes in food industry. Food Technol. Biotechnol. 56(1): 16–30.

Reed C.J., H. Lewis, E. Trejo, Winston V. and C. Evilia. 2013. Protein adaptations in archaeal extremophiles. Archaea 2013: 1–14.

Roberts M.F. 2005. Organic compatible solutes of halotolerant and halophilic microorganisms Mary. Saline Systems 30: 1–30.

Rojas-Contreras, J., A.P. de la Rosa and A. De Leon-Rodriguez. 2015. Expression and characterization of a recombinant psychrophilic Cu/Zn superoxide dismutase from *Deschampsia antarctica* E. Desv. [Poaceae]. Appl. Biochem. Biotechnol. 175: 3287–3296.

Rosenbaum E., F. Gabel, M.A. Durá, S. Finet, C. Cléry-barraud, P. Masson, et al. 2012. Effects of hydrostatic pressure on the quaternary structure and enzymatic activity of a large peptidase complex from *Pyrococcus horikoshii*. Arch Biochem Biophys. 517: 104–110.

Samoylova YV., K.N. Sorokina, M.V. Romanenko and V.N. Parmon. 2018. Cloning, expression and characterization of the esterase estUT1 from *Ureibacillus thermosphaericus* which belongs to a new lipase family XVIII. Extremophiles 22(2): 271–285.

Sari B., O. Faiz, B. Genc, M. Sisecioglu, A. Adiguzel and G. Adiguzel. 2018. New xylanolytic enzyme from *Geobacillus galactosidasius* BS61 from a geothermal resource in Turkey. Int. J. Biol. Macromol. 119: 1017–1026.

Saunders N.F., T. Thomas, P.M. Curmi, J.S. Mattick, E. Kuczek, R. Slade, et al. 2003. Mechanisms of thermal adaptation revealed from genomes of the antarctic archaea *Methanogenium frigidum* and *Methanacoccoides burtonii*. Genome Res. 13(7): 1580–1588.

Schwibbert K., A. Marin-Sanguino, I. Bagyan, G. Heidrich, G. Lentzen, H. Seitz, et al. 2011. A blueprint of ectoine metabolism from the genome of the industrial producer *Halomonas elongata* DSM 2581T. Environ. Microbiol. 13(8): 1973–1994.

Seeger M., G. Osorio and C.A. Jerez. 1996. Phosphorylation of GroEL, DnaK and other proteins from *Thiobacillus ferrooxidans* grown under different conditions. FEMS Microbiol. Lett. 138: 129–134.

Sleator R.D. and C. Hill. 2001. Bacterial osmoadaptation the role of osmolytes in bacterial stress and virulence. FEMS Microbiol Rev. 26(1): 49–71.

Smalas A.O., H.K., Leiros, V. Os and N.P. Willassen. 2000. Cold adapted enzymes. Biotechnology Annual Rev. 6: 1–57.

Takami H. 2002. Genome sequence of *Oceanobacillus iheyensis* isolated from the Iheya Ridge and its unexpected adaptive capabilities to extreme environments. Nucleic Acids Res. 30: 3927–3935.

Tang L., Y. Xia, X. Wu, X. Chen, X. Zhang, H. Le, et al. 2017. Screening and characterization of a novel thermostable lipase with detergent-additive potential from the metagenomic library of a mangrove soil. Gene 625: 64–71.

Tommonaro G., G. Pieretti, A. Poli, H. Kazak, B. Gürleyendag, E.T. Öner, B. Nicolaus, et al. 2009. High level synthesis of levan by a novel *Halomonas* species growing on defined media. Carbohydr. Polym. 78(4): 651–657.

Tutino M.L., G. di Prisco, G. Marino and D. de Pascale. 2009. Cold-adapted esterases and lipases: from fundamentals to application. Protein Pept Lett. 16: 1172–1180.

Wu C., L. Zhuang, S. Zhou, F. Li and J. He. 2011. Alkaliphilic, humic acid-reducing bacterium isolated from a microbial fuel cell. Int. J. Syst. Evol. Microbiol. 61(4): 882–887.

Yaakop A.S., K.G. Chan, R. Ee, Y.L. Lim, S.K. Lee, F.A. Manan, et al. 2016. Characterization of the mechanism of prolonged adaptation to osmotic stress of *Jeotgalibacillus malaysiensis* via genome and transcriptome sequencing analyses. Sci Rep. 6: 1–14.

Yong Y., Z. Cai, Y. Yu, P. Chen, R. Jiang, B. Cao, et al. 2013. Increase of riboflavin biosynthesis underlies enhancement of extracellular electron transfer of *Shewanella* in alkaline microbial fuel cells. Bioresour Technol. 130: 763–768.

Zhang T., L. Zhang, W. Su, P. Gao, D. Li, X. He, et al. 2011. The direct electrocatalysis of phenazine-1-carboxylic acid excreted by *Pseudomonas alcaliphila* under alkaline condition in microbial fuel cells. Bioresour Technol. 102: 7099–7102.

Zhang Y., J. An, G. Yang, X. Zhang, Y. Xie, L. Chen and Y. Feng. 2016. Structure features of GH10 xylanase from *Caldicellulosiruptor bescii*: implication for its thermophilic adaption and substrate binding preference. Acta. Biochim. Biophys. Sin. (Shanghai) 48: 948–957.

Zhao B., N.M. Mesbah, E. Dalin, L. Goodwin, M. Nolan, S. Pitluck, O. Chertkov, T.S. Brettin, J. Han, F.W. Larimer, et al. 2011. Complete genome sequence of the anaerobic, halophilic alkalithermophile *Natranaerobius* thermophilus JW/NM-WN-LF. J Bacteriol. 193: 4023–4024.

Zhou W., W. Guo, H. Zhou and X. Chen. 2016. Phenol degradation by *Sulfobacillus acidophilus* TPY via the meta-pathway. Microbiol. Res. 190: 37–45.

Zverlov V., K. Piotukh, O. Dakhova, G. Velikodvorskaya and R. Borriss. 1996. The multidomain xylanase A of the hyperthermophilic bacterium *Thermotoga neapolitana* is extremely thermoresistant. Appl. Microbiol. Biotechnol. 45: 245–247.

20

Extremophilic Isolates of *Pseudomonas aeruginosa* as Biomarkers of Presence of Heavy Metals and Organic Pollution and Their Potential for Application in Contemporary Ecotoxicology

Ivanka M. Karadžić[1,*], Milena G. Rikalović[2],
Lidija T. Izrael-Živković[1] and Ana B. Medić[1]

Introduction

As the environmental microorganisms, originated from diverse ecological conditions, have a particular ability to respond quickly and to degrade or transform pollutants into less toxic compounds in a specific, contaminated environment, they can serve as indicators of pollution. Metabolite and protein fingerprints of different species in an ecosystem, occurring as a response to pollutants, indicate the modes of action of pollutants and resistance mechanisms developed by the organisms. *Pseudomonas aeruginosa* is widely distributed on the Earth and inhabits all ecological niches, including polluted extreme environments. Its extraordinary potential to cope with a broad spectrum of toxic substances is related to its huge genome and well-adapted protein structures and metabolic pathways. In polluted environments, *P. aeruginosa* quickly develops specific metabolic patterns with nodes that can be used as an early warning signal of contamination, i.e., relevant pollution biomarkers. The search for environmental molecular biomarkers has been accelerated by the application of new, large-scale, high-throughput platforms of genomics, proteomics, and metabolomics, which solely or in combination can detect initial environmental contamination and its subtle effects on the ecosystem. These omics methods can be applied in microbial ecotoxicology not only to detect and monitor an early signal of pollution but to investigate complex pollutant-impacted ecosystems and the role of microorganisms in the removal of these pollutants. Nowadays, it is possible to monitor specific metabolic responses across multiple levels of function, to integrate data, and to obtain a holistic picture.

Environment Pollution and Impact on Living Systems

Pollution and its impact on the environment are correlated with increased anthropogenic activity, technological development, progression by the chemical industry, inadequate waste treatment and management, and poor education and social conscience (Massard-Guilbaud and Mathis 2017). Environmental matrices, especially waters and soils, are often loaded with toxic substances, such as heavy metals and organic compounds (petroleum derivatives, pesticides, plastic polymers, etc.)

[1] Department of Chemistry, Faculty of Medicine, University of Belgrade, Serbia.
[2] Department of Environment and Sustainable Development, Singidunum University, Belgrade, Serbia.
*Corresponding author: ivanka.karadzic@med.bg.ac.rs

(Chowdhury et al. 2018). These substances, due to their physicochemical properties, interact with the ecosystem and usually persist and circulate for a long (Chowdhury et al. 2018, Hazrat et al. 2019).

Polluted ecosystems present serious threats to the existence of all living organisms, especially through the processes of bioaccumulation and biomagnification when toxic substances enter the food chain and often chronically affect body functions (Hazrat et al. 2019, Thompson and Darwish 2019). These mechanisms are especially relevant in the cases of heavy metals and persistent organic pollutants (POPs) – organochlorinated pesticides and the petroleum derivatives, polycyclic aromatic compounds (PAH). Heavy metals interact with soil particles and water, enter root systems and upper parts of plants, further accumulate in fruits, and enter animals including humans via their diets. Organic compounds like PAH can alter soil properties and evaporate to air forms and spread through the ecosystem or accumulate in marine organisms (Kafel et al. 2012). Furthermore, most of these ecotoxicants are classified as xenoestrogens, mutagens, cancerogens, or teratogens (Soto and Sonnenschein 2010). Banks and Stark (1998) stated that ignorance of the presence of pollutants could lead to severe underestimates of toxin-related population declines.

Microbial Ecotoxicology and Microorganisms as Biomarkers

Ecotoxicology is a branch of environmental science that deals with toxic substances and their effects on living organisms. Considering both aspects of ecotoxicology, predictive, and retrospective, the importance of a toxicant's metabolic fate, its toxicokinetics and toxicodynamics, and especially any biological adaptation mechanisms, are crucial for understanding living systems' interactions with xenobiotics. Contemporary science demands interdisciplinary approaches in understanding the effects of xenobiotic compounds on organisms and populations in order to recognise hazard risks, prevent pollution, and/or remediate contamination (Ashauer and Escher 2010, Castro-Català et al. 2016). Biomarkers, such as *Mollusca* and some fish species, were recognised long ago as indicators of pollution in aquatic ecosystems due to their tendency to accumulate toxic substances (Gagnaire et al. 2008). On the other hand, local plant or animal species could serve as biomarkers for terrestrial ecosystems (Fontanetti et al. 2011). These biomarkers satisfy the demands of the traditional approach in environmental research in relation to basic methods (lethal doses LD_{50}, inhibitory concentration IC_{50}, or effective concentration EC_{50} determination) (Hook et al. 2014). On the other hand, the present state of the environment indicates that there is now a need to consider organisms with rapid metabolic rates that undergo rapid changes and to include in their investigation modern analytical methods and techniques (Richardson 2001, Castro-Català et al. 2016, Sauer et al. 2017, Xiaowei et al. 2018).

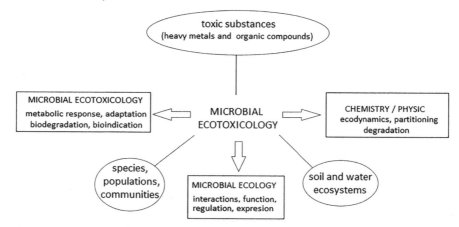

Figure 1. Schematic Overview of Microbial Ecotoxicology and Its Position in Environmental Science (Adapted from Ghiglione et al. 2016).

Microbial ecotoxicology, a relatively new area of ecotoxicology, deals with the ecological impacts of pollutants at a microbial level, and the role and activity of the organisms in the ecosystem (their distribution, transfer, degradation, and transformation) (Ghiglione et al. 2016, Shahsavari et al. 2017). Microbial ecotoxicology links microbial ecology, toxicology, physics, and chemistry (Figure 1). Its development coincides with progress in molecular methods, bioinformatics, and *in silico* research.

Microorganisms, due to their wide ecological valence, and metabolic and adaptation capacities, are important and relevant pollution biomarkers for metals and various organic pollutant compounds (Ye et al. 2017). Especially interesting are the extremophile microbes adapted to grow in extreme environments polluted with heavy metals and/or petroleum derivatives (Margesin and Schinner 2001, Olaniran et al. 2013, Di Donato et al. 2016, 2019). This type of biomarker provides not only a signal of pollutant present in the environment but also information on metabolic alterations and adaptations that could be further used in pollutant bioremediation (biotransformation and biodegradation) (Ghiglione et al. 2016, Shahsavari et al. 2017).

Position of *P. aeruginosa* in Microbial Ecotoxicology

In view of the above, microbial ecotoxicology can be regarded as an extension of microbiology in environmental sciences. Among the diverse microbial genera and species, *Pseudomonas*, and particularly *P. aeruginosa* as extremely abundant organisms in the environment, could be targeted for such research and be potential pollution biomarkers. *P. aeruginosa*, especially environmental isolates, qualify as biomarkers due to their interactions with all elements of the living and non-living environment, together with unique biological behaviours like biofilm formation, quorum sensing (*QS*) communication, diverse secondary metabolism, and their ability to switch oxygen metabolism pathways (Figure 2). Growing numbers of reports confirm the potential of *P. aeruginosa* to resist heavy metal and organic pollution and to bioremediate these pollutants (Rikalovic et al. 2014, Chellaiah et al. 2018, Izrael-Živković et al. 2018, 2019).

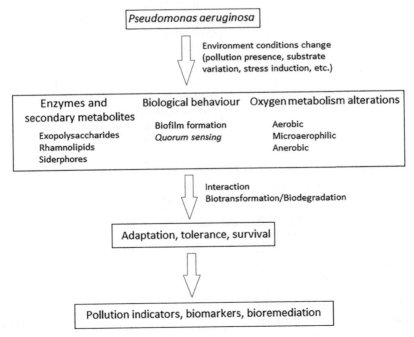

Figure 2. *P. aeruginosa* Biological Arsenal Important for Ecotoxicology Studies and Application in Environment Protection.

Extremophilic *P. aeruginosa*

Due to its great metabolic diversity, low nutritional requirements, pronounced production of secondary metabolites, and biofilm formation (Moradali et al. 2017), *P. aeruginosa* has the capability to survive in a broad range of conditions, including extreme environments. Extremophilic isolates of this bacterium have been classified as shown in Table 1.

TABLE 1: Classification of Extremophilic *P. aeruginosa*

Microorganism Designation	Extremophile Class	Carbon Sources Utilized by Hydrocarbonoclastic Microorganisms	Reference
P. aeruginosa	Psychrotrophic	/*	Molin and Ternstrom 1986
P. aeruginosa AP02-1	Thermophilic, Organotrophic	Alkanes (C5–C19), aromatic hydrocarbons (benzene, toluene, *o*-xylene), PAHs (anthracene, naphthalene), heterocyclic sulphurated hydrocarbons (thianaphthene, dibenzothiophene)	Perfumo et al. 2006
P. aeruginosa FRD1, RM4440	UV radiation-tolerant	/	Elasri et al. 1999
P. aeruginosa	Halotolerent	/	Varjani and Upasani 2019
P. aeruginosa Asph2	Halotolerent Oligotrophic	Petroleum hydrocarbons (saturates, aromatics, asphaltenes, and resins)	Ali et al. 2014
P. aeruginosa LVD-10	Halotolerent, Oligotrophic	Toluene, phenol, *p*-cresol, phenanthrene, 4-nitrophenol, indole, diesel	Drakou et al. 2015
P. aeruginosa san ai	Alkaliphilic, Metallotolerant Oligotrophic	2,6-di-*tert*-butylphenol, *n*-alkanes (C16, *n*-C19)**, PAHs (phenanthrene, fluorene, pyrene)**	Karadzic et al. 2004, Avramović et al. 2013, Izrael-Zivkovic et al. 2018, Medic et al. 2019.
P. aeruginosa	Oligotrophic	/	Tada et al. 1995
P. aeruginosa	Halotolerent, Oligotrophic	*n*-Alkanes (C12–C42), PAHs (naphthalene, fluorenephenanthrene, dibenzothiophene, fluoranthene, pyrene, anthracene) and their derivatives	Pasumarthi et al. 2013
P. aeruginosa RM1 and SK1	Halophile, Oligotrophic	Hydrocarbon fractions of the waste engine oil (*n*-alkanes C9-C26, anthracene, pyrene)	Salam 2016
P. aeruginosa BC15	Metallotolerant	/	Raja et al. 2006
P. aeruginosa ASU6a	Metallotolerant	/	Hassan et al. 2008
P. aeruginosa	Metallotolerant	/	Yamina et al. 2014
P. aeruginosa KP717554	Metallotolerant	/	Aka and Babalola 2016
P. aeruginosa	Metallotolerant	/	Lin et al. 2016
P. aeruginosa E1	Metallotolerant	/	Zeng et al. 2009
P. aeruginosa JP11	Metallotolerant	/	Chakraborty and Das 2014
P. aeruginosa	Metallotolerant	/	Filali et al. 2000
P aeruginosa	Metallotolerant	/	Teitzel and Parsek 2003

P. aeruginosa is well known for its ability to use different sources of carbon and at the same time, it is commonly present in almost all contaminated sites (Chellaiah et al. 2018). Furthermore, some of the *P. aeruginosa* isolates listed in Table 1 are polyextremophiles, while others are oligotrophic with hydrocarbonoclastic capacity, i.e., a particular ability to survive, grow, and degrade hydrocarbons, including their complex mixtures, in crude oil.

Due to accidental spills, even extreme environments may become contaminated with heavy metals and organic pollutants. In these conditions, conventional microorganisms are unable to efficiently remove or degrade pollutants. This problem can be solved by the use of extremophilic microorganisms that are already adapted to these circumstances. A wide range of hydrocarbon pollutants has been biodegraded in conditions of low (Arctic and Antarctica) or high (deserts) temperatures, acidic aquifers or forest soils, alkaline waste industrial effluent, and the high salinity niches of salt-containing soils or aquatic environments (Margesin and Schinner 2001). Microorganisms adapted to extreme conditions have a particular potential to adapt and respond rapidly to the contamination in these extreme conditions, thanks to sophisticated molecular strategies based on adapted biomolecules and biochemical pathways. Enzyme nanomachines remain catalytically active under extremes of temperature, salinity, pH, and solvent conditions, owing to several adaptation strategies. For example, thermophile proteins have disulphide bridges and multiple subunit structures; halophile proteins have enhanced numbers of acidic residues, while high flexibility around active sites that have low activation enthalpies facilitates the activity of psychrophile proteins at low temperatures (Karadzic et al. 2004, 2005, Brininger et al. 2018). Based on well-adapted enzyme structures, the adaptation of the microorganism occurs though a multifactorial response based on three main programmes: metabolic (global and the compound-specific pathways), stress-responsive (adaptation to sub-optimal growth conditions), and social (motility, chemotaxis, and biofilm, intercellular interactions) (Nogales et al. 2017). Quick and highly specific metabolic adaptation to a particularly toxic substance accompanied by rapid growth ensures *P. aeruginosa*, the very abundant microorganism, can provide a concrete early warning signal of pollutant present in the ecosystem.

Biological Mechanisms of *P. aeruginosa's* Interactions with Pollutants

The survival of eurivalent *P. aeruginosa* under broadly fluctuating environmental factors is due to its genome characteristics and metabolic capacity. *P. aeruginosa* isolates have a large gene repertoire (genome size ~5–7 Mbp) with a high proportion of regulatory genes available for response and adaptation to diverse environments (Moradali et al. 2017). The wide environmental distribution of *P. aeruginosa* is attributed to its extraordinary survival capability, conditioned by recruiting an arsenal of responsive mechanisms (Moradali et al. 2017, Izrael-Živković et al. 2019). The reported data substantiate the genetic potential of these microbes and their role in the bioremediation of heavy metals, especially Cd, Cr, Cu, and petroleum derivatives (Avramović et al. 2013, Rikalovic et al. 2014, 2015, Izrael-Živković et al. 2019). Additionally, the unique genetic and metabolic characteristics of *P. aeruginosa* and its potential for application in fields of biotechnology—medicine, pharmacy, agriculture—make this species unique (Karadzic et al. 2004, 2006, Dimitrijevic et al. 2011, Grbavcic et al. 2011, Wu et al. 2018).

As previously mentioned, *P. aeruginosa*'s responses to heavy metal and organic pollution events involve complex mechanisms at the molecular level, dependent on *QS* regulation (Zhang et al. 2016, Moradali et al. 2017, Wang et al. 2017). Precisely, *QS* regulates secondary metabolite production (exopolysaccharides (EPS), rhamnolipid biosurfactant (RL), elastase, Xcp secretion, lipase, exotoxin A, lectins, alkaline protease, hydrogen cyanide, pyocyanin, virulence factors, etc., biofilm formation, and swarming and twitching movement. These physiological roles of *P. aeruginosa QS* are an important part of its response to pollutant exposure through initiation and formation of biofilm, and optimisation of microbial community composition, which ultimately strengthen pathways of biodegradation and/or transformation (Zhang et al. 2016). The main constituent features of biofilm formation are the synthesis of EPS (the major components of biofilm matrix) and RL biosurfactant (provides optimal substrate availability and swarming motility due to its surfactant structure) (Wei and Ma 2013, Rikalovic et al. 2014). Furthermore, studies confirm the great impact of the surface-active compounds EPS and RL on biodegradation efficiency, based on their amphipathic structures, tension properties, and their capacity to bind heavy metal ions (Avramović et al. 2013, Rikalovic et al. 2014).

Contemporary Analytical Approaches in Microbial Ecotoxicology

Traditional ecotoxicological tests have severe drawbacks that can be overcome by the use of molecular biomarkers at the population level (Hook et al. 2014). The reductionist approach of using a single biomarker usually cannot provide insight into the state of whole ecosystems. Therefore, a holistic approach based on multiple biomarkers, i.e., both exposure and effect biomarkers, is desirable to determine the effects of a toxic substance and to distinguish between organism adaptive responses, which could be further used for monitoring and/or remediation, and harmful effect to a living system (Hook et al. 2014). In this regard, the newly developed omics platforms could fulfil all the requirements of the integrated biomarker approach in ecotoxicology (Xiaowey et al. 2018, Gouveia et al. 2019).

The fascinating progress in bioanalytical methods and bioinformatics has allowed the development of large-scale technological platforms of genomics, transcriptomics, proteomics and metabolomics for high-throughput research of genomes, transcriptomes, proteomes, and metabolomes, respectively. These omics methods enable the study of the early molecular responses of organisms to environmental stressors and as such, they can be used to identify specific metabolic responses to toxic substances (i.e., the production of novel biomarkers), to predict the impact of a pollutant on a broad diversity of biological species, and to assess the environmental risks of bioavailable chemicals (Gouveia et al. 2019). Most of these molecular omics methods can be applied in microbial ecotoxicology to investigate complex pollutant-impacted ecosystems and the role of microorganisms in the removal of these pollutants (Ghiglione et al. 2016).

The Genomes Online Database (GOLD) (https://gold.jgi.doe.gov), as an up to date catalogue of genomes and whole metagenome sequences (metagenome), reports 60,112 sequenced bacterial genomes and 8,644 metagenomes (2019). According to the GOLD, 723 genomes of *P. aeruginosa* isolates have been processed. In the latest release (2019), UniProt (https://www.uniprot.org/proteomes/) provides an impressive number of 4,990 fully characterised *Pseudomonas* proteomes of which 2,298 are *P. aeruginosa* proteomes, demonstrating our deepening knowledge and understanding of this organism at the molecular levels of genes and proteins. In less than two decades, from the first human genome sequencing to date, an enormous effort has been made to understand and connect genes to their functions and to systematically analyse their expression levels in relation to other biomolecules (mRNA, proteins, metabolites) that constitute a living system.

Although omics technologies have not yet been applied as standard methods in regulatory hazard assessment, their rapid progress, availability, and reliability will speed up their application in regulatory and legal issues (Sauer et al. 2017). The next section provides a succinct overview of the potential for using the state-of-the-art omics technologies (genomics, proteomics, metabolomics, lipidomics) as well metagenomics, metaproteomics, and integrative multiomics platforms to study physiological changes and adaptations of environmental microorganisms and microbial communities exposed to contaminants and to discover and use relevant biomarkers in ecotoxicology.

Genomics Studies of environmental *P. aeruginosa*

Advances in high-throughput sequencing methods, where particularly Illumina platforms have resulted in an explosion in the number of genomic and metagenomic studies, enabled identification of the microbial functions of both cultivated and uncultivated microorganisms present in a specific environment and better understanding of their adaptations to specific environmental conditions (Shahsavari et al. 2017). Genomes of bacteria, in particular genus *Pseudomonas* isolated from extreme conditions, revealed the existence of genes responsible for proteins that serve for adaptation under these conditions (Das et al. 2015, He et al. 2018, Izrael-Živković et al. 2019). The architecture of bacterial genomes is compact and shows a strong correlation between genome size and the number of functional genes (Koonin 2009). The availability of bacterial genomic data accompanied by the use of genomic computational tools has created a deeper understanding of bacterial survival capabilities in different ecological niches and the application of these organisms as bioremediation tools (Hivrale et al. 2016).

P. aeruginosa san ai, isolated from alkaline water-soluble mineral cutting oil (pH 10), is an alkaliphilic, heavy metal tolerant, hydrocarbon-degrading bacterium which has a collection of genes providing heavy metal resistance and the key enzymes responsible for the aerobic transformation of *n*-alkanes, alkanesulphonate, aromatic hydrocarbons, and halogenobenzene (Izrael-Živković et al. 2019) (Figure 3). As a heavy metal tolerant organism, *P. aeruginosa* san ai has genes to produce chelating compounds (siderophores) and heavy metal-binding proteins (Izrael-Živković et al. 2018). Furthermore, in oligophilic *P. aeruginosa* san ai, the presence of genes encoding proteins involved in the catabolism of aromatic compounds through the β-ketoadipate *ortho* degradation pathway (*catABC* of catechol and *pcaBCDG* of protocatechuate branch) has been identified (Medić et al. 2019).

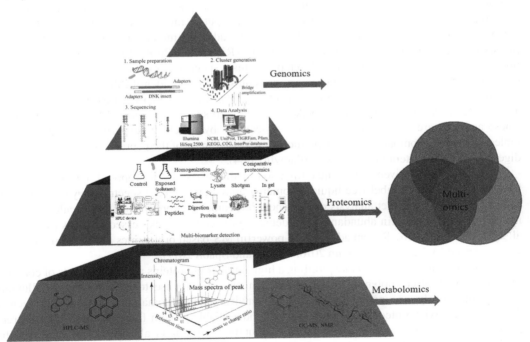

Figure 3: Multiomics Strategy to Investigate Extremophilic *P. aeruginosa* Isolates as Biomarkers of Heavy Metals and Organic Pollution

Whole-genome analysis of a hydrocarbonoclastic, *P. aeruginosa* N002 isolated from an environment contaminated with crude oil, showed this bacterium has an active genetic mechanism for crude oil degradation, which comprises genes encoding for hydrocarbon biodegradation and biosurfactant production similar to *P. aeruginosa* san ai (Das et al. 2015). The genes involved in the protocatechuate (*pcaBCG*) and catechol (*catABC*) pathways were identified in *P. aeruginosa* DN1, a microorganism with a strong ability to utilise fluoranthene and crude oil as sole carbon sources (He et al. 2018).

The genome sequencing of *P. aeruginosa* DQ8, isolated from petroleum-contaminated soil, showed the presence of key genes and enzymes involved in PAH (fluorene, phenanthrene, fluoranthene, and pyrene) biotransformation via two pathways (Zhang et al. 2011). All genes encoding the naphthalene upper pathway enzymes were identified in *P. aeruginosa* PaK1, isolated from sludge in a wastewater treatment plant (Takizawa et al. 1999). Eaton and Chapman (1992) reported that naphthalene degradation by *P. aeruginosa* PAO1 is associated with genes encoding the first three enzymes of the naphthalene pathway to salicylate. The key genes involved in degradation of aromatic compounds were identified in genomes of *P. aeruginosa* isolates: PAO1, KF702, N002, and DSM50071 (Dong et al. 2017).

Besides the bacterium's innate, although changeable genetic potential, the key to our understanding of adaptation strategies of *P. aeruginosa* to polluted extreme environments lies in proteomic changes and metabolic paths, which enable bacterial survival when exposed to stressors.

Proteomics Studies of Environmental *P. aeruginosa*

All of the proteins encoded by a genome make a proteome, which can now be studied by proteomics technology that includes the large-scale identification and characterisation of proteins, including their posttranslational modifications, quantification, cellular localisation, modifications, interactions, activities, and functions (Simpson 2003). In the post-genomic era, large-scale analysis of primary protein structure by sensitive, high-throughput techniques such as mass spectrometry (MS) has become routine. This has facilitated the study and monitoring of complex biological processes, including those in polluted environments (Richardson 2001). Two principal strategies—bottom-up and top-down—have been developed in mass spectrometry-oriented proteomics. In both cases, robust instrumentation is a necessary prerequisite. Protein identification in the widely used bottom-up approach is based on analysing peptides obtained by proteolytic digestion using matrix-assisted laser desorption/ionisation (MALDI) or electrospray ionisation (ESI) MS techniques. These analyses are accompanied by database searches by which the experimentally determined molecular mass is compared with the predicted molecular mass of the peptides obtained by *in silico* digestion of each protein in a database (Bladwin 2004).

A variety of MS-driven protein quantification methods have been proposed, but recent studies have revealed gel-free, label-free liquid chromatography-tandem mass spectrometry (LC-MS/MS) shotgun proteomics (direct analysis of the whole proteome) as a promising, large-scale method of measuring relative protein abundance, and detecting differentially expressed proteins (Vandera et al. 2015, Yu et al. 2016). Recent advances in powerful MS-based proteomics and software tools are well-reviewed (Aebersold and Mann 2016, Vidova and Spacil 2017).

Proteomic platforms that enable profiling a high number of proteins across large sets of complex samples provide great potential to study mechanisms of response to environmental pollutants and simultaneously, an opportunity to discover environmental biomarkers. Indeed, the recently developed ecotoxicoproteomics approach reveals identifiable protein fingerprints of the different organisms in an ecosystem as a response to pollutants (Gouveia et al. 2019). This indicates the modes of action of pollutants and the resistance mechanisms developed by organisms in response to the contamination (Gouveia et al. 2019).

The state-of-the-art technology now enables investigation of complex microbial communities in their specific environments by metaproteomics, a powerful method for multiple proteome investigation (Wilmes et al. 2006). To overcome limits and to avoid redundant protein identification in highly complex metaproteomics data analysis, powerful computational equipment, and improved protein databases with an enhanced number of taxonomic species are necessary (Heyer et al.

2017). Compared to standard proteomics, which analyses known proteins in a reference database, an integrative proteogenomics approach has been recently developed as an alternative to genome sequencing (Armengaud et al. 2014). Integrative proteogenomics combines RNA-sequencing of the transcriptome to rapidly identify protein-coding genes and shotgun proteomics to discover species-specific protein sequences (Armengaud et al. 2014). High-throughput shotgun, proteogenomics, and targeted proteomics have been pointed out as being the next-generation proteomics tools in ecotoxicoproteomics for the discovery of metabolic responses to toxic pollutants and species-specific biomarkers (Gouveia et al. 2019).

Although many publications report heavy metal resistance in *Pseudomonas*, only a few have studied metabolic changes and proteome expression triggered in *P. aeruginosa* by the presence of heavy metals. Such studies have been conducted on *P. aeruginosa* PAO1 (referent proteome), *P. aeruginosa* san ai (polyextremophile strain) (Table 1), and a *P. aeruginosa* isolated from wastewater containing chromium (Kilic et al. 2010, Poirier et al. 2013, Izrael-Živković et al. 2018, Wright et al. 2019). Comparative proteome analysis revealed metabolic alteration in *P. aeruginosa* exposed to heavy metal, which included up-regulation of proteins involved in energy production, amino acid metabolism, and carbohydrate metabolism together with up-regulation of metalloproteins (Kilic et al. 2010, Poirier et al. 2013, Izrael-Živković et al. 2018, Wright et al. 2019). Up-regulation of metalloproteins that are mostly located in the outer membrane, i.e., are the first line of defence, was detected in both *P. aeruginosa* and *P. fluorescens* (Poirier et al. 2013, Izrael-Živković et al. 2018). This signifies the importance and similarity of metalloprotein regulation in genus *Pseudomonas*. Remarkably, differentially up-regulated denitrification metalloproteins were found, even though the denitrification cascade was not active (Izrael-Živković et al. 2018).

Metabolomic Studies of Environmental *P. aeruginosa*

Metabolomics is a powerful molecular method that enables metabolites that serve as direct signatures of biochemical activity to be profiled. Aside from a global, untargeted approach designed to simultaneously measure a huge number of metabolites in biological samples, a defined set of metabolites can be investigated by a targeted approach. Although in contemporary studies of metabolites, highly sensitive and robust instrumentation based on liquid or gas chromatography coupled with MS (LC- or GC-MS) and nuclear magnetic resonance (NMR) techniques have become standard, many other analytical approaches can be applied for measuring metabolites (Patti et al. 2013). Targeted metabolomic approaches answer a specific biochemical question or hypothesis related to a particular pathway as in the case of the major compounds obtained after biodegradation of fluorene by *Pseudomonas* sp. NCIB 9816-4, which were identified by GC-MS, [1H] NMR, [13C]NMR analysis (Resnick and Gibson 1996), and which confirmed the proposed metabolic path of degradation (Seo et al. 2009). Using GC-MS analysis, phthalic acid was reported as a key intermediate during the growth of *P. aeruginosa* W10 on phenanthrene as a carbon source, leading to the conclusion that phenanthrene can be degraded via the protocatechuate pathway (Chebbi et al. 2017).

A global study of the *P. aeruginosa* metabolome by GC-MS analysis revealed 243 compounds, which indicates the existence of a basic, core metabolism that is independent of the growth phase and carbon source, aside from any specific adaptive response (Frimmersdorf et al. 2010). Similar induction of substrate-specific sets of metabolites in *P. putida* (van der Werf et al. 2008) and *P. aeruginosa* (Frimmersdorf et al. 2010) indicates the remarkable capacity for substrate utilisation based on powerful adaptive responses in genus *Pseudomonas*. The successful interplay of genome sequencing and metabolomics based on GC-MS of a fluoranthene-degrading *P. aeruginosa* DN1 revealed the degradation pathway of fluoranthene (He et al. 2018).

Production of EPS and biofilm formation are the most commonly described heavy metal resistance mechanisms in *Pseudomonas* (Moradali et al. 2017). EPS constitute the first line of defence, having ionisable functional groups engaged in metal chelation. Besides the complexation of heavy metals, retarding the diffusion of metals into, the biofilm also protects the bacteria from heavy metal toxicity (Chic-Ching et al. 2013). *P. aeruginosa* san ai exposed to the heavy metal cadmium produced EPS and accumulated 70% of the cadmium on the cell surfaces (Izrael-Živković et al. 2018). The increase of EPS production is related to its protective role for bacteria exposed to environmental stress and it can be regarded as an effect biomarker of broad specificity.

Siderophores are small secondary metabolites with high affinity for iron, the major limiting factor for microbial growth. Biosynthesis of siderophores is regulated by the amount of iron present in the environment and by the amount of iron already acquired by the bacteria (Schalk et al. 2011). Alteration of siderophore production and synthesis of proteins involved in the siderophore biosynthetic process was observed in *P. aeruginosa* exposed to heavy metals (Kilic et al. 2010, Izrael-Živković et al. 2018, Wright et al. 2019). Therefore, siderophores can be used as effect biomarkers.

The chemical diversity of the metabolome, which covers a broad spectrum of water-soluble and completely insoluble compounds, stimulated the development of lipidomics as a subomics approach. Comparative analysis of rhamnolipid mixtures from several environmental isolates of *P. aeruginosa* and one microbial consortium obtained from hydrocarbon-polluted areas was conducted by LC-MS and revealed isolate-specific rhamnolipid fingerprints (Rikalovic et al. 2013). To overcome limits related to the chemical diversity of the metabolome, a novel, promising, integrated metabolomics/lipidomics workflow has been just published, enabling simultaneous analysis of polar metabolites (hydrophilic sugars and organic acids) and non-polar lipids within one analytical run (Schwaiger et al. 2019). Merging metabolomics and lipidomics could be a key method for biomarker identification in large-scale global studies of microbial communities in polluted environments.

Multiomics

The broad and more frequent use of omics platforms in biological sciences has greatly increased the potential to conduct comprehensive multiomics studies and to collect multiomics data with the ultimate goal of creating an advanced, holistic understanding of living organisms and their adaptations. The data generated from multiple omics approaches can be analysed by post-analysis data integration, integrated data analysis, and systems modelling techniques (Pinu et al. 2019). So far, the most frequently applied technique in contemporary bioscience is post-analysis data integration, based on different omics data sets which are analysed separately and then combined to make a general model pathway. On the other hand, as the integrated data analysis approach demands specialised tools to merge different omics data, a growing number of databases, software tools, and approaches have been developed and made accessible (Biswapriya et al. 2019).

The post-analysis genomics and proteomics data integration approach revealed worthwhile insights into global metabolic and regulatory networks in an environmental isolate of *P. aeruginosa* san ai exposed to heavy metals and identified the degradation path of a phenolic pollutant (Izrael-Živković et al. 2018, 2019, Medić et al. 2019) Fig. 3.

Metallotolerant *P. aeruginosa*

The ability of various *P. aeruginosa* isolates to resist the most toxic heavy metals is presented in Table 2. Most of the listed resistant bacteria were isolated from metal contaminated environments.

TABLE 2: Minimal Inhibitory Concentrations (MICs) Determined in Heavy Metal Resistant *P. aeruginosa* Isolates for a Variety of Heavy Metals

Microorganism designation	Isolation site	MIC (mM)					
		Cd	Hg	Zn	Cr	Ni	Pb
P. aeruginosa BC15[*]	Oil mill-treated wastewater	4.5	/	/	7.7	11.9	3.9
P. aeruginosa ASU6a	Wastewater	4.4	/	9.2	5.8	6.8	3.1
P. aeruginosa	Hospital wastewaters	8.5	0.1	17.6	/	/	/
P. aeruginosa KP717554	Soil recovered from mine tailings	0.5	/	/	15	+	/
P. aeruginosa	Cd-contaminated soil	19.6	0.05	27.5	11.5	1	5.8
P. aeruginosa E1	Cd-contaminated soil	18	/	+	/	/	+
P. aeruginosa JP11	Coastal marine sediment	11.2	0.05	13.8	3.9	2.9	1.9
P. aeruginosa	Sewage wastewaters	2	1.2	/	/	/	/
P aeruginosa PAO1	Laboratory strain	/	/	8	/	/	0.125
P. aeruginosa san ai	Mineral cutting oil	7.2	/[**]	/[**]	5	/	/[**]

+ resistance detected but MIC not determined
[*]the same microorganisms have been already mentioned and references are given in Table 1
[**]data not published

The concentration of heavy metals is one of the main parameters monitored in ecotoxicological studies and is regarded as an effect biomarker (Fontanetti et al. 2011). As metals can trigger the production of reactive oxygen species, enzymatic antioxidants were found to be up-regulated in *P. aeruginosa* exposed to heavy metals (Kilic et al. 2010, Poirier et al. 2013, Izrael-Živković et al. 2018, Wright et al. 2019) and as such, these enzymes have been used as biomarkers of effect in ecotoxicological studies on the environmental impact of heavy metals (Hook et al. 2014). Our proteomics study indicated that the up-regulation of metalloproteins is also a biomarker of effect of heavy metal contaminated environments (Izrael-Živković et al. 2018).

Extremophilic *P. aeruginosa* in Organic Pollution

The recovery of the biosphere from organic contamination largely depends on the microbial population's ability to degrade these pollutants (Bara Caracciolo et al. 2013). However, the non-polar structure of most organic pollutants and related to that, their concentration in the oligotrophic range can limit microbial population growth and consequently limit biodegradation. Ubiquitously present in diverse, including extreme, environmental niches, *P. aeruginosa* is the most frequently isolated microorganisms from areas contaminated with organic pollutants. In fact, the genus *Pseudomonas* has been regarded as critical for recycling organic carbon on the planet (Kahlon 2016). Nevertheless, the low concentrations of organic pollutants that *P. aeruginosa* can use classify this microorganism into a group of oligotrophic bacteria.

The thermophilic bacterium *P. aeruginosa* AP02-1, a biosurfactant producer, has high multidegradative capacity, ranging from *n*-alkanes to PAH (Perfumo et al. 2006). Thermophiles are particularly useful for biodegradation of hydrocarbons as the biotransformation increases with temperature rise due to the increased solubility of organic pollutants. The halotolerant *P. aeruginosa* Asph2 successfully degrades different fractions of petroleum hydrocarbons in polluted seawater (Ali et al. 2014). Genes encoding important enzymes involved in biotransformation of aromatic hydrocarbons have been detected in *P. aeruginosa* LVD-10 under extreme conditions, indicating this isolate has a great ability to degrade aromatic hydrocarbons (Drakou et al. 2015).

Hydroxy-pyrene, a PAH metabolite, is a commonly used exposure biomarker of PAH pollution, while antioxidant enzymes serve as biomarkers of effect for these xenobiotics (Hook et al. 2014). In addition to these biomarkers, our omics study of phenolic derivatives and PAH biodegradation by *P. aeruginosa* san ai revealed several new, potential biomarkers. Specifically, biodegradation of the plastic additive, 2,6-di-*tert*-butylphenol (2,6-DTBP), a hazardous, toxic substance for aquatic life, was investigated by genomics and gel-based comparative proteomics (Fig. 3). The study showed up-regulation of aromatic ring cleavage enzymes, leading to the conclusion that 2,6-DTBP is degraded by *ortho*-ring cleavage. These cleavage enzymes and their coding genes can be used as exposure biomarkers of phenolic pollution. *P. aeruginosa* san ai contains key enzymes of the glyoxylate shunt that could serve as effect multimarkers of organic pollution (Medic et al. 2019). In addition to this, enhanced activities of enzymes that neutralise the production of reactive oxygen species were measured, accordingly to their reported use as biomarkers of effect (Hook et al. 2014).

Biotechnological applications in the bioremediation of aromatic compounds using several *P. aeruginosa* isolates: RW41, RS1, LY11, N6P6, NY3, and PaJC have been reported (Nogales et al. 2017). Unambiguously, *P. aeruginosa* can be regarded as an exceptional candidate microorganism for both bioremediations of environmental pollutants and early detection of contamination.

Conclusions

Contemporary studies in ecotoxicology and monitoring of environmental pollution are strongly dependent on the selection of biomarkers that can enable early detection of pollutants and their effective monitoring. However, only a few, mostly oxidative biomarkers of environmental pollutants are known, representing a general response to a wide variety of pollutants and as such, they lack specificity. Due to their rapid metabolism and agile adaptation strategies, microorganisms can serve as biomarkers for pollution of soil and water ecosystems with POPs and heavy metals. Especially, eurivalent *P. aeruginosa*, which is able to exist in polluted extreme habitats and is an excellent candidate for an environmental biomarker as it has a broad spectrum of biological mechanisms and strategies to cope with pollutants. Rather than seeking out singular biomarkers, the state-of-the-art omics methods, operating at the molecular level of genes, proteins, and metabolites provide holistic insight into microbial metabolic responses to specific pollutants. The omics methods produce specific biological fingerprints related to the contaminant in question, and these fingerprints serve as early warning signs of environmental damage. In particular, up-coming integrative omics approaches will provide comprehensive insight into the response of biosystems to stressors as well as the creation and development of bioremediation procedures.

Acknowledgment

This study was supported by the Ministry of Science and Technological Development of Serbia, Project III43004.

References

Aebersold, R. and M. Mann. 2016. Mass-spectrometric exploration of proteome structure and function. Nature 537: 347–355.

Aka, R.J.N. and O.O. Babalola. 2016. Effect of bacterial inoculation of strains of *Pseudomonas aeruginosa*, *Alcaligenes faecalis* and *Bacillus subtilis* on germination, growth and heavy metal (Cd, Cr, and Ni) uptake of *Brassica juncea*. Int. J. Phytoremediat. 8: 200–209.

Ali, H.R., D.A. Ismailand and N.S. El-Gendy. 2014. The biotreatment of oil-polluted seawater by biosurfactant producer halotolerant *Pseudomonas aeruginosa*. Energ. Source Part A 36: 1429–1436.

Armengaud, J., J. Trapp, O. Pible, O. Geffard, A. Chaumot and E.M. Hartmann. 2014. Nonmodel organisms, a species endangered by proteogenomics. J. Proteomics. 105: 5–18.

Ashauer, R. and B.I. Escher. 2010. Advantages of toxicokinetic and toxicodynamic modelling in aquatic ecotoxicology and risk assessment. J. Environ. Monit. 12: 2056–2061.

Avramović, N., S. Nikolić-Mandić and I. Karadžić. 2013. Influence of rhamnolipids, produced by *Pseudomonas aeruginosa* NCAIM(P), B001380 on their Cr(VI) removal capacity in liquid medium. J. Serb. Chem. Soc. 78: 639–652.

Banks, J.E. and J.D. Stark. 1998. What is ecotoxicology? An ad-hoc grab bag or an interdisciplinary science. Integr. Biol. 5: 196–204.

Bara Caracciolo, A., P. Bottoni and P. Grenni. 2013. Microcosm studies to evaluate microbial potential to degrade pollutants in soil and water ecosystems. Microchem. J. 107: 126–130.

Biswapriya, B.M., L. Carl, O. Michael and A.C. Laura. 2019. Integrated omics: Tools, advances and future approaches. J. Mol. Endocrinol. 62: R21–R45.

Bladwin, M. 2004. Protein identification by mass spectrometry. Mol. Cell Proteomics 3: 1–9.

Brininger, C., S. Spradlin, L. Cobani and C. Evilia. 2018. The more adaptive to change, the more likely you are to survive: Protein adaptation in extremophiles. Semin. Cell. Dev. Biol. 84: 158–169.

Castro-Català, N., M. Kuzmanovic, N. Roig, J. Sierra, A. Ginebreda, D. Barceló, et al. 2016. Ecotoxicity of sediments in rivers: Invertebrate community, toxicity bioassays and the toxic unit approach as complementary assessment tools. Sci. Total Enviro. 540: 297–306.

Chakraborty, J. and S. Das. 2014. Characterization and cadmium-resistant gene expression of biofilm-forming marine bacterium *Pseudomonas aeruginosa* JP-11. Environ. Sci. Pollut. Res. 21: 14188–14201.

Chebbi, A., D. Hentati, H. Zaghden, N. Baccar, F. Rezgui, M. Chalbi et al. 2017. Polycyclic aromatic hydrocarbon degradation and biosurfactant production by a newly isolated *Pseudomonas* sp. strain from used motor oil-contaminated soil. Int. Biodeterior. Biodegrad. 122: 128–140.

Chellaiah, E.R., K. Anbazhagan and G.S. Selvam. 2006. Isolation and characterization of a metal resistant *Pseudomonas aeruginosa* strain. World J. Microbiol. Biotechnol. 22: 577–585.

Chellaiah, E.R. 2018. Cadmium (heavy metal) bioremediation by *Pseudomonas aeruginosa*: a minireview. Appl. Water Sci. 8: 154.

Chic-Ching, C., L. Bo-Chou and W. Chun-Hsien. 2013. Biofilm formation and heavy metal resistance by an environmental *Pseudomonas* sp. Biochem. Eng. J. 78: 132–137.

Chowdhury, A.R., R. Datta and D. Sarkar. 2018. Heavy metal pollution and remediation. pp. 359–373. *In:* B. Török and T. Dransfield [eds.]. Green Chemistry an Inclusive Approach. Elsevier Inc. Netherlands.

Das, D., R. Baruah, A.S. Roy, A.K. Singh, H.P.D. Boruah, J. Kalita, et al. 2015. Complete genome sequence analysis of *Pseudomonas aeruginosa* N002 reveals its genetic adaptation for crude oil degradation. Genomics 105: 182–190.

Di Donato, P., A. Buono, A. De Castro-Català, M. Kuzmanovic, N. Roig, J. Sierra, et al. 2016. Ecotoxicity of sediments in rivers: Invertebrate community, toxicity bioassays and the toxic unit approach as complementary assessment tools. Sci. Total Environ. 540: 297–306.

Di Donato, P., A. Buono, A. Poli, I. Finore, G.R. Abbamondi, B. Nicolaus, et al. 2019. Exploring marine environments for the identification of extremophiles and their enzymes for sustainable and green bioprocesses. MDPI. 19: 149.

Dimitrijevic, A., D. Velickovic, M. Rikalovic, N. Avramovic, N. Milosavic, R. Jankov, et al. 2011. Simultaneous production of exopolysaccharide and lipase from extremophylic *Pseudomonas aeruginosa* san-ai strain: a novel approach for lipase immobilization and purification. Carbohyd. Polymers 83: 1397–1401.

Dong, W., C. He, Y. Li, C. Huang, F. Chen and Y. Ma. 2017. Complete genome sequence of a versatile hydrocarbon degrader, *Pseudomonas aeruginosa* DN1 isolated from petroleum-contaminated soil. Gene Reports 7: 123–126.

Drakou, F., M. Koutinas, I. Pantelides and I. Vyrides. 2015. Insights into the metabolic basis of the halotolerant *Pseudomonas aeruginosa* strain LVD-10 during toluene biodegradation. Int. Biodeterior. Biodegradation 99: 85–94.

Eaton, R. and P. Chapman. 1992. Bacterial metabolism of naphthalene: construction and use of recombinant bacteria to study ring cleavage of 1,2-dihydroxynaphthalene and subsequent reactions. J. Bacteriol. 174: 7542–7554.

Elasri, M.O. and R.V. Miller. 1999. Study of the response of a biofilm bacterial community to UV radiation. Appl. Environ. Microbiol. 65: 2025–2031.

Filali, B.K., J. Taoufik, Y. Zeroual, Z. Dzairi, M. Talbi and M. Blaghen. 2000. Waste water bacterial isolates resistant to heavy metals and antibiotics. Curr. Microbiol. 41: 151–156.

Fontanetti, C.S., L.R. Nogarol, R. Bastão de Souza, D.G. Perez and G.T. Maziviero. 2011. Bioindicators and biomarkers in the assessment of soil toxicity. pp. 143–168. *In:* S. Pascucci [ed.]. Soil Contamination. In Tech, Croatia.

Frimmersdorf, E., S. Horatzek, A. Pelnikevich, L. Wiehlmann and D. Schomburg. 2010. How *Pseudomonas aeruginosa* adapts to various environments: a metabolomic approach. Environ. Microbiol. 12: 1734–1747.

Gagnaire, B., O. Geffard, B. Xuereb, C. Margoum and J. Garric. 2008. Cholinesterase activities as potential biomarkers: characterization in two freshwater snails, *Potamopyrgus antipodarum* (Mollusca, Hydrobiidae, Smith 1889) and *Valvata piscinalis* (Mollusca, Valvatidae, Müller 1774). Chemosphere 71: 553–560.

Ghiglione, J-F., F. Martin-Laurent and S. Pesce. 2016. Microbial ecotoxicology: an emerging discipline facing contemporary environmental threats. Environ. Sci. Pollut. Res. 23: 3981–3983.

Gouveia, D., C. Almunia, Y. Cogne, O. Pible, D. Degli-Esposti, A. Salvador, et al. 2019. Ecotoxicoproteomics: A decade of progress in our understanding of anthropogenic impact on the environment. J. Proteomics 198: 66–77.

Grbavcic, S., D. Bezbradica, L. Izrael-Živković, N. Avramovic, N. Milosavic, I. Karadzic, et al. 2011. Production of lipase and protease from an indigenous *Pseudomonas aeruginosa* strain and their evaluation as detergent additives: compatibility study with detergent ingredients and washing performance. Bioresour. Technol. 102: 11226–11233.

Hassan, S.H.A., R.N.N. Gad Abskharon, S.M.F. El-Rab and A.A.M. Shoreit. 2008. Isolation, characterization of heavy metal resistant strain of *Pseudomonas aeruginosa* isolated from polluted sites in Assiut city, Egypt. J. Basic. Microbiol. 48: 168–176.

Hazrat, A., K. Ezzat and I. Ikram. 2019. Review article environmental chemistry and ecotoxicology of hazardous heavy metals: environmental persistence, toxicity, and bioaccumulation. Hindawi J. Chem. Article ID 6730305.

He, C., Y. Li, C. Huang, F. Chen and Y. Ma. 2018. Genome sequence and metabolic analysis of a fluoranthene-degrading strain *Pseudomonas aeruginosa* DN1. Front. Microbiol. 9: 2595.

Heyer, R., K. Schallert, R. Zoun, B. Becher, G. Saake and D. Benndor. 2017. Challenges and perspectives of metaproteomic data analysis. J. Biotechnol. 26: 24–36.

Hivrale, A.U., P.K. Pawar, N. Rane and S. Govindwar. 2016. Application of genomics and proteomics in bioremediation in toxicity and waste management using bioremediation. IGI Glob. https://doi.org/10.4018/978-1-4666-9734-8.ch005.

Hook, S., E. Gallagher and G. Batley. 2014. The role of biomarkers in the assessment of aquatic ecosystem health. Integr. Environ. Assess. Manag. 10: 327–341.

Izrael-Živković, L., M. Rikalović, G. Gojgić-Cvijović, S. Kazazić, M. Vrvić, I. Brčeski, et al. 2018. Cadmium specific proteomic responses of a highly resistant *Pseudomonas aeruginosa* san ai, RCS Advances. 8: 10549–10560.

Izrael-Živković, L., V. Beškoski, M. Rikalović, S. Kazazić, N. Shapiro, T. Woyke, et al. 2019. High-quality draft genome sequence of *Pseudomonas aeruginosa* san ai, an environmental isolate resistant to heavy metals. Extremophiles 23: 399–405.

Kafel, A., J. Gospodarek, A. Zawisza-Raszka, K. Rozpedek and E. Szulinka. 2012. Effects of petroleum products polluted soil on ground beetle *Harpalus rufipes*. Ecol. Chem. Eng. A. 19: 731–740.

Kahlon, R.S. 2016. Biodegradation and bioremediation of organic chemical pollutants by *Pseudomonas*. pp. 343–417. *In:* Kahlon R. [ed.]. *Pseudomonas:* Molecular and Applied Biology. Springer International Publishing Switzerland.

Karadzic, I., A. Masui and N. Fujiwara. 2004. Purification and characterization of a protease from *Pseudomonas aeruginosa* grown in cutting oil. J. Biosci. Bioeng. 98: 145–152.

Karadzic, I. and J. Maupin-Furlow. 2005. Improvement of two-dimensional gel electrophoresis proteome maps of the haloarchaeon *Haloferax volcanii*. Proteomics 5: 354–359.

Karadzic, I., A. Masui, L. Izrael-Živkovic and N. Fujiwara. 2006. Purification and characterization of an alkaline lipase from *Pseudomonas aeruginosa* isolated from putrid mineral cutting oil as component of metalworking fluid. J. Biosci. Bioeng. 102: 82–89.

Koonin, E.V. 2009. Evolution of genome architecture. Int. J. Biochem. Cell Biol. 41: 298–306.

Kılıc, N.K., A. Stensballe, D.E. Otzen and G. Dönmez. 2010. Proteomic changes in response to chromium (VI) toxicity in *Pseudomonas aeruginosa*. Bioresour. Technol. 101: 2134–2140.

Leahy, J.G. and R.R. Colwell. 1990. Microbial degradation of hydrocarbons in the environment. Microb. Rev. 54: 305–315.

Lin, X., R. Mou, Z. Cao, P. Xu, X. Wu, Z. Zhu and M. Chen. 2016. Characterization of cadmium-resistant bacteria and their potential for reducing accumulation of cadmium in rice grains. Sci. Total. Environ. 569–570: 97–104.

Liu, K., W. Han, W.P. Pan and J.T. Riley. 2001. Polycyclic aromatic hydrocarbon (PAH) emissions from a coal fired pilot FBC system. J. Hazard. Mater. 84: 175–188.

Margesin, R. and F. Schinner. 2001. Biodegradation and bioremediation of hydrocarbons in extreme environments. Appl. Microbiol. Biotechnol. 56: 650–663.

Massard-Guilbaud, G. and C.-F. Mathis. 2017. A brief introduction to the history of pollution: from local to global. pp. 3–15. *In:* C. Cravo-Laureau, C. Cagnon, B. Lauga and R. Duran [eds.]. Microbial Ecotoxicology. Springer International Publishing AG.

Medić, A., K. Stojanović, L. Izrael-Živković, V. Beškoski, B. Lončarević, S. Kazazić, et al. 2019. A comprehensive study of conditions of the biodegradation of a plastic additive 2,6-di-tertbutylphenol and proteomic changes in the degrader *Pseudomonas aeruginosa* san ai. RSC Adv. 9: 23696–23710.

Molin G. and A. Ternstrom. 1986. Phenotypically based taxonomy of psychrotrophic *Pseudomonas* isolated from spoiled meat, water, and soil. Inter. J. Syst. Bacteriol. 36: 257–274.

Moradali, M.F., S. Ghods and B.H.A. Rehm. 2017. *Pseudomonas aeruginosa* Lifestyle: A paradigm for adaptation, survival, and persistence. Front Cell Infect Microbiol. 7: Article 39.

Nogales, J., J.L. García and E. Díaz. 2017. Degradation of aromatic compounds in Pseudomonas: a systems biology view. pp. 1–35. *In:* F. Rojo [ed.]. Aerobic Utilization of Hydrocarbons, Oils and Lipids, Handbook of Hydrocarbon and Lipid Microbiology. Springer AG, New York, USA.

Olaniran, A.O., A. Balgobind and B. Pillay. 2013. Review bioavailability of heavy metals in soil: Impact on microbial biodegradation of organic compounds and possible improvement strategies. Int. J. Mol. Sc. 14: 10197–10228.

Pasumarthi R., S. Chandrasekaran and S. Mutnuri. 2013. Biodegradation of crude oil by *Pseudomonas aeruginosa* and *Escherichia fergusonii* isolated from the Goan coast. Mar. Pollut. Bull. 76: 276–282.

Patti, G.J., O. Yanes and G. Siuzdak. 2013. Metabolomics: the apogee of the omic triology. Nat. Rev. Mol. Cell. Biol. 13: 263–269.

Perfumo, A., I.M. Banat, F. Canganella and R. Marchant. 2006. Rhamnolipid production by a novel thermophilic hydrocarbon-degrading *Pseudomonas aeruginosa* AP02-1. Appl. Microbiol. Biotechnol. 72: 132.

Pinu, R., D. Beale, A. Paten, K. Kouremenos, S. Swarup and H.J. Schirra. 2019. Systems biology and multi-omics integration: viewpoints from the metabolomics research community. Metabolites 9: 76.

Poirier, I., P. Hammannb, L. Kuhn and M. Bertrand. 2013. Strategies developed by the marine *Pseudomonas fluorescens* BA3SM1 to resist metals: A proteome analysis. Aquat. Toxicol. 128–129: 215–232.

Raja, E.C., K. Anbazhagan and G.S. Selvam. 2006. Isolation and characterization of a metal resistant *Pseudomonas aeruginosa* strain. World J. Microbiol. Biotechnol. 22: 577–585.

Resnick, S.M. and D.T. Gibson. 1996. Regio- and stereospecific oxidation of fluorene, dibenzofuran, and dibenzothiophene by naphthalene dioxygenase from *Pseudomonas* sp. strain NCIB 9816–4. Appl. Environ. Microbiol. 62: 4073–4080.

Richardson, S.D. 2001. Mass spectrometry in environmental sciences. Chem. Rev. 101: 211–225.

Rikalovic M., A.M. Abdel-Mawgoud, E. Déziel, G. Gojgic-Cvijovic, Z. Nestorovic, M. Vrvic, et al. 2013. Comparative analysis of rhamnolipids from novel environmental isolates of *Pseudomonas aeruginosa*. J. Surfact. Deterg. 16: 673–682.

Rikalovic, M., M. Vrvic and I. Karadzic. 2014. Bioremediation by rhamnolipids produced by environmental isolates of *Pseudomonas aeruginosa*. pp. 299–333. *In:* J.B. Velazquez-Fernandez and S. Muniz-Hernandez [eds.]. Bioremediation Processes, Challenges and Future Prospects. Nova Science Publishers, New York.

Rikalovic, M., M. Vrvic and I. Karadzic. 2015. Rhamnolipid biosurfactant from *Pseudomonas aeruginosa* – from discovery to application in contemporary technology, J. Serb. Chem. Soc. 80: 279–304.

Salam, S.B. 2016. Metabolism of waste engine oil by *Pseudomonas* species. 3 Biotech. 6: 98.

Sauer, U.G., L. Deferme, L. Gribaldo, J. Hackermüller, T. Tralau, B. Ravenzwaay, et al. 2017. The challenge of the application of 'omics technologies in chemicals risk assessment: Background and outlook. Regul. Toxicol. Pharmacol. 91: S14eS2619.

Schalk, I., M. Hannauer and A. Braud. 2011. New roles for bacterial siderophores in metal transport and tolerance. Environ. Microbiol. 13: 2844–2854.

Schwaiger M., H. Schoeny, Y. El Abiead, G. Hermann, E. Rampler and G. Koellensperge. 2019. Merging metabolomics and lipidomics into one analytical run. Analyst 144: 220–229.

Seo, J.S., Y.S. Keum and Q.X. Li. 2009. Bacterial degradation of aromatic compounds. Int. J. Environ. Res. Public. Health. 6: 278-309.

Shahsavari, E., A. Aburto-Medina, L.S. Khudur, M. Taha and A.S. Bal. 2017. From microbial ecology to microbial ecotoxicology. pp. 17–40. *In*: C. Cravo-Laureau, C. Cagnon, B. Lauga and R. Duran [eds.]. Microbial Ecotoxicology. Springer International Publishing AG, Cham, Switzerland.

Simpson, R., Proteins and Proteomics: A Laboratory Manual. 2002. pp. 15–50. Cold Spring Harbor Laboratory Press, NY, USA.

Soto, A.M. and C. Sonnenschein. 2010. Environmental causes of cancer: endocrine disruptors as carcinogens. Nat. Rev. Endocrinol. 6: 363–370.

Tada Y., M. Ihmori and A. Yamaguchi. 1995. Oligotrophic bacteria isolated from clinical materials. J. Clin. Microbiol. 33: 493–494.

Takizawa, N., T. Iida, T. Sawada, K. Yamauchi, Y. Wang and M. Fukuda. 1999. Nucleotide sequences and characterization of genes encoding naphthalene upper pathway of *Pseudomonas aeruginosa* PaK1 and *Pseudomonas putida* OUS82. J. Biosci. Bioeng. 87: 721–731.

Teitzel, G.M. and M.R. Parsek. 2003. Heavy metal resistance of biofilm and planktonic *Pseudomonas aeruginosa*. Apply Environ. Microbiol. 64: 2313–2320.

Thompson, L.A. and W.S. Darwish. 2019. Environmental chemical contaminants in food: Review of a global problem. Hindawi J. Toxicol. Article ID 2345283.

Vandera, E., M. Samiotaki, M. Parapouli, G. Panayotou and A. Koukkou. 2015. Comparative proteomic analysis of *Arthrobacter phenanthrenivorans* Sphe3 on phenanthrene, phthalate and glucose. J. Proteomics 113: 73-89.

van der Werf, M.J., K.M. Overkamp, B. Muilwijk, M. Koek, van der Werff-van der Vat, R.H. Jellema et al. 2008. Comprehensive analysis of the metabolome of *Pseudomonas putida* S12 grown on different carbon sources. Mol. Biosyst. 4: 315–327.

Varjani S. and V.N. Upasani. 2019. Evaluation of rhamnolipid production by a halotolerant novel strain of *Pseudomonas aeruginosa*. Bioresour. Technol. 288: 121577.

Vidova, V. and Z. Spacil. 2017. A review on mass spectrometry-based quantitative proteomics: targeted and data independent acquisition. Anal. Chem. Acta. 964: 7–23.

Wang, M.Z., B.M. Lai, A.A. Dandekar, Y. Yang, N. Li, J. Yin, et al. 2017. Nitrogen source stabilization of quorum sensing in the *Pseudomonas aeruginosa* bioaugmentation strain SD-1. Appl. Environ. Microbiol. 83: e00870–17.

Wei, Q. and L.Z. Ma. 2013. Biofilm matrix and its regulation in *Pseudomonas aeruginosa*. Int. J. Mol. Sci. 14: 20983–21005.

Wilmes, P. and P.L. Bond. 2006. Metaproteomics: studying functional gene expression in microbial ecosystems. Trends Microbiol. 14: 92–97.

Wright, B., K. Kamath, C. Krisp and M. Molloy. 2019. Proteome profiling of *Pseudomonas aeruginosa* PAO1 identifies novel responders to copper stress. BMC Microbiology 19: 69.

Xiaowei, Z., Pu Xia, P. Wang, J. Yang and D.J. Baird. 2018. Omics advances in ecotoxicology. Environ. Sci. Technol. 52: 3842–3851.

Yamina, B., B. Tahar, M. Lila, H. Hocine and F.M. Laure. 2014. Study on cadmium resistant-bacteria isolated from hospital wastewaters. Adv. Biosci. Biotechnol. 5: 718–726.

Ye, H., Z. Yang, X. Wu, J. Wang, D. Du and J. Cai, et al. 2017. Sediment biomarker, bacterial community characterization of high arsenic aquifers in Jianghan Plain, China. Sci. Rep. 7: 42037.

Yu, F., F. Qiu and J. Meza. 2016. Design and statistical analysis of mass-spectrometry-based quantitative proteomics data. pp. 211–237. *In*: P. Ciborowski and J. Silberring [eds.]. Proteomic Profiling and Analytical Chemistry, Elsevier, Amsterdam, Netherland.

Zeng, X., J. Tang, X. Liu and P. Jiang. 2009. Isolation, identification and characterization of cadmium-resistant *Pseudomonas aeruginosa* strain E1. J. Cent. South. Univ. Technol. 16: 0416–0421.

Zhang, W. and C. Li. 2016. Exploiting quorum sensing interfering strategies in Gram-negative bacteria for the enhancement of environmental applications. Front. Microbiol. 6: Article 1535.

Zhang, Z., Hou, Z., Yang, C., Ma, C., Tao, F. and P. Xu. 2011. Degradation of *n*-alkanes and polycyclic aromatic hydrocarbons in petroleum by a newly isolated *Pseudomonas aeruginosa* DQ8. Bioresour. Technol. 102: 4111–4116.

21

Moroccan New and Extremophilic Actinobacteria and Their Biotechnological Applications

Ahmed Nafis[1,*], Anas Raklami[2], Brahim Oubaha[2], Salah-Eddine Samri[3], Lahcen Hassani[2] and Yedir Ouhdouch[2]

Introduction

Microorganisms are microscopic living beings that are widespread (ubiquitous) and exist in large numbers with a very immense diversity. This microbial diversity not only represents the diversity in terms of existing species' numbers, but it also represents the diversity of the properties of the strains within the species. In this regard, microbial biodiversity interest has grown very remarkably in recent years as judged by the increasing number of scientific reports devoted to scientific literature. Although among the most widely distributed group of microorganisms in nature, actinobacteria have attracted more attention because of its applications in terms of biotechnological uses. Actinobacteria represent a high proportion of microbial biomass and appear to be of importance among the microbial flora (Sardi et al. 1992). They can colonize a wide variety of natural habitats and can develop on a wide range of substrates. These bacteria are present in polar soils, permanently frozen, as in hot and dry desert soils, in crude oils, soils highly contaminated with heavy metals, extremely alkaline, and salt lakes. This group covers genera covering a wide range of morphology, extending from the coccus, and rod-coccus cycle through fragmenting hyphal forms to genera with permanent and highly differentiated branched mycelium (Piepersberg 2008).

Since their discovery, actinobacteria have provided many important bioactive compounds of high commercial value and continue to be regularly screened for new bioactive substances (Barakate et al. 2002). They are well-known for the production of a wide variety of secondary metabolites useful for humans, such as antibiotics, vitamins, enzymes, and many other products (Barakate et al. 2002). Besides, they are also known for expressing different PGPR activities that can promote plant growth and nutrient uptake as well as enhancing plants' systemic resistance. Besides their ability to promote plant growth, these filamentous and sporulating organisms are often able to colonize plant tissues and are natural producers of antibiotics and antifungal substances, in particular, that could protect the plants against various phytopathogens.

To date, more than 20,000 naturally occurring antibiotics have been discovered and the vast majority of them are produced by microorganisms. Actinobacteria are the main reservoir of these

[1] Biology Department, Faculty of Sciences, Chouaib Doukkali University El Jadida Morocco.
[2] Laboratory of Microbial Biotechnologies, Agrosciences, and Environment, Faculty of Sciences Semlalia, Cadi Ayyad University, Marrakech, Morocco.
[3] Biology Department, Polydisciplinary Faculty of Nador, Morocco.
* Corresponding author: a.nafis.palmes@gmail.com

antibiotics since more than 60% of known antibiotics, and more than 90% of antibiotics that have been applied as antimicrobials have been isolated from this group of microorganism (Běhal 2000). They produce a large number of antibiotics of varied chemical structures (aminoglycosides, anthracycline, glycopeptide, B-lactams, macrolides, peptides, etc.) which have many therapeutic applications. Among the species belonging to different kinds of actinobacteria, *Streptomyces* are the most important producers of antibiotics and other secondary metabolites (Lechevalier 2017).

The list of novel actinobacteria and products found in microbiologically unexplored areas around the world suggests that a careful exploration of new habitats might continue to be useful. The Moroccan actinobacterial microflora has been poorly explored to search for new species. This country presents many unexplored environments such as the rhizosphere of endemic plants, soils from the desert or snowy peaks of the Atlas, and seawater from bays (e.g., Essaouira, where the temperature is around 20°C throughout the year) that suggest a careful exploration of new habitats might be useful.

This chapter presents results briefly obtained in the search for new actinobacterial species isolated from different habitats along with their different biotechnological uses.

Actinobacteria Taxonomy

Actinobacteria taxonomy has been progressed in line with the development of knowledge, and it is based on morphological, chemotaxonomic, and molecular criteria. Morphological traits may be useful in differentiating actinobacterial genera, including the presence or absence of substrate mycelium and aerial mycelium (Shirling and Gottlieb 1966). The color of mycelium and pigments, diffusible in the culture medium, could also be used in separation between actinobacterial isolates (Shirling and Gottlieb 1966). Indeed, spores and sporangia are also structures used in classification. Actinobacterial spores can be differentiated by their shape, size, arrangement, presence or absence of sporophores, and the surface of spores. While, the presence or absence of sporangia on mycelium, shape, size, and number of spores per sporangium are among the characteristics to be examined (Breton et al. 1992).

Although it has been possible to classify a microbial isolate on the basis of these morphological criteria with further advancement in science, they were considered insufficient to establish the correct determination. Therefore, it became essential to consider other characters as well to classify a given microbial isolate. Chemotaxonomy relied on the use of stable chemical characters, based on the cell wall compositions such as glycoproteins and membrane lipids (Lechevalier et al. 1977). Later, the classification of actinobacteria was based on the determination of the percentage of G+C, DNA-DNA hybridization, and 16S rDNA sequencing (Patel 2001, Garrity et al. 2002). This molecular approach was considered a powerful and essential tool in taxonomy.

Actinobacteria: Biotechnological Applications

Actinobacteria are microorganisms of great practical importance. Not only they play a vital role in humification process and organic matter decomposition, but they are also known for the production of a wide variety of secondary metabolites useful for humans and plants, such as antibiotics, antifungals, vitamins, plant growth promoters, enzymes, and many other products (Barakate et al. 2002, Dairi 2005, Pizzul 2006). For instance, actinobacteria produce approximately two-third of the known antibiotics produced by all the microorganisms.

Actinobacteria are well recognized for the production of primary and secondary metabolites that have important applications in various fields', especially human health. Actinobacteria produce a large number of structural antibiotics, a variety of chemicals (aminoglycosides, anthracyclines, glycopepetids, beta-lactams, tetracyclines, macrolides, nucliosides, etc.) that have many therapeutic applications (Mangamuri et al. 2016). For the sake of argument, actinobacteria are the main producers of antibiotics since more than 70% of known antibiotics, which have been of application value, have been isolated from this bacterial group (Běhal 2000). The genus *Streptomyces* produces nearly 80% of the actinobacterial commercially and medically useful antibiotics with the genus *Micromonospora*, producing one-tenth as many as the *Streptomyces* (Jensen et al. 1991, Hassan et al. 2011). Thereby, *Streptomyces* has been the primary-antibiotics producing organisms exploited by the pharmaceutical industry (Ramesh and Mathivanan 2009). Actinobacteria produce a large number of antibiotics of varied chemical structures, such as aminoglycosides (e.g., streptomycin), ansamycins (e.g., rifampin), anthracyclines (e.g., doxorubicin), β-lactam (e.g., cephalosporins), macrolides (e.g., erythromycin), and tetracycline (Mangamuri et al. 2016).

Actinobacteria are also a promising source of a wide range of important enzymes produced on an industrial scale (Qin et al. 2011). The interest of microorganisms, including actinobacteria, as a source of enzymes is due to the high efficiency and low cost of their production. Currently, enzymes of microbial origin are widely used in pharmaceutical, therapeutic, and biotechnology industries (Loucif 2011). They produce enzyme inhibitors useful for cancer treatment and immunodifiers that enhance immune response (Qin et al. 2011). As an example, actinobacteria can secret amylase, which is of great importance in biotechnological applications (food industry, fermentation, and paper industry), because of its ability to degrade starch (Pandey et al. 2000). Moreover, actinobacteria are known to produce cellulase (collection of hydrolytic enzymes), lipase (used in detergent industries, foodstuff, and in industries of pharmaceutical fields), protease (used in detergent industries), phytase, keratinase, chitinase, urease, etc. (Yadav et al. 2013).

Sustainable agriculture requires new, promising, and safe strategies to increase or maintain the current food production state while reducing environmental damage caused by chemical compounds. The use of microbial plant growth promoters is one of the most strategic technologies discussed in the last decades; this strategy might be an alternative to conventional agricultural technologies (Nafis et al. 2019). Actinobacteria are agriculturally important as they are involved in various interactions known to affect plant fitness and soil quality, thereby increasing the productivity of agriculture and stability of soil (Ranjani et al. 2016). Actinorhizal plants comprise around 200 plant species belonging to 25 genera in eight families (Chaia et al. 2010). There are several ways in which different members of actinobacteria promote plant growth directly, such as nitrogen fixation, nutrient acquisition (phosphate and potassium solubilization), phytohormone (auxin, gibberellins), siderophores production, and ACC deaminases activity (Qin et al. 2011, Ranjani et al. 2016). Actinobacteria could also improve plant growth and fitness by indirect mechanisms, including the production of ammonia, hydrogen cyanides, antibiotics, antifungals, and lytic enzymes. Illustratively, *Streptomyces rochei* and *S. thermolilacinus* had led to an increase of 12.2-24.5% in shoot length and 1.8-2.3 folds in biomass of wheat (Jog et al. 2012). Several studies have indicated that actinobacteria application can also result in improving plants' natural systemic resistance, which confers plants' tolerance to biotic and abiotic stresses (Qin et al. 2011, Ranjani et al. 2016).

Another important aspect of actinobacteria in plant biotechnology is their antagonistic activity against phytopathogens (Franco-Correa and Chavarro-Anzola 2016). For instance, *Streptomyces kasugaensis* was able to produce Kasugamycin, which is a bactericidal as well as fungicidal that inhibit protein biosynthesis in microorganisms (Okuda and Tanaka 1992). *Pythium* damping-off in tomato plants and *Fusarium* wilt in cotton are inhibited and controlled by *S. ambofaciens*. Tubercidin produced by *S. Violaceusniger* controls *Phytophthora* blight of pepper; the literature is full of examples highlighting the importance of the role of actinobacteria to control plant diseases (Ranjani et al. 2016). In addition to diseases caused by microorganisms' phytopathogens, actinobacteria can produce several compounds that control weeds (herbicide) and insects (insecticide). Interestingly, 60% of new insecticides and herbicides were originated from *Streptomyces* (Tanaka and Omura 1993).

In the course of screening for metabolic compounds, several studies were oriented toward the isolation of actinobacteria from extreme habitats. Extreme conditions of nutrition and culturing may affect the ability of actinobacteria cultures to survive, hence selecting rare actinobacteria species and new metabolic products with great biotechnological importance. In the next part, we have focused on the new actinobacteria strains, purely Moroccan, isolated from different ecosystems.

New Actinobacteria Species Isolated From Extreme Ecosystems of Morocco

Streptomyces marokkonensis sp.nov.

Streptomyces marokkonensis (ma.rok.ko.nen 'sis. N.L. masc. adj. Marokkonensis belonging to Marokko, the Dutch name of Morocco, from which the type strain was isolated) with the code Ap1^T (= R-22003T = LMG23016T = DSM41918T). *S. marokkonensis* was isolated from the rhizosphere of Moroccan endemic argan tree (*Arganiaspinosa* L.) in the region of Essaouira (southern Morocco) which is characterized by semi-arid climate with an annual rainfall of around 300 mm and average temperatures of 20°C (Figure 1) (Bouizgarne et al. 2009). As is the case for arid and semi-arid regions, the Essaouira is encountered at a high risk of water scarcity due to climate change. The strain was isolated using the agar soil extract medium according to the method described by Barakate et al. (2002). Ap1^T is formed of a branched mycelium and a greyish-white aerial mycelium with long spiral spore chains and more than 20 non-flagellated spores with a smooth surface and a cell wall containing LL-diaminopimelic acid (Figure 2).

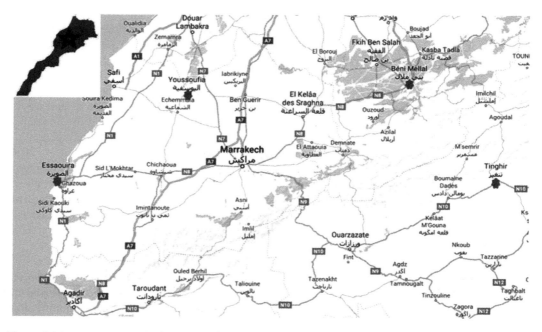

Figure 1. Moroccan new actinobacteria isolation sites.

The Ap1^T strain has been designated as a new producer of pentene polyene macrolide isochainin, which strongly inhibits the growth of pathogenic yeasts (*Candida albicans* and *C. tropicalis*) and phytopathogenic fungi (*Fusariumoxysporum* f.sp. *albedinis*, *Fusariumoxysporum* f.sp. *lycopersici*, and *Verticilliumdahliae*) (Bouizgarne et al. 2006).

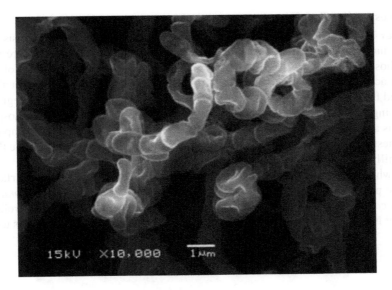

Figure 2. Microscopic observation of the spiralisspore chains of *Streptomyces marokkonensis* sp. nov.

Streptomyces thinghirensis sp. nov.

Streptomyces thinghirensis (thin.ghi.ren'sis.N.L. masc. adj. thinghirensis from Thinghir, the name of a city in southern Morocco from where the strain was isolated) carrying the code S10^T (= CCMM B35T = DSM41919T) (Bouizgarne et al. 2009). It has been isolated as part of the program actinobacteria screening of Moroccan origin that has activity against several phytophatogens (Loqman et al. 2009) (Figure 3). Strain S10^T was isolated on the medium extracted from agar sol described by Ouhdouch et al. (2001) from the soil of the rhizosphere of wild and healthy plants of *Vitisvinifera*, collected in Tinghir province of Ouarzazate, southern Morocco. Tinghirdesert has a warm arid climate with dry summers and waterless-chilly winters according to the Köppen-Geiger classification. The average temperature in Tinghir is 20.1°C and the average rainfall is 156.9 mm.

Strain S10^T has a yellow mycelium substrate and white-gray aerial mycelium with long spiral spore chains carrying smooth surface spores and produces a diffusible yellow pigment (Figure 3). It is active against several strains of yeasts and molds; moreover, it has a strong activity against Gram-negative strains mainly *Escherichia coli*. A patent (EP 09 290 240.2) was filed under the title "New actinobacteria strain compositions and their use for the prevention and/or control of microorganisms inducing plant diseases" by Loqman et al. (2010).

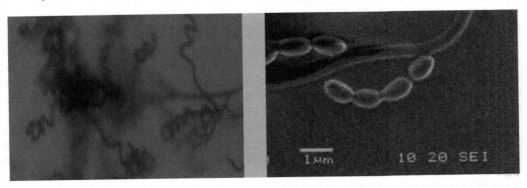

Figure 3. Microscopic observation of the mycelium and spore chains of *Streptomycesthinghirensis* sp. nov.

Streptomycesyoussoufiensis sp. nov.

Streptomyces youssoufiensis (yous.sou.fi.en' sis. N.L. masc. adj. Youssoufiensis belonging to Youssoufia, named after the phosphate rock mining town in Morocco from where the strain was isolated) whose code X4ᵀ (= CCMM B709T = DSM41920T) was found as part of the screening program for actinobacteria strains capable to solubilize phosphate rock (Hamdali et al. 2008). It has been isolated from a Moroccan phosphate mine in Youssoufia, in the Marrakesh region of southern Morocco using agar-extracted medium (Hamdali et al. 2011) (Figure 4). The phosphate mine of Youssoufia is a sedimentary type, which means that phosphate is presented in complex forms with sands or limestones. In this region, the summers are short, very hot, and arid, while the winters are generally chilly. During the year, the temperature can exceed 42°C.

X4ᵀ had white aerial mycelium with rectiflexibile spore chains with smooth surface spores and no diffusible pigments (Figure 4). It has an interesting activity against *Pseudomonas aeruginosa*, *Escherichia coli*, *Pythiumultimum*, *Mucorraman*, and *Staphylococcus aureus*. Virolle et al. (2012) have a patent No. US9206212 B2 entitled "Actinomycete strain composition and its use".

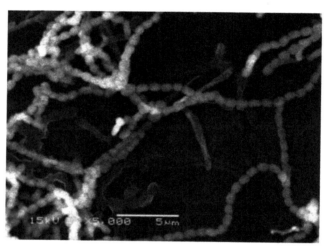

Figure 4. Microscopic observation of spore chains of *Streptomyces youssoufiensis* sp. Nov.

Streptomyces beta-vulgaris sp. nov.

Streptomyces beta-vulgaris is an actinobacteria strain deposited at the CNCM (Collection Nationale de Cultures de Microorganismes) under No. 1-3639 and L6M under No. P-23735. Its isolation was made from samples of the rhizosphere soil of sugar beet (*Beta-vulgaris L.*) from the Beni-Mellal region, Morocco (Figure 1). Beni-Mellal is a mountainous region characterized by a warm Mediterranean climate with dry summer according to the Köppen-Geiger classification.

The strain was isolated on medium extracted from agar sol or Olson medium according to two methods: treatment by heat-stirring or chemotaxis treatment. *Streptomyces beta-vulgaris* is characterized by a substrate mycelium, which varies from yellowish-brown to grayish-brown depending on the nature of the culture medium and a whitish-yellow aerial mycelium. It has a smooth spiral spores with an average of 10 to 50 spores (Lebrihi et al. 2009).

This strain produces bioactive substances used in plant protection and biological control of certain phytopathogenic agents and/or to stimulate plant growth. It has a particularly strong activity against fungi causing diseases of the vine wood, such as *Phaeomoniellachlamydospora*, *Eutypalata*, *Fomitiporiamediterranea*, and *Botrysphaeriaobstua*. This work is crowned by the filing of a patent No. WO2009156688 A3 in 2009 entitled "New *Streptomyces beta-vulgaris* strain, culture filtrate, derivative active compounds and their use in the treatment of plant".

Conclusions

Moroccan ecosystems represent a very rich source of microorganisms based on the results of research studies carried out in recent decades. In addition to that, innovation patents filed in this regard encourage Moroccan microbiologists to seek new habitats that have not been exploited before, to find other new species and therefore introduce new secondary metabolites of interest in agriculture as well as medical and food.

References

Barakate, M., Y. Ouhdouch, K. Oufdou and C. Beaulieu. 2002. Characterization of rhizospheric soil streptomycetes from Moroccan habitats and their antimicrobial activities. World J. Microbiol. Biotechnol. 18: 49–54. doi: 10.1023/A:1013966407890.

Bouizgarne, B., I. El Hadrami and Y. Ouhdouch. 2006. Novel production of isochainin by a strain of *Streptomyces* sp. isolated from rhizosphere soil of the indigenous moroccan plant *Argania spinosa* L. World J. Microbiol. Biotechnol. 22: 423–429. doi: 10.1007/s11274-005-9051-y.

Bouizgarne, B., B. Lanoot, S. Loqman, C. Sproer, H.-P. Klenk, J. Swings, et al. 2009. *Streptomycesmarokkonensis* sp. nov., isolated from rhizosphere soil of *Arganiaspinosa* L. Int. J. Syst. Evol. Microbiol. 59: 2857–2863. doi: 10.1099/ijs.0.011387-0.

Breton, A. M. Dusser, B. Gaillard-Martinie, J. Guillot and L. Millet. 1992. Caractérisation de champignons polycentriques du rumen observés *in vivo*. Annales de Zootechnie 41: 79–80.

Běhal, V. 2000. Bioactive products from *Streptomyces*. Adv. Appl. Microbiol. 47: 113–156.

Chaia, E.E., L.G. Wall and K. Huss-Danell. 2010. Life in soil by the actinorhizal root nodule endophyte *Frankia*: a review. Symbiosis 51: 201–226. doi: 10.1007/s13199-010-0086-y.

Dairi, T. 2005. Studies on biosynthetic genes and enzymes of isoprenoids produced by Actinomycetes. J. Antibiot. (Tokyo) 58: 227–243. doi: 10.1038/ja.2005.27.

Franco-Correa, M. and V. Chavarro-Anzola. 2016. Actinobacteria as plant growth-promoting rhizobacteria, in: actinobacteria – basics and biotechnological applications. InTech. doi: 10.5772/61291.

Garrity, G.M., H.G. Trüper, W.B. Whitman, P.A.D. Grimont, X. Nesme, W. Frederiksen, et al. 2002. Report of the ad hoc committee for the re-evaluation of the species definition in bacteriology. Int. J. Syst. Evol. Microbiol. 52: 1043–1047. doi: 10.1099/00207713-52-3-1043.

Hamdali, H., B. Bouizgarne, M. Hafidi, A. Lebrihi, M.J. Virolle and Y. Ouhdouch. 2008. Screening for rock phosphate solubilizing Actinomycetes from Moroccan phosphate mines. Appl. Soil. Ecol. 38: 12–19. doi: 10.1016/j.apsoil.2007.08.007.

Hamdali, H., M.J. Virolle, M. von Jan, C. Sproer, H.-P. Klenk and Y. Ouhdouch. 2011. *Streptomyces youssoufiensis* sp. nov., isolated from a Moroccan phosphate mine. Int. J. Syst. Evol. Microbiol. 61: 1104–1108. doi: 10.1099/ijs.0.023036-0.

Hassan, A.A., A.M. El-Barawy and E.M.M. Nahed. 2011. Evaluation of biological compounds of *Streptomyces* species for control of some fungal diseases. J. Am. 7: 752–760.

Jensen, P.R., R. Dwight and W. Fenical. 1991. Distribution of actinomycetes in near-shore tropical marine sediments. Appl. Environ. Microbiol. 57: 1102–1108.

Jog, R., G. Nareshkumar and S. Rajkumar. 2012. Plant growth promoting potential and soil enzyme production of the most abundant *Streptomyces* spp. from wheat rhizosphere. J. Appl. Microbiol. 113: 1154–1164. doi: 10.1111/j.1365-2672.2012.05417.x.

Lebrihi, A. R. Errakhi and M. Barakate. 2009. Nouvelle souche *Streptomyces* beta-vulgaris, filtrat de culture, composes actifs dérivés et leur utilisation dans le traitement des plantes. doi: Brevet numéro WO2009156688 A3.

Lechevalier, H. 2017. Rôle écologique des Actinomycètes. Encycl. Universalis 1–7.

Lechevalier, M.P., C. De Bievre and H. Lechevalier. 1977. Chemotaxonomy of aerobic actinomycetes: phospholipid composition. Biochem. Syst. Ecol. 5: 249–260. doi: 10.1016/0305-1978(77)90021-7.

Loqman, S., B. Bouizgarne, E.A. Barka, C. Clement, M. von Jan, C. Sproer, et al. 2009. *Streptomyces thinghirensis* sp. nov., isolated from rhizosphere soil of *Vitis vinifera*. Int. J. Syst. Evol. Microbiol. 59: 3063–3067. doi: 10.1099/ijs.0.008946-0.

Loqman, S., Y. Ouhdouch, J. Renault, J. Nuzillard, C. Clément and E. Ait-Barka. 2010. New actinomycetes strain compositions and their use for the prevention and/or the control of microorganism inducing plant diseases. doi: Brevet numéro EP 09 290 240.2.

Loucif, K., 2011. Recherche De Substances Antibactériennes À Partir D'une Collection De Souches D'actinomycètes. Caractérisation Préliminaire De Molécules Bioactives. Magister en Microbiol. appliqué Biotechnol. Microb. à l'Université Mentouri-Constantin, Algérie. 1–139.

Mangamuri, U.K., M. Vijayalakshmi, S. Poda, B. Manavathi, B. Chitturi and V. Yenamandra. 2016. Isolation and biological evaluation of N-(4-aminocyclooctyl)-3, 5-dinitrobenzamide, a new semisynthetic derivative from the Mangrove-associated actinomycete Pseudonocardia endophytica VUK-10. 3 Biotech 6: 158. doi: 10.1007/s13205-016-0472-0.

Nafis, A., A. Raklami, N. Bechtaoui, F. El Khalloufi, A. El Alaoui, B.R. Glick, et al. 2019. Actinobacteria from extreme niches in Morocco and their plant growth-promoting potentials. Diversity 11: 139. doi: 10.3390/d11080139.

Okuda, S. and Y. Tanaka. 1992. Fungicides and antibacterial agents, in: Springer Book Archive. pp. 213–223. doi: 10.1007/978-1-4612-4412-7_11.

Ouhdouch, Y., M. Barakate and C. Finance. 2001. Actinomycetes of moroccan habitats: isolation and screening for antifungal activities. Eur. J. Soil. Biol. 37: 69–74. doi: 10.1016/S1164-5563(01)01069-X.

Pandey, A. P. Nigam, C.R. Soccol, V.T. Soccol, D. Singh and R. Mohan. 2000. Advances in microbial amylases. Biotechnol. Appl. Biochem. 31: 135. doi: 10.1042/BA19990073.

Patel, J. 2001. 16S rRNA gene sequencing for bacterial pathogen identification in the clinical laboratory. Mol. Diagnosis 6: 313–321. doi: 10.1054/modi.2001.29158.

Piepersberg, W. 2008. Streptomycetes and Corynebacteria. pp. 433–468. *In*: Biotechnology. Wiley-VCH Verlag GmbH, Weinheim, Germany, doi:10.1002/9783527620821.ch13.

Pizzul, L. 2006. Degradation of Polycyclic Aromatic Hydrocarbons by Actinomycetes. Dr. thesis Swedish Univ. Agric. Sci. Sweden. 1–39.

Qin, S., K. Xing, J.-H. Jiang, L.-H. Xu and W.-J. Li. 2011. Biodiversity, bioactive natural products and biotechnological potential of plant-associated endophytic actinobacteria. Appl. Microbiol. Biotechnol. 89: 457–473. doi: 10.1007/s00253-010-2923-6.

Ramesh, S. and N. Mathivanan. 2009. Screening of marine actinomycetes isolated from the Bay of Bengal, India for antimicrobial activity and industrial enzymes. World J. Microbiol. Biotechnol. 25: 2103–2111. doi:10.1007/s11274-009-0113-4.

Ranjani, A., D. Dhanasekaran and P.M. Gopinath. 2016. An Introduction to Actinobacteria, in: Actinobacteria - Basics and Biotechnological Applications. InTech. doi: 10.5772/62329.

Sardi, P., M. Saracchi, S. Quaroni, B. Petrolini, G.E. Borgonovi and S. Merli. 1992. Isolation of endophytic Streptomyces strains from surface-sterilized roots. Appl. Environ. Microbiol. 58: 2691–2693.

Shirling, E.B. and D. Gottlieb. 1966. Methods for characterization of Streptomyces species. Int. J. Syst. Bacteriol. 16: 313–340. doi: 10.1099/00207713-16-3-313.

Tanaka, Y. and S. Omura. 1993. Agroactive compounds of microbial origin. Annu. Rev. Microbiol. 47: 57–87. doi: 10.1146/annurev.mi.47.100193.000421.

Virolle, M.-J., H. Hamdali, Y. Ouhdouch, M. Hafidi, A. Benharref, A. Lebrihi, et al. 2012. Actinomycete strain composition and its use. doi: Brevet numéro US9206212 B2.

Yadav, A.K., S. Vardhan, S. Kashyap, M. Yandigeri and D.K. Arora. 2013. Actinomycetes diversity among rRNA gene clones and cellular isolates from Sambhar Salt Lake, India. Sci. World J. doi: 10.1155/2013/781301.

Index